PURINE METABOLISM IN MAN—II
Regulation of Pathways and Enzyme Defects

ADVANCES IN EXPERIMENTAL MEDICINE AND BIOLOGY

PURINE METABOLISM IN MAN—II

Regulation of Pathways and Enzyme Defects

Edited by

Mathias M. Müller and Erich Kaiser

University of Vienna
Vienna, Austria

and

J. Edwin Seegmiller

University of California, San Diego
La Jolla, California

PLENUM PRESS • NEW YORK AND LONDON

Library of Congress Cataloging in Publication Data

International Symposium on Purine Metabolism in Man, 2d, Baden, Austria, 1976.
   Purine metabolism in man, II.

  (Advances in experimental medicine and biology; v. 76)
  Includes index.
  CONTENTS: Pt. A. Regulation of pathways and enzyme defects.—Pt. B. Physiology, pharmacology, and clinical aspects.
  1. Purine metabolism—Congresses. 2. Uric acid metabolism—Congresses. 3. Metabolism, Inborn errors of—Congresses. I. Müller, Mathias M. II. Kaiser, Erich, Dr. med. III. Seegmiller, J. E. IV. Title. V. Series. [DNLM: 1. Purines—Metabolism—Congresses. 2. Gout—Enzymology—Congresses. 3. Purine-pyrimidine metabolism, Inborn errors—Congresses. 4. Carbohydrates—Metabolism—Congresses. 5. Lipids—Metabolism—Congresses. W1 AD559 v. 76 1976 / QU58 I635 1976p]
QP801.P8I56 1976   612'.0157   76-62591
ISBN 0-306-39089-2 (vol. 76A)

Proceedings of the first half of the Second International Symposium
on Purine Metabolism in Man, held in Baden, Vienna, Austria,
June 20–26, 1976

©1977 Plenum Press, New York
A Division of Plenum Publishing Corporation
227 West 17th Street, New York, N.Y. 10011

Printed in the United States of America

PREFACE

The study of gouty arthritis has provided a common
meeting ground for the research interests of both the
basic scientist and the clinician. The interest of the
chemist in gout began 1776 with the isolation of uric
acid from a concretion of the urinary tract by the
Swedish chemist SCHEELE. The same substance was
subsequently extracted from a gouty tophus by the
British chemist WOLLASTONE in 1797 and a half century
later the cause of the deposits of sodium urate in
such tophi was traced to a hyperuricemia in the serum
of gouty patients by the British physician Alfred
Baring GARROD who had also received training in the
chemical laboratory and was therefore a fore-runner of
many of today's clinician-investigators.

The recent surge of progress in understanding of some
of the causes of gout in terms of specific enzyme
defects marks the entrance of the biochemist into this
field of investigation. The identification of the first
primary defect of purine metabolism associated with
over-production of uric acid, a severe or partial
deficiency of the enzyme hypoxanthine-guanine phosphoribosyltransferase was achieved less than a decade ago.
The knowledge of the mechanism of purine over-production
that it generated led shortly to the identification of
families carrying a dominantly (possibly X-linked)
inherited increase in the activity of the enzyme
phosphoribosylpyrophosphate synthetase as a cause of
purine over-production. Yet this is only a start as
these two types of enzyme defects account for less
than five per cent of gouty patients.

The rapid pace at which new knowledge of aberrations of
human purine metabolism is being acquired is adequate
reason for holding the Second International Symposium
on Purine Metabolism in Man (Baden, Austria, June 20 - 26,
1976) just three years after the first symposium was

convened. It also marks the bicentennial aniversary of the discovery of uric acid by SCHEELE. The table of contents shows a further consolidation of our understanding of the mechanisms involved in the synthesis and degradation of purines and the aberrations produced in regulation of these processes by well characterized defects in purine metabolism. In addition are reports of newly discovered defects in enzymes of purine metabolism not previously presented at the last symposium. Homozygousity for deficiency of adenine phosphoribosyl-transferase has now been identified in three children, two of whom presented with calculi of the urinary tract composed of 2,8-dihydroxyadenine thus setting at rest previous speculations based on studies of heterozygotes for this disorder.

On the basis of recent experiments the understanding of renal handling of urate has been further increased indicating a pre- and post-secretory reabsorption. The significance of protein-binding of urate is still open for discussion. However the knowledge of mechanisms regulating purine transport through membranes has improved by development of rapid micromethods.

A whole new area of considerable importance for the future is the association of an impaired function of the immune system in children with a gross deficiency of either of two sequential enzymes of purine interconversion, adenosine deaminase or purine nucleoside phosphorylase. Further investigation of the mechanism of this phenomenon gives promise of extending substantially our knowledge of the normal control of the immune response.

We wish to acknowledge the support of the Dean of the Faculty of Medicine of the University of Vienna, DDr. O. Kraupp and financial support from Dr. Madaus and Co. The contribution of the Organizing Committee and of the Scientific Committe in arranging the meeting and the details of the program is also gratefully acknowledged.

MATHIAS M. MÜLLER

ERICH KAISER

J. EDWIN SEEGMILLER

CONTENTS OF VOLUME 76 A

De Novo Synthesis: Phosphoribosylpyrophosphate and
Phosphoribosylpyrophosphate Synthetase

De Novo Synthesis:
Phosphoribosylpyrophosphate Amidotransferase

Nucleotide Metabolism

Catabolism

MUTATIONS AFFECTING PURINE METABOLISM IN MAN

General Aspects

Hypoxanthine-Guanine Phosphoribosyltransferase:
    Complete Deficiency (Lesch-Nyhan Syndrome)

Lipid and Purine Metabolism

METHODOLOGY

CONTENTS OF VOLUME 76 B

BIOCHEMISTRY OF PURINE TRANSPORT

BIOCHEMICAL PHARMACOLOGY

HYPERURICEMIA AS A RISK FACTOR

HISTORY OF GOUT.

INCLUDING COMMENTS FROM AN ILLUSTRIOUS TIMELESS GATHERING

Andre de Vries

Rogoff-Wellcome Medical Research Institute, Tel-Aviv
University Medical School, Beilinson Medical Center,
Petah Tikva, Israel

In the history of medicine gout stands out because of its du-
ration extending over thousands of years, because of its striking
features and consequences pertaining to the most diverse fields of
medicine, and not in the least because of its recent exciting development.

But why history? Because it enables us to "make the experience
of the past available to the present for the sake of the future
(Jevons, 1). And why history of medicine in particular? An answer
is given by Hippocrates: "I consider that clear knowledge of nature
can be derived from no source except medicine" (2). Did Hippocrates
foresee that so much of the development in modern basic biology was
triggered by clinical observation? I presume, if not by sophisti-
cation, then by intuition. Thus, in a sense, Hippocrates joins that
modern class of forecasters called "intuitive futurists" (3). Certainly, Hippocrates' presumption holds as well for gout, a disease
with which he appears to have been acquainted, and which is an
exemplification of modern clinical-biochemical interplay.

Thus, a review of the history of gout would seem most appropriate at this symposium on purine metabolism, were it not for the
indubitable fact, that all participants, addicted to purines as they
are, are familiar with the excellent writings of such honorable
modern gout-historians as John H. Talbott (4), J.P. Rodnan (5),
D.P. Mertz (6), Hartung (7) and others. Therefore, I shall limit
my discussion to specific aspects of the history of this disorder,
selected according their bearing on actual problems of "gouty" and
general medical interest. In addition, I shall, in an unorthodox

way, invite some outstanding historical personalities, to speak
for themselves.

Let me first, however, summarize some critical stages in the
development of our knowledge of gout. Gout is one of the oldest
diseases for which we have documentation. For ancient gout there
is definite proof, furnished by the remarkable discovery of urate
tophi in a mummy from Nubia, Upper Egypt, dating thousands of years
ago (8). The finding of urate-containing stones in Egyptian mummies
(9) and in a three-thousand years old mummy from Arizona (10) can
provide, of course, only circumstantial evidence for the ancient
prevalence of gout, since urate stones, as known at present, may
occur also, and apparently not infrequently, in the absence of gout,
at the least of clinically manifest gout. From these ancient times
we pass into the classical period with the description by Hippocrates in the fifth century B.C. (2,11) and that by Celsus and Galenus (12,13) in the first two centuries A.D., continuing through
the first millenium into the second, our own millenium, with the
biographies of a host of famous gouty personalities, many of intel-
lectual stature, including John Calvin, Desiderius Erasmus, Benjamin Franklin, William Harvey, and numerous others.

This symposium being held in this wonderful country of Austria,
I cannot refrain from mentioning a famous doctor of the sixteenth
century, who spent a part of his youth in Villach in Tirol and studied at Vienna, to return after a tumultuous life to practise in
Salzburg, where he died and was buried - Philippus Theophrastus
Bombastus von Hohenheim, Aureolus, Paracelsus (14,15,16). However
one judges this physician-philosopher-scientist-astrologer, quack
if you wish, one must admire his clinical acumen and his conviction
as to the prime importance of clinical observation as opposed to the
mechanical absorption of the classical texts: "The patients are
your textbook, the sick bed is your study". He knew the gout, and
proposed probably the first reasonable metabolic mechanism for the
acute arthritis-deposition in the joints of "Tartarus", an acid deposit resembling that in vessels containing old wine. The "Tartarus"
was also present in the "sand" in the urine of such patients, an
impressive foresight indeed. He was a great therapist and excelled
in esoteric prescriptions which,no doubt, were successful, whether
rationally or psychologically. In those times, however, diagnosis
of gout must have often been uncertain, and it was in the seven-
teenth century that the diagnosis of gout become more reliable when
Thomas Sydenham delineated its clinical characteristics from tho-
rough self-observation (17). His description is unsurpassed as to
accuracy and elegance, and, like Paracelsus, he was an independent
spirit accumulating experience in private praxis and distrusting
the university establishment: "Physick is not to be learned by going
to universities. One has as good send a man to Oxford to learn shoe-
making as practising Physick".

While gout continued to play its clinical role, occupying the
bodies and minds of patients and physicians, the breakthrough in
the understanding of the nature of the disease came in the late
eighteenth century with the identification of urate in bladder stones
and tophus by Scheele and Wollaston, and in mid-nineteenth century
with the demonstration of hyperuricemia by Garrod (18). We then
witness in the present century the struggle in classfication-pri-
mary, idiopathic, secondary, renal gout, and now in the very few
last decades four main exciting advances: the diagnostic break-
through-demonstration of urate crystals in joint fluid, leukocytes
and synovia (19) leading to the separation of pseudogout; then the
clinical and biochemical breakthrough with the discovery of enzyme
mutations causing purine overproduction - the hypoxanthine-guanine
phosphoribosyltransferase deficiency (20) associated with the Lesch-
Nyhan syndrome (21) and mutant phosphoribosylpyrophosphate superactivity (22,23);then, the decisive therapeutic breakthrough with
the uricosuric agent probenecid (24) and the xanthine oxidase inhibitor allopurinol (25). And, finally, there is the exciting immunological development linked to adenosine deaminase deficiency (44)
and purine nucleoside phosphorylase deficiency (45).

What do we learn from this amazing gouty history?

Firstly, the prime value of clinical observation, in that gout,
in its acute and chronic forms and associated lithiasis, became
remarkably well defined by the clinicians before biochemistry had
developed.

Secondly, that the value of accurate clinical observation is
not necessarily mitigated by wild speculation, to wit Thomas Sydenham's adherence in the explanation of gout to the Galenic humors and
his proposal that the disease was due to accumulation of heat produced by excessive "incineration" in the body not being able to
escape through a suitable "vent" (17).

Thirdly, as well demonstrated by the recent gouty events, that
the discovery of basic mechanisms in disease is often triggered by
the impact of clinical reality, to wit the Lesch-Nyhan syndrome.

Fourthly, and related to the previous argument, the importance
of the close association between clinical observation and basic research, whether by cooperating clinician and biochemist or, often
still more advantageous, by unification of the two research types
in one and the same person, as exemplified by not a few modern investigators.

I now turn to the fascinating history of the semantics of gout
and associated phenomena. The issue of semantics in medicine is
often looked upon with disdain since names obviously are less important than facts and principles. Nevertheless, a well-adjusted
terminology is valuable, both in scientific communication and in

education, in order to secure effective transmission of reality, and in order to prevent persistence of obsolete concepts. Here, I quote George Orwell (26): "Our language becomes ugly and inaccurate because our thoughts are foolish, but the slovenliness of our language makes it easier for us to have foolish thoughts".

The first relevant historical term we know of is "podagra" (from pous=foot, agra=hunting or chase, (27)) originally used for "an affection of the joints of beasts of burden" (28), and already used by the Hippocratic school (28,29) in the 5th century B.C., then later by Celsus and Galenus in the 1st and 2nd centuries A.D. (30)m and around that time by Talmudic scholars in Jerusalem (27). In the 2nd century B.C. the Cappadocian physician Aretaeus equalled podagra with "a trap for the feet" (30). The term tophus, derived from Greek, probably meaning" a rough crumbling rock", was used by Galenus (13). The word gout, from gutta, drops of unbalanced Galenic humors falling into the joint, came into use much later, around the beginning of the 13th century (29), and was not used by Hippocrates, although it is erroneously mentioned in some translatiors of his works.

An upheaval in gouty terminology is witnessed in recent years, resulting from spectular biochemical advances. Modern definitions such as by Wyngaarden and Kelley (31) and by Seegmiller (32) include the characteristic clinical picture, hyperuricemia and the presence of sodium urate crystals in and about the joints. The first two, clinical symptomatology and hyperuricemia, each alone or even in combination, do not suffice, since pseudogout due to pyrophosphate crystal deposition may mimick true gout, and hyperuricemia may be asymptomatic or in non-gouty arthritis a fortuitous association. But should the presence of urate crystals be an absolute criterion, if a recent report (33) on acute gouty arthritis in the absence of such crystals is valid?

And what about the terms primary, secondary, idiopathic, renal? Does primary gout include idiopathic gout? If primary gout is due to an inborn error of metabolism (31), should that not invalidate the designation idiopathic when there is normal uric acid excretion (31), or should the term idiopathic be reserved for gout without purine overproduction but due to renal urate retention of unknown etiology? Or should the term renal gout be used exclusively for gout due to known kidney disease or induced by drugs affecting renal urate handling? And should secondary gout include, in addition to that due to increased nucleic acid turnover (32), also gout due to renal urate retention (31)? Would it not be preferable to discontinue the use of these designations and apply semantics according to the measure of purine production as follows?

Gout:
        with purine overproduction due to
            inborn error : known
                          : unknown
            increased nucleic acid turnover
        without purine overproduction, due to renal urate retention
            specific for urate
            renal failure

Another confusion in terminology pertains to uric acid lithiasis. When the lithiasis appears together or following the first
gouty arthritic attack the term is easily justified. But how to
designate the lithiasis when, as not uncommonly occurs, it precedes
the arthritis ? Should it be called idiopathic, to be changed re-
trospectively to gouty lithiasis when classical gout appears? And
does idiopathic uric acid lithiasis include that associated with
hyperuricosuria as well as that with normouricosuria?

Then, the confusion as to the terminology of the socalled gouty or hyperuricemic nephropathy (34,35), an interstitial type of
nephropathy believed to be caused by inflammatory reaction to urate
deposits in the renal tissue. Is the designation for this entity as
gouty nephropathy justified when it may, although rarely, precede
gouty arthritis, and when there is no final proof, in many instances,
that the chronic renal condition is indeed initiated by the interstitial urate deposits but instead may be due to associated metabolic disturbances, hypertension or the consequences of accompa-
nying urolithiasis including urinary tract infection? And is the
designation hyperuricemic nephropathy, indicating a direct plasma-
tic source of the interstitial urate deposits, justified when their
derivation from intratubular uric acid deposits has not been excluded (36) ?

Finally, the remarkable history of gout therapy. It seems a
valid statement that as long as gout was a purely clinical observation, prevention and therapy were derived from simple clinical
experience of physicians and magi, whether guided by high intelligence or embellished by daring speculation. Thus colchicin-contai-
ning herbs and discretion in the excesses of living were long known.
Thomas Sydenham, as many others before him and contemporary with
him knew of the harm brought about by food and alcoholic spirits
and, as he believed, by mental exertion and coition (17). It was,
however, only with the modern developments in chemistry and biochemistry, that it became possible to synthetize and rationally apply
the uricosuric and the uric acid production-inhibiting compounds,
roughly in the same period when biochemistry provided the understanding of basic mechanisms in gout. Still, also regarding treatment,
the history of gout has not ended, since new drugs will be found
and indications have to be clarified and tightened.

I would like to illustrate this selective consideration of the history of gout by invoking your imagination for the participation in a fictitious round table devoted to the subject of gout, whereby we can listen, as it were at this very moment to past and present personalities involved with this disease. Let us invite to the panel an ancient Egyptian courtier, historian Herodotus, biblical King Asa, doctor Hippocrates, general Alexander the great, philosopher Aristotle, doctor Galenus, reformer John Calvin, humanist Desiderius Erasmus, doctor Paracelsus, scientist-diplomat Benjamin Franklin, doctors William Harvey and Thomas Sydenham, scientists-doctors Scheele, Wollaston and Garrod. Let us further invite some living honorable colleagues active in the field, whom I keep anonymous since they may hesitate to be considered historical already, and finally modern philosopher George Santayana and writer Andre Gide.

HISTORICAL PANEL

Chairman anonymous: Dear dead and living, those suffering from gout and those investigating and treating it, the subject of our panel discussion is gout or whatever term you are acquainted with. May I modestly ask you not to talk too long, as some of you are in the habit of doing, as for instance, you Paracelsus! Shall we start with our most ancient member?

Ancient Egyptian: My suffering is great. When my foot becomes swollen the pain is so severe that I can not fulfill my court duties as minister of embalming and burial. I am glad to know that now in the twentieth century you could make the exact diagnosis of my disease by examining my mummy. I think you call it gout.

Herodotus: Your excellency, you are of course outstanding in having this ailment, because as a courtier "you eat and drink so much, while the simple people are frugal by necessity" (37).

Twentieth century doctor: You had also stones?

Ancient Egyptian: I had; stones were indeed frequent at the court.

Twentieth century doctor: But that does not fit with the paucity of stones found in the thousands of recently excavated mummies.

Herodotus: Possibly, an explanation can be found by the method of embalming whereby "the body was placed in natrium for seventy days" ...... "nothing (being) left of the dead body but the skin and the bones" (37).

Twentieth century doctor: Certainly uric acid stones would have dissolved (18).

Herodotus: But who treated you, your Egyptian excellency? "Because Medicine (in Egypt) is distributed in the following way: every physician is for one disease, and not for several, and the whole country is full of physicians; for there are physicians of the eyes, others of the head, others of the teeth, others of the obscure diseases. There is even a special "guardian of the anus" and "one understanding the internal fluids" (37).

Ancient Egyptian: That is my difficulty, I do not know whom to approach, so many ranks of doctors we have: "general practitioners, high doctors, inspector doctors, chief doctors, supreme doctor for Upper and Lower Egypt, Director-general of the House of Life and Chief of the Secrets of Life in the Institute of Tout" (11).

Twentieth century doctor: We have the same problem. Shall the patient go to a doctor specialist in phosphoribosylpyrophosphate synthetase superactivity or to a doctor specialist in hypoxanthine-guanine phosphoribosyltransferase deficiency, or to a nucleic acid turnover specialist?

King Asa: I, the great-grandson of king Solomon, live in the ninth century B.C. and suffer from my feet, as is written some 500 years later in the Old Testament (38): "In the thirty-ninth year of his reign Asa became affected with gangrene in the feet; he did not seek guidance of the Lord but resorted to the physicians. He rested with his forefathers, in the forty-first year of his reign .....". My disease was diagnosed in the 3rd century A.D., by Talmudic scholars in Jerusalem as podagra, and accepted in the 18th century, i.e. some 2600 years later, to have been gout (39). Is that right?

Twentieth century doctor: No evidence for gout whatever from your case history. According to the symptomatology it must have been peripheral vascular disease (39).

Hippocrates: On my little island of Cos, when visiting the citizens with my students, I do see patients who have what I call "Walking Disease" or podagra. One of my favorite observations is that a young man does not suffer from it until he indulges in coition. Did not you, Thomas Sydenham, observe also that coition brings about the attack? In any event my system of education of the island of Cos is very effective, just as you have it now in Britain with the Open University, and then, of course, what I do is the equivalent of your modern home care par excellence.

Alexander the Great: I have podagra, I think because I eat too much out of the booty after each conquest, but I have great trouble with the colchicium causing diarrhea. I ask my mentor Aristotle how to behave.

Aristotle: "Never accept medicine from a single doctor, but employ many and only act on their unanimous advice".

Twentieth century doctor: "By the way, the mechanism of action of colchicine, still used today with great benefit, is still a matter of dispute, although there is recent talk on an action on microtubules.

Galenus: Like my colleague Celsus I also see patients with podagra. Many of them have also stones in the kidneys and the bladder, as well as around the joints. I call these stones poroi (13), which is a Greek word. I think you call it in Latin tophus, but I wish to stress that to me stones and tophi are the same. - "In arthritic patients the chalk-stones (poroi) are formed by a thick and sticky humor. They are not easily broken up into small pieces, but all at once they dry up under the influence of sharp medication. In the same manner kidney stones are formed when a sticky and heavy humor is overheated in these organs" (37). I think, like most of you, that on important etiological factor is gluttony and that drinking of much water may prevent stones. But there is certainly also a familial predisposition.

Desiderius Erasmus: My arthritis is bad, but my stones are worse: - "There was no end or let-up, birth followed birth.... two weeks ago I almost perished with birth-pangs. The stone was huge, my stomach collapsed and even yet has not recovered, your Erasmus is nothing but skin and bones" (40). Since I live now in Basel, I ask you doctor Paracelsus who dwells in the same house to treat me.

Paracelsus: "A helpful salve for the gout. When you venesect or cup a person who suffers from this disease, keep his blood, but without his knowledge of what use you will make of it. Distill it to water three times over a mild fire or a bath; take then of this water 14 lot, of human fat 1 ounce, of horse oil $1/2$ ounce, a little melted Venetian soap, also $1/2$ ounce bear fat, of the juice of cultivated leek 1 ounce, cattle marrow $1/2$ ounce. Place all this into a copper pan and let it boil slowly until it becomes a thick paste; stir it constantly with a spatula. When it has become as thick as a salve it is ready; then puncture the sole of the patient where the gout is located with a cupping iron every eight day as the moon is rising, and rub the warm salve thoroughly. Then his gout will disappear in nine weeks. The older this salve becomes the better it is. It maintains its strength and value for ten years when kept in a cool place" (41).

John Calvin, Benjamin Franklin, William Harvey in chorus: We all suffer from gout and stones.

John Calvin: I have many diseases: "Gout, malaria, stomach trouble, migraine, insomnia, hemoptysis, lung infection, hemorrhoids, renal colic and stone".

   William Harvey: "I am much and often troubled with the gout"
and "sit with my legs bare if it were frost on the leads of Cockaine
house, put them into a pail of water till I am almost dead with
cold, then betake myself to a stove, and so it is gone".

   Benjamin Franklin: My suffering is so severe that I recently
had a dialogue on the subject with gout itself.

   Thomas Sydenham: "Either men will think that the nature of
gout is wholly mysterious and incomprehensible, or that a man like
myself who has suffered from it thirty-four years, must be of a slow
and sluggish disposition not to have discovered something respecting
the nature and treatment of a disease so peculiarly his own. Be this
as it may, I will give a bona fide account of what I know. The difficulties and refinements relating to the disease itself, and the
method of its cure, I will leave for Time, the guide to truth, to
clear up and explain" (17). May I congratulate my younger colleagues
Scheele, Wollaston and Garrod to have done such a fine job on the
clarification of the disease, although you will agree that it is
good that the thread test for hyperuricemia has become obsolete
thanks to our 20th century colleagues.

   Chairman: "You will all agree that for the patient the most
important aspect is treatment. Would you express your opinions?

   All in chorus: We agree upon a frugal life, avoiding excessive
food, alcohol, coition, physical and mental exertion.

   Desiderius Erasmus: Burgundy wine is excellent. When my stones
get too bad, I travel to France. But beer is bad - " O felicitous
Burgundy, truly worthy to be called the mother of mankind, that has
such milk in her breasts. It is no wonder that the first mortals
worshipped you as a goddess.... and so I can foresee that with little
urging I shall depart for Burgundy. On account of the wine, you ask?
Indeed " (40).

   Thomas Sydenham: Apply bleeding and cooling of the body to
decrease the excessive incineration. For elderly gouty patients I
recommend a carriage" which "is a blessing to gouty people; inasmuch
as that very wealth which fostered the luxury which brought about
the disease supplies the means of keeping a vehicle, whereby those
can take the one sort of exercise when they could not take the other"
(horse riding) (17).

   Twentieth century doctor: We are proud of our enzyme mutations
but raise the difficulty in therapeutic indication, reflecting on
our attitude to future life style, in particular since in the majority

of the gouty subjects the pathogenetic mechanism is not known. With the advent of the new anti-gouty agents, uricosuric and uric acid production blockers, we have two extreme possibilities: to try to prevent gout by discontinuing our vices, that is to refrain from excess in eating, drinking, sex and other delights, or to persist in our vices but with a cover of medication. Our answer can not be final, since, on the one hand, our vices constitute an integral part of our life while, on the other hand, medication is beset by un-avoidable side-effects, including iatrogenic urolithiasis such as xanthine and oxipurinol stones (42,43).

Chairman: "Who is right? The future will show and at a certain stage of it our offspring will judge our decision historically. But for that they will have to study the history of gout.

George Santayana: "He who is ignorant of history is doomed to repeat it".

Andre Gide: "Everything has been said but it is necessary to say it again because some have not listened".

REFERENCES

1. Jevons, F.R. In the Future as an Academic Discipline, Ciba Foundation Symposium 36 (new series). Elsevier, Excerpta Medica, North Holland, Amsterdam, Oxford, New York, pp. 53-71, 1975.
2. Clendening, L. Source of Medical History, P. Hoeber, Inc. London, 1942.
3. Eldredge, H.W. In the Future as an Academic Discipline, Ciba Foundation Symposium in 36 (new series), Elsevier, Excerpta Medica, North Holland, Amsterdam, Oxford, New York, pp. 5-18, 1975.
4. Talbott, J.H. Gout (Third Ed.) Grune & Stratton, New York, London, 1967.
5. Rodnan, G.P. Arthritis & Rheumatism, 4:27, 176, 1961.
6. Mertz, D.P. Gicht, Grundlagen, Klinik und Therapie, Georg Thieme Verlag, Stuttgart, 1971.
7. Smith, G.E. and Jones, F.W. The archeological survey of Nubia. Report for 1907-8, vol. II. Cairo: National Printing Dept., pp. 44, 269, 1910. (from ref. 4).
8. Hartung, E.F. Metabolism, 6:1967, 1957.
9. Miller, J.L. Ann. Med. Hist., 1:400, 1929.
10. Williams, G.D. J. Amer. Med. Assoc., 87:941, 1926.
11. Sigerist, H.E. A History of Medicine, Vol. I.: Primitive and Archaic Medicine, Oxford University Press, 1951.
12. Talbott, J.D. A Biographical History of Medicine. Excerpta and Essays on the Men and their Work, Grune & Stratton, New York-London, 1970.

13. Siegel, R.E. Galen's System of Physiology and Medicine. An Analysis of his Doctrines and Observations on Bloodflow, Respiration, Humors and Internal Diseases. S. Karger, Basel, New York, 1968.

14. De Vries, A. Paracelsus, Sixteenth-century physician-scientist-philosopher, New-York State J. of Med. In press, 1976.

15. Lother, R. and Wollgast, S. Paracelsus. Das Licht der Natur. Philosophische Schriften. Phillip Reclam Verlag, Leipzig D.D.R., 1973.

16. Pagel, W., Paracelsus. An Introduction to Philosophical Medicine in the Era of the Renaissance. S. Karger, Basel, New-York, 1958.

17. Dewhurst, K. Dr. Thomas Sydenham (1624-1689). His Life and Original Writings. The Wellcome Historical Medical Library, London, 1966.

18. Atsmon, A., de Vries, A. and Frank, M. Uric Acid Lithiasis. Elsevier Publ. Co., Amsterdam, 1963.

19. McCarty, D.J. and Kozin, F. Arthritis & Rheumatism, 18 (No.6 Suppl.):757, 1975.

20. Seegmiller, J.E., Rosenbloom, F.M. and Kelley, W.N. Science, 155:1682, 1967.

21. Nyhan, W.L. Ann. Rev. Med. 24:4, 1973.

22. Sperling, O., Boer, P., Persky-Brosh, S., Kanarek, E. and de Vries, A. Europ. J. Clin. Biol. Res., 17:703, 1972.

23. Becker, M.A., Meyer, L.J., Wood, A.W. and Seegmiller, J.E. Science, 179:1123, 1973.

24. Fanelli, G.M. Arthritis & Rheumatism, 18 (No.6 Suppl.):853, 1975.

25. Hitchings, G.H. Arthritis & Rheumatism, 18 (No.6 Suppl.):863, 1975.

26. Orwell, G. Politics and the English Language, in the Orwell Reader, New York, 1946.

27. Liddle, H.G. and Scott, R. Greek English Lexicon, Oxford University Press, Oxford, 1968.

28. Skinner, H.A. The Origin of Medical Terms. The Williams & Wilkins Co. Baltimore, 1949.

29. Wain, H. The Story Behind the Word. Some Interesting Origins of Medical Terms. Ch. C. Thomas, Springfield, Ill. U.S.A., 1968.

30. Hammond, N.G. and Schullard, H.A. Oxford Classical Dictionary. Oxford University Press, 1970.

31. Wyngaarden, J.B. and Kelley, W.N. Gout, in The Metabolic Basis of Inherited Disease (Eds. J.B. Stanbury, J.B. Wyngaarden and D.S. Fredrickson). Third Edition, McGraw-Hill Book Co., A Blakiston Publication, New York, pp. 889-968, 1972.

32. Seegmiller, J.E. Diseases of Purine and Pyrimidine Metabolism, in Duncan's Diseases of Metabolism, Seventh Edition. (Eds. P.K. Bondy and L.E. Rosenberg) W.B. Saunders Co., Philadelphia, pp. 655-774, 1974.

33. Schumacher, H.R., Jimenez, S.A., Gibson, T., Pascual, E., Traycoff, R., Dorwart, R.B. and Reginato, A.J., Arthritis & Rheumatism, 18:603, 1975.

34. Brod, J. The Kidney. London, Butterworth & Co. Ltd., p. 523, 1973.

35. Klinenberg, T.R., Kippen, I. and Bluestone, R. In Influence of the Kidney upon Urate Homeostasis in man (Rieselbach, R.E. and Stede, Th., guest eds.) Nephron, 14:88, 1975.

36. De Vries, A. and Sperling, O. Implications of Disorders of Purine Metabolism for the Kidney and the Urinary Tract. Ciba Foundation Symposium No. 48 (new series):Purine and Pyrimidine Metabolism, June 1976.

37. The History of Herodotus. Transl. G. Rawlinson Ed. M. Komroff, Tudor Publ. C. New York, 1928.

38. New English Bible. II Chronicles 12-14, Oxford and Cambridge Univ. Press, 1970.

39. De Vries, A. and Weinberger, A. New York State J. of Med., 452:75, 1975.

40. O'Malley, C.D. J. Am. Med. Assn., 216:66, 1970.

41. Pachter, H.M. Paracelsus, Magic into Science. Henry Schumann, New York, 1951.

42. Greene, M.L., Fujimoto, W.N. and Seegmiller, J.E. New Engl. J. Med., 280:426, 1969.

43. Landgrebe, A.R., Nyhan, W.L. and Coleman, M. New Engl. J. Med., 292:626, 1975.

44. Giblett, E.R., Anderson, J.E., Cohen, F., Pollara, B. and Meuwissen, H.J. Lancet, 2:1067, 1972.

45. Giblett, E.R., Ammann, A.J., Wara, D.W., Sandman, R. and Diamond, L.K. Lancet, 1:1010, 1975.

A MULTIENZYME COMPLEX FOR DE NOVO PURINE BIOSYNTHESIS

Peter B.Rowe, Gemma Madsen, E.McCairns and Dorit Sauer

Department of Child Health, University of Sydney,

New South Wales, AUSTRALIA

Multienzyme complexes are defined as aggregates of different functionally related enzymes integrated into highly organized structures by non-covalent bonds. Such a structural arrangement is considered to confer a catalytic advantage in that an unstable intermediate may be transferred more efficiently from one active site to the next. A number of such complexes have been isolated in association with subcellular organelles or membranes. Examples of these include the glycogen phosphorylase particles of mammalian muscle and the cytochrome electron transport system of mitochondrial membranes. More significantly the cytosol of a wide variety of cell types has been shown to contain enzyme complexes e.g. the glutamine synthetase and the fatty acid synthetase complexes. Recent evidence would suggest however, that in some species the latter complex is, in fact, a single multifunctional polypeptide chain (1).

In general the isolation of enzyme complexes from mammalian cytosol has not been a rewarding area of research. Some success has however, been achieved with the pyrimidine biosynthetic enzymes (2) and certain folate interconversion enzymes (3). This lack of success is probably related to a number of factors such as (i) protein dilution following cell disruption (ii) the effects of high ionic strengths of agents such as ammonium sulfate used in routine enzyme purification procedures and (iii) the differential adsorption of proteins onto the solid phases used in gel filtration and ion exchange chromatography.

13

TABLE 1

ASSAY: DE NOVO PURINE BIOSYNTHESIS

Aspartic acid, glutamine, PP-ribose-P, ATP, KCl, Na formate...	2.0 mM
PEP..	10.0 mM
$MgCl_2$..	3.5 mM
$N^{5,10}$ methenyltetrahydrofolate...........................	1.5 mM
$KHCO_3$..	10.0 mM
Pyruvate kinase..	80.0 units
Glycine-1-^{14}C (sa 1.6 Ci/mole).........................	0.6 mM
Tris-Cl pH 7.8	50.0 mM
Enzyme - to final volume..................................	3.5 mM

1. Incubate 38°.
2. Stop reaction with 3.5 ml 2N PCA. Remove protein.
3. Divide supernatant into three fractions.
   (i)   Hypoxanthine extracted by silver salt procedure.
   (ii) Hydrolyse.
   (iii) Neutralize with KOH.
4. Fractions analysed by thin layer chromatography and high pressure
   liquid chromatography.

The de-novo purine biosynthesis pathway consists of ten
sequential enzymatic reactions with an overall high energy requirement
The pathway, in eukaryotes, is unbranched. The diversity of types of
reactions involved - one carbon transfers, amino transfers, cyc-
lization reactions - and the instability of some of the intermediates
and of some of the enzymes suggests the existence of an organized
enzyme system in the cell cytosol in order to facilitate vectorial
catalysis.

We have established an in-vitro assay for de-novo purine biosynthesis which measures the incorporation of glycine-1-^{14}C into
hypoxanthine, the major end product in pigeon liver extracts (Table 1)
Formate is not essential in the presence of 5,10 methenyltetrahydro-
folate but as the enzymes involved in the synthesis of this folate
derivative and of 10 formyltetrahydrofolate, the other one carbon
derivative folate donor, are present in the active purine synthetic
fractions, it serves to constitute a regeneration system for these
derivatives. The requirement for PP-ribose-P can also be adequately
met by ribose-5-Pi. Qualitative and quantitative evaluation of the
products of the reaction was undertaken by thin layer chromatography
and high pressure liquid chromatography. Table II summarises the
results achieved by our technique of partial purification of the
enzymes of purine biosynthesis from pigeon liver. The critical factors in achieving this purification has been the use of the non

TABLE II
PURIFICATION OF ENZYMES OF DE NOVO
PURINE BIOSYNTHESIS FROM PIGEON LIVER

Step	Specific Activity	Total Protein	Recovery	Purification
	nmoles/mg protein/hr	mg	% initial activity	-fold
1. Homogenate	0.76	5460		
2. High Speed supernatant	24	1589	935	31
3. 4% PEG supernatant	32	1079	843	42
4. 11% PEG precipitate	187	171	774	246
5. Column Pool	4848	33	3864	6379
6. 15% PEG precipitate	4800	20	3477	6316

ionic polymer, polyethylene glycol (PEG), of defined molecular weight ranges and the minimisation of the ionic strengths of buffer solutions.

Pigeon liver was homogenized in 1.5 volumes of a solution containing 0.25 M sucrose, 5.0 mM β mercaptoethanol, 3.0 mM $MgCl_2$, 1.0 mM ATP and 50 mM Tris-Cl, pH 8.0 at 4° and centrifuged for 6.0 x 10^6g min. The resulting supernatant solution was fractionated between 4% and 11% concentration (w/v) of PEG 4000 by the dropwise addition of a 50% (w/v) aqueous solution of PEG. The 4-11% precipitate was dissolved in a minimum volume of solution consisting of 2% (w/v) PEG 20 M in 5.0 mM β mercaptoethanol, 3.0 mM $MgCl_2$, 1.0 mM ATP and 50 mM Tris-Cl pH 8.0. 1.0 ml of this solution (approximately 50-60 mg protein) was applied to a controlled pore glass (CPG-10-240, average pore diameter $215A^\circ$, 80-120 mesh) column, 172cm x 1.5cm equilibrated in the above buffer solution.

The column was developed at 4° at a flow rate of 15 ml per hour and the purine biosynthetic enzymes eluted as a single peak behind the major leading protein peak (Fig 1). There was, on the average, a five-fold increase in total activity recovered from the column, The activity was recovered from the eluate pool by precipitation with PEG 4000 to a final concentration of 15% (w/v).

The activity values given in Table II for the first three purification steps are not accurate maximal values as substrate limitation is a significant problem with these crude fractions. ATP, for example becomes limiting very rapidly despite the presence of a very generous ATP regeneration system.

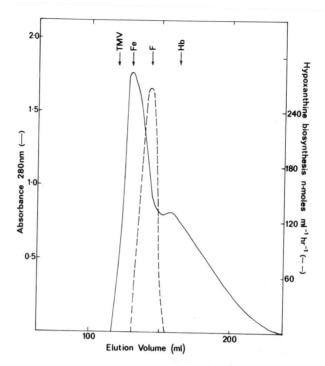

Figure 1. Chromatographic elution profile of the 4-11% PEG fraction
on CPG 10-240 glass column 172 cm x 1.5 cm. Continuous line
represents absorbance at 280 nm and broken line represents de novo
purine biosynthetic activity. Column markers (↓) are tobacco
mosaic virus (TMV), a void volume marker, horse spleen apoferritin
(Fe) MW 480,000, formyltetrahydrofolate synthetase (F) MW 240,000
and hemoglobin (Hb) MW 68,000.

Using this partially purified enzyme fraction a broad pH
optimum of between 7.5 and 8.5 and a temperature optimum of 50°
was established for de novo purine synthesis. Substrate deletion
experiments confirmed the requirements for purine synthesis. The
addition of 1.0 mM AMP to the assay system did not affect the
incorporation of glycine-1-^{14}C into hypoxanthine but at 5.0 mM AMP
isotope incorporation was completely inhibited. The analysis of
this inhibition is, however, quite complex as adenylate kinase is
present in the purified enzyme fraction and a number of labelled
intermediates are clearly present.

The elution volume of the peak of purine biosynthetic activity
on the glass chromatographic column raises a number of questions.

According to an internal enzyme marker, formyl tetrahydrofolate synthetase (EC 6.3.4.3) this corresponds to a molecular weight of 240,000 while the published molecular weight of different individual pathway enzymes cover a wide range from 34,000 (4) to 50,000 (5) for phosphoribosyl glycineamide synthetase (EC 6.3.1.3) from Aerobacter aerogenes, through 135,000 for avian liver phosphoribosyl formyl glycineamide synthetase (EC 6.3.5.3) (6) to 200,000 for phosphoribosyl pyrophosphate amidotransferase (EC 2.4.3.14) also from avian liver (7) and approximately 350,000 for the chicken liver protein containing phosphoribosylaminoimidazole carboxylase (EC 4.1.1.21) and phosphoribosylaminoimidazole succinocarboxamide synthetase (EC 6.3.2.6) activities (8). If the lower molecular weight enzymes were to polymerize or to be associated with one another it is possible that all enzyme activities would achieve closely overlapping elution profiles on the basis of molecular weight.

The existence of a large macromolecular complex cannot however, be entirely excluded. The relative elution volumes of many globular proteins on carefully prepared glass columns is a function of their molecular weights (or more exactly their Stokes' radii) but certain proteins such as cytochrome C have a strong tendency to adhere to the glass beads. A certain amount of information of the physical properties of some of the purine biosynthetic enzymes is available but not enough to be able to assess their potential chromatographic properties under all conditions.

While further studies are still required to identify the possible nature of an enzyme complex it is clear, however, that techniques have been developed to achieve partial purification in high yield of the enzymes of purine biosynthesis. This clearly has significant implications in terms of establishing the mechanisms of regulation of purine synthesis.

REFERENCES

1. Stoops,J.K., Arslanian,M.J.,Yang,H.OH., Kirk, A.C., Vanman,T.C. and Wakil,S.J. 1975. Presence of two polypeptide chain comprising fatty acid synthetase. Proc. Nat. Acad. Sci, (U.S.) 72, 1940-1944.

2. Shoaf,W.T. and Jones,M.E. 1973. Initial steps in pyrimidine synthesis in Ehrlich ascites carcinoma. Biochem. Biophys. Res. Comm. 45, 796-602

3. MacKenzie,R.E. 1973. Copurification of three folate enzymes from porcine liver. Biochem. Biophys. Res. Comm. 53, 1088-1095.

4. Nierlich,D.P. and Magasanik,B. 1965. Phosphoribosylglycinamide synthetase of Aerobacter aerogenes. J. Biol. Chem. 240,
   366-372.

5. Henrickson,K.D. 1967. PhD Thesis, Harvard University.

6. Mizobuchi,K. and Buchanan,J.M. 1968. Biosynthesis of the
   purines. XXIX Purification and properties of formylglycinamide ribonucleotide amidotransferase from chicken liver.
   J. Biol. Chem. 243, 4842-4850.

7. Rowe,P.B. and Wyngaarden,J.B. 1968. Glutamine phosphoribosylpyrophosphate Amidotransferase. Purification, substructure,
   amino acid composition and absorption spectra. J. Biol. Chem.
   243, 6373-6383.

8. Patey,C.A.H. and Shaw,G. 1973. Purification of an enzyme duet,
   phosphoribosylaminoimidazole carboxylase and phosphoribosylaminoimidazole succinocarboxamide synthetase, involved in the biosynthesis of purine nucleotides de novo. Biochem. J. 135,
   543-545.

COORDINATE REGULATION OF THE PROXIMAL AND DISTAL STEPS OF THE

PATHWAY OF PURINE SYNTHESIS DE NOVO IN WI-L2 HUMAN LYMPHOBLASTS

Michael S. Hershfield* and J. Edwin Seegmiller

Department of Medicine, University of California

San Diego, La Jolla, California 92093

In classic terms, there are potentially three sites at which *de novo* purine synthesis might be regulated: at the first step committed to IMP synthesis, and at the distal branch point where IMP is converted to adenine and guanine nucleotides. Since the late 1950's, numerous demonstrations in intact cells of inhibition of the overall rate of *de novo* synthesis by exogenous purines have been complemented by elegant mechanistic studies with partially purified phosphoribosyl amidotransferase (PAT)[1] from several sources, including humans (1,2). Regulation of the branch point has received less attention. IMP dehydrogenase and adenylosuccinate synthetase, the first enzymes committed to GMP and AMP synthesis, respectively, from IMP, have been partially purified and shown to be subject to inhibition by a variety of purine nucleotides (2). However, in the absence of studies in intact cells, schemes for regulation derived from these investigations can only be considered as potential mechanisms and their integration into regulation of the overall pathway cannot be established.

In attempting to characterize regulation of the IMP branch point in intact cells, we found most methods generally used in studying purine synthesis to be inadequate. The technique in which the fourth step of the pathway is inhibited with azaserine and incorporation of labeled formate or glycine into formylglycinamide
- - - - - - - - - -

* Present Address: Duke University Medical Center, Durham, North Carolina, 27710.

[1] Abbreviations: PAT, phosphoribosyl amidotransferase; PP-ribose-P, phosphoribosylpyrophosphate; FGAR, formylglycinamide ribonucleotide; 6-MMPR, 6-methylmercaptopurine riboside.

ribonucleotide (FGAR) is measured (3) gives no information regarding distal steps in the pathway. Furthermore, since azaserine blocks the synthesis of the endproducts of the pathway, the method potentially entails undesired and uncontrolled alterations in regulation of the proximal steps whose rate is being studied. Techniques currently used to measure labeling of soluble and nucleic acid purines are cumbersome because they require separate quantitation of a large number of soluble purine nucleotides as well as of label appearing in nucleic acid purines. Those methods which measure only the rate of labeling of soluble purine nucleotides or of nucleic acid purine nucleotides are unsuitable for characterizing IMP branch point regulation because of dilution effects due to the unequal size of the soluble adenine and guanine nucleotide pools. On the other hand, methods utilizing isotopically labeled hypoxanthine are unsuitable, first because hypoxanthine itself inhibits the *de novo* pathway; and second, because it competes for phosphoribosylpyrophosphate (PP-ribose-P) with other purine bases whose effects on the branch point are to be studied. We have developed a method (4) that provides for simple and simultaneous quantitation of the overall rate of *de novo* synthesis and of label entering *all* intracellular adenine and guanine nucleotides, whether in the soluble pools or nucleic acids. The method involves labeling cells with $\{^{14}C\}$formate for various periods of time, after which the cells are chilled, collected by centrifugation, and the cell pellet hydrolyzed in 0.4 N perchloric acid which converts all intracellular purine nucleotides to the three free purine bases adenine, guanine, and hypoxanthine. The hydrolysate is passed over a Dowex 50 column which is washed in 0.1 N HCl after which the three bases are eluted as a single fraction in 6 N HCl. An aliquot of this fraction is counted to give the overall rate of purine synthesis *de novo*. The remainder of the 6 N HCl fraction is evaporated, redissolved and the three purine bases separated from eachother by one-dimensional thin-layer chromatography. The distribution of counts in these three compounds permits quantitation of the fraction of newly synthesized IMP which is converted to guanine and adenine nucleotide. The ratio of total counts in guanine to total counts in adenine, termed the G/A ratio, provides a sensitive index of IMP branch point regulation.

In preliminary studies (4) of the kinetics of formate labeling of purines, we determined that the rate of labeling of total purines was linear for at least four hours at the cell concentrations and concentrations of sodium $\{^{14}C\}$formate used in our experiments. The apparent rate of purine synthesis depended critically on the concentration of $\{^{14}C\}$formate used. The rate of labeling of purines did not saturate even up to 5 mM formate. In addition, in comparing rates of purine synthesis in cells at different stages of growth, we found that at low concentrations of sodium $\{^{14}C\}$formate, e. g., 0.15 mM (Figure 1, lower lines) the rate of purine synthesis appeared to be faster in cells studied in medium which had been conditioned by growth of cells for about 30 or more hours compared to cells in

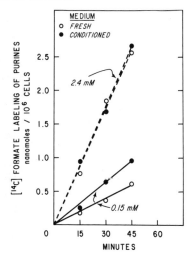

Figure 1. *Dependence of rate of labeling of purines on sodium-{[14]C}formate concentration in fresh and conditioned medium.* Mid-log phase cells were equilibrated for 4 hr in fresh medium (O) containing dialyzed serum or at the same density in medium in which cells had grown for 18 hr (conditioned medium, ●). Duplicate aliquots were then labeled with sodium {[14]C}formate at a concentration of either 0.15 mM (solid lines) or 2.4 mM (dashed lines) and labeling of total intracellular purines determined at 15, 30, and 45 min.

fresh medium. However, when the rates of purine synthesis in fresh and conditioned medium were compared using 2.4 mM sodium {[14]C}formate (Figure 1, upper lines), there was no difference in rate of labeling in the two media. Therefore, in experiments comparing relative rates of purine synthesis in different cell lines or in the same cell line at different stages of growth, rates were determined at more than one formate concentration. Since formate enters the nascent purine ring prior to IMP, the G/A labeling ratio was independent of formate concentration. In other experiments, we determined that the G/A labeling ratio became nearly constant after about 20 minutes so that in experiments where this ratio was determined, cells were pulsed for at least 30 minutes, by which time 93-97% of the counts incorporated into total intracellular purines were in guanine and adenine, the remainder in hypoxanthine.

Figure 2 shows the effects of exogenous purine bases on the overall rate of purine synthesis and on the guanine to adenine ratio, and also demonstrates the decrease in intracellular PP-ribose-P concentration which accompanies conversion of these bases to

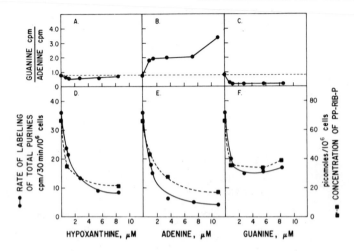

Figure 2. *Effect of exogenous purine bases on* de novo *purine synthesis and PP-ribose-P concentration.* Cultures were incubated with the indicated concentrations of hypoxanthine, adenine, or guanine. After 15 min, {^{14}C}formate was added and incubation continued another 30 min, at which time labeling of all intracellular purines and G/A labeling ratio were determined. PP-ribose-P concentrations were measured 30 min after addition of purine base.

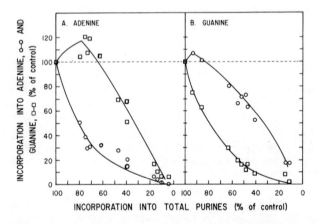

Figure 3. *Effects of exogenous adenine or guanine on IMP branch point.* Summary of several experiments similar to that described in legend to Figure 2, in which various concentrations of adenine or guanine were used to cause a range of degrees of inhibition of the overall rate of *de novo* synthesis (abscissa) at each of which percent inhibition of labeling of guanine (□) and adenine (0) compounds were determined (ordinate). (MSH and JES, submitted for publication.)

nucleotides. 50% inhibition of the overall rate of *de novo* synthesis occurs at concentrations of about 1-2 μM with each purine base. Hypoxanthine, which is converted to IMP, the common precursor of both adenine and guanine nucleotides, inhibits the rate of purine synthesis without altering significantly the G/A ratio. However, exogenous adenine and guanine each profoundly alter this ratio, even at the lowest concentrations which affect the overall rate of the *de novo* pathway. Adenine increases the G/A ratio while guanine has the opposite effect. Figure 3 demonstrates these results in a slightly different manner. Here, we summarize a large number of experiments in which various concentrations of exogenous adenine or guanine were used to produce different degrees of inhibition of the overall rate of *de novo* synthesis. It can be seen in Figure 3A that exogenous adenine not only inhibits the conversion of newly synthesized IMP to adenine nucleotides, but stimulates the synthesis of guanine nucleotides. This is most apparent at concentrations of adenine which cause less than about 40% inhibition of the overall rate of *de novo* activity. This results not from 40% inhibition of both adenine and guanine nucleotide synthesis but from about 70% inhibition of adenine and up to a 20% stimulation in guanine nucleotide synthesis. Exogenous guanine has the opposite effect, inhibiting guanine nucleotide synthesis and stimulating adenine nucleotide synthesis. Thus, in cells which derive their purine nucleotides from a combination of *de novo* synthesis and salvage of a particular purine base, there exists a coordinated and complementary mechanism for insuring an overall ratio of guanine and adenine nucleotides consistent with the cell's metabolic requirements.

The dual effects of adenine and guanine, i. e., selective inhibition of one and stimulation of the other branch of the *de novo* pathway, can be explained in either of two ways. Using adenine as an example, the first mechanism would simply involve inhibition of adenylosuccinate synthetase by AMP derived from salvage of the exogenous base. Inhibition of this branch of IMP utilization would permit the residual IMP synthesized by the *de novo* pathway to be utilized to a greater extent for guanine nucleotide synthesis simply because IMP normally is limiting for both branches of the pathway, being in very low intracellular concentration. Similarly, IMP dehydrogenase would be inhibited by guanine nucleotides formed from exogenous guanine, permitting a greater fraction of IMP derived from residual de novo synthesis to be converted to adenine nucleotides. That nucleotide formation is required for effects observed with exogenous purines is shown by the absence of these phenomena in mutants deficient in hypoxanthine-guanine phosphoribosyltransferase or adenine phosphoribosyltransferase activities. The alternative mechanism, suggested by Kelley *et al.* (2) would involve GTP-mediated regulation of the IMP branch point via inhibition of IMP dehydrogenase combined with stimulation by GTP of adenylosuccinate synthetase. This mechanism would not account for the effects caused by exogenous adenine. In addition, we

TABLE 1

EFFECTS OF 6-MMPR AND AZASERINE ON PURINE SYNTHESIS
VIA DE NOVO AND REUTILIZATION PATHWAYS

		LABELING OF INTRACELLULAR PURINES WITH			
		^{14}C FORMATE		20 uM ^{14}C HYPOXANTHINE	
ADDITION*		TOTAL PURINE	G/A RATIO	TOTAL PURINE	G/A RATIO
	µM	% OF CONTROL		% OF CONTROL	
6-MMPR	-	100	0.91	100	0.86
	.04	48	1.5	-	-
	.05	33	2.1	120	0.88
	0.10	18	3.5	126	0.82
	0.25	6	6.8	110	0.84
	-	-	-	100	0.72
	45	-	-	89	0.75
	90	-	-	95	0.75
AZASERINE	-	100	0.67	100	0.64
	4.5	93	0.73	121	0.68
	18	45	0.77	120	0.66
	45	11	0.54	130	0.69
	90	-	-	124	0.70

* CULTURES LABELED FOR 30 MIN FOLLOWING 15 MIN PREINCUBATION WITH INHIBITOR.

MSH and JES, submitted for publication.

favor the former mechanism rather than the latter because of the observation shown in Table 1 that the adenosine analog 6-methyl-mercaptopurine ribonucleoside (6-MMPR) mimics adenine by inhibiting adenine nucleotide synthesis *de novo* while causing a relative stimulation in guanine nucleotide synthesis. Since 6-MMPR is converted only to the nucleoside monophosphate derivative in WI-L2 cells (5), its effects are more easily explained in terms of its resemblance to AMP rather than by some mechanism involving GTP. Table 2 demonstrates the increase in intracellular PP-ribose-P concentration that accompanies the inhibition of *de novo* synthesis caused by 6-MMPR, even at very low concentration. This result, previously reported in other cell lines (6), is a clear example of purely nucleotide mediated inhibition dissociated from any PP-ribose-P limitation.

TABLE 2

EFFECT OF 6-MMPR ON INTRACELLULAR PP-RIBOSE-P CONCENTRATION AND PURINE SYNTHESIS DE NOVO

6-MMPR	De Novo Synthesis	PP-ribose-P Concentration
uM	% OF CONTROL	
0	100	100 (74 pmols/10^6 cells)
0.1	52	185
0.2	37	335
0.5	16	343
5.0	6	318

Protocol as described in legend to Figure 2 (M. S. Hershfield, R. Trafzer, and J. E. Seegmiller, in preparation).

Previous studies (7) showed no effect on conversion of $\{^{14}C\}$hypoxanthine to adenine nucleotides. We have made the same observation (Table 1). Also, in studies in intact lymphoblasts (8) and Ehrlich ascites cells (9) in which intracellular adenine nucleotide pools were expanded by preincubation with adenine, less than 10% inhibition of subsequent adenine nucleotide synthesis from exogenous $\{^{14}C\}$hypoxanthine could be demonstrated, in contrast to results we have obtained with our assay system. We initially thought the results we obtained with 6-MMPR, rather than indicating direct inhibition of adenylosuccinate synthetase by 6-MMPR-5'-PO$_4$, might simply be a consequence of some alteration in intracellular purine nucleotide pools caused by the inhibition of de novo synthesis by 6-MMPR which would not occur when hypoxanthine was available for nucleotide synthesis. To test this hypothesis, we examined the effects of azaserine, which also inhibits the de novo pathway, on the G/A ratio using our method or $\{^{14}C\}$hypoxanthine as precursor (Table 1). No effect on the IMP branch point was seen in either case, in contrast to results with 6-MMPR. Our results suggest that direct inhibition of adenylosuccinate synthetase by adenine nucleotides and their analogues does occur in intact cells but is quite sensitive to the intracellular concentration of IMP and does not occur when IMP synthesis from "salvage" of exogenous hypoxanthine is active. Our method appears to offer a useful alternative to the use of labeled hypoxanthine in identifying sites of action of purines and their analogs in intact cells.

Finally, I would like to mention some studies in which we examined changes in IMP branch point regulation during growth of cells in purine-free medium. This type of experiment is important in characterizing the inherent control mechanisms which operate in cells deriving their purine nucleotides solely from *de novo* synthesis rather than from a combination of *de novo* synthesis and reutilization of exogenous purines. In the experiment shown in Figure 4 parallel cultures at a range of cell densities were allowed to double in purine-free medium so that those at highest density would be approaching stationary phase. Fig. 4 (left) shows the overall rate of purine synthesis *de novo* declines sharply in these cultures. Fig. 4 (right) shows this decline involves a much greater inhibition of guanine than adenine nucleotide synthesis, producing a marked decrease in the G/A ratio from close to 1 at low and intermediate densities to about 0.25 at the highest cell density studied. An hour after returning these cells to fresh purine-free medium without changing cell density, the overall rate and G/A ratio had returned to the values observed at low cell densities. Parallel studies showed that the decrease in G/A ratio at higher cell densities was

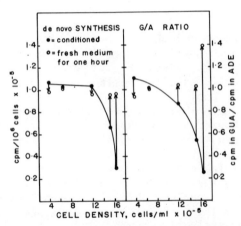

Figure 4. *Effect of growth to various cell densities on rate of purine synthesis* de novo *and G/A ratio.* Four parallel subcultures of WI-L2 at various initial cell densities were grown for 18 hr (one doubling) in medium containing 10% dialyzed calf serum. Then duplicate aliquots of each were centrifuged and cell pellets resuspended in the original volumes of either fresh medium (O) or the medium in which they had been growing (conditioned medium, ●). After an hour cultures were labeled for 60 min with {^{14}C}formate and labeling of total purines and G/A ratio determined. Arrows connect cells from the same original subculture. Four hr after resuspension aliquots from the subculture at highest density were again labeled for 60 min (most rightward data points in each panel). (MSH and JES, submitted for publication.)

TABLE 3

PURINE NUCLEOTIDE POOLS AND ENERGY CHARGE AFTER GROWTH
TO VARIOUS CELL DENSITIES IN PURINE FREE MEDIUM

CELL DENSITY (CELLS/ML x 10^{-5})	3.5	6.0	10.7	14.0
ATP + ADP + AMP (NANOMOLS/ 10^6 CELLS)	4.14	4.15	4.51	3.91
ADENYLATE ENERGY CHARGE	0.94	0.94	0.94	0.94
GTP + GDP + GMP (NANOMOLS/ 10^6 CELLS)	1.11	1.31	1.39	1.11
GUANYLATE ENERGY CHARGE	0.92	0.92	0.91	0.93

[*]ENERGY CHARGE =(NTP + 0.5 NDP) ÷ NTP + NDP + NMP

Aliquots from the four subcultures described in the legend to Figure 3, before changes of medium, were removed for determination of concentrations of intracellular purine nucleotides by high-pressure liquid chromatography. (MSH and JES, submitted for publication.)

not accompanied by any change in the concentrations of soluble adenine or guanine nucleotide pools or changes in energy charge (Table 3). The decrease in G/A ratio is not caused by a depletion of glutamine since the addition of glutamine to cells in conditioned medium does not change either the rate of *de novo* synthesis or the G/A ratio. It is possible that the equal rates of guanine and adenine nucleotide synthesis during early and mid-log phase reflect the requirements for DNA and RNA synthesis in actively growing cells, while the lower G/A ratio as cells enter stationary phase reflects a selective turn-off in synthesis of certain nucleic acid species, such as ribosomal RNA.

Another factor which might be involved in regulation of *de novo* purine synthesis during growth in purine free medium is PP-ribose-P concentration. In contrast to the constancy of purine nucleotide pools during growth (Table 3), we have repeatedly observed that PP-ribose-P concentration fluctuates during exponential growth in both WI-L2 and in mutants derived from WI-L2. This is demonstrated in Table 4 in which parallel cultures of WI-L2, an adenosine kinase deficient line, and a double mutant lacking both

TABLE 4

CHANGES IN PP-RIBOSE-P CONCENTRATION AND RATE OF
PURINE SYNTHESIS DE NOVO IN WI-L2 AND IN MUTANTS DEFICIENT
IN ADENOSINE KINASE (MTI^{r}107a) AND IN BOTH
ADENOSINE KINASE AND HPRT ACTIVITY (MTI-TG)

CELL LINE	CELL DENSITY CELLS/ML x 10^{-5}	LABELING OF INTRACELLULAR PURINES CPM/10^6 CELLS	PP-RIB-P PMOLES/10^6 CELLS
WI-L2	5.4	12050	21.5
	7.8	10710	44.8
	10.1	11240	95.2
	15.2	7760	105.7
MTI^{R}107A	5.0	11450	24
	7.2	10190	49.9
	9.7	8980	81.9
	13.6	5303	61.8
MTI-TG	5.0	14690	92.2
	6.9	14610	185.1
	9.0	15000	495.7
	12.4	17710	472.3

The indicated cell lines were subcultured in fresh purine free
medium containing 10% dialyzed, heated (5 hr at 62°) fetal calf
serum at initial densities of about 2.7 x 10^5 cells/ml and grown in
parallel. After 20, 27.5, 34, and 44 hrs growth aliquots were removed for determination of PP-ribose-P concentration and for pulse
labeling with {^{14}C}formate (2.17 mM, 4.6 μCi/μmol) for 60 min. Values shown are averages of triplicates.(MSH & JES, in preparation.)

adenosine kinase and hypoxanthine-guanine phosphoribosyltransferase
were grown in parallel and PP-ribose-P concentration and rate of
labeling of intracellular purines *de novo* monitored. Despite a
five-fold increase in PP-ribose-P concentration, no significant
increase in rate of *de novo* synthesis was observed. We conclude
that the rate of purine synthesis *de novo* is limited by factors
other than PP-ribose-P concentration under the growth conditions
studied. This is further suggested by the similarity in rates of
purine synthesis *de novo* in the normal and hypoxanthine-guanine
phosphoribosyltransferase-deficient lines despite over a four-fold
higher PP-ribose-P concentration in the mutant at all stages of
growth. This latter observation will be discussed in greater

length in another paper in this symposium (MSH, EBS, and JES, #41).

In summary, we describe a method for simultaneously assessing the overall rate of intracellular purine synthesis *de novo* as well as the ratio of newly synthesized guanine to adenine nucleotides. This method has been used to study IMP branch point regulation in the *de novo* pathway both in cells growing in purine-free medium and in cells deriving their purine nucleotide requirements from a combination of *de novo* synthesis and utilization of exogenous purine bases. In the former situation, the G/A ratio is close to 1 in actively dividing cells and falls precipitously as cells enter stationary phase, implying that regulation of the IMP branch point is attuned to the requirements for RNA and DNA synthesis in dividing or stationary cells. In the latter situation, the IMP branch point is regulated in such a way that the two routes of purine nucleotide synthesis, reutilization and *de novo*, complement each other in providing cells with the ratio of guanine to adenine nucleotides consistent with their metabolic activity. Purine nucleotide pools are more rigorously regulated than PP-ribose-P concentration during growth in purine free medium. The rate of purine synthesis *de novo* is not limited by PP-ribose-P concentration under these growth conditions.

ACKNOWLEDGEMENTS

This work was supported in part by United States Public Health Service grants AM-13622, AM-05646, and GM-17702, and by grants from the National Foundation and the Kroc Foundation. MH is the recipient of National Institutes of Health Research Fellowship AM-00710-01.

REFERENCES

1.   Henderson, J. F.: *Regulation of Purine Biosynthesis*. American Chemical Society, Washington, D. C., 1972.
2.   Kelley, W. N., Holmes, E. W., and Van Der Weyden, M. B.: *Arthritis Rheum.* 18 (Suppl.): 673-680, 1975.
3.   Henderson, J. F.: *J. Biol. Chem.* 237: 2631-2635, 1962.
4.   Hershfield, M. S., and Seegmiller, J. E.: Submitted for publication.
5.   Hershfield, M. S.: Unpublished observation.
6.   Henderson, J. F., and Khoo, M. K. Y.: *J. Biol. Chem.* 240: 3104, 1965.
7.   Shantz, G. D., Smith, C. M., Fontenelle, L. J., Lau, H. K. F., and Henderson, J. F.: *Cancer Res.* 33: 2867-2871, 1973.
8.   Astrin, K. H.: *Purine Metabolism in Cultured Human Lymphoid Cells*. Doctoral Thesis, University of California, San Diego, 1973.
9.   Snyder, F. F., and Henderson, J. F.: *Can. J. Biochem.* 51: 943-948, 1973.

PURINE TOXICITY IN HUMAN LYMPHOBLASTS

Floyd F. Snyder,* Michael S. Hershfield† and J. Edwin
Seegmiller
Department of Medicine, University of California, San
Diego, La Jolla, California 92093

The selective toxicity of adenosine to dividing lymphoid cells
is considered a possible basis for a type of severe combined immunodeficiency disease in which there is absence of adenosine deaminase (ADA)[1] activity, the enzyme which converts adenosine to inosine.
Treatment of lymphoid cells with adenosine has been reported to lower
intracellular pyrimidine nucleotide concentrations (1,2) and to
transiently increase adenosine 3',5'-cyclic phosphate (cAMP) concentration (3,4). We have recently shown that depletion of intracellular
phosphoribosylpyrophosphate (PP-ribose-P) concentration in cultured
human lymphoblasts is responsible for the decrease in concentration
of nucleotides dependent on PP-ribose-P for their synthesis (5). It
is not known which of these several effects are mediated by adenosine
directly or by phosphorylated products of adenosine. We now present
evidence that adenosine's toxicity may not require its phosphorylation
and does not depend upon change in cAMP concentration.

The growth inhibitory effect of adenosine on the diploid human
lymphoblast line WI-L2 is potentiated by inhibitors of adenosine
- - - - - - - - - -

* Present address: Biochemistry Group, Department of Chemistry, The
University of Calgary, Calgary, Alberta, Canada.

† Present Address: Duke University Medical Center, Durham, North
Carolina, 27710.

[1] Abbreviations: ADA, adenosine deaminase; AK, adenosine kinase;
HPRT, hypoxanthine-guanine phosphoribosyltransferase; EHNA, erythro-
9-(2-hydroxy-3-nonyl)adenine; 6-MMPR, 6-methylmercaptopurine ribonucleoside; APRT, adenine phosphoribosyltransferase; PP-ribose-P,
phosphoribosylpyrophosphate; MEM, minimal essential medium; cAMP,
adenosine 3',5'-cyclic phosphate.

deaminase activity. In their presence we find adenosine is still growth inhibitory to mutants of WI-L2 which have less than 1% of the normal adenosine kinase (AK) activity. Therefore, adenosine *per se*, rather than intracellular nucleotides to which it is converted, may be the primary inhibitor of cell division. In support of this conclusion, we have found some biochemical consequences of exposure to adenosine to be distinct from those caused by the related purine compounds, adenine and hypoxanthine, which can be converted to the same intracellular nucleotides as adenosine. We further report that mutants of WI-L2 severely deficient in adenine phosphoribosyltransferase (APRT) activity are still sensitive to growth inhibition by adenine, just as AK mutants are still sensitive to the effects of adenosine.

CULTURE AND CHARACTERIZATION OF PARENT AND MUTANT WI-L2 HUMAN LYMPHOBLASTS

The AK$^-$ mutant MTI^{r}107a was selected by resistance to the adenosine analog 6-methylmercaptopurine ribonucleoside (6-MMPR) and has less than 1% of residual AK activity (M. Hershfield, R. Trafzer, and J. E. Seegmiller, in preparation, and this symposium, #41) when measured using as substrate either {^{14}C}6-MMPR or {^{14}C}adenosine in the presence of an inhibitor of ADA activity, erythro-9-(2-hydroxy-3-nonvl)adenine (EHNA) (6). A series of mutants, severely deficient (<1%) or partially deficient (10-30%) in APRT activity were selected from WI-L2 by resistance to 2,6-diaminopurine (M. Hershfield,

TABLE 1

EFFECTS OF ADENINE AND HYPOXANTHINE ON PURINE SYNTHESIS
DE NOVO AND PP-RIBOSE-P CONCENTRATION IN WI-L2

ADDITIONS*	PURINE SYNTHESIS	PP-RIBOSE-P
	CPM/10 MIN PER 10^6 CELLS	PICOMOLES
NONE	26700	89
0.5 MM HYPOXANTHINE	1130	11.5
0.1 MM ADENINE	1200	3.2

* ADDED 24 HRS PRIOR TO ASSAY

TABLE 2

EFFECTS OF ADENINE, HYPOXANTHINE, AND URIDINE
ON SOLUBLE NUCLEOTIDE POOLS IN WI-L2

ADDITIONS* =	NONE	0.5 mM HYPOXANTHINE	0.5 mM ADENINE	1 mM URIDINE +0.5 mM ADENINE
		NANOMOLES/10^6 CELLS		
AMP/OROTIC ACID	0.07	0.15	0.05	0.1
AMP + ADP + ATP	4.3	4.4	6.2	4.9
GMP + GDP + GTP	1.3	1.1	1.3	0.8
UMP + UDP + UTP + UDP - SUGARS	2.9	1.4	1.5	7.0

* ADDED 24 HRS PRIOR TO ASSAY

E. B. Spector, and J. E. Seegmiller, this symposium #41). Lymphoblast lines conditioned to growth in medium supplemented with 10%
horse serum, which lacks adenosine deaminase activity, were used in
studies with adenosine. Similar results were obtained in medium
containing 10% fetal calf serum which had been dialyzed extensively
and heated to 62° for five hr to inactivate ADA activity. Mutants
and parent grew at the same rate in both media.

EFFECTS OF ADENINE AND HYPOXANTHINE ON THE GROWTH, PURINE
SYNTHESIS DE NOVO, INTRACELLULAR PP-RIBOSE-P CONCENTRATION,
AND PURINE AND PYRIMIDINE NUCLEOTIDE POOLS

Experiments with adenine, which is toxic, and with hypoxanthine,
which is not, demonstrate that inhibition of *de novo* purine synthesis
and depletion of PP-ribose-P concentration do not of themselves inhibit growth. Thus 10^{-4} M adenine, which just begins to slow growth,
and 5×10^{-4} M hypoxanthine both strongly inhibit *de novo* purine
biosynthesis throughout the log phase of growth and both cause prolonged reduction in the intracellular concentration of PP-ribose-P
(Table 1). Table 2 compares the effects of 5×10^{-4} M adenine or
hypoxanthine in the presence and absence of 1 mM uridine, on the intracellular concentrations of adenine, guanine, and uridine
nucleotides. Both purines cause about a 50% reduction in the concentration of uridine nucleotides, but neither causes an accumulation of orotic acid (which does not separate from AMP in the high-
pressure liquid chromatographic system used in these studies). In
the presence of 1 mM uridine, pyrimidine pools were expanded to
greater than their concentration in the control cells. Nevertheless, uridine has virtually no effect on the growth inhibition

caused by adenine (Table 3). Nor did various concentrations of the
deoxypurine or pyrimidine nucleosides diminish adenine toxicity
(which might have occurred if adenine toxicity resulted from inhibition of ribonucleotide reductase). These experiments indicate
that (1) diminished PP-ribose-P concentration *per se* is not growth-
inhibitory since adenine, which is toxic, and hypoxanthine, which
is not toxic, both lowered PP-ribose-P concentration to the same
degree; and (2) diminished pyrimidine nucleotide concentration is
not the basis of adenine toxicity since hypoxanthine causes a
similar lowering of intracellular pyrimidine pools, and repletion of
these pools does not alter inhibition of growth by adenine.

ADENINE MEDIATED GROWTH INHIBITION OF LYMPHOBLASTS DEFICIENT IN ADENINE PHOSPHORIBOSYLTRANSFERASE ACTIVITY

Cell lines selected from WI-L2 which were resistant to 200 µM
diaminopurine (compared to 50% inhibition of the parent strain at
less than 10 µM diaminopurine) had less than 1% of parental levels
of APRT activity. These APRT⁻ mutants, as well as others which had
between 10 and 30% of residual APRT activity, showed no increase in
resistance to the growth-inhibitory effects of adenine at any concentration compared to the parental strain (Table 3). The most
deficient strains showed less than 1% of the parental rate of
conversion of [^{14}C]adenine (0.5 mM) to intracellular nucleotides.
Thus, adenine is toxic even when it is not converted to intracellular nucleotides.

TABLE 3

ADENINE TOXICITY TOWARD GROWTH OF WI-L2 AND APRT⁻ LYMPHOBLASTS

A.

	ADENINE, mM			
	0.1	0.4	0.6	0.8
	GROWTH RELATIVE TO CONTROL*			
WI-L2	1.0	0.67	0.48	.36
APRT⁻**	0.7	0.52	0.40	0.31

B.

	0.5 mM HYPOXANTHINE	0.5 mM URIDINE	1 mM ADENINE	1 mM ADENINE + 0.5 mM URIDINE
	GROWTH RELATIVE TO CONTROL*			
WI-L2	1.4	0.92	0.08	0.11
APRT⁻**	-	1.0	0.11	0.13

* CONTROL = NO ADDITIONS, 72 HRS GROWTH.
** < 1% OF WI-L2 APRT SPECIFIC ACTIVITY.

EFFECTS OF ADENOSINE AND INHIBITION OF ADENOSINE
DEAMINASE ACTIVITY ON WI-L2 LYMPHOBLASTS

EHNA, 5 μM, which inhibits lymphoblast ADA activity by >95%, has little effect on cell growth in the absence of exogenous adenosine but enhances by greater than 10-fold the sensitivity of WI-L2 lymphoblasts to growth inhibition by adenosine (Table 4). Complete growth inhibition occurred only after 24 hr exposure, approximately one doubling (Figure 1A) and was reversible up to 72 hr culture with adenosine, 50 μM, and EHNA, 5 μM (Figure 1B), demonstrating that cell viability is retained despite growth inhibition. We have presented evidence elsewhere that adenosine in combination with ADA inhibitors is not irreversibly toxic to non-dividing human lymphocytes (7). The growth inhibitory effect of adenosine and EHNA is is reversed by 10 or 100 μM uridine (Fig. 1A). The first column in Table 5 shows the effect of 24 hr incubation with adenosine and inhibitors of ADA activity on PP-ribose-P concentration in WI-L2. Despite inhibiting lymphoblast ADA activity by more than 95%, coformycin or EHNA alone had little effect on reducing cell growth or PP-ribose-P concentration. In contrast, adenosine in combination with the ADA inhibitors lowered the PP-ribose-P concentration by

TABLE 4

POTENTIATION OF ADENOSINE GROWTH INHIBITION OF WI-L2

LYMPHOBLASTS BY EHNA

	GROWTH RELATIVE TO CONTROL*	
	EHNA (μM)	
Adenosine (M)	0	5
0	1.00	0.92
10^{-6}	0.99	0.91
10^{-5}	0.87	0.68
10^{-4}	0.62	0
10^{-3}	0.17	0

* Control: no additions, 0-72 hours growth.

FIGURE 1

ADENOSINE GROWTH INHIBITION IN PARENT
AND ADENOSINE KINASE DEFICIENT LYMPHOBLASTS

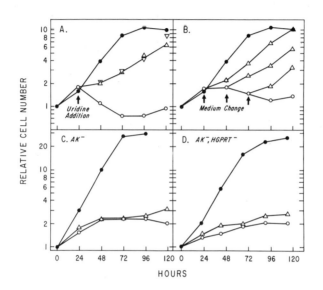

Lymphoblasts were cultured with no additions (●) or 5 µM EHNA plus 50 µM adenosine (0).

A.   10 µM (Δ) or 100 µM (▽) uridine was added to EHNA and adenosine treated lymphoblasts at 24 hours.

B.   Lymphoblasts exposed to EHNA plus adenosine were resuspended in fresh medium without additions at 24, 48 or 72 hours (Δ).

C.   EHNA and adenosine treated adenosine kinase deficient lymphoblasts were cultured in the absence (0) and presence (Δ) of 100 µM uridine.

D.   EHNA and adenosine treated adenosine kinase, HPRT deficient lymphoblasts were cultured in the absence (0) and presence (Δ) of 100 µM uridine.

TABLE 5

EFFECT OF ADENOSINE IN COMBINATION WITH ADENOSINE DEAMINASE

INHIBITORS UPON LYMPHOBLAST PP-RIBOSE-P CONCENTRATION

	RELATIVE PP-RIBOSE-P CONCENTRATION*	
ADDITIONS	WI-L2 24 HOURS	AK⁻ 48 HOURS
None	1.00	1.00
Coformycin	1.28	1.59
EHNA	0.93	0.88
Adenosine	0.32	0.22
Coformycin + Adenosine	<0.01	<0.01
EHNA + Adenosine	0.07	<0.01

Coformycin, 1μg/ml; EHNA, 5 μM; adenosine, 50 μM.
* PP-ribose-P concentrations with no additions: 133 and 190 pmoles/10^6 cells for WI-L2 and AK⁻ respectively.

greater than 90%. In other studies we have shown the consequence of this adenosine-mediated reduction in PP-ribose-P concentration to be depletion of UTP, CTP, UDP-sugars, NAD, and under certain conditions GTP concentrations and, unlike adenine, an accumulation of orotic acid (5).

EFFECT OF ADENOSINE ON MUTANTS OF WI-L2 DEFICIENT IN ADENOSINE KINASE ACTIVITY

The mutants MTI^{r}107a (AK⁻) and MTI-TG (AK⁻-HPRT⁻) grew normally in 1 mM 6-MMPR, whereas WI-L2 cells were inhibited completely by 0.5 μM (M. Hershfield, R. Trafzer, and J. E. Seegmiller, in preparation). Nevertheless, the combination of adenosine and EHNA remains growth inhibitory to both mutants (Fig. 1C and 1D). Uridine, in contrast to its effects on normal lymphoblasts, does not reverse adenosine growth inhibition in adenosine kinase deficient mutants (Fig. 1C,D). Table 5, column 2 shows PP-ribose-P concentrations in AK⁻ cells were reduced greater than 95% after 48 hr culture in the presence of adenosine plus either of the ADA inhibitors, coformycin or EHNA. At 24 hr, prior to complete growth inhibition (Fig. 1C), UTP and CTP concentrations were decreased to 40% of control while purine nucleotide concentrations were unchanged. The decrease in pyrimidine nucleotides and PP-ribose-P concentrations (also at 24 hr) in AK⁻ cells are therefore not produced by increased purine nucleotide concentrations found in AK⁺ lymphoblasts under these conditions.

The ability of uridine to reverse adenosine-mediated growth inhibition in the parental cell line but not in the AK⁻ mutants can best be explained if one assumes that PP-ribose-P depletion effectively inhibits both purine and pyrimidine nucleotide synthesis *de novo*. In the parental cell line, purine nucleotides may be provided via the phosphorylation of adenosine to AMP by adenosine kinase and pyrimidine nucleotides via phosphorylation of uridine to UMP by uridine kinase. Neither of these reactions utilizes PP-ribose-P, and we have shown lymphoblast adenylate energy charge (8) to be normal at a reversible stage of adenosine growth inhibition (5), indicating a capability for energy dependent processes such as nucleoside phosphorylation. Uridine does not reverse growth inhibition in the AK⁻ cells since these cells presumeably have no source of purine nucleotide: the absence of human inosine or guanosine kinase activities (9) and the requirement of PP-ribose-P for nucleotide

TABLE 6

EFFECT OF ADENOSINE ON {1-^{14}C}-GLUCOSE METABOLISM

IN HUMAN LYMPHOBLASTS

ADDITIONS	{^{14}CO$_2$}GENERATION
	PERCENT INHIBITION
None	0
Adenine	8
Adenosine	7
Coformycin	14
Coformycin + Adenine	7
Coformycin + Adenosine	38

Lymphoblasts were cultured for 24 hours with additions prior to incubation with {1-^{14}C}glucose.

Adenine, 50 μM; Adenosine, 50 μM; Coformycin, 1 μg/ml.

synthesis from purine bases effectively preclude the possibility of reversal of adenosinegrowth inhibition in AK⁻ cells by a second exogenous purine. Finally, the mechanism by which adenosine inhibits pyrimidine synthesis *de novo* differs from that of adenine and hypoxanthine in that adenosine, but not the purine bases, causes an accumulation of orotic acid, despite equivalent depression in PP-ribose-P concentration. These findings suggest that PP-ribose-P synthesis is inhibited by adenosine, whereas adenine and hypoxanthine merely depress the steady state concentration of PP-ribose-P, but not its availability for nucleotide synthesis.

In additional studies we have attempted to elucidate the biochemical site of adenosine-mediated reduction in PP-ribose-P. Adenosine, 50 or 500 µM, in the absence or presence of EHNA, 5 µM, had no effect on PP-ribose-P synthetase activity in lymphoblast extracts (5). The metabolism of {1-^{14}C}glucose to {^{14}CO$_2$} was then examined after 24 hr exposure to adenosine in the presence and absence of the ADA inhibitor, coformycin (Table 6). The growth inhibitory concentration of adenosine and coformycin produced the greatest inhibition of {^{14}CO$_2$} generation from {1-^{14}C}glucose. The specificity of inhibition for the nucleoside adenosine was demonstrated by adenine producing less than 10% inhibition, either alone or in combination with coformycin. Thus, an adenosine mediated inhibition of pentose phosphate synthesis may account in part for the reduction in lymphoblast PP-ribose-P synthesis.

The possibility of the adenosine effect being mediated by cAMP has also been considered, even though the adenosine mediated increase in glycogen content in HeLa cells is opposite to an effect of increased cellular cAMP (1). We found no significant difference in lymphoblast cAMP concentration after 24 hr incubation, a time of maximal but reversible growth inhibition, in the absence or presence of 50 µM adenosine and 5 µM EHNA. Taken together, these findings indicate the nucleoside adenosine may alter the relative routes of glucose-6-phosphate metabolism, increasing glycogenesis (1) and decreasing pentose phosphate synthesis, the ultimate consequence being a depletion of PP-ribose-P and thus nucleotides required for nucleic acid synthesis. The exact mechanism of adenine toxicity remains unclear.

ACKNOWLEDGEMENTS

This work was supported in part by United States Public Health Service grants AM-13622, AM-05646, and GM-17702, and by grants from the National Foundation and the Kroc Foundation. MH is the recipient of National Institutes of Health Research Fellowship AM-00710-01.

REFERENCES

1. Hilz, H., and Kaukel, E.: *Mol. Cell Biochem.* 1: 229-239, 1973.
2. Green, H., and Chan, T.-S.: *Science* 182: 836-837, 1973.
3. Wolberg, G., Zimmerman, P., Hiemstra, K., Winston, M., and Chu, L.-C.: *Science* 187: 957-959, 1975.
4. Zenser, T. V.: *Biochim. Biophys. Acta* 404: 202-213, 1975.
5. Snyder, F. F., and Seegmiller, J. E.: *FEBS Letters*, in press, 1976.
6. Schaeffer, H. J., and Schwender, C. F.: *J. Med. Chem.* 17: 6-8, 1974.
7. Snyder, F. F., Mendelsohn, J., and Seegmiller, J. E.: *J. Clin. Invest.*, in press, 1976.
8. Atkinson, D. E.: *Biochemistry* 7: 4030-4034, 1968.
9. Friedmann, T., Seegmiller, J. E., and Subak-Sharpe, J. H.: *Expt. Cell Res.* 56: 425-429, 1969.

STUDIES ON THE REGULATION OF THE BIOSYNTHESIS OF MYOCARDIAL

ADENINE NUCLEOTIDES

H.-G. Zimmer, E. Gerlach

Physiologisches Institut der Universität München

Pettenkoferstr. 12, 8000 München 2

    Stimulation of cardiac ß-adrenergic receptors with isoproterenol results in a considerable enhancement of the biosynthesis of adenine nucleotides (AN) in the rat heart in vivo (17). AN biosynthesis proves also to be increased when the isoproterenol-induced diminution of the concentration of myocardial AN is prevented (19). Thus, release of feedback inhibition of 5-phosphoribosyl-1-pyrophosphate-amidotransferase (EC 2.4.2.14) cannot be exclusively responsible for the acceleration of AN synthesis observed under these conditions (8,16). An alternative mechanism seems to be operative which affects de novo synthesis of AN through the available pool of 5-phosphoribosyl-1-pyrophosphate (PRPP) (9).

    Isoproterenol may influence the PRPP pool by the following sequence of metabolic events: The increased concentration of 3',5'-cyclic AMP (cAMP) resulting from the stimulation of adenyl cyclase (13) leads to an enhancement of glycogenolysis with elevated concentrations of glucose-1-phosphate and glucose-6-phosphate (12,15, 21). Provided that under these conditions the pentose phosphate pathway is also activated (14), PRPP may become available to a greater extent.

    The present investigations were designed in order to elucidate whether isoproterenol, cAMP and substrates of the pentose phosphate cycle actually influence myocardial de novo synthesis of AN by affecting the available pool of PRPP.

MATERIALS AND METHODS

Female Sprague-Dawley rats (200-220 g) fed a diet of Altromin were used in all experiments. $1-^{14}C$-glycine (specific activity 56 mCi/mmole) and $8-^{14}C$-adenine (specific activity 54.2 mCi/mmole) were purchased from the Radiochemical Centre, Amersham, England. Isoproterenol was obtained from C.H. Boehringer Sohn, Ingelheim. Propranolol was a gift from ICI-Pharma, Heidelberg; compound D 600 (α-isopropyl-α-[(N-methyl-N-homoveratry)γ-aminopropyl] 3,4,5-tri-methoxyphenylacetonitrile) was generously supplied by Knoll AG, Ludwigshafen. Adenine and Dibutyryl-cAMP were obtained from Boeh-ringer Mannheim, D(+)xylose, D(-)ribose, xylitol and ribitol were purchased from Sigma, München. All other chemicals were obtained from Merck AG, Darmstadt and were of analytical grade.

The substances were dissolved in saline (0.9%, w/v) and ad-ministered subcutaneously or intravenously. At various times after administration, measurements of de novo synthesis of myocardial AN were performed. The rats which had been fasted for 12 hours, were intravenously injected with $1-^{14}C$-glycine (0.25 mCi/kg in 1 ml of saline). At the end of a 60-minute exposure to $1-^{14}C$-glycine, they were anesthetized with diethyl ether and the hearts rapidly ex-cised. The ventricles were freed of adhering blood and quickly frozen in liquid nitrogen. In experiments designed to determine concentration changes of cardiac AN (ATP, ADP, AMP, cAMP), the hearts taken from artificially ventilated animals were rapidly fro-zen in Freon at a temperature of -158°C. Subsequently, the ventri-cles were freed of frozen blood under liquid nitrogen.

The concentrations of ATP, ADP and AMP were determined after separation by paper chromatography (4). Rates of de novo synthesis of myocardial AN were obtained by relating the total radioactivity of AN to the mean specific activity of the intracellular glycine (18). cAMP was determined according to known methods (2,5) utili-zing the radioisotope dilution test (Boehringer Mannheim).

RESULTS AND DISCUSSION

Table 1 summarizes the results of studies in which the con-centrations of cAMP and rates of de novo synthesis of AN were de-termined in rat hearts. It is evident that isoproterenol causes an increase of both cAMP concentration and synthesis de novo of AN. The isoproterenol-induced enhancement of these parameters is at-tenuated, when isoproterenol and compound D 600 (7) are admini-stered at the same time. Furthermore, propranolol leads to a dimi-nution of the concentration of cAMP and of the rate of de novo syn-thesis.

Table 1: Changes of the concentration of 3',5'-cyclic AMP (cAMP)
and of the rate of de novo synthesis of adenine nucleotides in rat
hearts in vivo. Measurements were performed 60 min after s.c. administration of the drugs. Data are expressed as percentage of
change of controls. n = number of experiments.

	cAMP	Synthesis de novo
Isoproterenol (5 mg/kg)	+ 134 (n=8)	+ 149 (n=9)
Isoproterenol (5 mg/kg) + D 600 (10 mg/kg)	+ 37 (n=11)	+ 53 (n=6)
Propranolol (50 mg/kg)	− 31 (n=6)	− 64 (n=3)

Additional experiments were carried out to determine whether
an increase of cAMP not mediated by an activation of myocardial
adenyl cyclase is also associated with an acceleration of de novo
synthesis of cardiac AN. Table 2 shows that dibutyryl-cAMP as well
as aminophylline which inhibits phosphodiesterase (3,10) stimulate
de novo synthesis by 130% and 148%, respectively. This increase is
of the same order of magnitude as that induced by a low dose of
isoproterenol. When isoproterenol and aminophylline are administered simultaneously, the effects of both drugs are additive. It
thus appears that there are parallel changes in the concentration
of cAMP and the rate of de novo synthesis in the myocardium under
the experimental conditions studied.

In order to elucidate whether in rat hearts the flow through
the pentose phosphate pathway may limit the availability of ribose-
5-phosphate and PRPP and thus the rate of biosynthesis of AN, several substrates of the pentose phosphate cycle were supplied. The
data in Table 3 indicate that the pentoses and pentitols tested
cause an enhancement of myocardial AN biosynthesis. These findings
suggest that the supply of ribose-5-phosphate and PRPP limited in
the normal heart can be increased by exogenously applied pentoses
and pentitols.

Table 2: Influence of dibutyryl-cAMP, aminophylline and isoproterenol on the synthesis de novo of myocardial adenine nucleotides
(nmoles/g/h). Doses applied, kind of administration and time of
exposure are given in parentheses. Mean values ± SEM; n = number
of experiments.

	nmoles/g/h
Control	5.8 ± 0.6 (n=11)
Dibutyryl-cAMP (50 mg/kg, i.v., 1 h)	13.4 ± 1.4 (n=6)
Aminophylline (50 mg/kg, s.c., 3 h)	14.4 ± 0.5 (n=2)
Isoproterenol (0.1 mg/kg, s.c., 3 h)	17.4 ± 1.0 (n=5)
Isoproterenol (0.1 mg/kg, s.c., 3 h) + Aminophylline (50 mg/kg, s.c., 3 h)	26.7 ± 1.9 (n=2)

To test whether such a limitation may also exist in the isoproterenol-stimulated heart, in which AN synthesis is already accelerated to about 40 nmoles/g/h, the effects of added xylitol
were studied under these conditions. As is evident from the data
in Table 4, AN synthesis proved to be further exaggerated, the increase amounting to 190 nmoles/g/h. It thus appears that also the
isoproterenol-induced enhancement of AN synthesis is most likely
restricted by the available pool of PRPP.

In Table 4 data are also included concerning the effects of
adenine on myocardial AN synthesis under various conditions. Adenine was used in an attempt to further evaluate the significance
of the PRPP pool for the regulation of AN synthesis. The approach
is based on the fact that PRPP is needed for both de novo synthe-

sis of nucleotides and salvage of purine bases. Provided the two
pathways compete for PRPP in the myocardium as they do in other
tissues (1,6,11), it should be possible to suppress de novo synthesis by increasing the supply of purine bases. Obviously, adenine suppresses markedly de novo synthesis in the normal heart and
prevents the acceleration of this process in the isoproterenol-
stimulated myocardium. Further experiments revealed that the concentrations of AN are not affected by adenine (20). The increase
of AN synthesis caused by isoproterenol and xylitol given either
alone or in combination proved also to be abolished by adenine.
Thus, the suppression of de novo synthesis in the normal heart and
the reduction of this process in isoproterenol-stimulated and xy-
litol-treated hearts can only be explained to result from a reduced availability of PRPP due to its intensified utilization in
the reaction catalyzed by adenine phosphoribosyltransferase
(EC 2.4.2.7). Additional support for the validity of this concept
was obtained in studies on the incorporation of radioactive adenine.

Table 3: Effects of some pentoses and pentitols (100 mg/kg, i.v.)
on the rates of de novo synthesis of adenine nucleotides
(nmoles/g/h) in rat hearts in vivo. Analyses were done 60 min after i.v. application of the substrates and of 1-^{14}C-glycine. Mean
values $\pm$ SEM; n = number of experiments.

	nmoles/g/h
Control (n=11)	5.8 $\pm$ 0.6
D-Ribose (n=4)	21.2 $\pm$ 1.0
Ribitol (n=3)	25.3 $\pm$ 6.7
D-Xylose (n=2)	23.2 $\pm$ 2.5
Xylitol (n=4)	22.7 $\pm$ 3.5

Table 4: Rates of biosynthesis of myocardial adenine nucleotides (nmoles/g/h) as influenced by isoproterenol and xylitol in the absence and presence of adenine (10 mg/kg, i.v.). Rats were exposed to these substances for 60 min. Only in experiments in which isoproterenol and xylitol were applied together, isoproterenol was administered 4 hours prior to xylitol and adenine, respectively. Mean values $\pm$ SEM; n = number of experiments.

	Without adenine	With adenine
Control	5.8 ± 0.6 (n=11)	0.4 ± 0.2 (n=8)
Isoproterenol (5 mg/kg, s.c.)	14.5 ± 0.7 (n=9)	3.0 ± 0.5 (n=7)
Xylitol (100 mg/kg, i.v.)	22.7 ± 3.5 (n=4)	5.6 ± 0.4 (n=3)
Isoproterenol (25 mg/kg, s.c.) + Xylitol (100 mg/kg, i.v.)	190.0 ± 29 (n=7)	26.5 ± 1.5 (n=4)

Fig. 1 shows the effect of isoproterenol on the incorporation of 8-^{14}C-adenine into myocardial AN. In isoproterenol-stimulated hearts 8-^{14}C-adenine is incorporated to a greater extent than in normal hearts. Under the influence of xylitol, the incorporation of adenine proved to be much greater. These findings as well as the observed inhibitory action of adenine on the isoproterenol- and xylitol-induced acceleration of de novo synthesis provide indirect evidence that isoproterenol and xylitól, respectively, actually cause an enlargement of the available pool of PRPP.

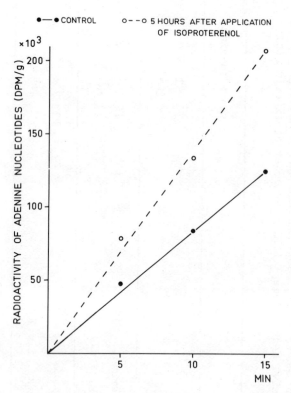

<u>Fig. 1</u>: Effect of isoproterenol (5 mg/kg, s.c.) on the incorporation of 8-^{14}C-adenine (0.25 mCi/kg, i.v.) into myocardial adenine nucleotides. Each point represents the mean of 2 experiments.

SUMMARY

   1) Changes in the rates of biosynthesis of adenine nucleotides
in rat hearts under various experimental conditions are paralleled
by corresponding alterations in the concentration of cyclic AMP.

   2) Pentoses and pentitols cause an acceleration of the de novo synthesis of adenine nucleotides in the normal heart and a further amplification of the increase of adenine nucleotide synthesis
in isoproterenol-stimulated hearts.

   3) The enhancement of de novo synthesis of adenine nucleotides
induced by isoproterenol as well as by pentoses and pentitols appears to be causally related to a greater availability of 5-phos-
phoribosyl-1-pyrophosphate.

REFERENCES

1) ABRAMS,R.: Some factors influencing nucleic acid purine renewal
   in the rat. Arch. Biochem. Biophys. 33, 436-447, 1951.

2) BROWN,B.L., ALBANO,J.D.M., EKINS,R.P., SGHERZI,A.M. and
   TAMPION,W.: A simple and sensitive saturation assay method for
   the measurement of adenosine 3':5'-cyclic monophosphate.
   Biochem. J. 121, 561-562, 1971.

3) BUTCHER,R.W. and SUTHERLAND,E.W.: Adenosine 3':5'-phosphate in
   biological materials. I. Purification and properties of cyclic
   3':5'-nucleotide phosphodiesterase and use of this enzyme to
   characterize adenosine 3':5'-phosphate in human urine.
   J. Biol. Chem. 237, 1244-1250, 1962.

4) GERLACH,E., DEUTICKE,B., DREISBACH,R.H. and ROSARIUS,C.W.: Zum
   Verhalten von Nucleotiden und ihren dephosphorylierten Abbauprodukten in der Niere bei Ischämie und kurzzeitiger postischämischer Wiederdurchblutung. Pflügers Arch. 278, 296-315, 1963.

5) GILMAN,A.G.: A protein binding assay for adenosine 3':5'-cyclic
   monophosphate. Proc. Nat. Acad. Sci. 67, 305-312, 1970.

6) GOLDTHWAIT,D.A. and BENDICH,A.: Effects of a folic acid antagonist on nucleic acid metabolism. J. Biol. Chem. 196, 841-852,
   1952.

7) FLECKENSTEIN,A.: Specific inhibitors and promoters of Ca^{++} action. In "Calcium and the Heart", edited by Harris and Opie.
   pp 135-188, Academic Press, London and New York 1971.

8) HENDERSON,J.F. and KHOO,M.K.Y.: On the mechanism of feedback
inhibition of purine biosynthesis de novo in Ehrlich ascites
tumor cells in vitro. J. Biol. Chem. 240, 3104-3109, 1965.

9) HENDERSON,J.F. and KHOO,M.K.Y.: Availability of 5-phosphoribo-
syl-1-pyrophosphate for ribonucleotide synthesis in Ehrlich
ascites tumor cells in vitro. J. Biol. Chem. 240, 2358-2362,
1965.

10) KUKOVETZ,W.R. and PÖCH,G.: The action of imidazole on the effects of methyl-xanthines and catecholamines on cardiac contrac-
tion and phosphorylase activity. J. Pharmacol. Exp. Ther. 156,
514-521, 1967.

11) MARKO,P., GERLACH,E., ZIMMER,H.-G., PECHAN,I., CREMER,T. and
TRENDELENBURG,Chr.: Interrelationship between salvage pathway
and synthesis de novo of adenine nucleotides in kidney slices.
Hoppe Seyler's Z. Physiol. Chem. 350, 1669-1674, 1969.

12) MAYER,S.E., WILLIAMS,B.J. and SMITH,J.M.: Adrenergic mechanisms
in cardiac glycogen metabolism. Ann. New York Acad. Sci. 139,
686-702, 1966/67.

13) ROBISON,G.A., BUTCHER,R.W., ØYE,I., MORGAN,H.E. and SUTHERLAND,
E.W.: The effect of epinephrine on adenosine 3':5'-phosphate
levels in the isolated perfused rat heart. Mol. Pharmacol. 1,
168-177, 1965.

14) SATO,K.: Stimulation of pentose cycle in the eccrine sweat gland
by adrenergic drugs. Am. J. Physiol. 224, 1149-1154, 1973.

15) WILLIAMSON,J.R. and JAMIESON,D.: Dissociation of the inotropic
from the glycogenolytic effect of epinephrine in the isolated
rat heart. Nature 206, 364-367, 1965.

16) WYNGAARDEN,J.B. and ASHTON,D.M.: The regulation of activity of
phosphoribosylpyrophosphate amidotransferase by purine ribonucleotides: A potential feedback control of purine biosynthesis. J. Biol. Chem. 234, 1492-1496, 1959.

17) ZIMMER,H.-G., STEINKOPFF,G. and GERLACH,E.: Veränderungen der
myokardialen Adenin-Nucleotid-Synthese durch Isoproterenol und
Propranolol. Verh. Dtsch. Ges. Kreislaufforschg. 39, 183-188,
1973.

18) ZIMMER,H.-G., TRENDELENBURG,Chr., KAMMERMEIER,H. and GERLACH,E.:
De novo synthesis of adenine nucleotides in the rat: Acceleration during recovery from oxygen deficiency. Circulation Res.
32, 635-642, 1973.

19) ZIMMER,H.-G. and GERLACH,E.: Effect of ß-adrenergic stimulation
    on myocardial adenine nucleotide metabolism. Circulation Res.
    35, 536-543, 1974.

20) ZIMMER,H.-G., STEINKOPFF,G. and GERLACH,E.: Studien über die
    Regulation der de novo-Synthese von Adenin-Nucleotiden im Her-
    zen. Verh. Dtsch. Ges. Kreislaufforschg. 40, 348-352, 1974.

21) ZIMMER,H.-G., BÜNGER,R. and GERLACH,E.: Influence of calcium,
    isoproterenol and compound D 600 on contractility, 3':5'-cyclic
    AMP and glucose-phosphates of the myocardium. Pflügers Arch.
    359, R 9, 1975.

EFFECT OF TRIIODOTHYRONINE ON THE BIOSYNTHESIS OF ADENINE NUCLEOTIDES AND PROTEINS IN THE RAT HEART

H.-G. Zimmer, G. Steinkopff, E. Gerlach

Physiologisches Institut der Universität München

Pettenkoferstr. 12, 8000 München 2

During the development of cardiac hypertrophy de novo synthesis of myocardial adenine nucleotides (AN) was found to be enhanced prior to the increase of protein synthesis (10,11,13,14). However, there are differences concerning time course, direction and extent of changes of AN and protein synthesis in various experimental models of myocardial hypertrophy. After aortic constriction, AN synthesis was reduced within the first 5 hours, increased after 24 hours and reached a maximum on the third day (11). Protein synthesis showed a similar biphasic pattern with a reduction during the first 5 hours and an enhancement on the second day (10). In contrast, during development of cardiac hypertrophy induced by a single high dose of isoproterenol, both AN and protein synthesis were stimulated considerably already within the first 12 hours (13,14).

In the present studies myocardial hypertrophy was induced by daily administrations of $3,3',5$-triiodo-L-thyronine (T_3). After various periods of time, rates of de novo synthesis and concentrations of cardiac AN as well as protein synthesis and heart weight were determined. The time course of changes of these parameters was compared with the respective alterations occurring in the other models of hypertrophy.

MATERIALS AND METHODS

Female Sprague-Dawley rats (200-220 g) fed a diet of Altromin were used in all experiments. 1-^{14}C-glycine (specific activity 56 mCi/mmole) and U-^{14}C-adenine (specific activity 225 mCi/mmole)

were purchased from the Radiochemical Centre, Amersham, England.
Isoproterenol was obtained from C.H. Boehringer Sohn, Ingelheim.
Propranolol was a gift from ICI-Pharma, Heidelberg. Adenine was
purchased from Boehringer Mannheim, 3,3',5-triiodo-L-thyronine from
Sigma, München. All other chemicals were obtained from Merck AG,
Darmstadt and were of analytical grade.

Rats were s.c. injected on 6 days with 3,3',5-triiodo-L-thyro-
nine (0.2 mg/kg per day). Measurements of the synthesis of AN and
proteins were performed 12 hours after the last application. At the
end of a 60 min exposure time to $1-^{14}C$-glycine, the rats were anesthetized with diethyl ether, and the hearts were rapidly excised
and quickly frozen in liquid nitrogen.

The concentrations of ATP, ADP and AMP were determined after
separation by paper chromatography (1). Rates of de novo synthesis
of cardiac AN were obtained from the total radioactivity of AN and
the mean specific activity of the intracellular glycine (12). Protein synthesis was expressed as relative rates of glycine incorporation into myocardial proteins calculated from the radioactivity
of proteins and the mean specific activity of intracellular glycine (10).

RESULTS AND DISCUSSION

In Fig. 1 data are summarized concerning changes of the rates
of myocardial AN and protein synthesis as well as of the weight of
heart ventricles which were observed during the development of T_3-
induced cardiac hypertrophy. As is evident, AN synthesis increases
immediately, but slowly within the first 12 hours and reaches a
plateau on the second day. Protein synthesis becomes significantly
enhanced on the second day when AN synthesis is already considerably augmented. The increase in heart weight almost parallels the
observed enhancement of protein synthesis.

The results obtained in this model of cardiac hypertrophy as
well as in the other models studied so far (14) reveal a striking
parallelism between de novo synthesis of AN and protein synthesis
(2,6,7). The changes of protein synthesis closely follow the pattern of alterations in AN synthesis: When nucleotide synthesis is
decreased, e.g. during the first phase after aortic constriction,
protein synthesis proves also to be depressed. Furthermore, the
greater the increase of AN synthesis, the more pronounced is the
enhancement of protein synthesis.

An interesting finding of these studies is the early onset of
the enhancement of AN synthesis during the development of cardiac
hypertrophy. When compared with other known metabolic alterations

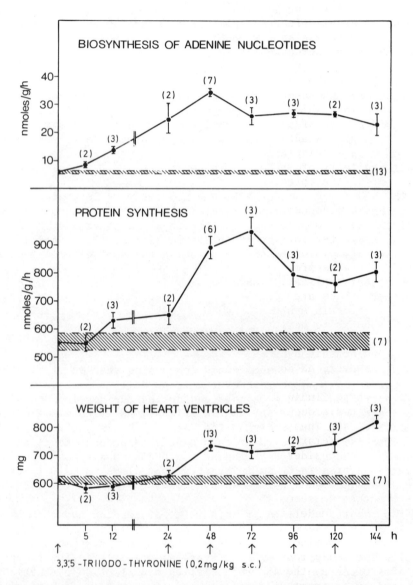

Fig. 1: Changes of the rates of myocardial adenine nucleotide and protein synthesis as well as of the weight of heart ventricles during the development of cardiac hypertrophy induced by 3,3',5-triiodo-L-thyronine. Mean values ± SEM. Number of experiments in parentheses.

(2,3,6-9), it is obvious that the stimulation of AN biosynthesis is one of the first changes which occur. It therefore seems conceivable that this process may well be involved in the enhancement of protein synthesis by providing greater amounts of precursor substances for the increased synthesis of RNA (3,8,9).

Another metabolic factor which has been proposed to be responsible for the enhancement of protein synthesis concerns the deficit of myocardial high energy phosphate compounds, particularly of ATP. The decrease of the ATP level was assumed to be an initial signal for somehow activating the genetic apparatus of the cell and thus inducing a stimulation of RNA synthesis (4,5). In order to test the validity of such an assumption, concentrations of AN were determined in the three models of myocardial hypertrophy.

In Table 1 concentration values of ATP and of the sum of ATP, ADP and AMP are compiled and compared with the rates of AN synthesis. 5 hours after aortic constriction, AN synthesis as well as the concentrations of AN are remarkably diminished. At this time, protein synthesis proved also to be depressed. On the other hand, 5 hours after application of a single high dose of isoproterenol, AN concentrations are considerably reduced, whereas AN synthesis as well as protein synthesis are maximally enhanced. A different pattern appears to be specific for the hypertrophying heart under the influence of T_3. 48 hours after the onset of T_3 treatment, both AN and protein synthesis (Fig. 1) proved to be strongly stimulated, however, AN concentrations are almost unchanged.

From these findings it appears that alterations of AN and protein synthesis occur independently of the concentration changes of ATP or of AN. Thus, the decrease of ATP levels seems neither to be causally related to the development of cardiac hypertrophy nor can it be considered to be the only stimulus for the acceleration of AN synthesis through the release of feedback inhibition of 5-phosphoribosyl-1-pyrophosphate amidotransferase (EC 2.4.2.14). Another mechanism seems to be operative particularly in the hypertrophying heart under the influence of T_3.

In order to elucidate whether a greater availability of 5-phosphoribosyl-1-pyrophosphate (PRPP) might be involved in the enhancement of AN synthesis under these conditions, adenine was applied in the T_3-treated animals. The T_3-induced increase of AN synthesis is abolished by adenine (Table 2) and the incorporation of ^{14}C-adenine into cardiac AN appears to be enhanced (Fig. 2). These results actually favour the concept that cardiac PRPP becomes available to a greater extent under the influence of T_3.

Table 1: Rates of biosynthesis and concentrations of adenine nucleotides (AN) in rat hearts after aortic constriction, application of isoproterenol and 3,3',5-triiodo-L-thyronine (T_3), respectively. Mean values $\pm$ SEM; n = number of experiments.

	AN biosynthesis (nmoles/g/h)	Concentration (nmoles/g)	
		ATP	$\sum$ ATP, ADP, AMP
Control	5.80 $\pm$ 0.6 (n=11)	4 480 $\pm$ 90 (n=21)	6 000 $\pm$ 110 (n=21)
Aortic constriction, 5 h	3.58 $\pm$ 0.5* (n=11)	4 070 $\pm$ 90* (n=7)	5 070 $\pm$ 140* (n=7)
Isoproterenol (25 mg/kg), 5 h	39.40 $\pm$ 2.3* (n=9)	3 370 $\pm$ 120* (n=9)	4 710 $\pm$ 100* (n=9)
T_3 (0.2 mg/kg daily), 48 h	34.61 $\pm$ 1.0* (n=8)	4 220 $\pm$ 90 (n=6)	5 720 $\pm$ 100 (n=6)

* $p < 0.005$

Table 2: Effects of adenine (10 mg/kg, i.v.) and of propranolol (50 mg/kg, s.c.) on the increase of adenine nucleotide synthesis (nmoles/g/h) in rat hearts in vivo caused by 3,3',5-triiodo-L-thyronine (T_3). The exposure time to the substances is indicated. Mean values $\pm$ SEM; n = number of experiments.

	n	nmoles/g/h
Control	11	5.8 $\pm$ 0.6
T_3, 48 h	8	34.6 $\pm$ 1.0
T_3, 48 h + adenine, 1 h	4	5.2 $\pm$ 1.5
T_3, 48 h + propranolol, 3 h	3	6.2 $\pm$ 0.2
T_3, 48 h + propranolol, 5 h	2	12.2 $\pm$ 1.9
T_3, 48 h + propranolol, 12 h	3	31.1 $\pm$ 4.3

Since ß-receptor blockade in the heart has been shown to reduce de novo synthesis of AN in the normal heart and to prevent the enhancement of AN synthesis in the isoproterenol stimulated heart (13), an attempt was made to influence cardiac AN synthesis by ß-blocking agents also in T_3-treated rats. It is evident from the data included in Table 2 that the T_3-induced acceleration of AN synthesis can be transiently reduced by propranolol. Thus, substances interacting with cardiac ß-adrenergic receptors seem to modulate the enhancement of AN synthesis under these experimental conditions.

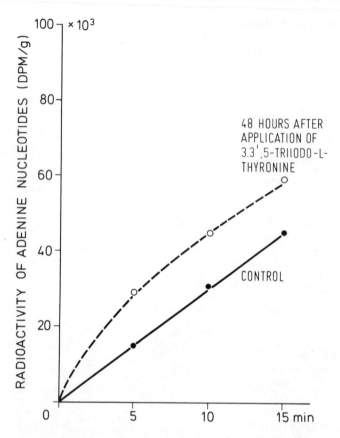

<u>Fig. 2:</u> Incorporation of U-^{14}C-adenine (0.05 mCi/kg, i.v.) into
myocardial adenine nucleotides in triiodothyronine-treated rats.
Each point represents the mean of 2 experiments.

SUMMARY

    1) During the development of myocardial hypertrophy induced
by 3,3',5-triiodo-L-thyronine the enhancement of de novo synthesis
of adenine nucleotides occurs very early and precedes the increase
of protein synthesis. In this respect there is a striking parallelism with other types of cardiac hypertrophy.

    2) The acceleration of adenine nucleotide synthesis under
these experimental conditions seems to be due to a greater availa-

bility of cardiac 5-phosphoribosyl-1-pyrophosphate.

3) The triiodothyronine-induced enhancement of adenine nucleotide synthesis can be attenuated by ß-receptor-blocking agents.

REFERENCES

1) GERLACH,E., DEUTICKE,B., DREISBACH,R.H. and ROSARIUS,C.W.:
   Zum Verhalten von Nucleotiden und ihren dephosphorylierten Abbauprodukten in der Niere bei Ischämie und kurzzeitiger postischämischer Wiederdurchblutung. Pflügers Arch. 278, 296-315, 1963.

2) GUDBJARNASON,S., TELERMAN,M. and BING,R.J.: Protein metabolism in cardiac hypertrophy and heart failure. Am. J. Physiol. 206, 294-298, 1964.

3) KOIDE,T., RABINOWITZ,M.: Biochemical correlates of cardiac hypertrophy. II. Increased rate of RNA synthesis in experimental cardiac hypertrophy in the rat. Circulation Res. 24. 9-18, 1969.

4) MEERSON,F.Z.: Possible mechanism of cardiac hypertrophy. Acta biol. med. germ. 29, 271-280, 1972

5) MEERSON, F.Z. and POMOINITSKI,V.D.: The role of high-energy phosphate compounds in the development of cardiac hypertrophy. J. Mol. Cell. Cardiol. 4, 571-597, 1972

6) MORKIN,E., KIMATA,S. and SKILLMAN,J.J.: Myosin synthesis and degradation during development of cardiac hypertrophy in the rabbit. Circulation Res. 30, 690-702, 1972.

7) MOROZ,L.A.: Protein synthetic activity of heart microsomes and ribosomes during left ventricular hypertrophy in rabbits. Circulation Res. 21, 449-459, 1967.

8) NAIR,K.G., CUTILETTA,A.F., ZAK,R., KOIDE,T. and RABINOWITZ,M.:
   Biochemical correlates of cardiac hypertrophy. I. Experimental model;changes in heart weight, RNA content, and nuclear RNA polymerase activity. Circulation Res. 23, 451-462, 1968.

9) RABINOWITZ,M. and ZAK,R.: Biochemical and cellular changes in cardiac hypertrophy. Ann. Rev. Med. 23, 245-262, 1972.

10) ZIMMER,H.-G., STEINKOPFF,G. and GERLACH,E.: Changes of protein synthesis in the hypertrophying rat heart. Pflügers Arch. 336, 311-325, 1972.

11) ZIMMER,H.-G., TRENDELENBURG,Chr. and GERLACH,E.: Acceleration
    of adenine nucleotide synthesis de novo during development of
    cardiac hypertrophy. J. Mol. Cell. Cardiol. 4, 279-282, 1972.

12) ZIMMER,H.-G., TRENDELENBURG,Chr., KAMMERMEIER,H. and GERLACH,E.:
    De novo synthesis of myocardial adenine nucleotides in the rat:
    Acceleration during recovery from oxygen deficiency. Circula-
    tion Res. 32, 635-642, 1973.

13) ZIMMER,H.-G. and GERLACH,E.: Effect of beta-adrenergic stimula-
    tion on myocardial adenine nucleotide metabolism. Circulation
    Res. 35, 536-543, 1974.

14) ZIMMER,H.-G. and GERLACH,E.: Comparative studies of myocardial
    adenine nucleotide and protein synthesis in two models of car-
    diac hypertrophy. International Symposium of the International
    Study Group for Research in Cardiac Metabolism. Brussels, p. 75
    1975.

ALTERED SYNTHESIS AND CATABOLISM OF PURINE NUCLEOTIDES IN REGULA-

TORY MUTANTS OF SACCHAROMYCES CEREVISIAE

J. Frank Henderson, P. W. Burridge and R. A. Woods

Cancer Research Unit and Department of Biochemistry

University of Alberta, Edmonton, Alberta, Canada

In 1970 Armitt and Woods (1) reported the isolation of a
series of mutants of Saccharomyces cerevisiae that did not require
purines for growth but which excreted substantial amounts of purines
into the growth medium. Genetic studies showed that these yeast
strains were associated with six unlinked genes (pur1, pur2, pur3,
pur4, pur5, pur6), and although most of the mutations were recessive, one, pur6, had both recessive and dominant (PUR6) alleles.

Because these mutants excreted purines into the medium at considerable rates but could depend entirely on purine biosynthesis
de novo to supply purines for growth, it has been thought that they
have accelerated rates of purine synthesis de novo and hence that
they are deficient in the regulation of this pathway. The excretion
of purines, mostly hypoxanthine and inosine, into the medium has been
thought to be merely a consequence of the accelerated purine synthesis de novo, a way of disposing of unneeded inosinate.

In order to test these hypotheses and to elucidate other
aspects of purine metabolism in these yeast strains, we have studied
the ways in which they metabolize radioactive glycine, adenine,
guanine and hypoxanthine.

Wild type and mutant yeast strains were grown up on yeast minimal medium until they were in mid log phase, then suspended in fresh
medium all at the same cell density, and incubated with each radioactive precursor for 20, 40 and 60 minutes. In each case the metabolism of the precursor was linear for this period. Incubations
were terminated by addition of perchloric acid, and various purine
derivatives were isolated and their radioactivity was measured; these
included nucleic acid adenine and guanine, adenosine and guanosine

mono-, di-, and triphosphates, inosinate and xanthylate, as well as adenine and adenosine, hypoxanthine and inosine, xanthine and xanthosine, guanine and guanosine, and uric acid and allantoin. In these experiments intracellular and extracellular purines were not separated or distinguished.

Qualitatively the metabolism of the radioactive precursors was similar in most of the eight mutants studied, although there certainly were quantitative differences. Only results for two of the mutants will be presented here, and only results for the 60 minute period will be reported.

The first point to be studied was the rate of purine biosynthesis de novo. This was considered to be equal to the total incorporation of radioactivity from glycine into all purine derivatives. The data of Table 1 show that purine biosynthesis de novo was indeed accelerated in these two mutants; hence the original hypothesis seems to have been correct in this regard. However, these results do not indicate the basis for the increased rate.

Because one possible cause of accelerated purine biosynthesis de novo was increased availability of phosphoribosyl pyrophosphate (PP-ribose-P), and because this might also result in accelerated nucleotide synthesis from purine bases, it was important to measure these processes. Nucleotide synthesis from the purine bases was considered to be equal to the incorporation of radioactivity into nucleic acid purines, acid-soluble nucleotides, and those nucleosides and bases which were derived from nucleotides. As shown in Table 2, the rates of nucleotide synthesis from adenine, guanine and hypoxanthine were in fact much reduced in the mutant yeast strains, although nucleotide synthesis from guanine was affected less than that from hypoxanthine and adenine. As the cells had been suspended in fresh media with 100 μM radioactive purine bases, it seems unlikely that very much of the reduction in rate of nucleotide synthesis could be due to dilution of the precursor with non-radioactive purines produced de novo. These results suggest that excess PP-ribose-P is not the sole basis of accelerated purine synthesis de novo. Other possibilities are that limited PP-ribose-P is somehow being diverted preferentially to the de novo pathway or is not available to the purine phosphoribosyltransferases even when high concentrations of purine bases are supplied, that the purine phosphoribosyltransferases are inhibited, or that they are present in reduced activities.

We next studied the processes of nucleotide catabolism to nucleosides and bases, both when glycine was the precursor and when the purine bases were used. As shown in Table 3, wild type yeast uses the nucleotides synthesized de novo very efficiently, and very

Table 1

PURINE BIOSYNTHESIS DE NOVO

YEAST STRAIN	RATE OF SYNTHESIS (cpm)
Wild type	6,740
purl'	11,600
PUR6	17,740

Table 2

NUCLEOTIDE SYNTHESIS FROM PURINE BASES

STRAIN	RATE OF SYNTHESIS (cpm) FROM:		
	Adenine	Hypoxanthine	Guanine
Wild type	68,220	54,240	31,277
purl'	6,772	4,977	24,070
PUR6	6,079	5,347	22,830

Table 3

CATABOLISM OF NUCLEOTIDES SYNTHESIZED DE NOVO

YEAST STRAIN	TOTAL CATABOLISM (Percent of Nucleotides Formed)
Wild type	5.3
purl'	69.4
PUR6	41.8

Table 4

CATABOLISM OF NUCLEOTIDES SYNTHESIZED
FROM PURINE BASES

YEAST STRAIN	TOTAL CATABOLISM (Percent of Nucleotides Formed)		
	Adenine	Hypoxanthine	Guanine
Wild type	22.6	27.6	15.9
purl'	46.8	85.8	87.7
PUR6	58.5	70.2	54.5

little is broken down. As already determined in other ways, however, the mutant yeast strains break down nucleotides synthesized de novo at very much increased rates. As would be expected, most of the catabolic products come from inosinate, although some accelerated catabolism of adenylate, xanthylate and guanylate also was detected.

When the catabolism of nucleotides formed from purine bases was measured (Table 4), we were surprised to find that these processes too were accelerated, even though the actual rate of nucleotide synthesis, as shown above, was very much decreased. These results also show that the catabolic process that is accelerated, whatever it may be, affects not only inosinate but also xanthylate, guanylate and probably adenylate as well.

All of these results, taken together, are difficult to interpret. Genetic studies have indicated that only single mutations are involved, but that there are six different loci represented among the eight yeast strains studied. However, three general types of biochemical abnormalities have been detected: accelerated purine biosynthes de novo, depressed nucleotide synthesis from purine bases, and accelerated catabolism of nucleotides synthesized both de novo and from preformed purines. Which of these is primary, and how is it linked to the other two?

An interpretation that could account for most of these results is as follows. The mutations involved either affect the dephosphorylation of purine nucleoside monophosphates directly or more likely, affect energy metabolism in such a way as to indirectly accelerate nucleotide catabolism. (Such relationships between energy metabolism and nucleotide catabolism have been observed in some studies of mammalian cells, for example (2)). This acceleration of nucleotide catabolism would affect both nucleotides synthesized de novo and those synthesized from preformed purines. This primary biochemical alteration could have two consequences. First, the concentrations of ATP and GTP might be reduced; in cultured mammalian cells, for example, these concentrations can be reduced by 50 or 60% without affecting growth rate (3). Other studies have shown that the rate of PP-ribose-P synthesis increases in cells in which ATP and GTP concentrations are decreased, and this could lead to accelerated purine biosynthesis de novo (J. Barankiewicz and J. F. Henderson, unpublished). A second possible consequence of accelerated catabolism, especially if this were based on altered energy metabolism, might be elevated concentrations of nucleoside monophosphates. These compounds are inhibitors of the purine phosphoribosyltransferases (4, 5), and this might account for the observed decrease in rate of nucleotide synthesis from purine bases.

     This hypothesis may not be the only possible explanation of
these results, and it has not yet been tested by actual measurements
of purine nucleotide concentrations in wild type and mutant yeast
strains.   However, the relative concentrations of radioactive
purine nucleoside monophosphates were indeed elevated relative to
those of the di- and triphosphates, in the mutant cells.   If this
explanation is correct, however, then these yeast strains are not
"regulatory" mutants at all.

References

1.   Armitt, S. and Woods, R. A.   (1970) Genet. Res., 15, 7-17.
2.   Lomax, C. A., Bagnara, A. S. and Henderson, J. F.   (1975)
     Can. J. Biochem., 53, 231-241.
3.   Warnick, C. T. and Paterson, A. R. P.   (1973) Cancer Res.,
     33, 1711-1715.
4.   Henderson, J. F., Brox, L. W., Kelley, W. N., Rosenbloom, F. M.
     and Seegmiller, J. E.   (1968) J. Biol. Chem., 243, 2514-2522.
5.   Henderson, J. F., Gadd, R. E. A., Palser, H. M. and Hori, M.
     (1970) Can. J. Biochem., 48, 573-579.

PHOSPHORIBOSYLPYROPHOSPHATE SYNTHESIS IN HUMAN ERYTHROCYTES:

INHIBITION BY PURINE NUCLEOSIDES

Guy Planet and Irving H. Fox

Purine Research Laboratory, University of Toronto

Rheumatic Disease Unit, Wellesley Hospital, Toronto

The intracellular concentration of PP-ribose-P is a balance between its formation and its utilization. Purine nucleosides stimulate PP-ribose-P formation in mammalian cells when inorganic phosphate exceeds 20 mM (1,2). This occurs by the conversion to ribose-5-phosphate of the ribose-1-phosphate liberated from nucleo- sides by purine nucleoside phosphorylase. Recently Bagnara and Finch (3) showed that nucleosides decreased intracellular PP-ribose-P in Escherichia coli incubated in 1.4 mM inorganic phosphate. The reason for the latter effect has not been clear since there is no substantial direct effect of nucleosides on purified PP-ribose-P synthetase.

We have evaluated the effect of purine nucleosides on PP- ribose-P metabolism in intact erythrocytes (4). Figure 1 shows the effects of nucleosides and Pi (inorganic phosphate) on erythro- cyte PP-ribose-P. Increased Pi concentrations from 0 to 25 mM stimulated PP-ribose-P formation. Under these conditions 0.63 or 1.25 mM adenosine, inosine, guanosine, adenosine and EHNA or 6- methylmercaptopurine riboside (MMPR) decreased PP-ribose-P con- centrations.

In order to discern how nucleosides might decrease intra- cellular PP-ribose-P it is necessary to consider their metabolism (Figure 2). Nucleosides may either decrease PP-ribose-P formation or increase its utilization. The nucleoside mediated decrease of intracellular PP-ribose-P may have occurred in 4 different ways: (a) the formation of a purine base from the nucleoside by erythro- cyte purine nucleoside phosphorylase may subsequently lead to PP- ribose-P utilization in the synthesis of nucleotide, (b) there may

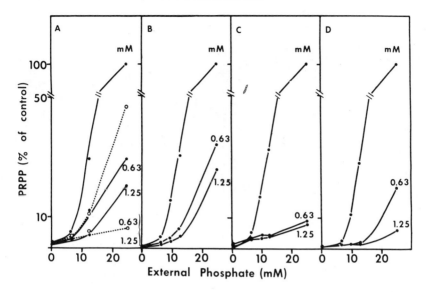

Fig. 1. Effect of nucleosides on PP-ribose-P formation. The values for PP-ribose-P are expressed as a percentage of the value at 25 mM Pi without nucleosides or erythro-9(2-hydroxyl-3-nonyl)adenine (EHNA), an adenosine deaminase inhibitor, which was a gift from Dr. G.B. Elion of Burroughs Welcome Company.

be a decrease in intracellular Pi with a consequent diminution of PP-ribose-P synthetase activity by the trapping of Pi (i) during the conversion of nucleoside to the purine base or (ii) during the phosphorylation of nucleoside, (c) the nucleoside is phosphorylated and the nucleotides produced inhibit the synthesis of PP-ribose-P, (d) the nucleosides themselves or an alteration of other intra-cellular metabolites inhibit PP-ribose-P synthetase. Purine nucleo-sides 1 mM only demonstrated 0 to 27% inhibition of PP-ribose-P synthetase and would not account for the decrease of PP-ribose-P observed (5).

Experiments were performed to assess for the relative importance of these mechanisms. Evidence for increased utilization of PP-ribose-P was sought by evaluating the effects of nucleosides and Pi on erythrocyte available PP-ribose-P (Figure 3). This assay esti-mates the capacity for PP-ribose-P production within the cell by allowing adenine to trap PP-ribose-P as AMP using adenine phos-phoribosyltransferase (APRT) (6). PP-ribose-P in this assay is preferentially utilized by APRT. Thus if reaction of PP-ribose-P with hypoxanthine or guanine accounted for the nucleoside effects observed, no inhibition would have been evident by the measurement of available PP-ribose-P. PP-ribose-P synthesis was stimulated by

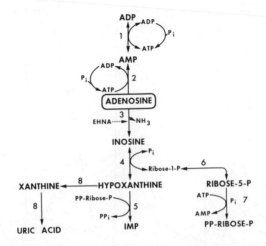

Fig. 2. Adenosine metabolism. Adenosine can be phosphorylated to
AMP. The AMP formed can be converted to ADP. Adenosine can be
deaminated to inosine, a reaction which is inhibited by EHNA.
Inosine is degraded to hypoxanthine in a reaction which requires Pi
and releases ribose-1-phosphate. Hypoxanthine can be synthesized
to IMP. Guanosine has a metabolism similar to inosine while MMPR
is mainly phosphorylated in a manner similar to adenosine. 1,
adenylate kinase; 2, adenosine kinase; 3, adenosine deaminase,
4, purine nucleoside phosphorylase; 5, hypoxanthine-guanine phosphoribosyltransferase; 6, phosphoribomutase; 7, PP-ribose-P
synthetase; 8, xanthine oxidase, not found in human erythrocytes.

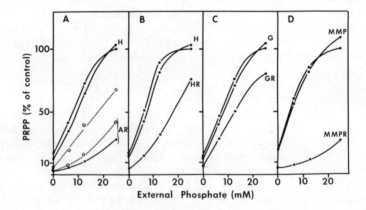

Fig. 3. Effect of nucleosides and bases on available PP-ribose-P.
The concentrations of PP-ribose-P are expressed as a percentage of
the values at 25 mM Pi without nucleoside or base.

increasing extracellular Pi up to 25 mM. The purine bases hypoxanthine, guanine or 6-methylmercaptopurine 1.25 mM did not change
available PP-ribose-P concentrations. In contrast adenosine,
inosine, guanosine or MMPR 1.25 mM decreased "available" PP-ribose-P
concentrations. Thus the mechanism of increased PP-ribose-P
utilization does not account for the nucleoside effect.

   The possible role of intracellular Pi concentration was next
evaluated. The effects of nucleosides and extracellular Pi on
intracellular erythrocyte Pi concentrations are indicated in Figure
4. There was a gradient between intracellular and extracellular Pi.
At 0, 10 or 20 mM Pi, erythrocyte Pi was 0.5, 1.5 or 2.4 mM respectively. Nucleosides at 0.63 or 1.25 mM decreased the intracellular concentration of Pi. Pi depletion occurred by (i) adenosine,
inosine or guanosine, which are primarily degraded to the purine
base by nucleoside phosphorylase and (ii) MMPR or adenosine and EHNA
which are mainly phosphorylated. PP-ribose-P synthesis and PP-
ribose-P synthetase are known to be regulated by small changes in
Pi. Therefore, PP-ribose-P synthesis could be substantially reduced
by the amount of Pi decrease observed.

   When the erythrocyte membrane and thereby the Pi gradient was
eliminated different effects were observed on PP-ribose-P formation.
Available PP-ribose-P in hemolysate was stimulated by increasing Pi
concentrations. Endogenous Pi in hemolysate was 3.6 mM after a 30
minute incubation. Above 1 mM added Pi, adenosine, adenosine with
EHNA, inosine, guanosine, but not MMPR increased "available" PP-
ribose-P.

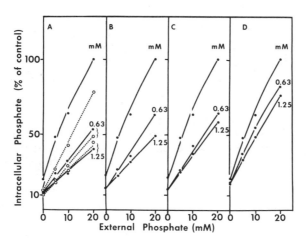

Fig. 4 Effect of nucleosides on erythrocyte Pi. The concentrations
of erythrocyte Pi are expressed as a percentage of 2.4 mM, the value
observed when extracellular Pi was 20 mM and no nucleosides were
present.

TABLE 1

EFFECT OF NUCLEOSIDES ON ERYTHROCYTE METABOLIC INTERMEDIATES

Compound mM	Pi	ATP	ADP	AMP
		(micromoles/ml)		
None	1.8	1.08	0.14	0.02
EHNA 0.021	1.3	0.98	0.16	0.03
Adenosine 1.25	0.7	1.19	0.16	0.02
Adenosine 1.25 + EHNA 0.021	0.7	1.10	0.18	0.03
Inosine 1.25	0.9	1.12	0.13	0.02
Guanosine 1.25	1.0	1.08	0.14	0.02
MMPR 1.25	1.3	0.98	0.15	0.02

The metabolic fate of nucleosides was next evaluated since the phosphorylation of adenosine or MMPR could potentially influence the synthesis of PP-ribose-P. Nucleotide synthesis accounted for only 2.4 to 4.6% of total radioactive adenosine, inosine or guanosine. Nucleotides accounted for 17.3% of total adenosine utilized with EHNA and 44.4 to 54.2% of total MMPR utilized. Table 1 illustrates effect of 1.25 mM nucleosides on erythrocyte adenine nucleotide concentrations. Alterations of intracellular concentration of ATP, ADP or AMP were not great enough to account for inhibition of PP-ribose-P formation. In contrast a large quantity of MMPR monophosphate up to 1.3 mM was formed from MMPR. We have shown that this compound inhibits partially purified PP-ribose-P synthetase. Thus MMPR may inhibit PP-ribose-P formation by both phosphate depletion and nucleotide inhibition.

The final possibility was that other metabolites could modify the synthesis of PP-ribose-P. Table 2 shows the intermediates measured. There was no elevation of 2,3-diphosphoglycerate or decrease of ribose-5-phosphate, which could decrease PP-ribose-P formation. The small diminution observed in cyclic AMP levels have no significance for PP-ribose-P synthesis. Therefore, no altered intracellular metabolite measured accounts for the decrease of PP-ribose-P synthesis.

Our studies demonstrate that when Pi ranges from 0 to 20 mM nucleosides decrease intracellular PP-ribose-P by inhibiting its synthesis. The decreased PP-ribose-P synthesis is related mainly to decreased intracellular Pi and may occur by nucleotide inhibition of PP-ribose-P synthesis in the case of MMPR. Regulation of

PP-ribose-P formation by alteration of Pi levels may be an important control mechanism. It also may account for a number of nucleoside related biological phenomena described recently including pyrimidine starvation and modification of certain component parts of the immune response. Disordered PP-ribose-P metabolism may have a role in explaining the pathophysiological basis for the immune abnormalities found with the deficiencies of adenosine deaminase or nucleoside phosphorylase. However, before extrapolation of the effects of nucleosides on PP-ribose-P metabolism to other systems, the mechanisms delineated in erythrocytes must be investigated in other tissues.

TABLE 2

EFFECT OF NUCLEOSIDES ON ERYTHROCYTE METABOLIC INTERMEDIATES

Compound mM	2,3-DPG	lactate	Ribose-5-P	Pyruvate	CAMP
	micromoles/ml			nanomoles/ml	
None	3.39	1.81	6	17	1.02
EHNA 0.021	3.45	2.62	16	13	0.83
Adenosine 1.25	3.56	2.33	60	12	0.90
Adenosine 1.25 + EHNA	3.60	2.86	42	12	1.01
Inosine 1.25	3.54	2.18	82	18	0.92
Guanosine 1.25	3.53	2.07	49	12	0.78
MMPR 1.25	3.52	2.43	5	13	1.10

REFERENCES

1. Henderson, J.F. and Khoo, M. 1965. Availability of 5-phospho-ribosyl-1-pyrophosphate for ribonucleotide synthesis in Ehrlich ascites tumor cells in vitro. J. Biol. Chem. 240: 2358-2362.

2. Hershko, A., Razin, A. and Mager, J. 1969. Relation of the synthesis of 5-phosphoribosyl-1-pyrophosphate in intact red blood cells and in cell free preparations. Biochim. Biophys. Acta. 184:64-76.

3. Bagnara, A.S. and Finch, L.R. 1974. The effects of basis and nucleosides on the intracellular contents of nucleotides and 5-phosphoribosyl-1-pyrophosphate in Escherichia coli. Eur. J. Biochem. 41:421-430.

4.   Planet, G. and Fox, I.H. 1976. Inhibition of phosphoribosyl-
        pyrophosphate synthesis by purine nucleosides in human
        erythrocytes. J. Biol. Chem. (in press).

5.   Fox, I.H. and Kelley, W.N. 1972. Human phosphoribosyl-
        pyrophosphate synthetase: kinetic mechanism and product
        inhibition. J. Biol. Chem. 247:2126-2131.

HUMAN PHOSPHORIBOSYLPYROPHOSPHATE SYNTHETASE: RELATION OF ACTIVITY AND QUATERNARY STRUCTURE

Michael A. Becker, Laurence J. Meyer, William H. Huisman, Cheri S. Lazar and William B. Adams
Department of Medicine, University of California, San Diego and San Diego Veterans Administration Hospital, La Jolla, California 92161 U.S.A.

Evidence from a variety of biochemical, pharmacological and clinical studies indicates that the intracellular concentration of 5-phosphoribosyl 1-pyrophosphate (PP-ribose-P)[1] is an important determinant of the rate of purine nucleotide and thus uric acid synthesis (Reviews, references 1,2). PP-Ribose-P formation (Figure 1) from ATP and ribose-5-phosphate is catalyzed by the enzyme PP-ribose-P synthetase in a reaction requiring inorganic phosphate (Pi) and magnesium. Small molecule inhibitors also affect PP-ribose-P synthetase activity and include purine, pyrimidine and pyridine nucleotides as well as 2,3-diphosphogly-cerate (2,3-DPG) (3). The significance of regulation of the activity of this enzyme is apparent in several families in whom purine overproduction and clinical gout result from different structural mutations in PP-ribose-P synthetase which lead to excessive enzyme activity and PP-ribose-P generation (4-6).

The present studies were undertaken in order to determine the relationship between the structure and activity of PP-ribose-P synthetase and the mechanisms of alteration of enzyme activity by effectors. These studies show that the enzyme is composed of a single repeated subunit and, in confirmation of the findings of Fox and Kelley (7), indicate that the enzyme can reversibly assume aggregated forms of molecular weights varying from 65,000 to greater than one million. Direct measurement of the enzyme activity

----

[1]Abbreviations: PP-ribose-P, 5-phosphoribosyl 1-pyrophosphate; ribose-5-P, ribose-5-phosphate; Pi, inorganic phosphate; 2,3-DPG, 2,3-diphosphoglycerate; SDS, sodium dodecylsulfate; DTT, dithio-threitol.

$$\text{ATP + RIBOSE—5—P} \xrightarrow[\text{Mg}^{++},\ \text{Pi}]{\text{PP—RIBOSE—P SYNTHETASE}} \text{PP—RIBOSE—P + AMP}$$

INHIBITORS: PURINE   ⎫
               PYRIMIDINE ⎬ NUCLEOTIDES
               PYRIDINE   ⎭

               2,3—DIPHOSPHOGLYCERATE
               PP—RIBOSE—P
               AMP

Figure 1. The PP-ribose-P synthetase reaction and inhibitors of
PP-ribose-P synthetase activity.

of the aggregated forms confirms the suggestion (7) that smaller
aggregates of PP-ribose-P synthetase are inactive while aggregates
composed of 16 and 32 subunits are active. Finally, the effects
of small molecule activators and inhibitors of PP-ribose-P synthetase are correlated with their effects on the state of aggregation of the enzyme.

    Human erythrocyte PP-ribose-P synthetase was purified 5000-fold
(7) to electrophoretic homogeneity (8) and was studied as summarized
in Table 1. Electrophoresis of PP-ribose-P synthetase in SDS-polyacrylamide (9) showed a single protein band corresponding to a
molecular weight of approximately 33,500. Polyacrylamide electrophoresis in 10 M urea (10) also showed a single protein band.
These results suggest that the enzyme contains only a single subunit
species, a suggestion supported by the results of amino acid

TABLE 1

SUBUNIT ANALYSIS OF PP-RIBOSE-P SYNTHETASE

ELECTROPHORESIS	
SDS-POLYACRYLAMIDE:	Single Protein Band, MW 33,500
10 M UREA-POLYACRYLAMIDE:	Single Protein Band
AMINO ACID ANALYSIS:	18 Arginine + 20 Lysine Residues
	per 33,500 MW Subunit
TRYPTIC PEPTIDE MAPPING:	37 Peptides Identified
	(Fluorescamine Reagent)
AMINO - TERMINAL ANALYSIS:	Threonine Only
	(Dansyl Chloride Method)
ANALYTICAL ULTRACENTRIFUGATION IN 0.01 N HCl:	Molecular Weight 33,200 (Sedimentation Equilibrium)

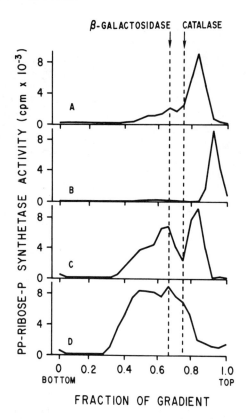

Figure 2. Sucrose density gradient ultracentrifugation of PP-ribose-P synthetase. Linear sucrose gradients (12 ml; 5 to 20% sucrose) were prepared containing the following additions: A. 1 mM sodium phosphate, 1 mM DTT (pH 7.4); B. 1 mM sodium phosphate, 1 mM DTT, 150 mM KCl (pH 7.4); C. 1 mM sodium phosphate, 1 mM DTT, 0.3 mM magnesium ATP (pH 7.4); D. 10 mM sodium phosphate, 1 mM DTT, 0.3 mM ATP, 6.0 mM magnesium chloride (pH 7.4). Partially purified (250-fold) PP-ribose-P synthetase (50 µg of enzyme) was chromatographed on Sephadex G-25 equilibrated with 1 mM phosphate, 1 mM DTT (pH 7.4) and divided into 4 fractions. Each fraction was then diluted in the appropriate solution (containing catalase and β-galactosidase) to give an enzyme sample corresponding in effector composition to A through D above. After 30 minutes incubation at 37°, samples (100 µl) were applied to the appropriate gradients which were centrifuged at 5° for 17 hours at 24,000 rpm in a Beckman L3-50 preparative ultracentrifuge (SW-41 rotor). After centrifugation, gradients were fractionated and samples (500 µl) were assayed for PP-ribose-P synthetase, catalase and β-galactosidase activities. Dotted lines correspond to peaks of standard protein activities.

analysis, tryptic peptide mapping, amino-terminal amino acid
determination and analytical ultracentrifugation. The amino acid
composition of PP-ribose-P synthetase calculated on the basis of
a 33,500 molecular weight subunit contained 38 arginine and lysine
residues, and 37 tryptic peptides were identified in the mapping
studies. The only amino-terminal amino acid residue identified
by the SDS-dansyl chloride method (11) was threonine. Finally,
the molecular weight of purified PP-ribose-P synthetase determined
by sedimentation equilibrium (12) in 0.01 N HCl was 33,200.

Despite the evidence for a single subunit species, a variety
of sedimentation profiles and elution patterns of PP-ribose-P
synthetase activity were identified by sucrose density gradient
ultracentrifugation and Sephadex gel filtration studies respectively. In Figure 2, comparison is made of sedimentation profiles
of PP-ribose-P synthetase activity on 5 to 20% linear sucrose
gradients under a variety of conditions. For each profile (Figures
2 A-D), partially purified PP-ribose-P synthetase, as well as internal standards catalase and β-galactosidase, were layered on the
top of preformed sucrose gradients poured in the presence of the
compounds indicated in the figure legend. In the presence of 1 mM
Pi and 1 mM dithiothreitol (DTT) (Figure 2A), the major peak of
enzyme activity sedimented to the right or nearer to the top of
the gradient than either of the protein standards and had a relative sedimentation coefficient of 7.1. A small shoulder of
enzyme activity sedimenting at 9.7 s was frequently found as well.
Addition of 150 mM KCl (Figure 2B) resulted in a shift of the
activity profile even further toward the top of the gradient to a
sharp peak corresponding to a relative $S_{20,w}$ value of 4.8. In
contrast, as shown in Figure 2C, addition of magnesium ATP, a
substrate of PP-ribose-P synthetase, yielded an activity profile
containing at least two distinct components, one to the right of
catalase and the other a heavier component sedimenting in the
region of β-galactosidase. The broadness of these two components
contrasts with the sharp peaks seen in the above two profiles and
is indicative of heterogeneity within each component. Even more
striking heterogeneity in the distribution of enzyme activity is
obvious in Figure 2D where the magnesium and Pi concentrations
were increased as indicated. Under these latter conditions, there
was very little enzyme activity corresponding to the lighter forms
of PP-ribose-P synthetase noted in Figures 2A and 2B.

Each of these activity profiles was highly reproducible with
recovery of 75 to 100% of the enzyme activity applied   to the
gradients. The broad patterns of activity in the heavy regions of

the gradients were not the result of disruption of the gradients
since in all cases activities of the standard enzymes were
identified in sharp peaks. Thus, as previously described (7),
PP-ribose-P synthetase can exist in multiple forms dependent on the
presence and concentration of a variery of effectors.

In Table 2, physical and catalytic characteristics of the
aggregated forms of PP-ribose-P synthetase detected in this study
are presented. Relative sedimentation coefficients for each form
were determined by sucrose gradient analysis (13) after further
resolution into discrete components was accomplished by centrifugation at higher velocities. Stokes radii estimations were derived
from gel elution studies (14) and in combination with relative
sedimentation coefficients were used to calculate provisional
molecular weights (15) for these forms. Subunit number was then
estimated by dividing the provisional molecular weight by the
33,500 molecular weight of the subunit. The smallest aggregated
form of the enzyme was seen in the presence of 150 mM KCl and was a
dimer. A tetrameric form of PP-ribose-P synthetase predominated
in the presence of Pi and DTT without KCl although some aggregates
of 8 (and also of 16) subunits were present as well. Aggregates of
16 and 32 subunits were observed upon addition of magnesium ATP to
the enzyme and, with increasing concentration of magnesium ATP,
increase in the proportion of heavier aggregates of the enzyme was
accompanied by virtual disappearance of the 4 and 8 subunit forms.
Aggregation to 16 and 32 subunits occurred with increasing free
magnesium as well as increasing magnesium ATP concentration. In
none of the above circumstances did Pi concentration alone (varied
over a range of 1 to 50 mM) determine the state of aggregation of
the enzyme.

Measurements of the activity of the monomer and of the
aggregates of PP-ribose-P synthetase isolated by sucrose gradient
centrifugation were made utilizing concentrations of magnesium ATP
(50 μM) and free magnesium (1 mM) at which interconversion of the
forms was insignificant during the course of the assay. As seen
in the last column of Table 2, aggregates of 8 subunits or less
showed less than 3% of the activity of the heavier aggregates per
mole of subunit. The apparent enzymatic activity of these smaller
forms of PP-ribose-P synthetase when measured in the standard assay
with higher concentrations of magnesium ATP and free magnesium was
shown to result entirely from a time, temperature and enzyme concen-
tration-dependent reaggregation of the enzyme which occurred under
standard assay conditions. The heavier forms of PP-ribose-P synthetase appeared equally active.

TABLE 2

AGGREGATED FORMS OF PP-RIBOSE -P SYNTHETASE

Relative S20, w	Stokes Radius Å	Molecular Weight	No. of Subunits	Enzyme Activity
4.8	31	65,000	2.0	<3%
7.1	43	133,000	4.0	<3%
9.7	61	258,000	7.7	<3%
15.9	74	513,000	15.4	100%
22.1	108	1,042,000	31.1	100%

The demonstration that activity of PP-ribose-P synthetase resided in the larger aggregates but not the smaller forms of the enzyme suggested that alteration in the state of aggregation of the enzyme could be a mechanism by which activity of this enzyme is controlled. In order to assess this possibility, the effect of inhibitors of PP-ribose-P synthetase on subunit association was studied. In Figure 3, the sedimentation profile of enzyme activity in 1 mM Pi, 1 mM DTT, and 0.3 mM ATP (Figure 3A) is compared with profiles observed in the presence of inhibitors of enzyme activity (Figures 3B - D). Addition of purine nucleotide inhibitors (ADP shown in Figure 3B , but also GDP and AMP) resulted in further aggregation of the enzyme to heavier forms. Aggregation of the smaller forms of the enzyme was also found with addition of pyrimidine and pyridine nucleotides as well as PP-ribose-P. These compounds appear to exert their inhibitory effects on PP-ribose-P synthetase activity by direct binding to and classical kinetic inactivation of the heavier forms of the enzyme. In contrast, 2,3-DPG, which shows competitive inhibition of PP-ribose-P synthetase with respect to ribose-5-P (3,8) caused disaggregation of the enzyme to the inactive smaller forms as shown in Figure 3C, and this effect was not overcome in the presence of 0.3 mM magnesium ATP (Figure 3D) or, in fact, magnesium ATP concentrations up to at least 5 mM. Disaggregation of PP-ribose-P synthetase by 2,3-DPG is apparent even at concentrations of 50 μM and provides a plausible mechanism for the inhibitory effects of this compound.

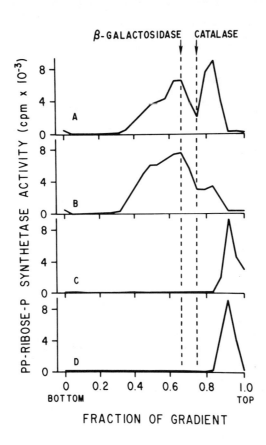

Figure 3.   Sucrose density gradient ultracentrifugation of PP-ribose-P synthetase.   Methods are those described in Figure 2 legend except that in this case sucrose gradients and samples contained the following additions: A.   1 mM sodium phosphate, 1 mM DTT, 0.3 mM magnesium ATP (pH 7.4); B. 1 mM sodium phosphate, 1 mM DTT, 0.3 mM magnesium ATP, 0.5 mM ADP (pH 7.4); C. 1 mM sodium phosphate, 1 mM DTT, 5.0 mM 2,3-DPG (pH 7.4); D. 1 mM sodium phosphate, 1 mM DTT, 0.3 mM magnesium ATP, 5.0 mM 2,3-DPG (pH 7.4).

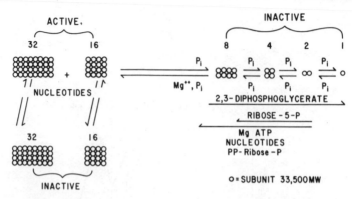

Figure 4.   Schematic representation of the relationship between
PP-ribose-P synthetase structure and activity and the alteration of
enzyme activity by effectors.

These findings suggest a model (Figure 4) which relates the
complex control of PP-ribose-P synthetase activity to the subunit
structure of the enzyme. PP-Ribose-P synthetase is composed of a
single repeated subunit of molecular weight 33,500 capable of
reversible aggregation to a number of polymeric states. While
the 16 and 32 subunit forms have enzymatic activity, the monomer
and aggregates containing two, four, or eight subunits are inactive
or nearly inactive. Neither Pi nor ribose-5-P alone cause enzyme
aggregation but these compounds appear to exert a permissive
effect on aggregation. Both magnesium ATP and nucleotide inhibitors induce aggregation, the latter exerting their inhibitory
effects on the aggregated enzyme. Another inhibitor, 2,3-DPG,
suppresses PP-ribose-P synthetase activity by disaggregating the
enzyme to the inactive small aggregated or monomer forms and
antagonizing aggregation by magnesium ATP. These separable mechanisms for the control of PP-ribose-P synthetase activity suggest a
variety of potential genetic aberrations that may lead to excessive
enzyme activity with consequent PP-ribose-P, purine and uric acid
overproduction. Such abnormalities might arise in several distinct
ways, such as: disordered carbohydrate metabolism with increased
ribose-5-P or decreased 2,3-DPG availability; depleted nucleotide
inhibitor pools; or structurally altered PP-ribose-P synthetases
defective in either inhibitor binding or subunit disaggregation.

ACKNOWLEDGEMENTS

This work was supported in part by grant AM-19187 from the
National Institutes of Health, a Veterans Administration Clinical
Investigatorship and a Veterans Administration Grant (MRIS 0865).

REFERENCES

1. Fox, I.H. and Kelley, W.N.: *Ann. Intern. Med.* 74: 424-433 (1971).
2. Becker, M.A. and Seegmiller,: *Annu. Rev. Med.* 25: 15-28 (1974).
3. Fox, I.H. and Kelley, W.N.: *J. Biol. Chem.* 247: 2126-2131 (1972).
4. Sperling, O., Persky-Brosh, S., Boer, P. and DeVries, A.: *Biochem. Med.* 7: 389-395 (1973).
5. Becker, M.A., Kostel, P.J., Meyer, L.J. and Seegmiller, J.E.: *Proc. Natl. Acad. Sci. U.S.A.* 70: 2749-2752 (1973).
6. Becker, M.A.: *J. Clin. Invest.* 57: 308-318 (1976).
7. Fox, I.H. and Kelley, W.N.: *J. Biol. Chem.* 246: 5739-5748 (1971).
8. Becker, M.A., Kostel, P.J. and Meyer, L.J.: *J. Biol. Chem.* 250: 6822-6830 (1975).
9. Neville, D.M., Jr.: *J. Biol. Chem.* 246: 6328-6334 (1971).
10. Reisfield, R.A. and Small, P.A.: *Science* 152: 1253-1255 (1966).
11. Weiner, A.M., Platt, T., and Weber, K.: *J. Biol. Chem.* 247: 3242-3251 (1972).
12. Yphantis, D.: *Biochemistry* 3: 297-317 (1964).
13. McEwen, C.R.: *Anal. Biochem.* 20: 114-149 (1967).
14. Ackers, G.K.: *J. Biol. Chem.* 242: 3237-3238 (1967).
15. Siegel, L.M. and Monty, K.J.: *Biochim. Biophys. Acta* 112: 346-362 (1966).

TRANSFER OF RESISTANCE TO SELECTIVE CONDITIONS FROM FIBROBLASTS WITH MUTANT FEEDBACK-RESISTANT PHOSPHORIBOSYLPYROPHOSPHATE SYNTHETASE TO NORMAL CELLS. A FORM OF METABOLIC COOPERATION

E. Zoref, A. de Vries and O. Sperling

Tel-Aviv University Medical School, Department of Chemical Pathology, Tel-Hashomer, and the Rogoff-Wellcome Medical Research Institute, Beilinson Medical Center, Petah Tikva, Israel

A mutant phosphoribosylpyrophosphate (PRPP) synthetase, recently found in our laboratory to be the primary abnormality underlying the excessive purine production in a family affected with primary metabolic gout (1,2), was used as a marker for the study of metabolic cooperation between cultured human fibroblasts (3). In physiological cellular milieu, the mutation is manifest in superactivity of the enzyme which is due to decreased sensitivity to feedback inhibition by several physiological intracellular inhibitors such as adenosine-5'-diphosphate, guanosine-5'-diphosphate and 2,3-diphosphoglyceric acid. The superactivity of the enzyme was shown to cause increased availability of its reaction product PRPP, a key substrate for both the de novo and salvage pathways of purine nucleotide synthesis. Accordingly, the mutant cell exhibits excessive de novo synthesis of purines as well as an improved capacity to synthesize purine nucleotides by the salvage pathway. Both properties render the mutant cell a suitable marker for the study of metabolic cooperation, the increased salvage capacity allowing selection between normal and mutant cells, and the excessive de novo purine synthesis allowing determination of the proportion of mutant cells in culture containing a mixture of mutant and normal cells.

The fibroblast growth medium was modified (4) to allow survival of only the mutant cells, i.e. cells possessing increased capacity to produce purine nucleotides. 6-Methyl-mercaptopurine-riboside (6-MMP-riboside) (0.2 mM) was added to block purine synthesis de novo, and hypoxanthine (0.2 mM) and uridine (0.5 mM) were added to allow salvage synthesis of inosinic acid and uridylic acid, respectively. Normal cells, under the selective conditions being

unable to produce a sufficient amount of purine nucleotides to sustain life, die within one week. In contrast, under the same selective conditions and during the same time period, the mutant cells
with the feedback-resistant superactive PRPP synthetase survive
the selection and multiply normally following transfer to the normal growth medium.

Experiments were designed to clarify if, and under what conditions, the capacity to resist selection could be transferred from
the mutant to the normal cells. Normal and mutant cells were mixed
in ratios of 1:1 and 1:3, respectively, and co-cultured for 3-10
generations. The mixed cell cultures were then exposed for seven
days to the modified selective growth medium, both at low and at
high cell densities. The cell cultures were then allowed to grow
in fresh regular growth media until confluency, and subsequently
trypsinized and propagated for one more generation. Subsequently,
the proportion between normal and mutant cells among the surviving
cells was established by measuring the rate of de novo purine synthesis (3). It was found (Table 1) that the capacity of the mutant
cells to resist selection could be transferred to the normal cells,
the degree of transfer depending on the degree of contact between
the normal cells and the mutant cells during selection; the higher
the proportion of mutant cells in the cell mixture and the greater
the density of the cells in culture during selection, the greater
the probability of contact between the mutant and normal cells.
Following exposure to selection of a normal:mutant cell mixture at
1:1 ratio and at high cell density, the rate of purine synthesis
de novo, as gauged by the rate of [^{14}C]formate incorporation into
purines excreted by the cells into the incubation medium (2), was
similar to that found before selection. As expected, these values
constitute the average of the synthesis rates of separate normal
and mutant cell cultures. On the other hand, when the same 1:1
mixed cultures were exposed to selection at low cell density, the
rate of purine synthesis following selection increased to approximately 2-fold, indicating survival of the mutant cells only. The
acquired resistance of the normal cells against selection was markedly reduced when normal cells were mixed with mutant ones at a
3:1 ratio. Following exposure to selection of such cell mixture at
low cell density, the surviving cells were too few to allow propa-
gation to confluency, suggesting the death of most of the normal
cells. However, when the same 3:1 cultures were exposed to selection at high cell density there were enough survivors to allow pro-
pagation, and the rate of purine synthesis de novo in the culture
was increased to 2.5-fold as compared to the preselection value.
These results indicate that in this latter experiment a small proportion of the normal cells survived the selective conditions, since
if all normal cells would have been killed, the rate of purine synthesis should have increased to approximately 4-fold.

TABLE 1

RATE OF PURINE SYNTHESIS DE NOVO IN MIXED NORMAL–MUTANT FIBROBLAST CULTURES FOLLOWING SELECTION
AGAINST NORMAL CELLS [E. Zoref et al. (3)]

Source of cells	Rate of incorporation of [14C]formate into purines excreted by the cells into the incubation medium (c.p.m./mg protein/6 h)		
	Pre selection	Post selection at	
		dense cultures[a]	dilute cultures[b]
Normal subjects			
J.B.	8,120	All cells killed	
J.Be.	12,050	All cells killed	
Subject O.G. with mutant enzyme	89,293	106,301	86,668
Mixture of normal and mutant cells			
O.G. + J.B. 1:1	59,294	70,447	94,567
O.G. + J.Be. 1:1	58,717	61,163	90,116
O.G. + J.Be. 1:3	27,269	66,246	too few survivors to allow propagation

a   < 250 cells/cm² growth surface

b   > 20,000 cells/cm² growth surface

In mutant cells, similarly exposed to the selective conditions, the rates of purine synthesis prior and following selection were found to be the same.

The transfer of the resistance to MMP-riboside from the mutant to the normal fibroblast is a new form of contact dependent metabolic cooperation. Metabolic cooperation is a form of intracellular communication by which cells in contact exchange molecules, a process providing multicellular organisms with an important mechanism for control of metabolic activity (5). This interesting phenomenon was originally described by Subak Sharpe, Burk and Pitts (6) who observed contact-dependent transfer of purine nucleotides from normal cells to cells mutationally incapable of producing inosinic acid due to deficiency in hypoxanthine-guanine phosphoribosyltransferase (7). Metabolic cooperation of this type was later also demonstrated with other enzymic markers, such as adenine phosphoribosyltransferase and thymidine kinase (8,9). The contact-dependent metabolic cooperation demonstrated with these markers is characterized by the normal cells being the donor and the mutant cell being the recipient, the former transferring to the latter a mutationally lacking metabolite. The type of metabolic cooperation described by us is unique in that the transfer of a metabolite occurs from a mutant donor cell to a normal recipient cell.

The dependence of the occurrence of the metabolic cooperation on the presence, during the selection, of close contact between the mutant and the normal cells, and the observation that the prolonged contact between the normal and the mutant cells prior to their exposure to the selective medium did not render the normal cells resistant to selection, indicate that the transferable molecule has a short life-span in the recipient cell. In view of the above and considering what is known from other reported observations on contact-dependent metabolic cooperation (5-9), the molecules which transferred the resistance to selection, may be either the reaction product PRPP or purine nucleotides, or both.

This investigation was partially supported by a U.S.A.-Israel Binational Science Foundation grant (No. 78).

REFERENCES

1. Sperling, O., Persky-Brosh, S., Boer, P. and de Vries, A. Biochem. Med., 7:389-395, 1973.
2. Zoref, E., de Vries, A. and Sperling, O. J. Clin. Invest., 56:1093-1099, 1975.
3. Zoref, E., de Vries, A. and Sperling, O. Nature, (London) 260:786-788, 1976.

4. Green, E.D. and Martin, D.W. Jr., Proc. Natl. Acad. Sci. U.S.A.,
    70:3698-3702, 1973.
5. Cox, R.P., Krause, M.R., Balis, M.E. and Dancis, J. in Cell
    Communication (edit. by Cox, R.P.), 67-95, (John Wiley & Sons,
    New York, 1974).
6. Subak-Sharpe, J.H., Burke, R.R. and Pitts, J.D. J. Cell Sci.,
    4:353-367, 1969.
7. Seegmiller, J.E., Rosenbloom, F.M. and Kelley, W.N. Science,
    155:1682-1684, 1967.
8. Cox, R.P., Krause, M.R., Balis, M.E. and Dancis, J. Exp. Cell
    Res., 74:251-268, 1972.
9. Pitts, J.D., in Growth Control in Cell Cultures, Ciba Founda-
    tion Symposium, (edit. by Wolstenholme, G.E.W., and Knight, J.),
    89-98 (Churchill & Livingstone, London, 1971).

PHOSPHORIBOSYLPYROPHOSPHATE DEGRADATION IN HUMAN TISSUES

Irving H. Fox and Pamela J. Marchant

Purine Research Laboratory, University of Toronto

Rheumatic Disease Unit, Wellesley Hospital, Toronto

Recent advances in the understanding of human purine metabolism have emphasized the essential role of phosphoribosylpyrophosphate (PP-ribose-P), an important substrate for purine, pyrimidine and pyridine metabolism. The intracellular concentration of this compound controls purine biosynthesis de novo in man such that increases and decreases in PP-ribose-P levels respectively accelerate and diminish the activity of this pathway. Pathological and pharmacological effects on the rate of purine biosynthesis de novo have been associated with altered cellular levels of PP-ribose-P. The intracellular concentration of PP-ribose-P is modulated by the rate of its formation as compared to its utilization. This compound is synthesized from ATP and ribose-5-phosphate. Degradation of PP-ribose-P has been believed to occur entirely by phosphoribosyltransferase reactions with purine, pyrimidine or pyridine bases.

Our studies of PP-ribose-P metabolism in human tissues have recently elucidated an alternative pathway for degradation of this compound which does not involve a phosphoribosyltransferase reaction. In this pathway PP-ribose-P is hydrolyzed with the release of inorganic phosphate and an unknown product. The potential importance of this reaction in controlling the intracellular PP-ribose-P concentration has prompted the investigation of the characteristics and distribution of this activity in man (1).

In the initial studies to prove the existence of an alternative pathway of metabolism, PP-ribose-P was incubated at $37^{\circ}C$ with dialyzed placental extracts at pH 7.0 and 9.5. Under these conditions PP-ribose-P disappeared. Measurement of inorganic phosphate release allowed a more direct assay of PP-ribose-P degradation which was

TABLE I

STOICHIOMETRY OF PP-RIBOSE-P HYDROLYSIS

Placental microsomal supernatant 1.85 μg was incubated at $37^{o}C$
with PP-ribose-P 2.4 mM, Tris-HCl 40 mM pH 9.5 from 10 minutes to
40 minutes. Assays were performed for Pi and PP-ribose-P. There
was non-linearity after 20 minutes.

Time (min)	PP-ribose-P Disappearance (nanomoles)	Pi Production (nanomoles)
10	15.4	13.4
20	24.5	28.1
30	36.7	30.8
40	41.6	36.9

linear with protein and with time up to 20 minutes. Table 1
illustrates the stoichiometry of PP-ribose-P hydrolysis at pH 9.5
with partially purified placental alkaline phosphatase. The
disappearance of PP-ribose-P approximately equals Pi production.

Our studies have delineated the characteristics of PP-ribose-P
hydrolysis. PP-ribose-P was degraded in a pH range from 5.0 to 10.5
with the optimum between 9.0 and 10.5 (Figure 1). There was no
requirement for divalent cations. However, 10 mM of many divalent
cations at $37^{o}C$ caused non-enzymatic hydrolysis of PP-ribose-P
from as low as 10% for $CaCl_2$ up to 100% of all substrate for $HgCl_2$.
Actual enzymatic hydrolysis of PP-ribose-P was inhibited by all
divalent cations tested except $CaCl_2$ and $MnCl_2$. The substrate
specificity for the enzyme preparation included numerous phosphorylated intermediates of purine, pyridine, pyrimidine and carbohydrate metabolism. The Km for PP-ribose-P was 0.3 mM as compared
to the Km of 0.5 mM for p-nitrophenylphosphate for alkaline phosphatase. The maximum specific activity was in the microsomal fraction of human placenta. All tissues assayed showed evidence of
PP-ribose-P hydrolysis at pH 7.0 and 9.5. The specific activity of
tissue homogenates at pH 7.0 ranged from as low as 0.15 in skeletal
muscle up to 1.37 micromoles/hr/mg in uterus. There was a greater
capacity for PP-ribose-P utilization than synthesis at neutral pH
in every tissue assayed.

PP-ribose-P hydrolyzing activity was localized mainly to the
microsomal fraction (Table 2). This activity may have been a

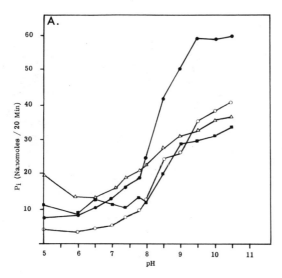

Fig. 1. pH Curves. 5 μg microsomal supernatant was assayed for 20 minutes using 40 mM Tris-HCl from pH 6 to pH 10.5 and 50 mM sodium acetate pH 5.0. Substrate concentrations were PP-ribose-P 2.4 mM (■—■), PPi (●—●), AMP (△—△), and p-nitrophenol-P (o—o) 5 mM. Pi assay was used for all substrate except p-nitrophenol-P where p-nitrophenol production was measured.

property of 5'-nucleotidase, non-specific phosphatase, or some other unique enzyme system. Studies were performed to distinguish between these possibilities. Human placental microsomal supernatant was fractionated by ion exchange chromatography on DE-52 in which 5'-nucleotidase was separated from alkaline phosphatase. 5'-nucleotidase did not bind to the column, while alkaline phosphatase was eluted with 0.3 M KCl. Table 3 illustrates that PP-ribose-P hydrolyzing activity copurified with alkaline phosphatase. The DE-52 wash which contained purified 5'-nucleotidase shows almost no increase in specific activity for PP-ribose-P hydrolysis. In the DE-52 KCl eluate, containing partially purified alkaline phosphatase, PP-ribose-P hydrolyzing activity increased from 0.75 to 80.7 micromoles/hr/mg representing a 108 fold purification.

Further studies provide evidence that alkaline phosphatase and PP-ribose-P hydrolysis were the same. The pH curve and optimum are similar. Both activities are lost together during thermal inactivation at 80°C. Both activities chromatograph together on 4% agarose during conformational changes induced by AMP and MgCl$_2$. Both are inhibited by 2 mM ZnCl$_2$ and 10 mM 1-phenylalanine but not by 10 mM sodium fluoride. The latter is a known inhibitor of 5'-

TABLE 2

SUBCELLULAR LOCALIZATION OF PP-RIBOSE-P HYDROLYZING

ACTIVITY

90 grams of human placenta were homogenized and fractionated. Dialyzed fractions were assayed at pH 9.5 using Pi assay for PP-ribose-P hydrolysis and p-nitrophenol production for alkaline phosphatase.

	Specific Activity (micromoles/hr/mg)	
	PP-ribose-P Hydrolysis	Alkaline Phosphatase
Homogenate	0.75	4.8
700 g x 20 min.	0.59	1.8
5000 g x 20 min.	3.20	11.4
8000 g x 480 min.	8.40	30.6
Supernatant	<.02	0.5

TABLE 3

SEPARATION OF PLACENTAL 5'-NUCLEOTIDASE FROM

NON-SPECIFIC PHOSPHATASE

Assays were performed at the pH indicated using the Pi assay for 5'-nucleotidase and PP-ribose-P hydrolysis and p-nitrophenol formation for non-specific phosphatase.

	Specific Activity (micromoles/hr/mg)			
	AMP pH 7.0	PP-ribose-P pH 7.0	p-nitrophenyl-P pH 9.5	PP-ribose-P pH 9.5
Homogenate	0.54	0.67	4.80	0.75
DE-52 Wash	14.84	1.00	4.40	0.18
DE-52 KCl eluate	–	34.30	102.20	80.70

nucleotidase. Commercially prepared alkaline phosphatase from calf intestinal mucosa degraded PP-ribose-P enzymatically, while 5'-nucleotidase from Crotalus ademateus did not have this activity. Although non-specific phosphatases rather than 5'-nucleotidase catalyzed the enzymatic hydrolysis of PP-ribose-P in these studies, the observations do not rule out the possible existence of a unique enzyme for PP-ribose-P degradation.

This large capacity to degrade PP-ribose-P clearly must be regulated to protect the PP-ribose-P present for other essential reactions. Many intracellular compounds were found to modify PP-ribose-P hydrolysis. Table 4 illustrates the effect of inorganic phosphate. This compound seems to be particularly important since it has a potent effect at a physiological 1 mM concentration and appears to have an increased inhibitory effect as PP-ribose-P levels decrease. Many nucleoside mono-, di- and triphosphates inhibit PP-ribose-P hydrolysis (1). Other phosphate compounds including phosphorylated sugars and inorganic pyrophosphate have an inhibitory effect. Many of these compounds are substrates for alkaline phosphatase. An example of inhibition of the hydrolysis of one phosphorylated compound by another is illustrated in Figure 2. The kinetic mechanism of inhibition by a diphosphonate analog of inorganic pyrophosphate[a] was also studied (Figure 3). The inhibitory effect on PP-ribose-P hydrolysis by these many compounds suggest that this pathway may be inhibited under normal conditions within the cell by the total nucleotide pool, the phosphorylated sugars, and inorganic phosphate.

The potential significance of the alternative pathway for PP-ribose-P degradation must be considered in the context of PP-ribose-P metabolism (Figure 4). PP-ribose-P is a rate limiting substrate for purine biosynthesis de novo and is an essential substrate for the salvage pathways. Therefore, a pathway which alters the cellular levels of PP-ribose-P may have a regulatory role in these important enzymatic steps. This may be relevant to PP-ribose-P hydrolysis, since our data suggests that this reaction can potentially utilize substantial quantities of PP-ribose-P. In all tissues assayed the capacity to degrade PP-ribose-P exceeded the synthetic ability at neutral pH. In addition the Km of PP-ribose-P for non-specific phosphatase is quantitively similar to its Km for the phosphoribosyl-transferase enzymes. However, the importance of PP-ribose-P hydrolysis in the intact cell awaits definitive proof.

---

a. Disodium ethane-1-hydrolyl-1,1-diphosphonate was generously supplied by Dr. I. Rosenbloom of Proctor and Gamble Company Limited.

TABLE 4

EFFECT OF INORGANIC PHOSPHATE ON PP-RIBOSE-P HYDROLYSIS

5.5 µg of microsomal supernatant was incubated for 20 minutes at 37°C with varying concentrations of PP-ribose-P and inorganic phosphate and Tris-HCl 40 mM pH 9.5. The PP-ribose-P assay was used to measure hydrolysis of PP-ribose-P.

| | Percent Inhibition | | |
| | Inorganic Phosphate | | |
PP-ribose-P	1 mM	2 mM	4 mM
0.75 mM	68	65	74
0.33 mM	80	84	100
0.25 mM	92	100	100

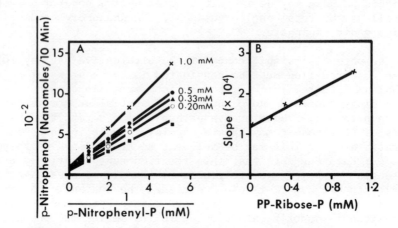

Fig. 2. Inhibition of alkaline phosphatase by PP-ribose-P. This (A) is a double reciprocal plot of inhibition of alkaline phosphatase with variable p-nitrophenylphosphate concentrations and fixed PP-ribose-P concentrations ranging from 0 to 1.0 mM. A competative mechanism of inhibition was observed. A secondary plot of slope (B) against PP-ribose-P concentrations gives a Ki of 0.8 mM.

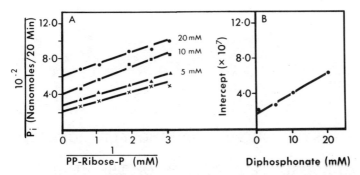

Fig. 3. Inhibition of PP-ribose-P hydrolysis by a diphosphonate.
This is a double reciprocal plot of inhibition of PP-ribose-P
hydrolysis with variable PP-ribose-P concentrations and fixed
diphosphonate concentrations ranging from 0 to 20 mM (A). A non-
competative mechanism of inhibition was evident (B). A secondary
plot of intercept against diphosphonate concentration gives a Ki
of 10 mM.

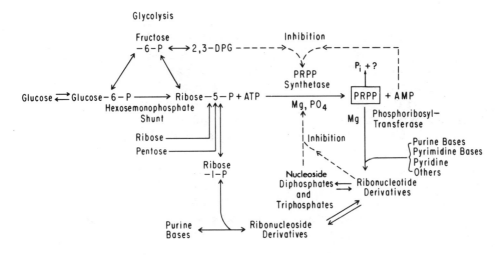

Fig. 4. Regulation of intracellular PP-ribose-P concentration.

REFERENCES

1. Fox, I.H. and Marchant, P.J. 1974. Phosphoribosylpyrophosphate
     degradation in human tissues. Can. J. of Biochem. 52:
     1162-1166.

PROPERTIES OF THE PHOSPHORIBOSYLPYROPHOSPHATE-GLUTAMINE AMIDOTRANSFERASE OF THE HUMAN LEUCEMIC CELLS

H. Becher and W. Gerok

Department of Medicine

University of Freiburg, Germany

The biosynthesis of purine nucleotides may be produced from two alternative routes, de novo and "salvage" pathway. From the investigations of SCOTT et al. (1) it is known that the novo pathway of purine synthesis is incomplete in leucocytes, and the incorporation of (^{14}C)-formate only served to close the purine ring of the otherwise complete precursor. Therefore the statements about the existence of such a pathway in the normal and leukemic blast blast cells are different. Recently, REEM (2) demonstrated two alternative pathways of purine de novo synthesis in normal spleen and leukemic lymphoblasts by determination of two enzyme activities which catalyse phosphoribosylamine.

The present study was undertaken to obtain more precise information about the purine biosynthetic pathway in human blood cells 5-Phosphoribosyl-1-pyrophosphate (PP-ribose-P) amido-transferase (E. G. 2.4.2.14) was assayed in cell free extracts of normal and leukemic blood cells, and some of the properties of the partially purified enzyme were determined.

Material and Methods

Enzyme assay: Glutamine PP-ribose-P amidotransferase was assayed by determining the PP-ribose-P dependent conversion of (^{14}C)-glutamate according to the method of PRUSINER and MILLNER (3). The standard assay was performed as described by HOLMES et al. (4) except that the reaction was stopped by the addition of 1 ml of ice-cold 20 mM imidazole at pH 7, 2 and immediately passed over a Dowex 1 x 8 (acetate form) column ((O, 5 x 4) cm). The (^{14}C)-glutamic acid was eluted from the column with 3 ml O, 5 M HCl and the eluate was collected in a counting vial and the radioactivity counted.

Preparation of cells

Only patients with a leukocyte count in excess of 1OO OOO cells/mm^3 were included in the study. The separation in lymphocytes and granulocytes was performed according to the method of BÖYUM (5). After the final wash the cells were suspended in the 5 mM potassium phosphate buffer, pH 7, 5 containing 1O mM $MgCl_2$ and 1O mM ß-mercapto-ethanol and after standing for 3O min, the cells were disrupted three times by freezing and thawing in liquid nitrogen and centrifuged at 1OO OOO x g for 6O min. The supernatant was used for the enzyme assay or purification procedures. The 1OO OOO x g supernatant was adjusted to pH 5, 2 with 1 M HCl and stirred for 15 min. After centrifugation solide ammonium sulphate was added to the supernatant and the protein precipitating between 35 and 6O % saturation was retained and dialyzed against 5O mM phosphate buffer, pH 7, 5 containing 1O mM $MgCl_2$ and 1O mM ß-mercapto-ethanol for 6O min prior to the assay. The spezific activity increased 2O fold and no phosphate dependent glutaminase activity was detectable in the resulting enzyme fraction.

Results and Discussion

Table 1 shows the PP-ribose-P amidotransferase activity in leukemic and normal blood cells, measured in the 1OO OOO x g supernatant.

Table 1

cells investigated	n	enzyme activity (nmoles glutamate/3O min/mg prot.)	
normal leukocytes	4	< O,O1	
normal bone marrow cells	4	< O,O1	
peripheral lympho- cytes	6	1,35 - 1,8	x̄ 1,55
spleen lymphocytes	5	1,5 - 2,55	x̄ 2,10
peripheral blast cells			
a) acute myeloblastic leucemia	12	2,4 - 10,8	x̄ 8,6
b) acute lymphocytic leucemia	8	3,5 - 9,6	x̄ 7,1
chronic myeloid leucemia	7	O,6 - 4,5	x̄ 2,7
chronic lymphatic leucemia	6	3,0 - 7,8	x̄ 5,1

In the normal blood cells enzyme activity was only detected in
peripheral and spleen lymphocytes. In comparison with the enzyme
activities in cultered lymphocytes and lymphoblasts, measured by
REEM (6) and WOOD (7), we found a relatively slow activity of
PP-ribose-P amidotransferase.
The further separation of spleen lymphocytes by a discontinuous
Ficoll density gradient (8) gave subpopulations with a three to four
times higher enzyme activity. The histological examination of the
cells with the lowest density shows lymphoblasts and other immature
cells. Therefore we can assume, that the de novo synthesis of
purine nucleotides in immature cells is of special importance, as
you can see in the leukemic cells. We found the highest enzyme
activity in the peripheral blast cells of patients with acute leukemia.
Although all investigated patients with acute leukemia contained
more than 75 % blast cells, the measured enzyme activity had a
considerable range and no correlation was found between the amount
of blast cells and enzyme activity.
The enzyme of acute myeloblastic cells was further purified
(20 fold) and substrate kinetic with PP-ribose-P and glutamine are
shown on the next slide. Substrate velocity curves for PP-ribose-P
and glutamine show hyperbolic kinetic and the reciprocal plots were
linear with an apparent Km for PP-ribose-P of 0, 5 mM and for
glutamine 1, 25 mM (Figure 1).

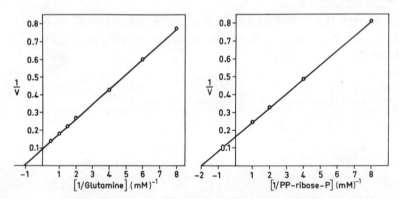

Figure 1   Lineweaver-Burk plots for Glutamine and PP-ribose-P
             Fraction III enzyme of leucemic blast cells was used
             for the test. All assays were performed under standard
             conditions as described in "Methods", except for
             variable Glutamine (A) and PP-ribose-P concentra-
             tions (B).

A higher Km for PP-ribose-P was noted in phosphate buffer. The well known importance of the intracellular PP-ribose-P concentrations on the rate of purine biosynthesis is shown in Figure 2.

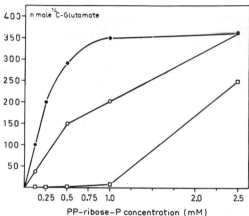

Figure 2   Michaelis-Menten plot illustrating the sigmoid kinetics with respect PP-ribose-P. Fraction III enzyme of leucemic blast cells (AML) was used for the test. All assays were performed under standard conditions as described in "Methods" except for variable PP-ribose-P concentrations.

(•——•) No added AMP, (o——o) O, 5 mM AMP and (□——□) 1, O mM AMP

In the absence of purine ribonucleotides the leukemic blast enzyme exhibits Michaelis-Menten-Kinetics for the substrate PP-ribose-P. The initial velocity patterns for saturation by PP-ribose-P in the presence of different fixed concentrations of AMP show that AMP acts as a competitive inhibitor of the PP-ribose-P binding. The molecular understanding for this regulation is recently given by the studies of HOLMES (9).

Table 2 shows the inhibition of enzyme activity in the 100 000 x g supernatant obtained from human spleens and leukemic blast cells by various purine and pyrimidine nucleotides.

Table 2   Inhibition of glutamine PP-ribose-P amidotransferase from leucemic cells (acute myeloblastic
leucemia) by purine and pyrimidine compounds.
All assays were performed under standard conditions as described in "Methods", containing
O, 8 mM PP-ribose-P.

Inhibitor added		Glutamine-PP-ribose-P-amidotransferase activity % control	
None	mM	leucemic blast cells 100%	spleen lymphocytes 100%
AMP	1	64	77
	4	48	9
GMP	1	100	64
	4	60	23
IMP	1	100	44
	4	43	18
6-Thioguanosine-5'-phosphate	1	91	44
	4	22	9
8-Azaguanosine-5'-phosphate	1	91	42
	4	14	0
6-Mercaptopurine-riboside-5'-P	1	56	36
	4	42	16
Allopurinol	1	100	100
Allopurinol ribonucleotide	1	88	56
	4	42	21
6-Azauracil	1	100	79
	4	100	81
8-Azaguanine	1	100	100
	4	100	83

The specific activity of spleen lymphocyte enzyme with 2, 1 nmole
^{14}C-glutamate/3O min/mg Protein was considerably lower than
the enzyme activity obtained from acute lymphoblastic cells with
9, 4 nmole (^{14}C)-glutamate/3O min/mg Protein and the inhibition
by the various purine nucleotides and derivates was significantly
lower. Of those tested the ribonucleotides of the purine analogs
6-mercaptopurine and 6-thioguanine were the most potent inhibitors.
This may be due to differences between the cell maturities rather
than expression of an alteration of the feedback sensitivity by the
neoplastic transformation but the significance of this observation
must be investigated on the purified enzyme.
The last Figure 3 demonstrates the interactions between de novo
and "salvage" pathway in the various blood cells.

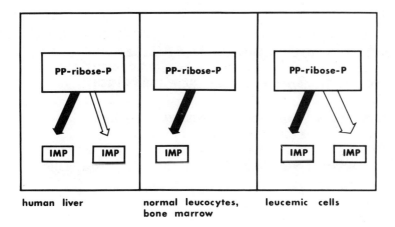

Figure 3 Interactions between de novo (⟹) and "Salvage" (➡) pathway

Amidotransferase activity was found in leukemic cells and normal lymphocytes and therefore these cells could synthesize the first intermediate of the purine biosynthetic pathway. Normal leukocytes and bone marrow cells lack this enzyme system and have an absolute requirement for externally supplied purines via salvage pathway. Leukemic blast cells show different enzyme activities independent of their cell count. In three patients with acute myeloblastic leukemia, we found very low values for the amidotransferase although there were more than 90 % peripheral blast cells. It was remarkable that these patients were primarily resistent against the usual zytostatic therapy.

ACKNOWLEDGEMENTS

I wish to thank Miss Beatrix Osswald for her excellent assistance with this study.

REFERENCES

1. Scott, J. L. J. Clin. Invest. 41, 67 (1962)

2. Reem, G.H. J.Biol. Chem. 249, 1693 (1974)

3. Prusiner, S., Milner, L. Analyt. Biochem. 37, 429 (1970)

4. Holmes, E. W., J. A. Mc Donald, J. M. Mc Cord, J. B. Wyngaarden, and W. N. Kelley J. Biol. Chem. 248, 144 (1973)

5. Böyum, A. Scand. J. Clin. Lab. Invest. 21 (1968) Suppl. 97

6. Reem, G. H. J. Clin. Invest. 51, 1058 (1972)

7. Wood, A W. and J. E. Seegmiller J. Biol. Chem. 248, 138 (1973)

8. Bach, M. K. and J. R Brashler Exp. Cell Res. 61, 387 (1970)

9. Holmes, E. W., J. B. Wyngaarden, and W. N. Kelley J. Biol. Chem. 248, 6035 (1973)

A PURINE AUXOTROPH DEFICIENT IN PHOSPHORIBOSYLPYROPHOSPHATE AMIDO-

TRANSFERASE AND PHOSPHORIBOSYLPYROPHOSPHATE AMINOTRANSFERASE ACTI-

VITIES WITH NORMAL ACTIVITY OF RIBOSE-5-PHOSPHATE AMINOTRANSFERASE

E. W. Holmes, G. L. King, A. Leyva and S. C. Singer

Duke University Medical Center, Durham, N. C.   27710

INTRODUCTION

The synthesis of phosphoribosylamine (PRA) is the first committed reaction unique to purine biosynthesis de novo (Wyngaarden,
1972). Traditionally the catalysis of this reaction has been attributed to the enzyme glutamine phosphoribosylpyrophosphate amidotransferase [E.C.2.4.2.14; PP-ribose-P amidotransferase (reaction 1)].

PP-ribose-P amidotransferase

(1)   Glutamine + PP-ribose-P + $H_2O \longrightarrow$ PRA + Glutamate + PPi

PP-ribose-P aminotransferase

(2) NH_3 + PP-ribose-P + $H_2O \longrightarrow$ PRA + PPi

Ribose-5-phosphate aminotransferase

(3)   Ribose-5-phosphate + ATP + $NH_3 \longrightarrow$ PRA + ADP + Pi

However, recent studies have suggested that two other enzymatic
activities also catalyze the synthesis of PRA in eukaryotic cells
(Reem, 1968; Reem, 1972; Reem, 1974; Reem and Friend, 1975). The
first of these (reaction 2) has been called ammonia PP-ribose-P
aminotransferase (PP-ribose-P aminotransferase) (Reem, 1972; Reem,
1974; Reem and Friend, 1975). This enzyme utilizes ammonia rather
than glutamine as substrate and has been separated from PP-ribose-P
amidotransferase on gel filtration chromatography (Reem, 1974).
This activity may represent a distinct protein or a subunit of

PP-ribose-P amidotransferase. A third enzyme, ammonia Ribose-5-phosphate aminotransferase (Ribose-5-phosphate aminotransferase), has also been reported to catalyze the synthesis of PRA (reaction 3) (Reem, 1968; Reem, 1972; Reem, 1974; Reem and Friend, 1975). However, the determination of PRA in this reaction has required an assay coupled with the second enzyme in the purine biosynthetic pathway. Since other studies have suggested that PRA can be synthesized non-enzymatically from NH_3 and ribose-5-phosphate (Westby and Gots, 1969; Henderson, 1963; Nierlich and Magasanik, 1965; Malloy, Sitz and Schmidt, 1973), the physiological significance of the Ribose-5-phosphate aminotransferase reaction in eukaryotic cells has been questioned. The recent isolation by Chu et al (Chu, Sun and Chang, 1972) of a eukaryotic cell line deficient in PP-ribose-P amidotransferase activity (Feldman and Taylor, 1973) provided the unique opportunity to evaluate the potential role of each of these three reactions in purine biosynthesis de novo.

MATERIALS AND METHODS

Cell Lines

Cells were routinely grown in monolayer in Falcon plastic Petri dishes or glass roller bottles using Eagle's minimum essential medium (F-15, Gibco) supplemented with 10% fetal calf serum (Irvine) and 10^{-4}M hypoxanthine. Experiments performed in purine-free medium utilized fetal calf serum that had been dialyzed twice against 40 volumes of 0.15 M NaCl for 12 hours.

Enzyme Assays

PP-ribose-P amidotransferase was assayed in a 100 µl reaction mixture which contained the following: 5mM PP-ribose-P, 4mM $[^{14}C]$-glutamine, 5mM $MgCl_2$, 0.75mM dithiothreitol (DTT) and 50 µl of cell extract (0.49 to 0.94 µg of protein) in 37.5mM KPi buffer, pH 7.4. This assay which has been previously described utilized a PP-ribose-P blank to determine PP-ribose-P amidotransferase activity (Holmes, et al. 1973). The PP-ribose-P independent conversion of $[^{14}C]$-glutamine to $[^{14}C]$-glutamate was attributed to glutaminase (Holmes, et al. 1973). PP-ribose-P aminotransferase was assayed in a 100 µl reaction mixture which contained the following: 5mM PP-ribose-P, 100mM NH_4Cl (1.26mM NH_3), 5mM $MgCl_2$, 1.4mM DTT, 40mM $[^{35}S]$-cysteine and 50 µl of cell extract (0.49 to 0.94 µg protein) in 25mM KPi buffer, pH 8.4. An NH_4Cl blank was used to determine the PP-ribose-P aminotransferase activity. This assay for PRA utilized a newly described reaction between $[^{35}S]$-cysteine and PRA (King and Holmes, 1975). The NH_3, ribose-5-phosphate and ATP dependent production of PRA was arbitrarily attributed to Ribose-5-phosphate aminotransferase activity, since it is not known whether the synthesis of PRA under these conditions is an enzymatic or non-enzymatic process. Since the newly described direct assay for PRA could not be used in the presence of ribose-5-phosphate (King and Holmes, 1975), the assay for Ribose-5-phosphate aminotransferase was performed in a 100 μl reaction mixture which contained the following: 27mM ribose-5-phosphate, 22mM NH_4OH (1.1mM NH_3), 2mM ATP, 2mM [^{14}C]-glycine, 10mM $MgCl_2$, 1mM DTT and 40 μl of cell extract (0.5 to 1.2 μg of protein) in 50mM Tris-HCl buffer, pH 8.0. The ribose-5-phosphate and NH_4OH were preincubated at 37° for 60 min. in 50mM Tris-HCl buffer, pH 8.0. The blank for this assay omitted the ribose-5-phosphate and NH_4OH, and the [^{14}C]-glycine was separated from the [^{14}C]-phosphoribosylglycinamide (PRG) on a Dowex column (Malloy, Sitz and Schmidt, 1973). Preliminary studies indicated that PRG synthetase activity from the cell lysate was not limiting and consequently an exogenous source of this enzyme was not added to the reaction mixture.

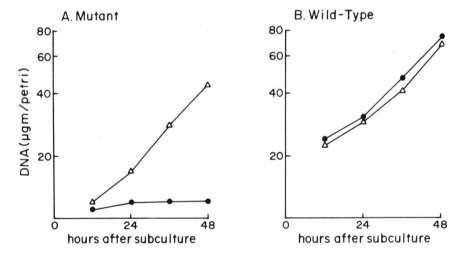

Fig. 1. Growth requirements of mutant and wild-type cells. Cells were grown in a purine-free medium without supplementation (o--o) or with 10^{-4}M hypoxanthine (Δ--Δ). Panel A, mutant cells; Panel B, wild-type cells.

Hypoxanthine-guanine phosphoribosyltransferase (HGPRT) (Kelley and Meade, 1971), adenine phosphoribosyltransferase (APRT) (Thomas, Arnold and Kelley, 1973), inosinic acid dehydrogenase (IMP-dehydrogenase (Holmes, Pehlke and Kelley, 1974), adenylosuccinate (S-AMP) synthetase (Van Der Weyden and Kelley, 1974), adenosine deaminase (ADA) (Van Der Weyden, Buckley and Kelley, 1974), xanthine oxidase, (Holmes, et al., 1974), and PP-ribose-P synthetase (Leyva, 1974) were determined as previously described. All of the above assays were linear with respect to time of incubation and protein concentration. 5-Aminoimidazole-4-carboxamide ribonucleotide (AICAR) was determined by the method of Ravel et al. (Ravel, et al.,1948).

RESULTS

Growth Requirements

Figure 1 demonstrates that after 24 hours of subculture in a purine-free medium the mutant cells were unable to replicate, while the wild-type cells continued to grow well. As shown, both cell lines grew equally well when the medium was supplemented with 10^{-4}M hypoxanthine. Although not presented in Fig. 1, 10^{-4}M adenine also supported growth of the mutant cells.

TABLE 1

ACTIVITY OF PRA SYNTHESIZING ENZYMES IN MUTANT
AND WILD-TYPE EXTRACTS

Enzyme	Wild-type*	Mutant*
	(nmoles/ hr/mg)	(nmoles/ hr/mg)
PP-ribose-P Amidotransferase[+]	88.5	<1
PP-ribose-P Aminotransferase[+]	256	<5
Ribose-5-Phosphate Aminotransferase[‡]	4.32	5.82

*Cells were grown in regular medium supplemented with 10^{-4}M hypoxanthine.
[+]Cell extracts were dialyzed against 50mM KPi buffer, pH 7.4, containing 1mM DTT.
[‡]Extracts were dialyzed against 50mM Tris-HCl buffer, pH 7.4, containing 1mM DTT.

Synthesis of PRA

Table 1 lists the three activities in mutant and wild-type extracts reported to synthesize PRA. Neither PP-ribose-P amidotransferase nor PP-ribose-P aminotransferase activity was detected in extracts from the mutant cells. In mixing experiments of extracts from mutant and wild-type cells there was no evidence for the presence of an inhibitor of PP-ribose-P amidotransferase or PP-ribose-P aminotransferase (Table 2).

In contrast to these findings, Tris-HCl dialyzed extracts from both the mutant and wild-type cells demonstrated an equal ability to synthesize PRA and PRG from ribose-5-phosphate, NH_3, ATP and glycine (Table 1). The synthesis of PRG in this reaction was linear with respect to time of incubation and extract protein concentration (Fig. 2).

TABLE 2

MIXING OF MUTANT AND WILD-TYPE EXTRACTS*

Mutant[+] Extract	Wild-type[+] Extract	PP-ribose-P Amidotransferase (nmoles/hr)	PP-ribose-P Aminotransferase (nmoles/hr)
–	25 µl	4.41	3.01
25 µl	–	<0.50	<0.50
25 µl	25 µl	4.86	2.81

*Cells were grown in regular medium supplemented with 10^{-4}M hypoxanthine.
Extracts were dialyzed against 50mM KPi buffer, pH 7.4, containing 1mM DTT for PP-ribose-P amidotransferase; and 50mM KPi buffer, pH 7.4, containing 5mM $MgCl_2$ and 60mM beta-mercaptoethanol for PP-ribose-P aminotransferase.
[+]The protein concentrations of the mutant cell extracts were 14.4 mg/ml and 12.5 mg/ml for the PP-ribose-P amidotransferase and PP-ribose-P aminotransferase experiments, respectively. The protein concentrations of the wild-type extracts were 18.8 mg/ml and 9.8 mg/ml, respectively.

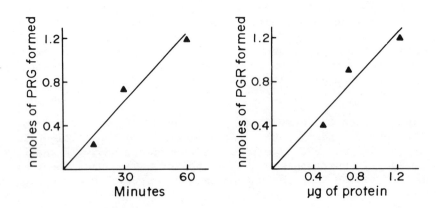

Fig. 2. Ribose-5-phosphate aminotransferase activity. Extracts
from the mutant cell line were dialyzed against 50mM Tris-HCl buffer,
pH 7.4, and the assay for PRG synthesis from ribose-5-phosphate,
NH_3, ATP and ^{14}C-glycine was performed as described in the methods.
The left hand panel depicts product formed versus time of incubation at 37° with 1.2 μg of extract protein; the right hand panel
depicts product formed versus protein concentration for a 60 min.
incubation.

TABLE 3

GROWTH OF MUTANT AND WILD-TYPE CELLS
IN SUPPLEMENTED MEDIUM*

| | % change in DNA/Petri dish | |
Supplement	Wild-type	Mutant
None	+ 400%	- 16%
1mM Ribose	+ 340%	- 10%
1mM NH_4Cl	+ 330%	- 5%
1mM Ribose + 1mM NH_4Cl	+ 340%	- 14%

*Cells were grown in purine-free medium with the indicated supple-
ment for 48 hours. The percent change in DNA is expressed as the
μg of DNA/Petri dish at 48 hours relative to that present at time 0.

Intracellular Ribose-5-phosphate Aminotransferase Activity

Table 3 lists the results of cell growth studies performed in purine-free medium which was supplemented with potential substrates for ribose-5-phosphate aminotransferase. Wild-type cells grew under all of the experimental conditions, but the mutant cells did not replicate even when the medium was supplemented with 1mM ribose, 1mM NH_4Cl, or the combination of both of these agents. Higher concentrations of NH_4Cl inhibited growth of the wild-type cells and did not support growth of the mutant cells.

TABLE 4

SYNTHESIS OF INTERMEDIATES OF PURINE METABOLISM
BY MUTANT AND WILD-TYPE CELLS

A. Synthesis of 5-Aminoimidazole-4-Carboxamide Ribonucleotide (AICAR)
(ng AICAR formed/90 min./mg of protein)[+]

Mutant extract	46.7
Wild-type extract	81.3

B. Utilization of 5-Aminoimidazole-4-Carboxamide (AIC)
(% change in DNA/Petri dish)

Mutant cells	+ 510%
Wild-type cells	N.D.*

In part A, the undialyzed extracts from freshly harvested cells were incubated with 27mM ribose-5-phosphate, 1.1mM NH_3, 2mM ATP, 10mM $MgCl_2$, 2mM glycine, 2mM formate, 10mM glutamine, 1mM aspartate, 10mM KCl, and 25mM bicarbonate in 50mM Tris-HCl buffer, pH 7.4, containing 1mM DTT for 90 min. at 37°. A ribose-5-phosphate/NH_3 blank was used to calculate AICAR produced during the 90 min. incubation. [+]The actual concentration of AICAR in the extract varied from 1.26 to 2.5 μg/ml.

In part B, the mutant cells were grown for 48 hours in a purine-free medium that was supplemented with 1mM 5-aminoimidazole-4-carboxamide (AIC). The percent change in DNA is expressed as the μg of DNA/Petri dish at 48 hrs. relative to that present at time 0. *N.D. = not determined since wild-type cells grew in purine-free medium.

PRA Utilization

The results presented in Table 4 indicate that the mutant cells were capable of catalyzing the remainder of the reactions in the pathway of purine biosynthesis de novo, if they were supplied with PRA. PRA was synthesized by incubating the extract from mutant or wild-type cells with ribose-5-phosphate and NH_3. As shown in part A of Table 4, extract from mutant, as well as wild-type cells, was capable of synthesizing AICAR, the 8th intermediate in the pathway of purine biosynthesis de novo. Part B of this table demonstrates that mutant cells grew well when the purine-free medium was supplemented with AIC. AIC is metabolized to AICAR by APRT (Thomas, Arnold and Kelley, 1973), and AICAR is then converted in two enzymatic reactions to the parent purine ribonucleotide, inosinic acid.

Enzyme Activities

The activities of a number of other enzymes important to purine biosynthesis are listed in Table 5. The activities of IMP dehydrogenase, S-AMP synthetase, HGPRT, APRT, PP-ribose-P synthetase, xanthine oxidase and glutaminase were comparable in the extracts from mutant and wild-type cells.

TABLE 5

ENZYME ACTIVITIES
IN MUTANT AND WILD-TYPE EXTRACTS*

Enzyme	Wild-type	Mutant
	(nmoles/hr/mg)	(nmoles/hr/mg)
HGPRT	252	303
APRT	452	583
IMP-dehydrogenase	1.55	1.68
S-AMP Synthetase	6.48	9.09
ADA	596	575
Xanthine Oxidase	54.6	48.3
PP-ribose-P Synthetase	154	167
Glutaminase	31.7	30.1

*All extracts were dialyzed against 50mM KPi buffer, pH 7.4, containing 1mM DTT, except those used for the xanthine oxidase assays and these were dialyzed against 50mM Tris-HCl buffer, pH 7.4 with 1mM DTT.

DISCUSSION

Ribose-5-phosphate aminotransferase activity was comparable in the extracts from mutant and wild-type cells. Since the mutant cells were demonstrated to be deficient in PP-ribose-P amidotransferase and PP-ribose-P aminotransferase activities, the Ribose-5-phosphate aminotransferase reaction represents the only known mechanism for PRA synthesis in these cells. However, the mutant cells were confirmed to be strict purine auxotrophs even when grown in medium supplemented with maximal concentrations of ribose and NH_3, potential substrates for Ribose-5-phosphate aminotransferase. The failure of the mutant cells to grow under these conditions cannot be explained by additional enzymatic defects in the pathway of purine biosynthesis de novo, since these cells can utilize PRA for the synthesis of AICAR and AICAR can be converted to purine ribonucleotides. Thus, the Ribose-5-phosphate aminotransferase activity observed in cell lysates does not play a significant role in the intracellular synthesis of PRA in this eukaryotic cell line under the in vitro conditions studied.

The present studies clearly establish a role for PP-ribose-P amidotransferase or PP-ribose-P aminotransferase or both in the synthesis of PRA, since the mutant cells, deficient only in the activity of these two enzymes, are strict purine auxotrophs. The relative contribution of each of these enzyme activities to PRA synthesis is not known, but probably depends upon the intracellular concentration of glutamine relative to that of NH_3 as well as the affinity of the protein(s) for each of these substrates.

In the absence of other demonstrable abnormalities of purine biosynthesis, the deficiency of both PP-ribose-P amidotransferase and PP-ribose-P aminotransferase activities in the mutant cells suggests a close relationship between these two enzyme activities. It is possible that these two reactions are catalyzed by a single protein (Hartman, 1963). However, Reem has reported that human PP-ribose-P amidotransferase can be separated from PP-ribose-P aminotransferase on gel filtration chromatography (Reem, 1974) and it could be postulated that these two proteins were structurally related through a common subunit, such as demonstrated for several glutamine utilizing enzymes (Hartman, 1973; Trotta, et al., 1973). It is also possible that PP-ribose-P amidotransferase and PP-ribose-P aminotransferase are distinct proteins whose synthesis or inactivation are closely coordinated at the genetic level.

REFERENCES

Chu, E.H.Y., Sun, N. C., and Chang, C. C. 1972. Proc. Nat. Acad. Sci. USA 69: 3459-3463.

Feldman, R. I., and Taylor, M. W. 1973. Biochem. Genetics 13: 227-234.

Fox, I. H. and Kelley, W. N. 1971. J. Biol. Chem. 246: 5739-5748.

Hartman, S. C. 1963. J. Biol. Chem. 238: 3024-3035.

Hartman, S. C. 1973. The enzymes of Glutamine Metabolism, ed. by Prusiner, S. and Stadtman, E. R. (Academic Press, New York) pp. 319-330.

Henderson, J. F. 1963. Biochim. Biophys. Acta 76: 173-180.

Holmes, E. W., McDonald, J. A., McCord, J. M., Wyngaarden, J. B. and Kelley, W. N. 1973. J. Biol. Chem. 248: 144-150.

Holmes, E. W., Pehlke, D. M., and Kelley, W. N. 1974. Biochim. Biophys. Acta 364: 209-217.

Holmes, E. W., Mason, D. H., Goldstein, L. I., Blount, R. E., Kelley, W. N. 1974. Clin. Chem. 20: 1076-1079.

Kelley, W. N., and Meade, J. C. 1971. J. Biol. Chem. 246: 2953-2958.

King, G. L., and Holmes, E. W. 1975. Anal. Biochem., in press.

Leyva, A. 1974. Thesis for Doctorate in Biochemistry, Duke University Medical Center.

Leyva, A. and Kelley, W. N. 1974. Anal. Biochem. 62: 173-179.

Lowry, O. H., Rosebrough, N. J., Farr, A. L., and Randall, R. J. 1951. J. Biol. Chem. 193: 265-275.

Malloy, G. R., Sitz, T. O., and Schmidt, R. R. 1973. J. Biol. Chem. 248: 1970-1975.

Nierlich, D. P., and Magasanik, B. 1965. J. Biol. Chem. 240: 366-37.

Ravel, J. M., Eakin, R. E., and Shive, W. 1948. J. Biol. Chem. 172: 67-70.

Reem, G. H. 1968. J. Biol. Chem. 243: 5695-5701.

Reem, G. H. 1972. J. Clin. Invest. 51: 1058-1062.

Reem, G. H. 1974. J. Biol. Chem. 249: 1696-1703.

Reem, G. H., and Friend, C. 1975. Proc. Nat. Acad. Sci. USA 72:
    1630-1634.

Thomas, C. B., Arnold, W. J., and Kelley, W. N. 1973. J. Biol. Chem.
    248: 2529-2535.

Trotta, P. P., Pinkus, L. M., Wellner, V. P., Estis, L., Haschemeyer,
    R. H., and Meister, A. 1973. The Enzymes of Glutamine
    Metabolism, ed. by Prusiner, S. and Stadtman, E. R. (Academic
    Press, New York) pp. 431-482.

Van Der Weyden, M. B., and Kelley, W. N. 1974. J. Biol. Chem. 249:
    7282-7289.

Van Der Weyden, M. B., Buckley, R. H., and Kelley, W. N. 1974.
    Biochem. Biophys. Res. Commun. 57: 590-595.

Westby, C. A., and Gots, J. S. 1969. J. Biol. Chem. 244: 2095-2102.

Wyngaarden, J. B. 1972. Current Topics in Cellular Regulation 5:
    135-176.

BIOCHEMICAL STUDIES OF PURINE AUXOTROPHS OF DROSOPHILA MELANOGASTER

Merrie M. Johnson, David Nash and J. Frank Henderson

Department of Genetics, and Cancer Research Unit, University of Alberta, Edmonton, Alberta, Canada

Axenic culture of Drosophila on sterile Sang's defined medium has enabled investigators to isolate mutants of this organism which require dietary supplementation with purine nucleosides. Five such mutants are the object of this study: two require either adenosine or guanosine (pur 1-1 and pur 1-2), one requires adenosine (ade 1-1), one requires adenosine or inosine (ade 2-1), and one requires guanosine (gua 2-1). Pur 1-1, pur 1-2 and ade 1-1 were isolated by Darrel Falk, ade 2-1 by Fardos Naguib, and gua 2-1 by David Nash. Genetic studies have indicated that each of these, with the possible exception of ade 2-1, have single mutations, and that pur 1-1 and pur 1-2 are allelic. Ade 2-1 may be a double mutant, since mapping data are ambiguous. Ade 1-1 is not a lethal mutant, but requires three days longer than wild type to develop without adenosine.

These results have been interpreted to suggest that pur 1-1 and pur 1-2 are blocked in purine biosynthesis de novo, ade 1-1 is blocked between inosinate and adenylate and gua 2-1 is blocked between inosinate and guanylate. Ade 2-1 may have two mutations, with one block in purine biosynthesis de novo and another in the utilization of guanosine. Alternatively, it may have a mutation in adenylosuccinate lyase, which is necessary for two steps in the pathway, as suggested by Naguib. The purpose of this investigation was to test these hypotheses as well as to examine some aspects of purine metabolism in wild type Drosophila using radioactive precursors.

The study of nutritional mutants of Drosophila requires that cultures be free of any microbial contamination. This is accomplished by the maintenance of flies sterilized as embryos in sterile culture.

Newly hatched larvae were transferred to a defined culture medium to which a radioactive precursor had been added. Cultures were incubated at 25° and samples of larvae were removed for analysis after two and four days. Larvae were immersed in perchloric acid and sonicated. Acid soluble and nucleic acid purine derivatives were separated and their radioactivity measured. Metabolism of radioactive ^{14}C hypoxanthine (1 mM), inosine (.3 mM), guanine (1 mM), guanosine (1 mM), adenosine (.3 mM), formate (1 mM), and glycine (1 mM) was studied. Since the differences between samples taken on the second and fourth days of incubation were generally quantitative rather than qualitative, only data collected on the second day are presented here.

All the radioactive precursors were incorporated into nucleotides by wild type larvae. As shown in Table I, guanine was converted into nucleotides at the lowest rate. Most of it was degraded, with high accumulation of radioactivity in uric acid, xanthine, and allantoin.

Hypoxanthine, when used alone, was largely degraded, with a smaller fraction incorporated into nucleotides than when most other precursors were used. However, when allopurinol was added, a significant decrease in catabolism was observed, although the fraction of unused base was higher than that of any other precursor.

Table I

UTILIZATION OF RADIOACTIVE PRECURSORS

BY WILD TYPE DROSOPHILA

Precursor	(Percent of radioactivity recovered) Converted to Nucleotides	Catabolized	Unused
Guanine	11	49	40
Hypoxanthine	32	30	38
Hypoxanthine (+ allopurinol)	22	1	77
Inosine	51	4	42
Guanosine	33	19	48
Adenosine	70	28	2

Since both guanine and hypoxanthine were converted to nucleotides, it may be concluded that hypoxanthine-guanine phosporibosyl
transferase is present in Drosophila. The extensive degradation of
guanine and hypoxanthine when allopurinol was not used indicate that
the three catabolic enzymes, xanthine dehydrogenase, guanine
deaminase, and uricase are quite active in this organism.

Inosine was readily incorporated into nucleotides, with very
little degradation. This difference in the utilization of the
nucleoside and base could indicate the presence of inosine kinase.
Alternatively, it could simply mean that when the nucleoside is
slowly converted to the base, hypoxanthine-guanine phosphoribosyl
transferase can compete more successfully for the substrate than
xanthine dehydrogenase.

Guanosine, like inosine, was incorporated into nucleotides
much more readily than its corresponding base and the same interpretations could apply in this case, except that hypoxanthine-
guanine phosphoribosyl transferase would be competing with guanine
deaminase. Although guanosine was utilized for nucleotide synthesis
to a much greater extent than guanine, more of it was catabolized
than in the case of inosine. Fifty-one percent of the radioactivity
recovered in the larvae when inosine was used was converted to
nucleotides, with only four percent catabolized. Only thirty-three
percent of the guanosine was converted to nucleotides, with nineteen
percent catabolized.

Adenosine was metabolized at the highest rate of any of the
precursors, with only two percent of the radioactivity recovered
as adenosine. Most of it was converted to nucleotides. The major
catabolic product was inosine, indicating a very active adenosine
deaminase.

Hypoxanthine, inosine, and adenosine were readily converted
to both adenine and guanine nucleotides. The adenine/guanine ratio
in nucleic acid was consistently about 1.6 in all three cases.
Guanosine, however, was not utilized for adenine nucleotides at all,
indicating the absence of guanosine monophosphate reductase. Incorporation of guanine into nucleic acid was so low that reliable
adenine/guanine ratios were not obtainable.

It appears that guanosine has an inhibitory effect on either
inosine monophosphate dehydrogenase or xanthosine monophosphate
aminase. When it was added to medium with radioactive inosine or
formate, nucleic acid adenine/guanine ratios were much higher,
about 38 with inosine and about 13 with formate.

Radioactive formate was used with aminoimidazole carboxamide
to study the latter portion of the de novo pathway of purine
synthesis, and it was readily converted by wild type Drosophila

into nucleotides. Nucleic acid adenine/guanine ratios were similar to those found with the other precursors used. Radioactive glycine in the presence of azaserine was substantially incorporated into phosporibosyl formylglycineamide.

The data obtained for the mutants indicate a much more complicated biochemical situation than the simple hypotheses suggested by the genetic data. One mutant, ade 1-1, showed no significant differences from wild type and so will not be discussed further. Pur 1-1, pur 1-2, and ade 2-1, the three strains which ostensibly have mutations in the de novo pathway, all incorporated both glycine and formate (with aminoimidazole carboxamide added) into nucleotides. Formate incorporation was approximately the same as wild type in all three cases, as was glycine incorporation (with azaserine) into phosphoribosyl formylglycineamide. However, the fraction of the radioactive nucleotides which were catabolized was about 21 percent in the mutants, compared to about 11 percent in the wild type.

The fact that all three of these mutants also show substantial incorporation of glycine into nucleotides makes it appear that they do not lack the ability to produce purines. In fact, the ade 2-1 mutant showed relatively greater incorporation into nucleic acid adenine than into nucleic acid guanine in two instances. The nucleic acid adenine/guanine ratio for pur 1-1 and pur 1-2 when glycine was used was about 6.6, whereas that for ade 2-1 was 27. The normal ratio for wild type when formate (with aminoimidazole carboxamide added) is used, 1.6, was decreased to 1.2 when adenosine was added to the medium. When adenosine was added to formate medium in the case of ade 2-1, the adenine/guanine ratio was increased from 1.6 to 2.4. There was no significant difference between these mutants and wild type with respect to radioactivity recovered in adenylosuccinate or succinyl adenosine.

The guanosine mutant, gua 2-1, showed only a slightly elevated adenine/guanine ratio in nucleic acid when inosine was used as a precursor, although this ratio in the acid soluble nucleotides was significantly higher than that in the wild type. The inhibition of incorporation of inosine into guanine nucleotides by added guanosine was more exaggerated than in wild type. No difference was found between the mutant and wild type in the levels of radioactivity in either xanthosine or xanthosine monophosphate. These data support the hypothesis that these mutants are deficient in the conversion of adenylate to guanylate, but do not make it possible to distinguish which of the two enzymes, inosine monophosphate dehydrogenase or xanthine monophosphate aminase, is affected by the mutation.

Inevitably in working with higher organisms, there are a number of complicating factors in the interpretation of data such as these. Thus the mutants are not, in a strict sense, comparable to the wild

type controls because they do not grow unless supplemented with
high concentrations (3.5 mM) of either adenosine or guanosine. In
such circumstances, measurements of specific activity (incorporation
per microgram of DNA) might emphasize changes in residual metabolism,
while overlooking major effects on net biosynthetic activity. It is
true that much less DNA is synthesized by most unsupplemented mutants
than by wild type. It is not a simple matter to compensate for the
inherent difficulties in this system. High concentrations of adenosine or guanosine are, in fact, quite toxic to the wild type and so
experiments carried out at any precursor concentration must have a
detrimental effect on either the wild type or the mutant.

Also, this system cannot distinguish differences in purine
metabolism in different tissues; it only measures the sum of purine
metabolism in all tissues. It is quite possible either that the
mutations are expressed to a greater extent in some tissues, or more
likely, that some tissues need more purines than others. Finally,
different stages of development may have different requirements for
purines, and small changes in the metabolism of early larvae may
become consequential only later.

The data presented here do not permit firm conclusions about
the nature of the mutations involved in these strains of Drosophila.
It has been shown that in three of the mutants, pur 1-1, pur 1-2,
and ade 2-1, the rate of catabolism of nucleotides is increased, and
since this does not seem to be matched by an increase in de novo
synthesis, it could account for the dietary deficiency which is the
phenotype of the mutant. The fact that the adenosine mutant, ade
2-1, converts more precursor into adenine nucleotides than wild type
suggests that rather than having a deficiency associated with the
synthesis of adenylate, the mutation is associated with an increased
metabolism of adenine nucleotides.

In conclusion, it seems likely that the lethality of the mutant
phenotype occurs either because a particular tissue or tissues has
an especially high requirement for purines or because purines are
required especially at a particular time in development. The studies
of purine metabolism presented here do not indicate sites of mutation
and further study is required to define them in detail.

References

Falk, D. and D. Nash (1974). Sex-linked auxotrophic and putative
auxotrophic mutants of Drosophila melanogaster. Genetics, 76,
755-766.

Naguib, F. N. M. (1976). Auxotrophic mutants of the second
chromosome of Drosophila melanogaster. Ph.D. Thesis, University of
Alberta.

INOSINE TRIPHOSPHATE METABOLISM IN HUMAN ERYTHROCYTES

J. Frank Henderson

University of Alberta Cancer Research Unit (McEachern
Laboratory) and Department of Biochemistry
Edmonton, Alberta, Canada   T6G 2H7

Inosine triphosphate (ITP) is a nucleotide which is not usually thought of as occuring in cells. However, beginning with the studies of Vanderheiden in 1965 (1), several investigators have shown that ITP can accumulate in human erythrocytes. The extent of erythrocyte ITP accumulation has appeared to vary from individual to individual, and also with the particular experimental conditions used by each investigator.

In our own studies, 2% suspensions of human erythrocytes in either Fischer's tissue culture medium containing 25 mM phosphate or Krebs-Ringer medium containing glucose and 25 mM phosphate, have been incubated for two hours with 100 μM [^{14}C]hypoxanthine. The hypoxanthine is first converted to inosinate by hypoxanthine-guanine phosphoribosyltransferase, using phosphoribosyl pyrophosphate (PP-ribose-P) formed from glucose. The enzyme that converts inosinate to inosine diphosphate (IDP) is not known for certain, although it may be guanylate kinase or at least a particular isozyme of guanylate kinase (2). Nucleoside diphosphokinase is capable of converting IDP to ITP. ITP can be converted back to inosinate by pyrophosphorolysis; this process is catalyzed by nucleoside triphosphate pyrophosphohydrolase (NTPH), which cleaves ITP, GTP and some other nucleoside triphosphates, but not ATP (3).

We have previously reported (4) that under a particular set of conditions (i.e., 2% cell suspensions, [^{14}C]hypoxanthine, incubation for 2 hours), erythrocytes from each of the several hundred individuals studied accumulated at least some radioactive ITP. Furthermore, the amount of ITP accumulated was a relatively constant characteristic of each individual. In the vast majority of cases,

115

ITP constituted less than 10% of the total radioactive nucleotides synthesized from [^{14}C]hypoxanthine, the remainder being principally inosinate. (Appreciable amounts of IDP were never found). However, a small number of individuals accumulated larger amounts of ITP under these conditions, and in such cases ITP constituted as much as 40% or more of the radioactive nucleotides. Again, this is a relatively constant characteristic of these particular individuals.

Although variation in ITP synthesis could in theory be due to variation either in its rate of synthesis or in its rate of breakdown, Vanderheiden several years ago suggested that breakdown was the more important process (5). In collaboration with Dr. Allan Morris (Michigan State University), we have also studied the relation between radioactive ITP accumulation and NTPH activity in ca. 100 individuals. The data of Figure 1 completely support Vanderheiden's proposal that low NTPH activity leads to high ITP accumulation, and that erythrocytes with high NTPH activity accumulate only low amounts of ITP. (There are some differences between our results and those of Vanderheiden, but these may perhaps be due to differences in methodology.)

We are now studying the inheritance of low NTPH activity and high ITP accumulation, but so far we are not able to draw firm conclusions regarding this matter.

In addition to these studies of ours and of Vanderheiden which have emphasized individual variation and in which accumulation of large amounts of ITP have been rare occurances, there have been other published reports (6, 7, 8) that suggest that all human erythrocytes can accumulate large amounts of ITP under certain conditions. Recently we have attempted to study the basis of these observations.

Because [^{14}C]hypoxanthine must first be converted to inosinate in order to be incorporated into ITP, we first studied factors that were likely to affect the amount of inosinate formed. Figure 2 shows that inosinate synthesis varied considerably in erythrocytes from different individuals, and there even appear to be two populations of cells with respect to this process; one synthesizes between 400 and 800 nmoles/10^{10} cells in 2 hours, whereas the second synthesizes between 800 and 1200 nmoles/10^{10} cells. However, other studies have shown that the proportion of total radioactive nucleotides that is comprised of ITP is not related to the total amount of nucleotides synthesized from hypoxanthine; i.e., variation in inosinate synthesis and variation in ITP accumulation are independent problems.

Although we have not studied the basis of individual variation in inosinate synthesis, effects of altering certain experimental conditions have been determined. Thus Table 1 shows that pH and

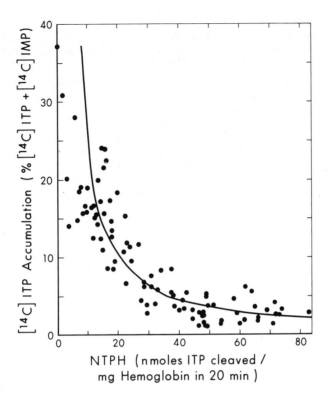

FIG. 1. Relationship between [14C] ITP accumulation in erythrocytes incubated with [14C]hypoxanthine, and nucleoside triphosphate pyrophosphohydrolase (NTPH) activity in erythrocyte lysates

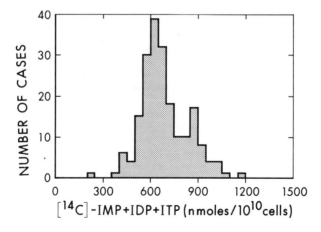

FIG. 2. Individual variation in inosinate synthesis in human erythrocytes

phosphate concentration both influence the rate of inosinate
formation, and this may be due at least in part to the effects of
these factors on PP-ribose-P synthesis from glucose. Table 2
shows that inosinate synthesis also varies with pO_2 and is stimulated
if pyruvate is added. Again, PP-ribose-P concentration may be the
basis for these changes in rate of inosinate synthesis. However,
none of these changes in experimental conditions resulted in very
marked changes in the accumulation of ITP.

TABLE 1

EFFECT OF pH AND PHOSPHATE CONCENTRATION

pH	$[P_i]$, mM	INOSINATE SYNTHESIS	PP-RIBOSE-P SYNTHESIS
		(Percent of control)	
7.4	25	100	100
7.4	50	151	137
7.0	25	149	161
7.0	50	173	193

TABLE 2

EFFECT OF pO_2 AND PYRUVATE

GAS PHASE	PYRUVATE (10mM)	INOSINATE SYNTHESIS	PP-RIBOSE SYNTHESIS
		(Percent of control)	
Air	–	100	100
N_2	–	59	43
Air	+	138	169
N_2	–	117	91

Further studies showed that the time of incubation was more important for ITP accumulation than were the other conditions of incubation. Figure 3 shows that when human erythrocytes were incubated for periods longer than 2 hours, ITP began to accumulate at increased rates. Thus after 24 hours incubation, ITP accumulation was 100-fold greater than after 2 hours incubation. Even cells that accumulate little ITP during 2 hours incubation will accumulate large amounts if incubated for 4 hours or more.

Another method used to study the influence of time was to incubate erythrocytes for 24 hours under different conditions in the absence of hypoxanthine, and then to measure radioactive ITP accumulation during a 2 hour period under our standard conditions. In these experiments, sterile 10% suspensions of erythrocytes were prepared and stored without shaking for 24 hours either at 4°, 22° or 37°. Table 3 shows that cells stored in this manner, particularly those stored at 22°, accumulated several times as much ITP as did fresh erythrocytes.

At the present time we unfortunately have no explanation for these results. NTPH activity in erythrocyte lysates does not change either during the course of 24 hours incubation with [^{14}C]hypoxanthine, or during storage without shaking. Of course, the possibility must still be considered that such measurements do not reflect the activity of the enzyme within cells under these conditions. Alternatively, the rate of ITP synthesis may increase, although nothing is known about this. We have also found that the concentrations of one possible effector molecule, 2,3-diphosphoglycerate, cannot be correlated with the observed changes in rate of ITP accumulation. Further studies are required to elucidate the basis of the change of rate of ITP accumulation with time.

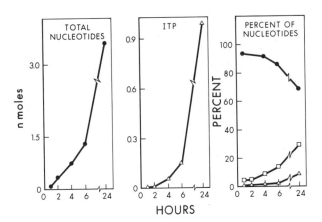

FIG. 3. Effect of incubation time on inosinate synthesis and ITP accumulation.

TABLE 3

EFFECT OF STORAGE FOR 24 HOURS

TEMP.	INOSINATE SYNTHESIS	ITP ACCUMULATION
	(Percent of Control)	
4°	126	113
22°	276	438
37°	251	274

References

1. Vanderheiden, B. S. (1965). Proc. 10th Congr. Int. Soc. Blood Transf., pp. 540-548.

2. Agarwal, K. C., and Parks, R. E., Jr. (1972). Mol. Pharmocol., 8, 123-138.

3. Chern, C. J., MacDonald, A. B., and Morris, A. J. (1969). J. Biol. Chem., 244, 5489-5495.

4. Fraser, J. H., Meyers, H., Henderson, J. F., Brox, L. W., and McCoy, E. E. (1975). Clin. Biochem., 8, 353-364.

5. Vanderheiden, B. S. (1969). Biochem. Genet., 3, 289-297.

6. Tatibana, M., and Yoshikawa, H. (1962). Biochim. Biophys. Acta, 57, 613-616.

7. Blair, D. G. R., and Dulmadge, M. (1969). Transfusion, 9, 198-202.

8. Zachara, B., and Lewandowski, J. (1974). Biochim. Biophys. Acta, 353, 253-261.

OXYPURINE AND 6-THIOPURINE NUCLEOSIDE TRIPHOSPHATE FORMATION IN HUMAN ERYTHROCYTES

Donald J. Nelson, Christoper Buggé, and Harvey C. Krasny

Wellcome Research Laboratories

Research Triangle Park, N.C., USA, 27709

Human erythrocytes are known to form a variety of purine nucleoside triphosphates from the bases and nucleosides, through the action of nucleoside kinases, phosphoribosyltransferases and nucleotide kinases. In addition to ATP and GTP, a very low concentration (6μM) of inosine triphosphate (ITP) was identified in fresh erythrocytes by Vanderheiden (1). More recently, after a 4 hour incubation of stored RBC's with inosine, Zachara reported that a large amount of ITP was formed (2). A specific ITP pyrophosphohydrolase (3) probably contributes to the low steady state levels of ITP in circulating cells. Inosine has been suggested as a purine source in blood preservation studies (4) and can augment adenine nucleotide pools in studies with ischemic kidneys (5).

Our work with oxypurines and 6-thiopurine analogs was initiated some time ago when it became clear that longer term incubations of red cells and the technique of high performance liquid chromatography (HPLC) could reveal the presence of nucleoside triphosphates not previously identified (6-9). The formation of even low levels of analog triphosphates may result in altered nucleotide metabolism which could be relevant to various therapeutic regimens.

FORMATION OF INOSINE TRIPHOSPHATE

RBC's incubated with hypoxanthine or inosine at 0.5 mM rapidly formed a large amount of IMP (>1mM) which, with longer incubation times, was converted to IDP and ITP (Figure 1). Glucose, 25mM, and phosphate, 50mM, were required to keep ATP levels normal and promote PRPP synthesis in extended incubations. By 24 hrs., the

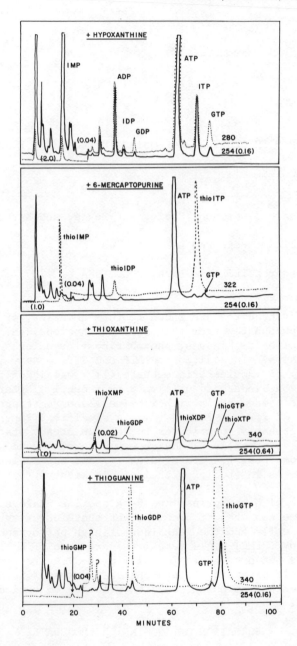

Figure 1. HPLC Chromatography of Nucleotides in Erythrocytes
Incubated with Various Purines for 24 hrs. The effluent was
monitored continuously at 254nm (———) and at a second wavelenth
(.....), with the full scale O.D. ranges shown in parentheses.
Conditions as in Fig. 3.

concentration of ITP reached 0.4mM (Figure 2). When care was taken to purify hypoxanthine-^{14}C, and erythrocytes were freed of WBC's and platelets, radioactivity was not found in either adenine or guanine nucleotides. This appears to confirm the absence of adenylosuccinate synthetase (10) and the low level of inosinate dehydrogenase (11) in human erythrocytes.

FORMATION OF 6-THIOINOSINE TRIPHOSPHATE

Incubations with 6-mercaptopurine (6-MP) or 6-mercaptopurine riboside resulted in the formation of 6-thioIMP and two new nucleotide metabolites, which appeared just before GDP and GTP, respectively, in the HPLC elution profile (Figure 1). Both exhibited an ultraviolet absorption spectrum indistinguishable from that of 6-MP riboside, λmax μ322nm, pH1, and both were hydrolysed to 6-MP riboside by alkaline phosphatase and to 6-thioIMP by venom phosphodiesterase. When mixed with a commercially available sample of authentic 6-MP riboside di- and triphosphate (PL-Biochemicals, Inc.), these two new 6-MP metabolites were found to have identical retention times on the HPLC chromatograph. It was concluded that erythrocytes are capable of phosphorylating thioIMP, probably by the same mechanism as in the case of IMP. The nucleotide kinase responsible for this may have been adenylate kinase, or possibly guanylate kinase. The very slow reaction velocity would require long incubation times to demonstrate these activities in vitro with the appropriate purified enzymes.

After a rapid formation of 6-thioIMP, the di- and triphosphates accumulated in erythrocytes at a linear rate over the 24 hr incubation period (Figure 2). 6-ThioIDP was always at a low level presumably as a result of its efficient conversion to 6-thioITP by nucleoside diphosphate kinase. There was no significant difference between 6-MP and 6-MP riboside in the rates of formation of the three nucleotide derivatives. There was no indication of the formation of 6-thioxanthine or 6-thioguanine derivatives which would have been easily detected on the HPLC elution profiles.

FORMATION OF XANTHOSINE TRIPHOSPHATE AND
6-THIOXANTHOSINE TRIPHOSPHATE

Both xanthine and 6-thioxanthine were converted to nucleoside monophosphates, XMP and 6-thioXMP, respectively, via a phosphoribosyltransferase reaction. XMP was aminated to GMP which then was phosphorylated to a large extent to GTP as shown in experiments with xanthine-^{14}C. A small amount of XTP-^{14}C was eluted 2 min. ahead of ITP in the HPLC chromatogram and co-chromatographed with authentic XTP. 6-ThioXMP was the major nucleotide metabolite of

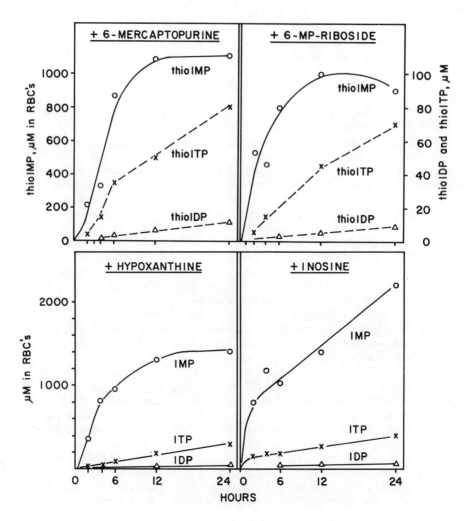

Figure 2.   Time Course of Formation of Nucleotides from Hypoxanthine, Inosine, 6-Mercaptopurine and 6-Mercaptopruine Riboside in
Erythrocytes.   A 10 ml incubation contained Na_2HPO_4 (50 mM, pH
7.4), glucose (25 mM), penicillin G and streptomycin (0.3 mg/ml),
the respective purine (0.5 mM), and washed human erythrocytes,
hematocrit = 25.   Nucleotides were quantitated as described in
Figure 3.

6-thioxanthine; however, it was possible to identify 6-thioGTP as well as two other compounds with absorption maxima at 338 nm in the elution profile (Figure 1). The first appeared just beyond ATP and the second appeared well beyond 6-thioGTP; both were sensitive to venom diesterase. These two components have been tentatively identified as 6-thioXDP and 6-thioXTP.

S-METHYLATION OF 6-THIOPURINE NUCLEOTIDES IN RBC's

A 200ml incubation of RBC's and 6-MP was done to obtain enough 6-thioITP for characterization. Nucleotides were separated on a large DEAE-Sephadex column (12) as described previously. In addition to 6-thioIMP a new peak was found which was eluted just ahead of 6-thioIMP and which had an absorption maximum at 293 nm. This compound was identified positively as 6-methylmercaptopurine riboside monophosphate, MMPRMP, and suggested that RBC's are capable of forming S-methylated derivatives of 6-MP. Further proof of this was obtained in an experiment with 6-MP-[^{35}S], 0.5mM, 20,100 dpm/nmole, and methionine-[CH_3-^{3}H], 0.06mM, ∿12,100 dpm/nmole. The methionine concentration and sp. act. were not fixed precisely because of the presence of an endogenous pool of unlabeled methionine, which was estimated to be 0.06mM in cells and plasma, for the purpose of the calculations. The results presented in Figure 3 show that MMPR-triphosphate was labeled with both ^{35}S and ^{3}H in a constant molar ratio of 1:1.09, which was very close to that calculated from the specific activities of the 6-MP-[^{35}S] and methionine-[CH_3-^{3}H]. The methylated thiopurine nucleotides were present at approximately 1/10th the concentrations of the respective free 6-thiol forms; this was also the ratio of methionine/6-MP in the incubation mixture. Analysis of the media revealed that over 90% of the radioactivity had been taken up by the cells and most of that was in the form of S-methylated thiopurines.

In a similar experiment with 6-thioguanine and methionine-^{3}H, 6-thioguanine nucleotides were methylated; 6-methylthioGTP reached a concentration of about 20 nmoles/ml packed cells while 6-thioGTP was present at about 1200 nmoles/ml cells. In an experiment with 6-thioxanthine and methionine-^{3}H, very low levels of tritiated metabolites appeared in the HPLC elution profile near thioXMP and thioXTP, and were presumed to be the respective S-methylated thioxanthylate derivatives.

Thiouric acid, 6-thio-8-hydroxypurine, uric acid-^{14}C, methyl-mercaptopurine riboside-[CH_3-^{14}C], and 6-methylthioguanosine were incubated in vitro with erythrocytes and methionine-^{3}H. The 8-hydroxy compounds were neither converted to nucleotides nor methylated. The S-methylated precursors were phosphorylated but did not undergo methyl exchange with methionine-[CH_3-^{3}H].

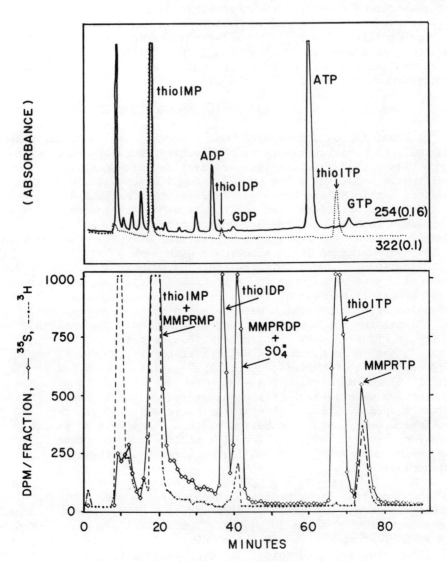

Figure 3. Acid-soluble Nucleotides from Human RBC's Incubated 24 hrs with [^{35}S]-6-Mercaptopurine and [^{3}H]-Methionine. The incubation media contained sodium phosphate 50 mM; glucose 25 mM; streptomycin and penicillin-G 0.3 mg/ml each; [^{35}S]-6-MP 0.5 mM, 21,100 dpm/nmole; [^{3}H-CH$_3$] methionine 0.06 mM, ∿12,100 dpm/nmole; 50% plasma and washed red cells, HCT 25. A Partisil-SAX (Whatman, Inc.) column, 0.46 x 25 cm, was eluted with a linear gradient of KH$_2$PO$_4$, 0.01-1 M, at pH 3.50, at a flow rate of 30 ml/hr.

Specific S-methyl transferases have been implicated in the metabolism of 6-MP since 8-hydroxy-6-methylthiopurine and 8-hydroxy-6-methylsulfinylpurine have been identified as urinary metabolites (13, 14). Bennett and Allen (15) have shown that a rapid non-enzymatic transfer of the methyl group of S-adenosyl-methionine to 6-thioIMP can occur. In the studies reported here, we have not attempted to distinguish between an enzymatic or chemical mechanism of methyl transfer although it would appear that S-adenosylmethionine is the methyl donor.

AZATHIOPRINE AND 2-AMINOAZATHIOPRINE

Azathioprine and 2-aminoazathioprine were rapidly cleaved to 6-MP and 6-thioguanine, which were readily converted to their nucleotide forms in amounts as described above for these bases. We failed to see any evidence for the possible existence of aza-thioprine nucleotide or 2-aminoazathioprine nucleotide which would have been evidenced by new peaks in the HPLC profiles.

PYRAZOLOPYRIMIDINES

Allopurinol-^{14}C and thiopurinol were found to be converted only to the respective 5'-monophosphates, in a 24 hr incubation. Oxipurinol-^{14}C and thiooxipurinol were not converted to measurable amounts of nucleotide in erythrocytes. These results are similar to the studies previously reported (16) which dealt with the formation of allopurinol and oxipurinol nucleotides in the rat.

OTHER COMPOUNDS

Several other purines and thiopurines were tested in the RBC incubation system with a 24 hr incubation and were found not to be converted to nucleotide forms. These were:

6-methylmercaptopurine, 6-thio-8-hydroxypurine, 6-methylmercapto-8-hydroxypurine, 6-methylsulfinyl purine, 6-methylsulfinyl-8-hydroxypurine, 6-methyl-sulfonyl-8-hydroxypurine, 6-thiouric acid, uric acid-2-^{14}C

SUMMARY

A variety of oxypurines and 6-thiopurines could be trans-formed by intact erythrocytes to their nucleoside triphosphate forms when incubations were extended for up to 24 hrs. The specific nucleotide monophosphate kinases which accomplish these reactions in erythrocytes were not identified but their ability to utilize 6-thioIMP, 6-thioXMP and 6-methylthioGMP as substrates, albeit very slowly, is clearly implied by these results. S-methylation of 6-thiopurines was demonstrated in erythrocytes incubated with physiological amounts of methionine-$[CH_3-^3H]$. 6-Methylthioguanosine triphosphate and 6-methylmercaptopurine riboside triphosphate were formed in micromolar amounts, probably from the corresponding thiopurine nucleotides by methyl transfer from S-adenosylmethionine.

REFERENCES

1. B. S. Vanderheiden, Proc. Xth Congress Internal. Soc. Blood Transfusion, Stockholm, 1964, 544 (1964).
2. B. Zachara, J. Biochem. 76, 891 (1974).
3. B. S. Vanderheiden, J. Cell. Physiol. 86, 167 (1974).
4. D. Rubinstein and E. Warrendorf, Can. J. Biochem. 53, 671 (1975).
5. A. R. Fernando, J. R. Griffiths, E. P. N. O'Donoghue, J. P. Ward, D. M. G. Armstrong, W. F. Hendry, D. Perrett, and J. E. A. Wickham, The Lancet, p. 555 (1976).
6. T. P. Zimmerman, L. C. Chu, C. J. L. Buggé, D. J. Nelson, R. L. Miller and G. B. Elion, Biochem. Pharmacol. 23, 2737 (1974).
7. T. P. Zimmerman and L. C. Chu, Biochem. Pharmacol. 23, 2473 (1974).
8. R. E. Parks and P. R. Brown, Biochemistry 12, 3294 (1973).
9. P. R. Brown and R. E. Parks, Jr., Anal. Chem. 45, 948 (1973).
10. C. Bishop, J. Biol. Chem. 235, 3228 (1960).
11. B. A. Lowy, M. K. Williams and I. M. London, J. Biol. Chem. 237, 1622 (1962).
12. D. J. Nelson, C. J. L. Buggé, H. C. Krasny and T. P. Zimmerman, J. Chromatog. 77, 181 (1973).
13. C. N. Remy, J. Biol. Chem. 238, 1078 (1963).
14. G. B. Elion, S. Callahan, R. W. Rundles and G. B. Hitchings, Cancer Research 23, 1207 (1963).
15. L. L. Bennett and P. W. Allen, Cancer Research 31, 152 (1971).
16. D. J. Nelson, C. J. L. Buggé, H. C. Krasny and G. B. Elion, Biochem Pharmacology 22, 2003 (1973).

ACTIVITY OF THE SALVAGE-PATHWAY IN ERYTHROCYTES OF NEWBORN INFANTS, CHILDREN AND ADULTS

M. M. Müller and P. Wagenbichler

1st Department of Medical Chemistry and

1st Department of Gynaecology

University of Vienna, Austria

INTRODUCTION

Since the important investigations from WILLIAMS (20) and GOLDWATER (4) in 1953 it is well known that besides the de novo synthesis of purines there exists a one-step reaction for the formation of purine nucleotides from purine bases. In this salvage-pathway (Figure 1) phosphoribosylpyrophosphate (PP-ribose-P, PRPP), which is catalyzed by phosphoribosylpyrophosphate synthetase (PP-ribose-P synthetase, PRPP-synthetase, E.C. 2.7.6.1) from ribose-5-phosphate and ATP, acts as ribose-5-phosphate donor. Cells of mammals contain two different enzymes for this synthesis of purine nucleotides from bases: the adenine phosphoribosyltransferase (APRT, E.C. 2.4.2.7), which catalyzes the formation of adenosinemonophosphate (AMP) from adenine and the hypoxanthine-guanine phosphoribosyltransferase (HGPRT, E.C. 2.4.2.8) converting hypoxanthine, guanine and xanthine to their nucleotides.

The importance of this pathway for cell metabolism is the reutilization of purine bases, formed by the breakdown of endogen and exogen nucleotides, particularly in cells without or with a limited de novo purine synthesis like erythrocytes, leucocytes, bone marrow cells and nerve cells (2, 9, 15). Furhtermore the purine phosphoribosyltransferases regulate the intracellular concentration of purine nucleotides which are feed-back inhibitors

129

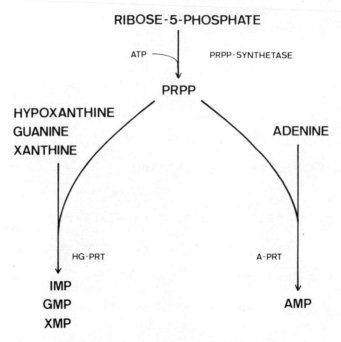

<u>Figure 1</u>.   Salvage-pathway

of de novo purine synthesis.

In 1967 SEEGMILLER and coworkers described the
complete deficiency of HGPRT in erythrocytes and fibroblasts of patients with Lesch-Nyhan syndrome (17, 18).
The activity of APRT is elevated in erythrocytes of
these patients as well as in erythrocytes of patients
with a partial HGPRT defect (7) and may be elevated in
patients with overproduction of uric acid (12) and no
abnormality of HGPRT.

It is well known that some enzymes in erythrocytes
of newborn infants show different activities compared to
those of adults. Activities of glucose-6-phosphate and

and 6-phosphogluconate dehydrogenase, hexokinase,
aldolase, pyruvate kinase, enolase and lactate dehydrogenase are higher, but activities of adenylate kinase,
phosphofructokinase, glutathione peroxidase and acetyl-
cholinesterase are in newborn infants' erythrocytes
lower (8, 21). WITT et al. (21) concluded on the basis
of density gradient separation of blood from neonates
and adults into younger and older fractions that
increased activities of pyruvate kinase and glucose-
6-phosphate dehydrogenase in erythrocytes of neonates
depend upon the reticulocyte content of cord blood. On
the other hand OSKI (13) and KONRAD (8) have emphasized
that the degree of increase in enolase and phosphoglycerate kinase activities in red blood cells of newborn infants is not explainable by reticulocyte content
alone.

Since there is only an information on APRT activity
in erythrocytes of neonates (1), an investigation on
enzymes and subatrates of salvage-pathway in erythrocytes
of neonates, children and adults was performed.

MATERIALS AND METHODS

Cord blood samples were collected in tubes which
contained heparin from 55 newborn infants. From 44
healthy children 1 to 10 years old and from 51 healthy
adults venous blood was used. After the preparation of
hemolysates APRT, HGPRT and PRPP-synthetase activities
were determined by radiochemical assays(3, 10). The
reactionproducts AMP or IMP were separated by means of
high voltage-electrophoresis.

Temperature dependence of purine phosphoribosyltransferases was examined by preincubating hemolysates
of adults and newborn infants at $56^{o}C$ for APRT and at
$80^{o}C$ for HGPRT respectively before performing the
enzyme assay.

The assay for PRPP was based on the utilization of
PRPP in a system containing 7-14C-orotic acid, orotate
phosphoribosyltransferase and orotidine-5-phosphate
decarboxylase (11).

RESULTS

APRT activities were highest in erythrocytes of newborn infants; lowest activities were found in adults' erythrocytes. (Table 1) The differencies in enzyme activities were statistically significant between newborn infants and the two other groups.

HGPRT was lowest in newborn infants and highest in adults. The differencies between all three groups could be statistically confirmed (Table 2).

No significant variations in the activities of PRPP-synthetase could be demonstrated (Table 3).

Evaluation of PRPP concentrations in erythrocytes of all three groups behaved like APRT activities: Erythrocytes of neonates contained significantly more PRPP than erythrocytes of children and adults(Table 4).

	Neonates	Children	Adults
N	55	44	51
$\bar{x}$	32.1	25.5	24.1
S.D.	8.3	9.0	8.6
p_1	<0.001		
p_2		<0.1	
p_3	<0.001		

Table 1. Activities of adenine phosphoribosyltransferase (nM/mg/hr). p_1 = significance neonates/children p_2 = significance children/adults p_3 = significance neonates/adults

	Neonates	Children	Adults
N	55	44	51
x̄	71.7	84.6	104.0
S.D.	14.1	19.6	26.2
p_1	<0.001		
p_2		<0.001	
p_3	<0.001		

Table 2. Activities of hypoxanthine-guanine phospho-ribosyltransferase (nM/mg/hr).
p_1 = significance neonates/children
p_2 = significance children/adults
p_3 = significance neonates/adults

	Neonates	Children	Adults
N	55	41	32
x̄	25.7	24.5	22.5
S.D.	12.3	8.4	7.3
p_1	<0.6		
p_2		<0.2	
p_3	<0.1		

Table 3. Activities of phosphoribosylpyrophosphate synthetase (nM/mg/hr).
p_1 = significance neonates/children
p_2 = significance children/adults
p_3 = significance neonates/adults

	Neonates	Children	Adults
N	47	28	30
$\bar{x}$	16.2	9.8	8.7
S.D.	8.1	5.6	4.5
p_1	<0.001		
p_2		<0.6	
p_3		<0.001	

Table 4. Concentrations of phosphoribosylpyrophosphate
          (nM/mg).
          p_1 = significance neonates/children
          p_2 = significance neonates/adults
          p_3 = significance neonates/adults

       The investigation of temperature dependence of the
two purine phosphoribosyltransferases revealed no
differences of HGPRT properties but a striking increase
in temperature stability of new born infants' APRT when
compared with erythrocyte enzyme of adults (Figure 2).
Preincubation at 56°C showed a decrease of activity to
30 % and approximately to 50 % in adults and newborn
infants respectively.

                              DISCUSSION

       The data obtained in this study indicate an increase
in APRT activity and PRPP concentration in accordance
with BORDEN et al. (1), but a decrease in HGPRT activity
in erythrocytes of neonates compared to the values
obtained with erythrocytes of adults. According to
OSKI et al. erythrocytes of neonates show an increased
activity of phosphogluconate pathway and glycolysis (14)
and possess a higher level of ATP and adenine nucleotides
(6). An increased influx of phosphogluconate pathway
should result in a higher availability of ribose-5-

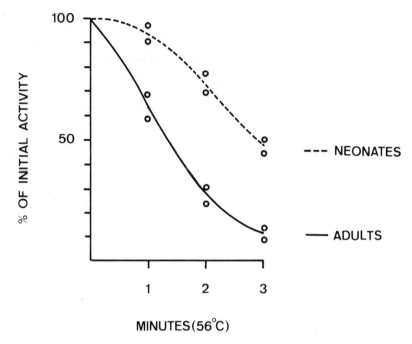

<u>Figure 2.</u>   Temperature sensitivity of adenine phospho-
             ribosyltransferase of neonates' and adults'
             erythrocytes.

phosphate. This compound together with the observed
increase in ATP concentration in erythrocytes of neonates
(6, 19) should enhance an elevated synthesis of PRPP.

    In erythrocytes of patients with Lesch-Nyhan syndrome
the absence of HGPRT activity is accompanied by an
elevation of APRT activity (18). The biological half-
life of the Lesch-Nyhan APRT in erythrocytes is about
2.5 times longer than that found in normal erythrocytes
due to increased PRPP concentrations (5, 16). Nevertheless
the amount of enzyme protein from erythrocytes of
Lesch-Nyhan patients or normals lies in the same range
and the enzymes are immunological indistinguishable from
each other (22). PRPP is able to stabilize APRT in vitro
(5). The demonstrated increase in temperature stability
of APRT could be due to the observed elevated PRPP

concentration in neonates. Therefore inactivation of APRT is slowed down.

On the other hand APRT decreases with age of the erythrocytes in its catalytic and antigenic activity at the same rate (23). Furthermore a larger proportion of young erythrocytes in neonates found by BORDEN and coworkers (1) could also explain the increase in APRT activity. This increased APRT activity could result in the described high levels of adenine nucleotides in newborn infants' red cells (6).

YIP and coworkers described no different amount of HGPRT protein in fractions of erythrocytes of different ages (23). In accordance with this the demonstrated low HGPRT activity in erythrocytes of neonates could be due to a slightly changed catalytic site of this enzyme which might be altered during maturation or due to an inhibitor not present in erythrocytes of children and adults.

The pattern of salvage-pathway enzymes determined in neonates' red blood cells could furthermore represent a persistence of a unique fetal erythropoiesis which disappeares after birth in a manner analogous to fetal hemoglobin. The altered enzyme activities of APRT and HGPRT could furthermore result from repression or derepression of the synthesis of specific enzymes according to the metabolic conditions of intrauterine development.

ACKNOWLEDGEMENT

This study was supported by a grant of "Fonds zur Förderung der wissenschaftlichen Forschung (Austria)" (project 1804).

REFERENCES

1.   Borden M., Nyhan W. L., Bakay, B. (1974)
      Pediatr. Res. 8, 31.

2.   Fontenelle L. J., Henderson J. F. (1969)
      Biochim. Biophys. Acta 177, 175.

3.   Fox I. H., Kelley W. N. (1971)
      J. Biol. Chem. 246, 5739.

4.   Goldwater E. (1954)
      Biochim. Biophys. Acta 13, 341.

5.   Greene M. L., Bayles J. R., Seegmiller J. E. (1970)
      Science 167, 887.

6.   Gross R. T., Schroeder E. A. R., Brounstein S. A.
      (1963)
      Blood 21, 755.

7.   Kelley W. N., Levy R. J., Rosenbloom F. M.,
      Henderson J. F., Seegmiller J. E. (1967)
      Proc. Natl. Acad. Sci (USA) 57, 1735.

8.   Konrad P. N., Valentin W. N., Paglia D. E. (1972)
      Acta Haematol. 48, 193.

9.   Lajtha L. G., Vane J. R. (1958)
      Nature 182, 191.

10.   Müller M. M. (1974)
      J. Clin. Chem. Clin. Biochem. 12, 28.

11.   Müller M. M., Fuchs H., Pischek G., Bresnik W. (1975)
      Therapiewoche 25, 514.

12.   Nyhan W. L., James J. A., Teberg A. J., Sweetman L.,
      Nelson L. G. (1969)
      J. Pediatr. 74, 20.

13.   Oski F. A. (1969)
      Pediatrics 44, 84.

14.   Oski F. A., Naiman J. L. (1965)
      Pediatrics 36, 104.

15.   Raivio K. O., Seegmiller J. E. (1970)
      In "Current Topics in Cellular Regulation", eds.
      Horecker B. L., Stadtman E. R., vol.2, p.201.,
      Academic Press, New York.

16.   Rubin C. S., Balis M. E., Piomelli S., Berman P. H.,
      Dancis J. (1969)
      J. Lab. Clin. Med. 74, 732.

17.   Sass J. K., Itabashi H. H., Dexter R. A. (1965)
       Arch. Neurol. 13, 639.

18.   Seegmiller J. E., Rosenbloom F. M., Kelley W. N.
       (1967)
       Science 155, 1682.

19.   Stave U., Cara J. (1961)
       Z. Kinderheilk. 86, 184.

20.   Williams W. J., Buchanan J. M. (1953)
       J. Biol. Chem. 203, 583.

21.   Witt I., Müller H., Künzer W. (1967)
       Klin. Wschr. 45, 262.

22.   Yip L. C., Dancis J., Balis M. E. (1973)
       Biochim. Biophys. Acta 293, 359.

23.   Yip L. C., Dancis J., Methieson B., Balis M. E. (1974)
       Biochemistry 13, 2558.

PURINE PHOSPHORIBOSYL TRANSFERASES IN

HUMAN ERYTHROCYTE GHOSTS

C.H.M.M. de Bruyn and T.L.Oei

Dept. Hum. Genetics, University of Nijmegen,The Netherlands

It is generally assumed that mammalian hypoxanthine-guanine phosphoribosyl transferase (HG-PRT; EC 2.4.2.8) and adenine phosphoribosyl transferase (A-PRT; EC 2.4.2.7) are soluble, cytoplasmic enzymes. All the isolation procedures for these enzymes are based on purification from cell free supernatant fractions (1-6). Little attention has been paid to the subcellular localisation of purine phosphoribosyl transferases. With respect to isolated cell membranes it has been reported that human erythrocyte and fibroblast membranes do not display HG-PRT activity (7).

In a previous study intact normal human erythrocytes were incubated with radioactive purine bases in the presence of exogenous phosphoribosylpyrophosphate (PRPP) and both cell content and incubation medium were analysed. Interestingly, labeled mononucleotides were found in the medium, suggesting the action of purine phosphoribosyl transferases in the intact cell membrane (8).

In the present paper further experiments are reported concerning effects of osmolarity and pH on the retention of HG-PRT and A-PRT activities associated with erythrocyte ghost preparations and the effect of fragmentation and solubilisation.

MATERIALS AND METHODS

Preparation of erythrocyte ghosts. Freshly drawn venous human blood was heparinized and allowed to stand for 10 minutes in an ice bath. After centrifugation (800 g; 10 min.) plasma and buffy coat were removed and the erythrocytes were washed three times with 0.9% NaCl. One volume of packed cells was lysed in 14 volumes of Tris buffer (pH 7.4) with varying osmolarities (10-80 mOsm) or in water. After three additional washings in the corresponding hemolysing media the various ghost preparations were suspended in 5 volumes of the corresponding media.

To investigate the effect of the pH of the hemolysing medium on the retention of enzyme activities, 10 and 30 mOsm Na-phosphate buffers with pH's ranging from 5.7 to 8.2 were employed. After three washes the erythrocyte ghosts were suspended in 5 volumes of the corresponding buffer.

To study eventual "cryptic" activities (increase of membrane bound enzyme activity by fragmentation), erythrocyte ghosts prepared at various osmotic strength were mixed with 5 volumes of water. The effect of Triton X-100, a non-ionic detergent, was also studied by suspending 1 volume of erythrocyte ghosts prepared at different osmotic strength in 5 volumes of a Triton X-100 solution (endconcentration 0.2%). After 10 minutes in an ice bath, these suspensions were centrifuged (30,000 g; 30 min.) and supernatants and pellets were assayed separately for HG-PRT and A-PRT activity.

The 30,000 g hemolysate supernatants, obtained after lysis of the erythrocytes at varying osmolarities and pH values, were treated in the same way.

Enzyme assays. HG-PRT activity was determined by measuring the conversion of 8-^{14}C-hypoxanthine (spec.act.59 mCi/mmole; radiochemical centre, Amersham) to 5'-IMP in the presence of the co-substrate 5'-phosphoribosyl-1-pyrophosphate (PRPP; Na-Salt from Sigma). The A-PRT activity was assayed by measuring the PRPP-dependent conversion of 8-^{14}C-adenine to 5'-AMP. The reaction mixture (final volume 50 ul) contained 4 nmoles of labeled purine base, 50 nmoles PRPP, 5 umoles Tris (pH 7.4), 0.5 umole MgCl$_2$ and 20 γ protein (ghost- or supernatant fractions). After incubation for 1 hour at 37°C, the mixtures were analysed by means of high-voltage electrophoresis and paperchromatography as described elsewhere (8).

Protein determinations. Total protein was determined by the method of Lowry et al. (9) using bovine serum albumin as standard. Hemoglobin was estimated with the pyridine-haemochromagen method (10). Non-hemoglobin content was calculated by subtracting the value obtained for hemoglobin from that obtained for total protein.

RESULTS

Incubations with erythrocyte ghosts. Thouroughly washed vertually hemoglobin free ghosts showed distinctly purine phosphoribosyl transferase activities. In fig.1 results of HG-PRT and A-PRT measurements

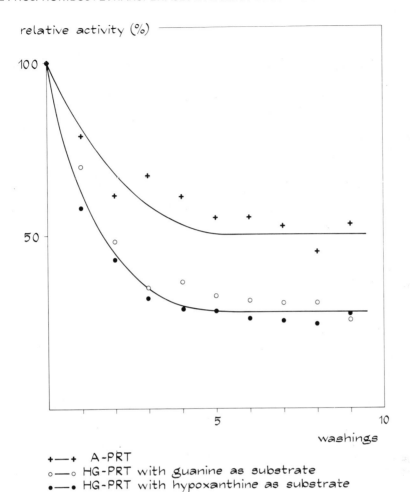

Figure 1.
Relative HG-PRT and A-PRT activities in erythrocyte ghosts (prepared in 20 mOsm phosphate buffer, pH 7.4) after various washings.
Values given are mean values of a typical experiment carried out in
triplicate.

are shown after repeated washings. The initial drop in activities
leveled off after 3-4 washings, and after nine washings still about
the same levels of activity were measured. In further experiments
always 5 times washed ghosts were used.

Table I shows that about 10% of the total erythrocyte protein
was associated with the first pellet (P_0). Subsequent washings did
not substantially reduce the protein content of the ghost fraction.

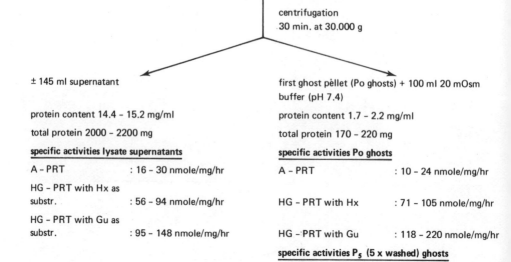

Table I.
Distribution of HG–PRT and A–PRT activities in 30,000 g hemolysate
supernatants and erythrocyte ghost preparations as determined in 5 independent isolations.

The specific activities in the hemolysate supernatant and the P_o
ghosts were in the same range, so that approximately 10% of total
erythrocyte purine phosphoribosyl transferase activity was associated with the P_o ghost fraction. In P_5 ghosts 3–5% of the total
erythrocyte HG–PRT and A–PRT activities were found. These P_5 ghosts
contained less than 0.3% of the hemoglobin originally present in
the intact erythrocyte (for P_o ghosts this value was 1.0%).

Effects of ionic strength. The retention of hemoglobin was markedly affected by the osmolarity of the hemolysing medium (fig.2). After
hemolysis in water, the erythrocyte ghost preparation contained
about 5% of the hemoglobin originally present in the red blood cell.
In erythrocyte ghosts prepared in buffers with osmolarities between
10 and 30 mOsm very little hemoglobin was retained: at 20 mOsm less
than 0.3% of the original hemoglobin. Increasing the osmolarity from

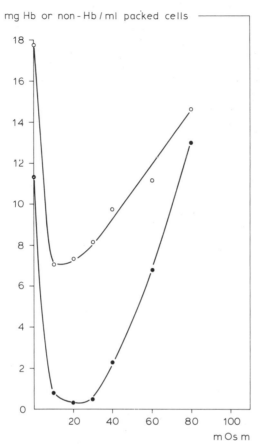

Figure 2.
The effect of the osmolarity of the hemolysing buffer on hemoglobin
retention in erythrocyte ghost preparations. o ———— o non-hemoglo-
bine protein (non-Hb). ● ———— ● hemoglobin (Hb).
Mean corpuscular hemoglobin: 260 mg/ml packed erythrocytes.

30 mOsm to 80 mOsm resulted in increasing amounts of hemoglobin
retained. The curve for non-hemoglobin protein had a comparable
shape. The non-Hb/Hb ratio, however, varied with ionic strength:
at 20 mOsm this ratio was about 14, at 80 mOsm it was approximately 1.

     The retention of HG-PRT and A-PRT was also influenced by the
osmolarity of the hemolysing buffer: at low concentrations low
activities were measured,whereas at intermediate osmolarities marked
optima of HG-PRT and A-PRT activity were observed (figs. 3 and 4).

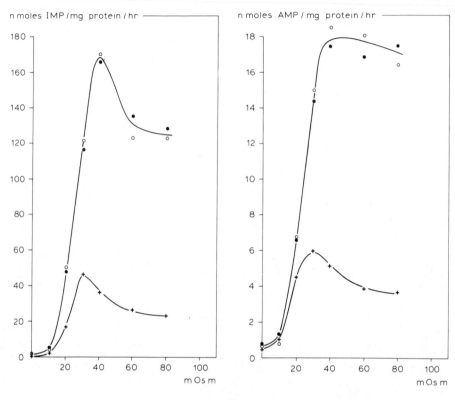

Figure 3   Figure 4

The effect of the osmolarity of the hemolysing buffer on HG-PRT
(fig.3) and A-PRT (fig.4) activities associated with erythrocyte ghost preparations. Aliquots of the suspensions were diluted
sixfold in the corresponding buffer (+ ——— +), water (o ——— o)
or in water containing 0.2% Triton X-100 prior to the enzyme assays
(● ——— ●).

    Phase contrast microscopy revealed structural differences
between the various erythrocyte ghost preparations. Intact ghosts
were observed after preparation at osmolarities between 20 and 80
mOsm, whereas at 10 mOsm there was much fragmentation. Preparation
with water resulted in complete fragmentation, since no intact ghost
could be observed.

    Dilution of the various erythrocyte ghost preparations with
water resulted also in an increase of HG-PRT activity when the
ghosts were prepared at osmolarities above 10 mOsm (fig.3). Addi-

tion of Triton X-100 to an endconcentration of 0.2% resulted in a similar curve (fig.3).

The effect of dilution with water in the absence or in the presence of 0.2% Triton X-100 on A-PRT activity in erythrocyte ghost preparations is depicted in fig.4. Here again, a stimulation of enzyme activity was seen when the ghosts were prepared at osmolarities higher than 10 mOsm. Maximal stimulation of HG-PRT and A-PRT activities were found with 40 mOsm erythrocyte ghost preparations.

Following treatment with 0.2% Triton X-100 and after centrifugation practically all the HG-PRT and A-PRT activities were recovered in the supernatant; only traces of enzyme activities were left in the 30,000 g pellet (fig.5);

The HG-PRT and A-PRT activities in the original 30,000 g hemolysate supernatants were neither affected by the different osmolarities used for hemolysis and dilution, nor by treatment with Triton X-100.

Effects of pH. Erythrocyte ghosts were prepared in phosphate buffers at pH values from 5.7 to 8.2 and at an ionic strength of 10 mOsm and 30 mOsm respectively. In this way both fragmented (10 mOsm) and intact (30 mOsm) erythrocyte ghosts were studied. HG-PRT

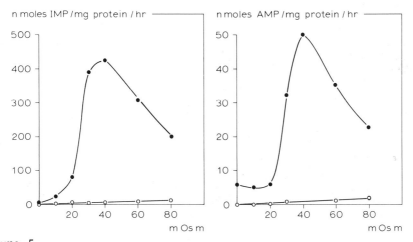

Figure 5.
HG-PRT (left) and A-PRT (right) activities in 30,000 g supernatants (● —— ●) and 30,000 g pellets (o —— o) from erythrocyte ghost suspensions, prepared at different osmolarities, after treatment with 0.2% Triton X-100.

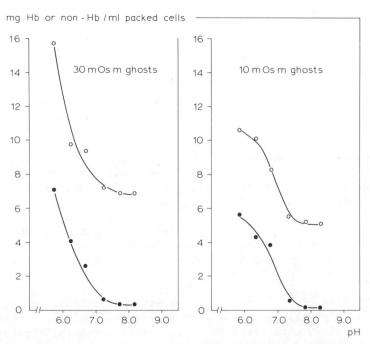

Figure 6.
The effect of pH on the retention of hemoglobin in erythrocyte
ghost preparation prepared in 30 mOsm buffer (left) and in 10 mOsm
buffer (right). o ——— o non hemoglobin protein (non-Hb). ● ——— ●
hemoglobin (Hb). Mean corpuscular hemoglobin: 260 mg/ml packed erythrocytes.

and A-PRT activities were assayed at pH 7.4 in the presence or absence of 0.2% Triton X-100. The treatment in the pH range between
5.7 and 8.2 did not produce loss of enzyme activities.

     In fig.6 the retention of Hb in 30 mOsm and 10 mOsm erythro-
cyte ghosts is shown. At low pH values relatively much Hb was
retained in 30 mOsm erythrocyte ghosts: at pH 5.7 about 3% of Hb
originally present in the erythrocyte; at pH 7.5 the Hb retention
was 0.3%. At pH 5.7 some 3% of original Hb was retained in 10 mOsm
erythrocyte ghosts, whereas at pH values higher than pH 7.3 hardly
any Hb could be detected. The curves for non-Hb protein have a comparable shape, but the pH of the hemolysing buffer significantly
influenced the non-Hb/Hb ratio. For instance at pH 5.7 this ratio
was about 2 in both 30 mOsm and 10 mOsm ghost preparations. At pH
7.8 the non-Hb/Hb ratio was about 14 in 30 mOsm ghosts, whereas in
10 mOsm ghost preparations practically all protein was non-Hb (fig.
6).

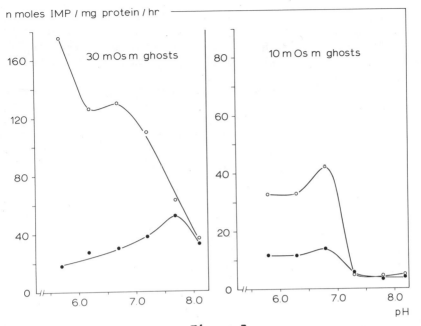

Figure 7

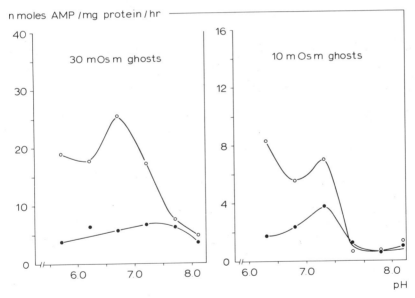

Figure 8
The effect of pH on the retention of HG-PRT (fig.7) and A-PRT (fig. 8) activities in erythrocyte ghosts prepared at 30 mOsm and at 10 mOsm. Enzyme activities were determined in the corresponding buffer (● —— ●) and in the presence of 0.2% Triton X-100 (o —— o).

The effect of pH upon HG-PRT activity was marked in both types
of erythrocyte ghosts. In 30 mOsm buffer preparations more HG-PRT
activity was retained at pH values higher than pH 7, reaching a
maximum at pH 7.8 in the absence of 0.2% Triton X-100. Addition of the
detergent resulted in an 8-fold increase of enzyme activity in suspensions of erythrocyte ghosts prepared at pH 5.7, but in suspensions
of erythrocyte ghosts prepared at pH 8.2 hardly any stimulation of
activity was noticed (fig.7,left). Ghosts prepared at 10 mOsm showed
lower HG-PRT activities and the differences between incubations with
and without 0.2% Triton X-100 were less pronounced. At higher pH value
again no stimulation of HG-PRT activity in the presence of Triton
X-100 was observed (fig.7, right).

A-PRT was also susceptible to the effect of pH. In preparations
at 30 mOsm and at lower pH values, treatment with Triton X-100 resulted in a considerable increase of activity; incubations without
detergent showed a rather constant retention of A-PRT activity
(fig.8, left). At higher pH values, no increased A-PRT activities
were found, as was the case with HG-PRT. Erythrocyte ghosts prepared
at 10 mOsm showed A-PRT activity curves comparable to those with
HG-PRT. At lower pH values there was a demonstrable increase in
activity in Triton X-100 treated samples, but with erythrocyte
ghosts obtained with 10 mOsm buffer at pH values higher than 7.3, no
differences between Triton X-100 treated and non-treated samples
were observed (fig.8, right).

A characteristic peak in the stimulated activities of HG-PRT
and A-PRT was observed between pH 6.3 and 6.8 with 30 mOsm as well
as 10 mOsm erythrocyte ghost preparations (fig.8, right).

The enzyme activities in the original 30,000 hemolysate
supernatants were not influenced by the pH of the hemolysing buffers and treatment with Triton X-100 did not result in stimulation
or inhibition of the soluble enzymes.

DISCUSSION

The studies with intact human erythrocytes and with practically hemoglobin-free erythrocyte ghosts presented in this
communication demonstrate purine phosphoribosyl transferase activities associated with red blood cell membranes.

During the preparation of hemoglobin free erythrocyte ghosts
the pink colour of the first ghost pellet disappeared after 3 or
4 washings. The initial loss in HG-PRT and A-PRT activities was
parallel to this phenomenon: after 3 to 4 washings the activities
remained on the same level (fig.1.). The absolute activities of both
enzymes varied according to source and preparation (table I), but

in all cases the same type of curves was noticed: after more than
4 times washing the activity levels remained unaffected (fig.1); 3-5%
of total erythrocyte purine phosphoribosyl transferase activities
were found to be associated with the practically hemoglobin-free
ghost fractions. The quantitative and qualitative significance
of this amount of activity for red cell purine metabolism requires
further investigation. Especially, compartmentation in the membrane
should be taken into consideration.

In the study of membrane structure and function, membranes
of erythrocytes have been widely used, because the so-called erythrocyte "ghost"-preparations are relatively easy to obtain by osmotic
lysis of washed erythrocytes. However, most erythrocyte ghost preparations show considerable variability with respect to hemoglobin
content and enzyme activities (10). It has been shown that the chemical
and enzymological composition of erythrocyte membranes varies with
ionic strength and pH of the hemolysing buffer (11,12). The term "ghost"
is a description of the discoid bodies obtained after removal of the
hemoglobin from the erythrocytes. Nevertheless, the limitations of the
use of erythrocyte ghosts should be realised: the erythrocyte membrane
obtained by osmotic lysis should be considered as a derivative only
of the intact erythrocyte membrane (13,14).

The present experiments on the effect of osmolarity on erythrocyte ghost HG-PRT and A-PRT activities showed that under certain conditions both enzymes were stimulated when the erythrocyte ghosts were
diluted with water. This increase in activity upon fragmentation of the
erythrocyte ghosts has been called "cryptic" activity, in contrast
to the "basic" activity of untreated erythrocyte ghosts (15).

The "basic" activities for HG-PRT and A-PRT were maximal in
erythrocyte ghosts prepared at 30 mOsm (fig.3 and 4). These erythrocyte
ghosts also contained very low amounts of Hb: 0.2% to 0.3% of Hb originally present in the intact erythrocytes (fig.2). The marked maxima
of the enzyme activities could be due to release of inhibitor, but
experimental evidence supporting this explanation is lacking. Similar
optima at this osmolarity were observed for other erythrocyte ghost
enzyme activities (16).

Ghosts prepared at low osmotic strength did not display "cryptic"
activity, whereas preparation at higher concentrations resulted in
rapidly increasing levels of enzyme activities (fig.3 and 4). Experiments on several enzymes in carbohydrate metabolism, such as lactic
acid dehydrogenase and phosphoglycerate kinase associated with
erythrocyte ghosts, showed similar profiles (16). During fragmentation,
the enzymes apparently become more accessible to substrates or other
factors.

HG-PRT and A-PRT were both released from the erythrocyte ghosts

by addition of a Triton X-100 solution: hardly any activity could
be measured in the 30,000 g pellet. Practically all HG-PRT and
A-PRT activity was found in the 30,000 g supernatant fraction
(fig.5).

The effect of solubilisation on HG-PRT and A-PRT activities
was comparable to that of fragmentation of the erythrocyte ghosts
by dilution with water.

The present results show that HG-PRT and A-PRT activities
associated with erythrocyte ghosts are also markedly affected by
changes in pH as is shown in figs. 7 and 8. Fragmented and intact
erythrocyte ghost fractions (prepared at 10 mOsm and 30 mOsm,
respectively) showed varying amounts of Hb depending on the pH
of the hemolysing buffer: at pH values higher than pH 7.3, 30 mOsm
and 10 mOsm ghosts contained less than 0.3% of the original erythrocyte Hb (fig.6) Strinking was the disappearance of the stimulation of enzyme activities at higher pH values in both 30 mOsm
and 10 mOsm ghost preparations. The typical maxima in both types
of ghost prepared at about pH 6.8 has also been observed by other
workers (16) but is not possible to give a plausible explanation
for this phenomenon with the present state of knowledge.

It should be noticed that in contrast to the membrane associated
purine phosphoribosyl transferase activities, the corresponding
soluble enzymes in the 30,000 g hemolysate supernatants are not
affected by the various treatments applied in this study. Whether
or not this difference in enzyme behavior is due to separate classes of enzyme in cytoplasm and membrane or to configurational alterations caused by the association of identical enzymes with the
membrane structure remains to be established.

Since loosely bound proteins are readily removed from the
red blood cell membranes as the ionic strength of the hemolysing
medium is lowered, a clear distinction can be made between loosely
bound (extrinsic) and firmly bound (intrinsic) proteins (12,16).
Also, the association of loosely bound enzymes with the membranes
is markedly affected by changes in pH, whereas that of firmly
bound enzymes is not. Therefore, HG-PRT and A-PRT may be classified
as loosely bound enzymes. Thus far, membrane associated purine
phosphoribosyl transferases have been reported from bacteria only:
they seem equally loosely bound, since they can also be released
by osmotic shock (17).

In studies on membrane-enzyme interactions most attention has
been paid to red blood cell membrane fractions with as little Hb
as possible. However, it has been suggested recently that hemoglobin may exert some effect, conceivably conformational in nature,
on the behavior of red blood cell membranes (14). From this point

of view the study of membrane associated enzymes should not be limited to almost hemoglobin free erythrocyte ghosts.

ACKNOWLEDGEMENTS

This study was supported by FUNGO, Foundation for Medical Scientific Research in the Netherlands and the Medical Prevention Fund. The skillfull assistance of Mr.C. van Bennekom and J. van Laarhoven is gratefully acknowledged.

REFERENCES

1. Krenitsky T.A., Papaioannou R. and Elion G.B. (1969).
   J.Biol.Chem. 244, 1263-1270.

2. Krenitsky T.A., Neil S.M., Elion G.B. and Hitchings G.H.(1969).
   J.Biol.Chem. 244, 4779-4784.

3. Craft J.A., Dean B.M., Watts R.W.E. and Westwick W.J. (1970).
   Eur.J.Biochem. 15, 367-373.

4. Arnold W.J. and Kelley W.N. (1971).
   J.Biol.Chem. 246, 7298-7304.

5. Rubin C.S., Dancis J., Yip L.C., Nowinski R.C. and Balis M.E.(1971).
   Proc.Nat.Acad.Sci. USA 68, 1461-1464.

6. Gutensohn W. and Guroff G. (1972).
   J.Neurochem. 19, 2139-2150.

7. Benke P.J., Herrick N. and Herbert A.(1973).
   Biochem. Med. 8, 309-322.

8. de Bruyn C.H.M.M. and Oei T.L. (1974).
   In: Purine Metabolism in Man (O.Sperling, A. de Vries and
   J.B. Wijngaarden Eds.) Plenum Press, New York, pp.223-227.
   de Bruyn C.H.M.M. and Oei T.L. (1976).
   These proceedings.

9. Lowry D.H., Rosebrough N.J., Farr A.L. and Randall R.J. (1951).
   J.Biol.Chem. 193, 265-275.

10. Hanahan D.J. (1973).
    Biochim.Biophys. Acta 300, 319-340.

11. Dodge J.T., Mitchell C. and Hanahan D.J. (1963).
    Arch.Biochem.Biophys. 100, 119-129.

12. Mitchell C.D., Mitchell W.B. and Hanahan D.J.(1965).
    Biochim. Biophys. Acta 104, 348-358.

13. Bramley T.A., Coleman R. and Fineman J.B. (1971).
    Biochim. Biophys. Acta 241, 752-769.

14. Hanahan D.J. and Ekholm J. (1972).
    Biochim. Biophys.Acta 255, 413-419.

15. Zamudio L., Celline M. and Canessa-Fische (1969).
    Arch.Biochem.Biophys. 129, 336-345.

16. Duchon G. and Collier H.B. (1971).
    J. Membrane Biol. 6, 138-157.

17. Hochstadt-Ozer J. and Stadtman E.R. (1971).
    J.Biol.Chem. 246, 5304-5311.

THE LOCALISATION OF THE PURINE-PHOSPHORIBOSYLTRANSFERASE IN RAT LIVER ORGANELLS

G. Partsch, I. Sandtner and R. Eberl

Ludwig Boltzmann-Institute for Rheumatology and Bal-

neology, Kurbadstrasse 1o, A-1107 Vienna, Austria

The liver is one of the most important organs where purine bases will be accumulated during catabolic processes and re-utilized. There exists a lot of information on preparing the different enzymes of the purine-phosphoribosyltransferase of cow and pigeon livers (1) or microorganism but there is less information about the localisation of these enzymes within cell organells. We therefore investigated rat liver organells with regard to their purine-phosphoribosyltransferase activities.

MATERIALS AND METHODS

The preparation of rat liver organells followed a combined method of Widnell (2), Sottocasa (3) and Chandra (4). 12 female Sprague Dawley rats starved for 12 hours were decapitated, the livers prepared and perfused with ice cold isotone sodiumchloride. 15o grams of liver were pressed through a sieve of steel and sus-pended in looo ml o,24 M sucrose complemented by $3 \cdot 1o^{-4}$ M Ca Cl$_2$ and homogenized (5). The homogenate was centrifuged at 6oo g for 15 minutes. From this sediment containing cell wall fragments, nuclei and whole cells the nuclei were prepared according to the method of Widnell (2). The supernatant of the 6oo g centrifugation was centrifuged at 6.5oo g for 2o minutes. The sediment contained the mitochondria fraction. To separate the light mitochondria from the heavy part the sediment was suspended in o,45 M saccharose gently homogenized and the suspension overlayed on 1,18 M sucrose (1:1/V/V). The cell material was centrifuged in a Beckman SW 4o rotor at 24.ooo ppm for 3 hours. The mitochondria fraction divided into four parts. On the top there was a light yellow soluble

153

fraction, then a yellow one, representing the light mitochondria,
then a colourless one, only gradient and on the bottom a dark brown
sediment, the heavy mitochondria.

The supernatant of the 6.5oo g centrifugation was centrifuged
at 27.ooo g for 15 minutes. The sediment was discharged. The supernatant was run at 1o5.ooo g for 6o minutes resulting in a sediment
containing the microsomal fraction. The ribosomes were purified
according to Chandra (4) from the microsomes.

The resting supernatant was the cytoplasmic part. The cytoplasm
the fraction of the soluble mitochondria and the ribosomes were
dialysed against TRIS-HCl buffer (o,1 M, pH 7,4 supplemented by
$5 \cdot 1o^{-3}$ M $MgCl_2$) overnight at 4° C. The sediments of separated
organells were washed twice in the same buffer and homogenized.
After centrifugation the cell free extracts were used for the enzyme
determination.

A part of the original squeezed liver material was homogenized
in buffer and the enzyme activity of the whole liver estimated.
Protein was measured according to Lowry (6). The determination of
the purine-PRT activities followed a radio isotope method described
earlier (7). To get information about the contamination of microsomal particles in the different fractions the specific glucose-
6-phosphatase was measured in each portion (8).

RESULTS AND DISCUSSION

As summarized in Table 1 the purine-phosphoribosyltransferases
are ubiloquaer in rat liver organells. As expected the main activity
is located in the cytoplasm. The A-PRT activity showed the highest
rate followed by the G-PRT and the H-PRT activity. This result is
divergent to the values obtained from the cell free extract of the
whole liver. The possible reason for this and also in other samples
is the different protein content of the preparations.

A very high enzyme activity could be determined in the 1o5.ooo
g sediment. This is a very heterogenious fraction. The so-called
microsomal part of the preparation contains parts of the endoplas-
matic reticulum, ergastoplasm, but also parts of the lysosomes and
microbodies. The distribution of the enzyme activities in the whole
microsomal fraction is completely different to the cytoplasm. The
A-PRT shows the highest activity followed by H-PRT and G-PRT. The
high enzyme activities lead to the conclusion that the microsomal
fraction is contaminated by parts of the cytoplasm. We believe that
due to the performed preparation this result is not a false one but
reflects the natural enzyme activity.

	A-PRT	G-PRT	H-PRT
LIVER TOTAL	1476,5	1148,o	846,6
NUCLEI	1o9,5	6o,8	54,7
MITOCHONDRIA TOTAL	125,9	7o,9	32,o
SOLUBLE FRACTION	o	o	o
LIGHT MITOCHONDRIA FRACTION	o	o	o
HEAVY MITOCHONDRIA FRACTION	288,7	93,4	71,3
MICROSOMAL FRACTION	882,6	511,4	768,6
RIBOSOMES	o	o	o
CYTOPLASM	1544,3	713,8	144o,8

Table 1. The purine-phosphoribosyltransferase activities in rat
liver organells derived from regression analysis of the
enzyme kinetics (nMol nucleotide/mg protein/h).

As shown in the table the ribosomal fraction prepared by
further gradient centrifugation did not show a measurable enzyme
activity.

Different enzyme activities were also obtained in the mitochondrial fraction. The total fraction (collected as the sediment
of the 16.oooxg centrifugation) shows a descending enzyme activity
from A-PRT to G-PRT and H-PRT.

The total mitochondria fraction was separated a second time
to obtain light and heavy mitochondria. With the gradient centrifugation in sucrose we got a soluble and a light fraction which
did not show any enzyme activity. As shown in the Table only the
heavy part of the mitochondria showed purine-phosphoribosyltransferase activity. The activity of the heavy part of the mitochondria
is approx. two times of the whole mitochondria fraction. Because of
the lower protein content of the separated heavy mitochondria sub-
fraction the enzyme activity is higher than in the total.

In the nuclei the A-PRT is also the dominant enzyme whereas
the G-PRT and H-PRT are of lower activity.

This study should elucidate the question as to whether the
purine-phosphoribosyltransferases are uniloquaer or ubiloquaer.
In the light of the present data we are able to favour the meaning
that there are two distinct main locations of the enzyme activities,
namely the microsomes and the cytoplasm. In view of the measurable
glucose-6-phosphatase activity in the preparations (around lo %
of microsomal fraction) which is reported to be specific for microsomes (8) the purine-PRT activities within other organells should
be considered with some caution. Otherwise it seems obvious that
the soluble and light fraction of mitochondria which had to be also
contaminated by microsomal particles did not exhibit any purine-
PRT activity and this also allows for the ribosomal fraction.
Thinès-Sempoux (9) on the other hand, mentioned that glucose-6-
phosphate may also be dephosphorylated by other phosphatases than
the specific glucose-6-phosphatase. The glucose-6-phosphatase
activity measured in our preparation therefore does not possibly
reflect microsomal contamination only.

It would be of some interest to find out for example if the
nuclei exhibit their own purine-PRT. To get more information about
the connection between purine-PRT activities and cell organells
this study had to be repeated on much more cell material especially
from the human liver.

REFERENCES

1. Flaks, J.G.: Methods in Enzymology. Academic Press, New York,
   Vol. VI, 136 (1963).

2. Widnell, C.C., Tata, J.R.: Biochem.J. $\underline{92}$, 33 (1964).

3. Sottocasa, G.L.: Methods in Enzymology. Academic Press, New
   York, Vol. X, 448 (1967).

4. Chandra, P.: Methoden der Molekularbiologie. Gustav Fischer
   Verlag, Stuttgart, p. 5 (1973).

5. Potter, V.R., Elvehjem, C.A.: J.Biol.Chem. $\underline{114}$, 495 (1936).

6. Lowry, O.H., Rosenbrough, N.J., Farr, A.L., Randall, R.J.:
   J.Biol.Chem. $\underline{193}$, 265 (1951).

7. Partsch, G., Altmann, H., Eberl, R.: Purine Metabolism in man.
   Advance in experimental medicine and biology. Plenum Press,
   New York, Vol. 41A, lo3 (1974).

8. Swanson, M.A.: Methods of Enzymology. Academic Press, New York,
   Vol. II, 541 (1955).

9. Thinès-Sempoux, D.: Frontiers of Biology. North Holland
   Publishing Company, Amsterdam, Vol. 29, Lysosomes 3, p. 281
   (1973).

REVERSIBLE AND IRREVERSIBLE INHIBITION OF HYPO-XANTHINE-PHOSPHORIBOSYLTRANSFERASE

Wolf Gutensohn

Institut für Anthropologie und Humangenetik der

Universität, D 8000 Munich, FRG

For some years our laboratory has been concerned with the purification and characterization of hypoxanthine-phosphoribo-syltransferase (HPRT, EC 2.4.2.8) from rat brain. By a new method (which will be discussed in detail in the methodology-section of this meeting) we now can easily obtain pure HPRT from rat brain or human erythrocytes and use these enzyme preparations to produce antisera in rabbits. It has been reported by several authors, also in this symposium 3 years ago, that specific antisera do inhibit HPRT-activity and we cannot add any exciting new evidence to this point. So only the anti-rat and anti-human serum shall shortly be compared with one another:
- Both sera inhibit their respective enzymes.
- Best results are obtained when the serum after having reacted with the enzyme is precipitated by goat-anti-rabbit-IgG.
- Anti-rat-serum does not crossreact with human HPRT.
- Anti-human-serum does not crossreact with rat HPRT.
- As expected both sera do not crossreact with rabbit HPRT (for example with the enzyme of the animal that produced the antiserum).
- Especially in the human system we found when - after reaction of the antiserum with the enzyme - samples are not carefully centrifuged, or are not precipitated with goat-anti-rabbit-IgG, or when an excess of antiserum was used, soluble enzyme-antibody-complexes are formed which are still enzymatically active.

The main part of this contribution is devoted to inhibition experiments which have been derived from our older studies on reversible inhibition of HPRT by natural nucleotides (1, 2). The fact was that in the guanine series of nucleoside-phosphates inhibition decreases with increasing number of phosphates, whereas in the series of the non-substrate-bases adenine, cytosine and uracil inhibition increases with increasing number of phosphates. Kinetic studies have shown that all these inhibitions, at least in the lower concentration range, up to 2 mM, are competitive with respect to phosphoribosyl-pyrophosphate (PRPP). ATP at higher concentrations seems to inhibit by a different mechanism. This is not a competition for Mg^{++}, since equimolar concentrations of this ion do not abolish the effect. In the concentration range above 2 mM ATP shows a curved up Dixon-plot suggesting a different kind of binding to the HPRT-molecule. Additional evidence to this point will be given below.

Of all the natural nucleotides GMP is by far the best inhibitor (K_i about 4 uM). Several structural features of this molecule seem to be important for tight binding to HPRT:
- The 2-aminogroup, since IMP has a much higher K_i than GMP
- The 5´-phosphate-group, since guanosine does not inhibit HPRT at all.
- A smaller contribution to binding comes from the 2´-hydroxyl, since dGMP inhibits the enzyme slightly less than GMP.

For theoretical reasons a specific irreversible inhibitor of HPRT should be a derivative of GMP with a reactive group in one of the structural areas just mentioned. That we found such a compound was more or less by chance. Our interest was directed towards periodate-oxidized nucleotides by studies of Cory and George (3) on irreversible inhibition of CDP-reductase. So we tried periodate-oxidized nucleotides on HPRT. The basic principles of the type of inhibition of HPRT observed with these compounds have been published (4) and will be shortly summarized.

Since we had difficulties in purifying and isolating the oxidized compounds the following method was adopted. The nucleotides were freshly prepared and immediately used. Three types of redox-reactions were set up:
- Nucleotide was oxidized with twice the concentration of $NaIO_4$. The excess of periodate was then reduced by an excess of glycerol (designated "oxidized nucleotide").
- The periodate was first completely reduced by glycerol and

the nucleotide (same concentration as above) was added after-
wards (designated "nucleotide-control").
- The periodate was reduced by glycerol without any further
  addition (designated "oxidized glycerol"). This solution - as
  the two others mentioned above - contains aldehydes as oxidation products from glycerol and iodate as a reduction product from periodate.

The basic effect seen with HPRT from rat and human is as
follows:
- Oxidized-GMP and oxidized-IMP in the concentration range of
  10 - 100 uM inhibit HPRT to over 90%.
- This effect is not observed with oxidized-AMP.
- It is not observed with all the nucleotide-controls and oxidized
  glycerol in the same concentration range.
- At higher concentrations (1 mM) unspecific effects also given
  by the controls superimpose upon the specific inhibition.

This type of inhibition, being complete after about 1 h of preincubation is irreversible and is
- not reversed by dialysis
- not reversed by Sephadex-gelfiltration
- not reversed by polyacrylamide-gelelectrophoresis: Two enzyme samples (identical amount of protein) treated with 10 uM
  oxidized GMP and 10 uM GMP-control respectively are compared in their enzyme-activity. They are then subjected to
  electrophoresis on a nondenaturing polyacrylamide-gel. After
  separation the two enzymes are extracted from the appropriate
  positions in the gels and their activities determined again. The
  ratio of the activities has remained unchanged.

This inhibition is active-site-directed:
- Enzyme can be protected against inactivation by 10 uM oxidized GMP by an excess of unoxidized GMP (1 mM), but only
  when this is given simultaneously with the inhibitor. Adding
  the GMP after inactivation has taken place does not restore
  enzyme activity.
- HPRT cannot be protected by 2.5 mM PRPP alone, however,
  in the presence of equimolar amounts of Mg^{++} PRPP provides excellent protection against inhibition by oxidized GMP.
  This shows again that the mono-Mg^{++}-salt of PRPP is the
  true substrate recognized by the binding-site of HPRT.
- ATP (5 mM) even in the presence of Mg^{++} does not protect
  the enzyme and this demonstrates that ATP must have a separate binding-site on HPRT.

When we use labeled inhibitors we can show binding of the inhibitor to the enzyme protein after gel-electrophoresis. This is observed when oxidized-^{14}C-GMP is allowed to act
- on HPRT in a crude preparation, i.e. within a mixture of proteins.
- on purified HPRT
The effect is observed
- on nondenaturing gels in a discontinuous system (Davis)
- on SDS-gels (Weber, Osborn-system)
Binding to HPRT cannot be demonstrated
- with oxidized-^{14}C-AMP
- with ^{14}C-GMP-control.
In the case of purified HPRT the stoichiometry of this binding can be calculated on the basis of the known molecular weight of the enzyme. This stoichiometry is still very poor and for reasons unknown not enough inhibitor is bound.

To test for the specificity of oxidized GMP for HPRT we have started to try this inhibitor on other enzymes of purine metabolism which bind or react with guanyl-nucleotides. Inhibitors were applied in the concentration-range of 10 - 100 uM where we observe effects on HPRT.
- Adenine-phosphoribosyltransferase (EC 2.4.2.7) is not inhibited by oxidized nucleotides, even not by oxidized AMP.
- 5´-nucleotidase (EC 3.1.3.5) is not inhibited.
- PRPP-synthetase (EC 2.7.6.1) which has regulatory binding-sites for nucleotides is not affected by oxidized nucleotides.
- Purine-nucleoside-phosphorylase (EC 2.4.2.1) is not inhibited by oxidized nucleotides, which was expected. But there was also no influence of oxidized nucleosides and oxidized inosine, even at substrate concentrations (0.5 mM) did not inhibit the enzyme nor was it accepted as alternative substrate.
- The effect of oxidized GMP on guanylate-kinase (EC 2.7.4.8) was studied in a crude brain-preparation. HPRT-activity of this preparation was inhibited in the usual way, whereas at the same time guanylate-kinase was unaffected.

Although oxidized GMP has so far proven a fairly specific inhibitor for HPRT in vitro the compound will most probably not be applicable in vivo. Oxidized GMP, still carrying the phosphate-group, is so polar that it cannot penetrate or act across an intact biological membrane. This is demonstrated in the last experiment. Oxidized GMP is allowed to act upon a human erythrocyte sample in three different ways:

- With an erythrocyte-lysate the effect of oxidized GMP on HPRT-activity is as described above.
- Oxidized GMP (10 - 100 uM) is preincubated with intact erythrocytes under isotonic conditions. It is then washed off, the cells are lysed and HPRT-activity is determined. In this case enzyme-activity compared with a control-sample is not diminished.
- Hypoxanthine-uptake and -retention in appropriately pretreated erythrocytes (a system first used by Mager and coworkers (5) and extended in our laboratory (6)) is a measure of intracellular HPRT-activity. This uptake-system is not affected by 10 - 100 uM oxidized GMP. At higher concentrations the system is inhibited, but in a completely unspecific way, which is as well observed with GMP-control and oxidized glycerol. This effect is probably due to a membrane damage by the aldehydes or iodate or both.

To summarize one can say that this type of affinity label might one day help in elucidating the structure of the active center of HPRT. But when when we want an agent blocking the enzyme in vivo, as we would like to do, we will have to look for other means.

Literature

1) Gutensohn, W. & Guroff, G. (1972) J. Neurochem. 19, 2139 - 2150

2) Gutensohn, W. (1974) in Advances in Experimental Medicine and Biology (Sperling, O., DeVries, A. & Wyngaarden, J. B. eds.) Vol. 41A, pp 19 - 22, Plenum Press, New York, London.

3) Cory, J. G. & George, C. B. (1973) Biochem. Biophys. Res. Commun. 52, 496 - 503.

4) Gutensohn, W. & Huber, M. (1975) Hoppe-Seyler's Z. Physiol. Chem. 356, 431 - 436.

5) Hershko, A., Razin, A., Shoshani, T. & Mager, J. (1976) Biochim. Biophys. Acta 149, 59 - 73.

6) Gutensohn, W. (1975) Hoppe-Seyler's Z. Physiol. Chem. 356, 1105 - 1112.

PURIFICATION AND CHARACTERIZATION OF HYPOXANTHINE-GUANINE PHOSPHO-RIBOSYL TRANSFERASE FROM CULTURED HTC CELLS

Vassilis I. Zannis, Lorraine J. Gudas, Deborah Doyle
and David W. Martin, Jr.
University of California, San Francisco
Departments of Medicine and Biochemistry and Biophysics
San Francisco, California 94143

Hypoxanthine-guanine phosphoribosyl transferase (HGPRTase) from $|^{35}S|$-methionyl-labelled rat hepatoma (HTC) cells has been purified more than 600-fold to near homogeneity by making use of conventional and affinity chromatographic techniques (Table 1). The conventional purification procedure utilizes a pH 5.0 precipitation, an ammonium sulfate precipitation, DEAE-cellulose chromatography, and heating at 80-85° according to the method of Olsen and Milman, 1974. Using this procedure, a partially pure enzyme preparation is obtained. Affinity chromatography according to Hughes et al., 1975, can be used either before or after the DEAE step to yield pure enzyme.

The DEAE-cellulose chromatographic procedure, used in both the conventional and affinity chromatographic purification, reveals two peaks of HGPRTase activity (Figure 1 I). The first peak of activity is eluted with a decreasing pH gradient of phosphate and the second eluted with a gradient of increasing potassium chloride concentration, as explained in the legend of Figure 1 I. We have used several techniques to study the HGPRTase molecules which can be resolved by this procedure. Figure 1 II shows on the left panel the SDS acrylamid gel electrophoresis pattern of the heated material eluted in peak A and in peak B. Shown in the right panel is the autoradiogram of SDS acrylamide electrophoretic gel of the material eluted from DEAE-cellulose in peaks A' and B' following the affinity chromatographic step. In all cases there was present a protein with a molecular weight of 26,000.

To verify that the 26,000 M.W. component was the HGPRTase sub-unit, heat treated $|^{35}S|$-labelled material from peak B, in which the 26,000 M.W. component was the major band, was isoelectric focused

Table I. Purification of HGPRTase from cultured HTC cells

Fraction	Total Protein mg	Total Activity units X 10⁻⁴	Specific Activity units/mg	Overall Yield	Purification Fold
1) Crude extract	1,330	59	444	100	-
2) pH 5.0 supernatant	665	59	887	100	2
3) 50%-70% ammonium sulfate pellet	262	51	1,950	86	4.4
4) DEAE cellulose:					
phosphate elution (Fraction A)	14.6	31	21,160	52	48
KCl elution (Fraction B)	6.6	9	13,200	15	30
5A) Concentrate 10X Heat at 80° for 6 min. (Fraction A)	2.1	25	116,500	42	262
5B) Concentrate 10X Heat at 85° for 10 min. (Fraction B)	.65	8.7	134,500	15	303
5C) Affinity chromatography of unheated (Fraction A)	.057	1.6	279,000	11	628
3') Affinity chromatography of pH 5.0 supernatant	.98	4.4	45,400	30	102
4') DEAE cellulose after affinity chromatography:					
Phosphate elution (Fraction A')	.067	.25	37,300*	1.5	>80
KCl elution (Fraction B')	.039	.47	120,500*	3.2	>275

*The low specific activity of HGPRTase is the result of severe inactivation of the enzyme due to its very dilute state.

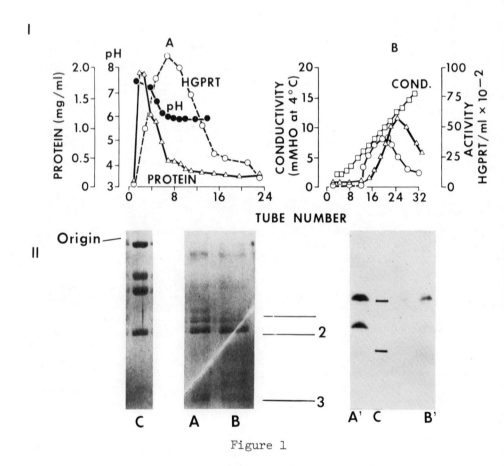

Figure 1

Figure 1 I. DEAE-cellulose column chromatography of HGPRTase from HTC cells.

Δprotein mg/ml; o activity units HGPRT/ml; ● pH; ☐conductivity mMHO at 4°C.

One unit is the amount of HGPRTase required to produce 1 nmole of IMP per hour under the assay conditions (50 mM Tricine, 6mM $MgCl_2$, 0.1 mM DTT, pH 7.4). The enzyme was loaded in the assay buffer and the column was washed with the same buffer. Peak A was eluted with potassium phosphate buffer 60mM, pH 5.8 containing 6mM $MgCl_2$, 0.1 mM DTT. Peak B was eluted with 10-300 mM KCl gradient in the assay buffer. Fractions of 6 ml were collected.

Figure 1 II. SDS acrylamide gel electrophoresis of purified HGPRTase.

A and B contain the fractions A and B, respectively, of DEAE-cellulose chromatography after heat treatment. The observed major protein bands in A and B are numbered in order to facilitate comparison with the two-dimensional electrophoresis gel Fig. 4A. Channel C contains from top to bottom the following protein markers: bovine serum albumin, ovalbumin, aldolase, and chymotrypsinogen. The right panel is the autoradiogram of SDS acrylamide gel electrophoresis of fractions A' and B' obtained by affinity chromatography followed by DEAE-cellulose chromatography (See Table 1). The position of the marker proteins chymotrypsinogen and RNase is marked in ink on the autoradiogram.

---

under native conditions on a Sephadex G-75 superfine gel with 2.1% ampholines at 4°C.

Following isoelectric focusing, the protein was eluted from 0.8 cm sections of the G-75 gel slab and the eluted material assayed for HGPRTase catalytic activity and $|^{35}S|$ radioactivity. Figure 2 demonstrates the coincidence of the $|^{35}S|$-labelled material and catalytic activity of HGPRTase, thus confirming that the major component of radioactive protein detected in the autoradiogram of the denaturing SDS polyacrylamide gel contains the HGPRTase catalytic activity. The native molecular weight of HGPRTase determined by Sephadex G-100 column chromatography was estimated to 77,000.

We further characterized and compared the HGPRTase molecules eluted in various fractions of peak A and peak B by isoelectric focusing on polyacrylamide slab gel under nondenaturing conditions. The catalytic activity of these fractions was then radiochemically assayed and the product detected by autoradiography according to Chasin, 1976. As seen in Figure 3, the isoelectric points of the

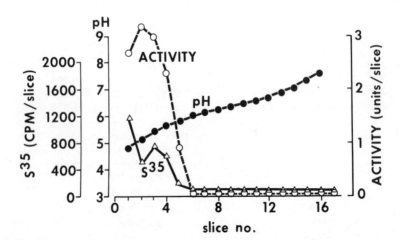

Figure 2. Native isoelectric focusing of HGPRTase on a Sephadex
G-75 slab containing ampholines in a final concentration of 2.1%
(0.9% pH 3.5-10, 0.7% pH 5-7, and 0.5% pH 6-8).

The track of the focused material was sliced in segments of 0.8 cm
and equivalent slices were tested for $|^{35}S|$ radioactivity, HGPRTase
activity and pH.

(Δ) $|^{35}S|$ counts; (o) activity units/slice; (•) pH.

isoenzymes eluted in peak A increase as the pH of the gradient increases. Peak B contains only the acidic isoenzymes. A mixture of
fractions A and B reveals the existence of at least seven distinct
isoenzymes with a pattern similar to that observed in the crude extract of HTC cells. The isoelectric points of the major isoenzymes
are between pH 5.2 and 6.2.

In order to characterize further the isoenzymes of peaks A and
B, equal catalytic activities of heat treated fractions A and B
were mixed with a small amount of crude extract of $|^{35}S|$-labelled
HTC cells, and the mixture was subjected to a high resolution two-
dimensional separation system according to O'Farrell, 1975. This
system employs as the first dimension isoelectric focusing in 9.5 M
urea and 14 mM mercaptoethanol in a disc gel. This disc gel is then

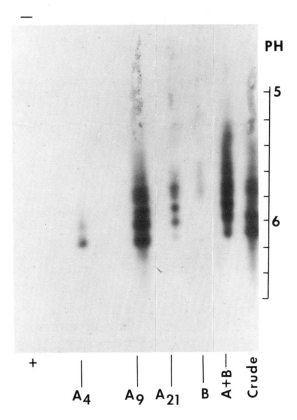

Figure 3. Autoradiogram of a native isoelectric focusing of HGPRT-ase on 5% acrylamide gel obtained by a modified method of Chasin, 1976.

The gel contained 2.4 % total ampholines (pH 3.5-10 1%; pH 4-6 0.6%; pH 5-8 0.8%). The gel after it was focused was covered with 2 ml assay mixture containing 20 mg BSA, 0.1 M Tricine pH 7.4, 0.01 M $MgCl_2$, 0.01 M 5-phosphoribosyl-1-pyrophosphate and 10 μC $|^{14}C|$-hypoxanthine (50 mC/mM) and was left to react at 37°C for 45 minutes. The gel was covered with a cellulose PEI-F sheet and remained at 25°C for 30 minutes to absorb the $|^{14}C|$-IMP produced. The sheet was then washed, dried, and autoradiographed. "A_4", "A_9", and "A_{21}" are the fractions 4, 9, and 21 of the DEAE cellulose peak A of Figure 1 I. "B" is fraction B of that same Figure. "A + B" is the mixture of fractions A and B and "crude" is a crude extract of HTC cells.

equilibrated with sodium dodecyl sulfate and mercaptoethanol and
layered on top of a slab of polyacrylamide containing 0.1% SDS for
a second dimensional electrophoresis which separates by molecular
weight. The autoradiographic patterns obtained from crude extracts
of $|^{35}S|$-labelled HTC cells run in this system are highly reproduc-
ible, as demonstrated in Figures 4B and 5A.

Figure 4A shows the two dimensional acrylamide gel electrophoresis of the mixture of heat treated fractions A and B of Figure 1 I.

The Rf's of the bands 1, 2, and 3 of the one dimensional gel of
Figure 2 II correlate well with the Rf's of the spots 1, 2, and 3
observed in Figure 4A.

On the basis of the Rf, the position of HGPRTase was identified
on the two dimensional gel as spot 2. The HGPRTase spot focuses in
the same position when heat treated fractions A and B are electrophoresed separately in the two dimensional system. The doublet
observed to the left of the HGPRTase spot is not observed in the
autoradiogram of a crude extract of HTC cells. Furthermore, its
removal by further purification doesn't affect the catalytic activity and therefore may represent an artifact generated during the
purification procedure.

The position of HGPRTase on the autoradiogram of a two dimen-
sional gel of an $|^{35}S|$-labelled crude extract of HTC cells was es-
tablished by superimposing an autoradiogram of the gel shown in
Figure 4B on its own protein stained gel (Figure 4A). Correct align-
ment was assured by the overlap of the protein spots 1, 2 and 3 of the
slab gel and the corresponding radioactivity spots of the autoradio-
gram. (Note that proteins 1, 2 and 3 are $|^{35}S|$-labelled.)

The spot assigned to HGPRTase is identifiable in the autoradiogram of two-dimensional gel electrophoresis of a crude extract of
$|^{35}S|$-labelled wild-type HTC cells as shown in Figure 5A. The identification of the HGPRTase was made by superimposing the autoradiograms of Figures 4B and 5A. Further confirmation of the identity
of the HGPRTase spot is provided by its absence in the autoradiogram
of a two-dimensional gel of a crude extract of an $|^{35}S|$-labelled
HGPRTase deficient clone (Tg 5) (Graf et al, 1974) of HTC cells.

This is demonstrated by comparing Figures 5B and C which represent enlarged appropriate portions of the autoradiograms of the
two-dimensional gels of the wild-type and mutant (Tg 5) $|^{35}S|$-labelle
crude extracts of HTC cells.

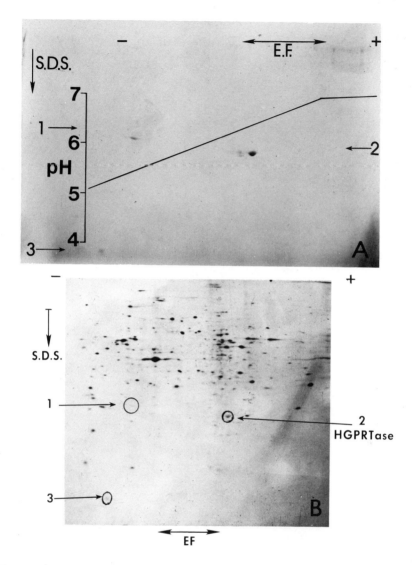

Figure 4A two-dimensional gel electrophoresis. The gel contains 8,000 cpm of an |^{35}S|-labelled crude extract of HTC cells mixed with equal catalytic activities of the heat treated fractions A and B of Figure 1. The observed spots were detected by protein stain and contain 55 cpm/μg protein. The position of HGPRTase is indicated on the picture. The pH gradient of the isoelectric focused disc gel is superimposed on the slab gel. The gel was run according to O'Farrell, 1975.

Figure 4B. Autoradiogram of the two-dimensional slab gel 4A; the film was exposed for 12 days. The position of the HGPRTase is indicated.

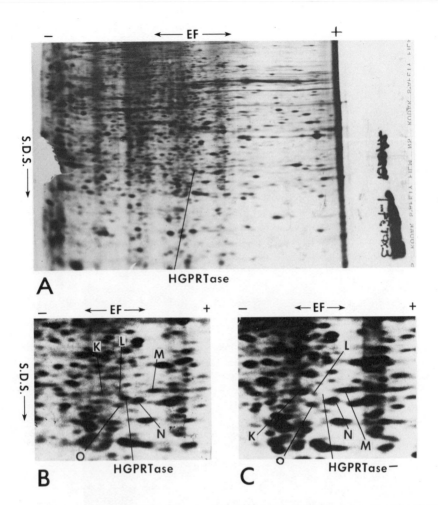

Figure 5A. Autoradiogram of two-dimensional gel electrophoresis of a crude extract of $|^{35}S|$ -labelled wild-type HTC cells. The sample contained 100,000 cpm of $|^{35}S|$, and the film was exposed for 12 days. The position of HGPRTase was identified as explained in the text and is indicated by the arrow.

Figure 5B. Enlarged portion of the autoradiogram of Figure 5A in the vicinity of HGPRTase. The proteins surrounding HGPRTase are identified by letters K, L, M, N and O.

Figure 5C. Enlarged portion of the autoradiogram in the vicinity of HGPRTase of a two-dimensional gel of a crude extract of $|^{35}S|$-labelled HGPRTase deficient HTC cells (Tg 5). The sample contained 200,000 cpm of $|^{35}S|$ and the film was exposed for 12 days. Note that HGPRTase spot is missing. The proteins surrounding HGPRTase are identified as in Figure 5B.

The HGPRTase spot is present in Figure 5A (wild-type) but absent in Figure 5B (Tg 5 mutant). It is worth noting that all the isoenzymes observed in the native isoelectric focusing Figure 3 focus as a single component in a two-dimensional gel electrophoresis system. A single major isoenzyme with a Pi of 5.9 is also observed in a native gel when the enzyme is preincubated for 30 minutes in 4.7 M urea, 7 mM mercaptoethanol before applying to the native gel. Having identified the HGPRTase subunit in the two-dimensional electrophoresis system we are able to screen a variety of HGPRTase mutants for altered Pi's and deficiencies of the HGPRTase molecules.

References

1. Chasin, L.A. and Urlaub, G. (1976) Somatic Cell Genetics 2, 453.
2. Graf, L.H., McRoberts, J.A., Harrison, T. and Martin, D.W. (1976) J. Cell. Physiol. 80, 331.
3. Hughes, S.H., Wahl, M. and Capecchi, M.R. (1975) J. Biol. Chem. 250, 120.
4. O'Farrell, P.H. (1975) J. Biol. Chem. 250, 4007.
5. Olsen, A.S. and Milman, G. (1974) J. Biol. Chem. 249, 4030.

EVIDENCE FOR THE EXISTENCE OF DIFFERENT TYPES OF METABOLIC COOPERATION

C.H.M.M.de Bruyn[*], T.L.Oei[*], M.P.Uitendaal[*] and P.Hösli[o]

[*]Dept.Hum.Genetics, University of Nijmegen, The Netherland

[o]Dept.Molecular Biology, Institut Pasteur,Paris,France

INTRODUCTION

Intercellular communication is supposed to be involved in a number of biological processes. For example, control of embryonic differentiation seems to require some form of close cellular interaction (1); the same has been suggested with respect to cell division and immune response (2,3).

One particular form of cell communication which has received relatively much attention during the past few years is metabolic cooperation. As a result of this phenomenon the mutant phenotype of certain enzyme deficient cells is corrected by contact with normal cells (4,5). Such a correction has been observed with hypoxanthine-guanine phosphoribosyl transferase deficient (HG-PRT⁻) cells in culture. Normal HG-PRT fibroblasts incorporate radioactive hypoxanthine or guanine into intracellular nucleotides demonstrable at the cellular level by autoradiography. Skin fibroblasts from patients with severe HG-PRT deficiency show a marked reduction in incorporation of these nucleotides under similar conditions. However, HG-PRT⁻ fibroblasts grown in close contact with normal fibroblasts become labeled (6,7,8).

In most studies published thus far, intercellular exchange has been studied with cultured skin fibroblasts. Normal and HG-PRT cells were cocultured in a ^{3}H-hypoxanthine containing medium and after incubation the cells were monitored for radioactivity by means of autoradiography.

In the present study metabolic cooperation has been analysed in two different ways: one method made use of normal donor cells (lymphocytes or erythrocytes) which were preincubated in ^{3}H-hypoxanthine containing medium (9). These cells were brought into contact with HG-PRT$^-$ recipient cells (fibroblasts or lymphocytes) and eventual transfer of label was studied by autoradiography. The second way was to use a recently developed HG-PRT activity measurement (10,11) of single fibroblasts isolated from a 1:1 mixture of normal and HG-PRT$^-$ cells which had previously been growing in close physical contact. In this way eventual appearance of HG-PRT activity in the HG-PRT$^-$ fibroblasts could be monitored.

MATERIALS AND METHODS

Autoradiography. These experiments were carried out with HG-PRT$^-$ fibroblasts and HG-PRT$^-$ lymphocytes as recipient cells. Fibroblasts were cultured in HAM F-10 medium containing antibiotics (penicillin 100 U/ml; streptomycin 100 γ /ml) and 15% fetal calf serum. Lymphocytes were isolated from fresh heparinised blood according to Roos and Loos (12). As donor cells normal lymphocytes and normal erythrocytes were used. Erythrocytes, retained after removal of plasma and buffy coat from heparinised blood, were washed three times with isotonic phosphate buffer (pH 7.4) before being used.

The normal donor cells were preincubated in a medium which contained per ml: 0.3 ml isotonic Na, K-phosphate buffer (pH 7.4), 0.45 ml of 0.9% NaCl, 0.05 ml of 2.35% $MgCl_2$ and 0.2 ml of 1% glucose; 8-^{3}H-Hypoxanthine (Radiochemical Centre, Amersham; spec.act.500 mCi/mmol) was included in a concentration of 0.02 mM. Lymphocytes were preincubated for 180 min. and erythrocytes for 60 min. at 37°C. Subsequently the cells were washed out untill the washing fluid contained no more radioactivity (normally after 2-3 washings). The preincubated and washed lymphocytes or erythrocytes were mixed with untreated HG-PRT$^-$ lymphocytes from a patient with the Lesch-Nyhan syndrome and spun down to assure cellular contact. In addition, preincubated control erythrocytes were allowed to sediment on top of HG-PRT$^-$ fibroblasts grown on coverslips. After incubation for 16 to 18 hrs, erythrocytes were removed by osmotic shock and the remaining cells (fibroblasts or lymphocytes) were submitted to autoradiography (13).

In a number of experiments the incorporation of label added to the medium as ^{3}H-IMP (2 μM) was studied in the presence of crude particulate fractions from normal and HG-PRT deficient erythrocytes and fibroblasts. These fractions were obtained after disruption of the cells by repeated freezing-thawing and centrifugation (1,000 g; 20 min) of the lysate. After several washings the pellets were suspended (10 γ /ml) in culture medium, containing ^{3}H-IMP, and allowed to sediment on top of HG-PRT deficient fibroblasts growing in mono-

layer. Following incubation for 16-18 hrs at 37°C autoradiography
was performed as described above.

   Measurement of HG-PRT in single cells. The method to assay activities at the single cell level has originally been developed for
fluorogenic substrates (14) and adapted recently for radiochemical
assays (10,11). Briefly, a 1:1 mixture of normal and HG-PRT⁻ fibroblasts is grown in confluency during 3 days in a Plastic Film Dish
(PFD; 14). After trypsinisation, the cell suspension is diluted and
replated; 8 hours later the cell culture is lyophilised and some 100
individual cells are cut out, free-hand under a stereomicroscope, from
the bottom of the PFD. Enzyme activities are measured in Parafilm
Micro Cuvettes (PMC's), small disposable incubation vessels, containing 0.3 μl of incubation medium and a plastic film leaflet carrying
the single fibroblast to be tested. Substrate and product are separated by paper chromatography and enzyme activity per cell is calculated after quantification of radioactivity in a liquid scintillation counter.

                              RESULTS

   Table I shows the autoradiographic experiments with the experimental mixtures. Intercellular exchange of label has taken place
in the combinations lymphocyte ⟶ lymphocyte; erythrocyte ⟶
lymphocyte; and erythrocyte ⟶ fibroblast. Most label in the recipient (HG-PRT⁻) cells was noticed in the nuclear regions.

   High voltage electrophoresis and paper chromatography has shown
that the radioactivity in the donor erythrocytes immediately after
preincubation was predominantly present as ^{3}H-IMP (> 95%). Therefore, the material transferred to the HG-PRT⁻ lymphocytes or fibroblasts might be IMP or a derivative which could be incorporated by
the recipient cells. ^{3}H-IMP added to the medium did not produce labeling of the HG-PRT⁻ cells (table I). However, addition to the
medium of crude particulate fractions from fibroblasts or erythrocytes (both HG-PRT⁺ or HG-PRT⁻), along with ^{3}H-IMP, induced incorporation in HG-PRT⁻ fibroblasts (figure 1). Heavy labeling was seen
in both nucleus and cytoplasm.

   The frequency distribution of HG-PRT activities in individual
control fibroblasts and fibroblasts from a patient with the Lesch-
Nyhan syndrome did not show overlap (fig.2). The activities of
cells from a 1:1 mixture of normal and mutant fibroblasts which
had been growing in close contact showed a completely different
distribution, suggesting a communicating cell system (fig.2). The
appearance of HG-PRT activity in HG-PRT⁻ cells is indicative of
the transfer of the enzyme itself, or an informational molecule,
or a regulatory molecule.

TABLE I

Metabolic cooperation in experimental cell mixtures.

a. intact donor and recipient cells

preincubated donor cells:	recipient cells:	appearance of label in recipient cells:
normal lymphocytes	HG-PRT⁻ lymphocytes	yes
normal erythrocytes	HG-PRT⁻ fibroblasts	yes
normal erythrocytes	HG-PRT⁻ lymphocytes	yes

b. only intact recipient cells

added to the medium:	recipient cells:	
^{3}H-IMP (20 μM)	HG-PRT⁻ fibroblasts	no
^{3}H-IMP (20 μM)	HG-PRT⁻ lymphocytes	no
^{3}H-IMP + crude particulate fraction (normal erythrocytes)	HG-PRT⁻ fibroblasts	yes
^{3}H-IMP + crude particulate fraction (normal fibroblasts)	HG-PRT⁻ fibroblasts	yes
^{3}H-IMP + crude particulate fraction (HG-PRT⁻ erythrocytes)	HG-PRT⁻ fibroblasts	yes

DISCUSSION

The evidence in favour of the transfer of enzyme product from normal to HG-PRT⁻ cells as a basis for metabolic cooperation is well documented (7,8,9,15), whereas far less evidence in favour of the possible transfer of other compounds have been published.

From present and previous (9) data it is evident that metabolic cooperation can occur not only between cultured normal and HG-PRT deficient fibroblasts, but also between normal and HG-PRT deficient lymphocytes and between normal red blood cells and HG-PRT deficient lymphocytes or fibroblasts. A basis for this phenotypic correction of mutant cells might indeed be the transfer of IMP or a derivative.

Ashkenazi and Gartler (16) added lysate from normal fibroblasts to a culture of HG-PRT⁻ fibroblasts and this caused incorporation of label originally added to the medium as ^{3}H-hypoxanthine. Auto-

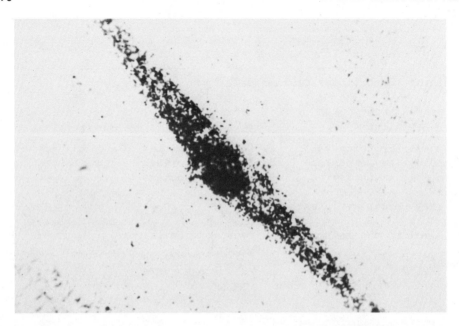

Figure 1.
TCA precipitable label in a HG-PRT deficient fibroblast after incubation with ^{3}H-IMP and a crude particulate fraction from normal
cultured fibroblasts.

radiographic examination revealed most label to be cytoplasmic, in
contrast to the nuclear labeling found by other investigators who
used intact donor cells. It has been inferred that metabolic cooperation is efficient when radioactive labeling is primarily located in the nucleus and inefficient if the label is cytoplasmic
(15). When incubating HG-PRT$^-$ fibroblasts with crude particulate
fractions and ^{3}H-IMP, we observed heavy labeling of both nucleus
and cytoplasm. Therefore, on the basis of the findings obtained
with autoradiography, there might be considerable heterogeneity
with respect to the mechanisms and the compounds involved in intercellular communication.

    The appearance of HG-PRT activity in HG-PRT$^-$ fibroblasts after
coculturing with normal fibroblasts, as judged from single cell
enzyme measurements, is the first direct evidence of exchange of
material leading to enzyme activity in mutant cells. Eventual
transfer of (unlabeled) enzyme products in these experiments
will not interfere with the enzyme assay, since the HG-PRT activity is determined with radioactive substrate after the cells

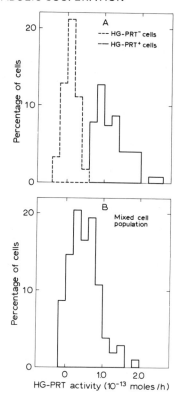

Figure 2.
Frequency distribution of HG-PRT activities in individual fibroblasts.
A: enzyme activities from non-interacting control and HG-PRT deficient
    cultures. About 100 cells of each culture were assayed.
B: enzyme activities from a mixed culture in which normal and HG-PRT
    deficient cells have been cocultured in a 1:1 ratio. About 100
    individual cells were assayed.

have been in contact (17), and isolated from the population. A possible
source of error could be a contamination of HG-PRT cells with membrane
associated HG-PRT from normal cells which might be released when the
confluent is trypsinised. However, the quantitative significance of
such a contamination as a result of a routine cell culturing procedure seems questionable.

    The presence of HG-PRT activity in part of the HG-PRT⁻ fibroblasts
was established eight hours after replating of the cells which had
been grown to confluency previously. Apparently this type of exchange
between intact cells differs from the mechanism observed when studying
the transfer of a low molecular weight metabolite: in that case HG-PRT⁻

cells promptly reverted to the mutant phenotype when separated
from normal donor cells (8,18).

It has been suggested that the ability of cells to communicate
might be correlated with the presence of specific structures in the
cell membrane. These structures, called low resistance- or gap
junctions, have been demonstrated in cells which were ionically
and metabolically coupled (19). On the other hand it has been reported that communication may also take place among cells that
apparently lack gap junctions: after prolonged interaction "low
efficiency cooperation" was observed (15). It could be possible
that exchange took place via some undetected gap junctions, but an
interesting alternative would be that it might represent a more
universal form of cell communication, i.e. the exchange of molecules located on or in the membranes of the apposed cells. In this
context it might be relevant to look for nucleotide metabolising
enzyme systems associated with the cellular membrane. An indication
in this direction is the formation of labeled extracellular IMP
or GMP when intact erythrocytes are incubated in a medium containing radioactive hypoxanthine or guanine and phosphoribosylpyrophosphate, the co-substrate of the HG-PRT reaction (20,21).

The concept has evolved that intercellular communication con-
stitutes an important mechanism for control of metabolic activity
and regulation of growth, differantiation and embryonic development. It must be recognised, however, that without being able
to discriminate between the various mechanisms of cell communication and the various molecules transferred, evidence for its
role in vivo is largely inductive.

ACKNOWLEDGEMENTS

This study was supported by a grant from FUNGO, Foundation
for Medical Scientific Research in the Netherlands. The skillfull
assistance of Miss E.Vogt, Mrs.C.Raymakers-Volaart and Mr.F.Oer-
lemans is gratefully acknowledged.

REFERENCES

1. Ebert J.D. (1965).
   Interacting systems in Development.
   Holt, Rinehart and Winston, pp.49-56.

2. Tadaro G.J., G.K.Lazar and H.Green (1965).
   J.Cell and Comp. Physiol. 66, 325-333.

3. Kreth H.W. and A.R. Williamsen (1971)
   Nature (Lond.) 234, 454-456.

4. Subak-Sharpe J.H., R.R. Bürk and J.D.Pitts (1966).
   Heredity 21, 342-343.

5. Subak-Sharpe J.H., R.R. Bürk and J.D. Pitts (1969).
   J.Cell Sci. 4, 353-367.

6. Friedman T., J.E.Seegmiller and J.H.Subak-Sharpe (1968).
   Nature (Lond.) 220, 272-274.

7. Cox R.P., M.R.Krauss, E.M.Balis and J.Dancis (1970).
   Proc.Nat.Acad.Sci. U.S.A. 67, 1573-1579.

8. Cox R.P., M.R.Krauss, E.M.Balis and J.Dancis (1972).
   Exp.Cell Res. 74, 251-268.

9. Oei T.L. and de Bruyn C.H.M.M. (1974).
   In: Purine Metabolism in Man (O.Sperling, A.de Vries and J.B.
   Wijngaarden,Eds.) Plenum Press, New York, pp.237-243.

10. Hösli P., de Bruyn C.H.M.M. and Oei T.L.(1974).
    In: Purine Metabolism in Man (O.Sperling, A.de Vries and J.B.
    Wijngaarden, Eds.). Plenum Press, New York, pp.811-815.

11. de Bruyn C.H.M.M., T.L.Oei and P.Hösli (1976).
    Biochem. Biophys. Res. Commun. 68, 483-488.

12. Roos D. and J.A.Loos (1970).
    Biochim. Biophys. Acta 222, 565-582.

13. de Bruyn C.H.M.M. and T.L.Oei (1973).
    Exp. Cell Res. 79, 450-452.

14. Hösli P. (1972).
    Tissue Cultivation Plastic Films.
    Tecnomara A.G., Zürich.

15. Cox R.P., M.R.Krauss, M.E.Balis and J.Dancis (1974).
    J.Cellul. Physiol. 84,237-252.

16. Ashkenazi Y.E. and S.M.Gartler (1971).
    Exp.Cell Res. 64, 9-16.

17. Uitendaal M.P., T.L.Oei, C.H.M.M. de Bruyn and P.Hösli (1976).
    Biochem.Biophys.Res.Commun., in press.

18. Pitts J.D.(1971).
    In: Growth Control in Cell Culture (G.E.W. Wolstenholme and
    J.Knight, Eds.) pp.89-96.

19. Gilula N.B., O.B.Reeves and A.Steinbach (1972).
    Nature (Lond.) 235, 262-265.

20. de Bruyn C.H.M.M. and T.L.Oei (1974).
    In: Purine Metabolism in Man (O.Sperling, A.de Vries and J.B.
    Wijngaarden,Eds.) Plenum Press, New York, pp. 223-227.

21. de Bruyn C.H.M.M. and T.L.Oei (1976).
    these Proceedings.

STABILITY OF THE AZAGUANINE RESISTANT PHENOTYPE IN VIVO

Gabrielle H. Reem and Charlotte Friend
Department of Pharmacology, New York University Medical
Center, New York, N.Y. 10016 and the Center for
Experimental Cell Biology, Mount Sinai School of Medi-
cine of the City University of New York, N.Y.C. 10029

Deficiency of hypoxanthine guanine phosphoribosyltransferase
(HGPRT EC 2.4.2.8) is one of the known causes of drug resistance
to 6-mercaptopurine and azaguanine in human leukemias and in
experimental tumors in animals. Tumor cells can develop resistance against these agents in the course of treatment; this may
reflect either a permanent mutational event or a change in gene
expression. Our experiments with virus-induced erythroleukemic
cells have shown that gene expression can be altered in vitro (1).
Erythroleukemic cells grown in tissue culture could be stimulated
by dimethylsulfoxide to differentiate along the erythroid pathway,
to synthesize hemoglobin while the activities of enzymes essential
to the biosynthesis of purines were repressed in the course of
differentiation in vitro.

Bakay and his associates suggested that a genetic factor from
human cells determined the expression of the rat structural gene
for HGPRT. These investigators observed reexpression of rodent
HGPRT after hybridization of HGPRT⁻ rodent cells with human cells.
Even HGPRT⁻ human cells, derived from a permanent cell line of a
patient with the Lesch Nyhan syndrome, when fused with rodent cells,
activated the rodent HGPRT gene. These experiments suggested the
existence of an as yet undefined factor which determined gene
activation (2,3).

Environmental factors can also cause cells genetically deficient in HGPRT activity to acquire the ability to synthesize
purine ribonucleotides from hypoxanthine (4). One possible
explanation for the increase in enzyme activity observed in HGPRT⁻
cells when grown in contact with HGPRT⁺ cells is the transfer of
enzyme activity by metabolic cooperation.

181

The present study was designed to test whether HGPRT deficient erythroleukemic cells formed tumors when injected into the
appropriate host and whether these tumors retained their original
phenotype. Since it seemed possible to change the HGPRT⁻ phenotype in vitro, the possibility whether a change in phenotype in
vivo could be effected was explored.

A thioguanine and azaguanine resistant cell line, selected by
Dr. Paul R. Harrison by exposure to thioguanine, was injected subcutaneously into a group of DBA mice. The azaguanine resistant
erythroleukemic cells were deficient in HGPRT and had elevated
concentrations of phosphoribosylpyrophosphate (P-Rib-PP) (Table 1).
These biochemical characteristics were used as cell specific
markers for the evaluation of the phenotype of tumors arising from
the cells. A second group of mice was injected with erythroleukemic cells of an HGPRT⁺ (5-86) cell line (Table 1).

Four weeks after the mice had been inoculated, tumors weighing between 1-2 gm were excised and their biochemical characteristics determined. Cells derived from one tumor of each group were
isolated and grown in tissue culture.

Enzyme activities were determined in freshly prepared, cell
free extracts. Tumors were homogenized in a Virtis homogenizer,
the homogenates centrifuged for 45 minutes at 15,000 x g and enzyme
activities and P-Rib-PP content determined as described (1,5).
The results of these determinations are recorded in Table 2.

Tumors derived from the HGPRT⁻ erythroleukemic cell line
(M-707) were found to have low HGPRT activity. The ratio between
HGPRT and adenine phosphoribosyltransferase (APRT EC 2.4.2.7)
activity was between 0.52 and 0.83, while the ratio in tumors
derived from the HGPRT⁺ cell line (5-86) was between 1.30 to 1.82.

TABLE 1. BIOCHEMICAL CHARACTERISTICS OF ORIGINAL CELL LINES

	ERYTHROLEUKEMIC CELL LINES	
	5-86	M-707
Azaguanine sensitivity	+	−
HGPRT activity	+	−
P-Rib-PP content	Control	Elevated

TABLE 2.   ENZYMATIC ACTIVITIES OF TUMORS DERIVED FROM AZAGUANINE
           SENSITIVE AND AZAGUANINE RESISTANT ERYTHROLEUKEMIC
           CELL LINES

ORIGIN OF	ENZYME ACTIVITIES		
TUMORS	HGPRT	APRT	RATIO
	nmoles/hr/mg		H/A
5-86 (HGPRT$^+$)	1165	840	1.39
	1060	815	1.30
	1000	550	1.82
M-707 (HGPRT$^-$)	496	817	.61
	450	771	.58
	303	486	.62
	330	400	.83
	275	307	.90
	264	507	.52

TABLE 3.   PHENOTYPE OF MURINE VIRUS-INDUCED ERYTHROLEUKEMIC CELL
           LINES BEFORE AND AFTER PASSAGE THROUGH DBA MICE

CELL LINE	ENZYME ACTIVITIES			P-RIB-PP
	HGPRT	APRT	RATIO	CONTENT
	nmoles/hr/mg		H/A	pmoles/10^6 cells
Original cultures				
5-86	341	397	1.0	28
M-707	62	160	.39	163
Cultures established from tumors				
5-86 III	157	186	.81	37
M-707 C	100	244	.34	142

        Cultures established from the two groups of tumors had the
biochemical characteristics of the cell lines of origin (Table 3).
HGPRT activities were low and P-rib-PP content elevated in the
original cell line (M-707) and in that derived from the tumor
(M-707C).

        Drug resistance was tested in the original cell lines and com-
pared with that of the two cell lines derived from the tumors
(Table 4).   When 1 γ/ml azaserine was added to the cultures growing in RPMI 1640 medium supplemented with 15% fetal calf serum,

TABLE 4. DRUG RESISTANCE BEFORE AND AFTER PASSAGE THROUGH MOUSE

CELL LINES	ADDITIONS	SURVIVAL
		%
Original culture	Azaguanine	
	γ/ml	
5-86	none	100
M-707	none	100
5-86	1.0	17
M-707	1.0	95
5-86	10.0	15
M-707	10.0	60
Cultures derived from tumors		
5-86 III	none	100
M-707 C	none	100
5-86 III	1.0	22
M-707 C	1.0	89
5-86 III	10.0	17
M-707 C	10.0	31
5-86 III	HAT	100
M-707 C	HAT	<1

95% of the cells of the original HGPRT$^-$ culture (M-707) survived and 89% of the cells of the culture derived from the tumor M-707(C) while only 17 to 22% of cells in the corresponding HGPRT$^+$ cultures survived. At 10 γ/ml azaguanine the rate of survival of the HGPRT$^-$ cultures was higher than that of the HGPRT$^+$ control cultures. On the other hand, when hypoxanthine, aminopterin, and thymidine were added to the medium (HAT medium), the cultures derived from the tumors of mice inoculated with HGPRT$^+$ cells survived, while those from the HGPRT$^-$ inocula did not.

These experiments show that inoculation of DBA mice with azaguanine resistant virus-induced erythroleukemic cells resulted in tumor growth. The tumors retained the phenotype of the cells of origin, and cultures from the tumors derived after passage through the mouse also retained the cell specific markers of the original cell line and remained drug resistant. The initial exposure of a relatively small number of cells to the host environment failed to change the gene expression of the growing tumors.

REFERENCES

1. Reem, G.H. and Friend, C., 1975 Proc. Nat. Acad. Sci. USA
    72, 1630-1634.

2. Bakay, B., Nyhan, W.L., Croce, C.M. and Koprowski, H., 1975
    J. Cell Sci. 17, 567-578.

3. Croce, C.M., Bakay, B., Nyhan, W.L. and Koprowsky, M., 1973,
    Proc. Nat. Acad. Sci. USA 70, 2590-2594.

4. Subak-Sharpe, H., Burk, R.R. and Pitts, J.D., 1969, J. Cell
    Sci. 4, 353-367.

5. Reem, G. and Friend, C., 1976, J. Cell Sci. 88, 193-196

FUNCTIONING OF PURINE SALVAGE PATHWAYS

IN ESCHERICHIA COLI K-12

Per Nygaard

University Institute of Biological Chemistry B

Sølvgade 83, Copenhagen, Denmark

INTRODUCTION

The increasing amount of data on purine and purine nucleoside metabolism in animals, microorganisms and plants suggest an important role of the purine salvage pathways in cellular metabolism: (i) in the reutilization of breakdown products of nucleic acids, (ii) in the interconversion of nucleotides and (iii) in the utilization of purine analogues in chemotherapy. It is known that not all tissues (cells) are dependent to the same extent on purine salvage pathways, and that they differ greatly with respect to enzymes involved and how the different reactions are regulated.

In the enteric bacterium Escherichia coli, the purine salvage pathways seem to be of importance for the cell: (i) the enzyme systems involved in the utilization of exogenous purines and purine nucleosides are efficient and highly regulated. (ii) certain of the purine salvage enzymes are involved in the interconversion of nucleotides, and (iii) mutants lacking certain purine salvage enzymes excrete purines and purine nucleosides into the growth medium. Such mutants show no apparant abnormality in their growth behavior, except when they contain an additional mutation inactivating the de novo pathway (1).

TABLE I

Bacterial strains. All strains are derived from E. coli
K-12 (λ),F⁻,metB,rel,strA = SØ OO3.
Nutritional growth requirements; all strains require
methionine 50γ/ml. (a) guanine or xanthine, (b) hypoxanthine, adenine or guanine, all 15 γ/ml. In addition
the medium contained thiamine 1γ/ml.

Strain	Relevant genotype	Enzyme not functioning
SØ OO3	wild type	
SØ O16	cya	Adenyl cyclase
SØ O93	crp	Cyclic AMP receptor protein
SØ 198	guaA a)	XMP-aminase
SØ 199	purE b)	Phosphoribosyl-aminoimidazole carboxylase*
SØ 312	purE,pup	Carboxylase* and purine nucleoside phosphorylase
SØ 505	apt,pup	Adenine phosphoribosyltransferase and purine nucleoside phosphorylase

PURINE NUCLEOTIDE SYNTHESIS

In E. coli purine nucleotides are synthesized via
the de novo pathway from simple precursors (5'-phospho-
ribosyl-l-pyrophosphate,glutamine,glycine,formate,CO_2
and aspartate) or from preformed purines or purine nucleosides,when these precursors are available. The endproducts of the de novo pathway, AMP and GMP, are formed
from IMP through reactions which may also operate when
nucleotides are synthesized via the salvage pathways,
Figure I. The control of the de novo synthesis of IMP is
complex. A major control is excerted by purine nucleotides which feed-back inhibits PRPP-amidotransferase and
PRPP-synthetase (2,3). A more specific regulation of the
synthesis of AMP and GMP from IMP occurs on the enzymes
located at the branch-point. This regulation involves
both control of enzyme synthesis and enzyme activity
(2,4,5). Although the regulatory enzymes have been subjected to numerous studies, no clear picture of the overall regulation of the de novo synthesis of AMP and GMP
has emerged.

When purine bases and nucleosides are made available

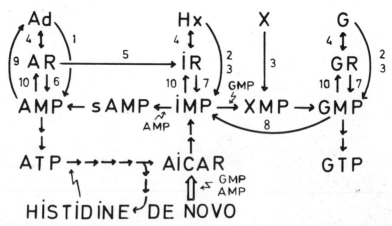

Figure I.   Salvage pathways for purine bases and purine
nucleosides in E. coli. Xanthosine is not included since
it cannot be metabolized, the purine deoxyribonucleo-
sides can only be metabolized via purine nucleoside phosphorylase and adenosine deaminase. Purines are not catabolized. (1) adenine phosphoribosyltransferase, (2) hypoxanthine phosphoribosyltransferase, (3) guanine phosphoribosyltransferase, (4) purine nucleoside phosphorylase, (5) adenosine deaminase, (6) adenosine kinase,
(7) guanosine kinase, (8) GMP reductase, (9) AMP hydrolase, (10) nucleotidases. ⇶ indicates feed-back inhibition.

to E. coli cells they are readily taken up and are utilized for nucleotide synthesis (6). Nucleosides are split
by purine nucleoside phosphorylase to yield the free purine base, while excess purine base is excreted into the
medium and the pentose moiety is catabolized. However,
most of the adenosine is deaminated to inosine which then
appears in the medium. Transport systems for nucleosides,
distinct from purine nucleoside phosphorylase, have been
characterized (7,8). So far the transport and the utilization of exogenous purines have only been related to
purine phosphoribosyltransferases. It appears that nucleotide synthesis from purines and purine nucleosides is
regulated by feed-back inhibition excerted by nucleoside
triphosphates. While the uptake of purine bases appears

TABLE II

Effects of exogenous purines on the de novo synthesis of
ATP and GTP in E. coli.
Cells (SØ OO3) were grown in minimal medium with glycerol as carbon source, in the presence of 0.33 mM U-14-C-
glycine (2.5 μC/μmole) for 4 generations. Nucleotide
pools were analyzed as described before (9).

Addition 100 μM	Incorporation of ^{14}C-glycine in per cent of incorporation in control cultures	
	ATP	GTP
No (control)	100	100
Hypoxanthine	19	15
Adenine	17	103
Guanine	77	50
Xanthine	66	19

to be controlled by their utilization, nucleosides are
constantly taken up and excess purine is excreted as bases.

The effects of exogenous purines on the de novo
synthesis of purine nucleotides in E. coli was investigated. The contribution of de novo pathway to the overall ATP and GTP synthesis was determined by measuring
the amount of incorporation of radioactivity from 14-C-
labelled glycine into ATP and GTP as affected by the presence of unlabelled purines in the growth medium. From
Table II it can be seen that in all cases both de novo
and the salvage pathways are functioning simultaneously.
The addition of hypoxanthine leads to the strongest inhibition of the de novo synthesis of both ATP and GTP,
while adenine, guanine and xanthine mainly inhibited the
de novo synthesis of their respective mononucleotide.

In order to determine whether purine bases and nucleosides are normally occurring metabolites in E. coli,
a mutant strain, SØ 505 (Table I), was constructed. This
mutant which lacks adenine phosphoribosyltransferase

and purine nucleoside phosphorylase activities is unable
to reutilize any endogenously formed adenine (Figure I).

Wild type (SØ 003) and mutant cells (SØ 505) were
incubated for several generations with 14-C-labelled
glycine (as described in Table II) and the medium was
analyzed for free radioactive purines and purine nucleosides by chromatographic methods (9). At a cell density
of 2 x 10^8 cells/ml less than 0.4 nmoles/ml purine compounds was excreted by wild type cells. However, the mutant strain did excrete appreciable amounts of adenine
(24 nmoles/ml) and of inosine (2 nmoles/ml). The excreted
quantities correspond roughly to the amount of purine
present in the cell as nucleic acids and nucleotides. It
is tentative to speculate that the purine compounds excreted into the medium are the products of mRNA turnover,
which in the wild type cells are reutilized.

PURINE SALVAGE ENZYMES

Studies on the nucleoside catabolizing enzymes in
E. coli have shown that the expression of the genes coding for these enzymes are regulated in an unusual and
complicated way (lo). So far two classes of regulatory
mutants have been identified. The regulatory genes mu-
tated in these mutants are the cytR and the deoR genes.
Both cytR and deoR code for repressor proteins (11).
According to their induction pattern, the nucleoside catabolizing enzymes can be divided into four different
regulatory units.

The first unit, under deoR control, comprises the
deo-enzymes (thymidine phosphorylase, tpp; purine nucleoside phosphorylase, pup; deoxyriboaldolase, dra; and
phosphodeoxyribomutase, drm). These enzymes catalyse the
following reactions (I), and they are all induced by de-
oxyribose-5-P.

(I)   pup
             tpp   drm   dra   acetaldehyde
     deoxy- ────→   deoxy- ────→   deoxy- ────→   + glyceral-
 nucleoside ↓ ribose-1-P   ribose-5-P   dehyde-3-P
             base

The second unit comprises two of the deo-enzymes,
purine nucleoside phosphorylase and phosphodeoxyribomutase. They catalyse reactions (II), and they are induced by inosine and guanosine.

(II)

$$\text{purine-ribonucleoside} \xrightarrow[\text{base}]{\text{pup}} \text{ribose-1-P} \xrightarrow{\text{drm}} \text{ribose-5-P} \longrightarrow \text{Pentose metabolism}$$

The third unit comprises the deo-enzymes (c.f. reaction I) and the cyt-enzymes, cytidine deaminase,cdd and uridine phosphorylase,udp. The cyt-enzymes catalyse the reactions, III:

(III)

$$\text{cytidine} \xrightarrow[NH_3]{\text{cdd}} \text{uridine} \xrightarrow[\text{uracil}]{\text{udp}} \text{ribose-1-P}$$

In addition cytidine deaminase catalyses the deamination of deoxycytidine to deoxyuridine which then may be catabolized via reaction (I). All the six enzymes are induced by cytidine and adenosine and are under the control of the cytR gene.

The forth unit consists of adenosine deaminase which catalyse the deamination of adenosine and deoxyadenosine to inosine and deoxyinosine respectively. The deamination is not required for the catabolism of adenosine and deoxyadenosine, since both compounds are substrates for purine nucleoside phosphorylase (12). The synthesis of adenosine deaminase is induced by adenine and hypoxanthine.

The conversion of guanine nucleotides into adenine nucleotides is mediated through GMP-reductase, Figure I, the enzyme is induced when E. coli is grown in the presence of exogenous guanine compounds (4). Most likely GMP is the endogenous inducer (13).

Very little is known about the regulation of the synthesis of the purine phosphoribosyltransferases and purine nucleoside kinases. If a purine auxotroph is starved for purine the synthesis of guanosine kinase is derepressed (1). While there is evidence that the purine phosphoribosyltransferases may utilize purine supplied either from the outside or endogenously formed, the kinases, most likely, act only on endogenously formed nucleosides, since exogenously supplied nucleosides are readily catabolized.

Table III summarizes our present knowledge of the regulation of the synthesis of the enzymes involved in the purine salvage pathways. Some of the genes coding

TABLE III

Regulation of the synthesis of purine salvage enzymes in E.Coli

Enzyme	Inducing compound	Fold of induction
Purine nucleoside-phosphorylase	deoxyribose-5-P inosine,guanosine cytidine,adenosine	3-6 2-4 2-4
Adenosine deaminase	adenine,hypoxanthine	10-30
GMP-reductase	GMP	10
Purine phosphoribosyl-transferases	not known	
Purine nucleoside-kinases	purine starvation	3

for purine salvage enzymes have been located on the E. Coli chromosome (1).

It has been shown that 3:5 cyclicAMP stimulates the synthesis of a variety of inducible enzymes in E.Coli and that cyclicAMP in conjunction with its binding protein (CR-protein) stimulates the initiation of the transcription of the mRNA for the inducible enzymes.

In order to determine the effect of cyclicAMP and the CR-protein on the synthesis of purine salvage enzymes,the levels of these enzymes were determined in mutants which were unable to synthesize either cyclicAMP or the CR-protein. The results given in Table IV show that none of the tested enzymes required an active cyclicAMP-CR-protein-complex for their synthesis, rather it was found that the level of one of the enzymes, adenosine deaminase was increased in these mutants. The induction of the synthesis of purine nucleoside phosphorylase and of adenosine deaminase does not require an active cyclicAMP-CR-protein-complex,(14). However, evidence has been presented that the induction of GMP-reductase by guanine compounds in mutants that lack adenyl cyclase requires the addition of cyclicAMP (13).

TABLE IV

Effects of cyclicAMP and the CR-protein on the synthesis of purine salvage enzymes in E.Coli
The cells were grown on glucose casamino acids and enzymatic activity, given as nmoles per min per mg protein, was determined as described before (1).

Enzyme	SØ003 wild	SØ016 cya	SØ093 crp
Adenosine deaminase	16	28	61
Purine nucleoside phosphorylase	155	206	182
Adenosine kinase	0.17	0.14	0.14
Guanosine kinase	0.7	0.7	0.7
Adenine phosphoribosyl transferase	95	63	64
Hypoxanthine phosphoribosyl transf.	102	95	106

INTERCONVERSION OF PURINE NUCLEOTIDES

The regulation of the synthesis of purine nucleotides must be directed toward a balanced supply of adenine and guanine nucleotides. In these processes IMP have a pivotal position, since IMP serve as a branchpoint for the formation of AMP and GMP, both when they are derived from preformed purine compounds and when they are synthesized de novo, Figure I .

AMP can be converted to GMP via two routes. The first involves the formation of ATP, which by a series of reactions in common with the histidine biosynthetic pathway is converted to aminoimidazole carboxamide ribotide (AICAR),(6). This pathway is feedback inhibited by histidine and is blocked if the cells are growing in the presence of histidine (6). The second pathway involves the formation of adenine which is converted to adenosine, followed by a deamination to inosine. Inosine is then either phosphorylated directly to IMP, or phosphorylized to hypoxanthine, which subsequently is converted into IMP, Figure I (15).

There seems to be only one route from GMP to AMP, namely via the reduction of GMP to IMP, which is catalysed by GMP-reductase, Figure I.

TABLE V

Purine nucleotide synthesis from exogenous purines in
E.Coli.
Cells were grown in glycerol minimal medium, 1 mM phosphate, in the presence of 32-P labelled phosphate ,30 μC
per μmole. Nucleotide were extracted and analyzed as described before, (9). Growth is given as generation time.

Strain	Addition (100μM)	Growth (min)	Nucleotide pool μmole/mg dryweight		Ratio ATP/GTP
			ATP	GTP	
S$\emptyset$ 003	No	76	8.5	3.8	2.2
-	Hypoxanthine	72	8.2	3.7	2.2
S$\emptyset$ 199	Hypoxanthine	67	5.9	2.5	2.4
- purE	Adenine	62	10.0	3.5	2.9
-	Guanine	140	4.4	4.2	1.1
-	Xanthine	245	2.6	2.9	0.9
S$\emptyset$ 312	Hypoxanthine	78	6.4	2.7	2.4
- purE	Adenine	180	14.1	1.5	9.4
- pup	Adenine + Histidine	>300	21.9	1.1	19.9
S$\emptyset$ 198 guaA	Guanine	88	6.5	3.5	1.9

The results shown in Table II, indicate indirectly
that nucleotide interconversion does occur in wild type
cells of E.Coli, when supplied with an exogenous purine
source. However, in a purine requiring mutant nucleotide
interconversion may be a prerequisite for growth.

The utilization of the four common purine bases for
adenine and guanine nucleotide synthesis in a purine requiring mutant, strain S$\emptyset$ 199, was investigated. The data
given in Table V, show that S$\emptyset$ 199 can utilize each of the
four purines, and that a correlation appears to exist between growth rate and the ATP/GTP ratio. Growth and nucleotide pool ratio in S$\emptyset$ 199 on adenine or hypoxanthine
as purine source were similar to that found for wild type
cells, S$\emptyset$ 003.

Next it was investigated to what extent the two pathways, mentioned above, for the conversion of adenine nu-

cleotides into guanine nucleotides are operating. A derivative of SØ 199, which lack purine nucleoside phosphorylase activity, SØ 312 was used. This strain can only synthesize guanine nucleotides from adenine via the pathway in common with the histidine biosynthetic pathway. The results given in Table V with SØ 312 show reduced growth rate and altered nucleotide pool ratio compared with SØ 199, likewise on adenine as purine source. If in addition histidine is added to SØ 312, and thereby the second pathway for the conversion of adenine nucleotides into guanine nucleotides is inhibited, growth almost stops.

If guanine or xanthine serve as the total purine source in SØ 199, growth is reduced compared with hypoxanthine and adenine, Table V. But, if guanine only is required for GMP synthesis, such as it is in SØ 198, which only can synthesize adenine nucleotides via the de novo pathway, growth and nucleotide pools tend to normalize, Table V. Most likely the synthesis of adenine nucleotides is limiting in the above experiments.

In summary, our studies indicate that purine salvage pathways are important for E.Coli and that these pathways, most likely are functioning in the cell not only when exposed to exogenous purine compounds.

(1) Jochimsen,B.Nygaard,P.Vestergaard,T.:Molec.Gen.Genet. 143(1975)85
(2) Gots,J.S.In Metabol.Regul.ed.Vogel,Acad.Press,Vol.V p225(1971)
(3) Baganara,A.S.Finch,L.R.:Eur.J.Biochem.41(1974)421
(4) Nijkamp,H.J.J.:J.Bacteriol.100(1969)585
(5) Eyzaguirre,J.Atkinson,D.E.:Arch.Bioch.Biophys.169 (1975)339
(6) Magasanik,B.Karibian,D.:J.Biol.Chem.235(1960)2672
(7) Peterson,R.N.Koch,A.L.:Biochim.Bioph.Acta 126(1966)129
(8) Mygind,B.Munch-Petersen,A.:Eur.J.Biochem.59(1975)365
(9) Jensen,K.F.Leer,J.C.Nygaard,P.:Eur.J.Biochem.40(1973) 345
(10) Munch-Petersen,A.Nygaard,P.Hammer-Jespersen,K.Fiil,N. Eur.J.Biochem.27(1972)208
(11) Hammer-Jespersen,K.Munch-Petersen,A.:Molec.Gen.Genet. 137(1975)327
(12) Jensen,K.F.Nygaard,P.:Eur.J.Biochem.51(1975)253
(13) Benson,C.E.Brehmeyer,B.A.Gots,J.S.:Biochem.Biophys. Res.Commun.43(1971)1089
(14) Olsen,F.Nygaard,P.:X[th]Int.Biochem.Congr.Hamburg(1976) Abstr.01-8-018
(15) Hoffmeyer,J.Neuhard,J.:J.Bacteriol.106(1971)14

METABOLISM OF INTRAVENOUS ADENINE IN THE PIG

*J.S. Cameron,*H.A. Simmonds,**A. Cadenhead,+D. Farebrother

* Department of Medicine, Guy's Hospital, London
** Rowett Research Institute, Bucksburn, Aberdeen
+ Wellcome Foundation, Beckenham, Kent

Although adenine is widely distributed throughout mammalian tissues; either in the form of the energy-rich adenine nucleotides, in combination as the nucleic acids, or the essential enzymes and co-factors such as NAD, FAD; free adenine is almost undetectable in body fluids and tissues under normal circumstances. Since there is no deamination system for adenine at the free base level in mammalian systems (1,6) it can be presumed that all metabolic transformations involving adenine must occur at nucleoside or nucleotide level. An extremely high Km for nucleoside phosphory-lase with adenine as substrate makes it extremely unlikely that direct utilisation or formation of adenine normally occurs via this route in mammalian systems (13) (although both the above trans-formations may be demonstrated in bacterial systems)(6). This being so, the main route for removal of adenine will be to AMP via PRPP and the enzyme APRTase, although endogenous production of free adenine appears unlikely by any of the above routes.

Our interest in adenine disposal was stimulated through two observations. First, in collaboration with Professor Van Acker in Belgium, we recently identified a family whose members suffer from complete deficiency of APRTase, reported elsewhere in this congress (12,16). In these affected siblings, the urine contained appreciable concentrations of adenine, and more importantly, its oxidation product 2,8-dihydroxyadenine formed via xanthine oxidase, for which adenine is a substrate (12). Second, when large amounts of adenine were administered to healthy individuals of a variety of mammalian species (1,3,4,7,8,14), an acute crystalline nephro-pathy resulted following precipitation in the urinary tract of the extremely insoluble 2,8-dihydroxyadenine.

EFFECTS OF ADENINE IN VARIOUS SPECIES

Route	Species	Dose mg/kg	2,8,DHA in urine	Crystals in kidney	Renal ↓ function
i.v.	Rabbit	100	yes	yes	yes
	Dog	80	n.d.	yes	n.d.
	Monkey	10-50	yes	no	yes
	Man	2-20	yes	no	?yes
		95	yes	yes	ARF
	Rat	40	n.d.	no	no
		60-220	n.d.	yes	ARF
Oral	Man	40	yes	n.d.	no
		80-100	yes	n.d.	ARF
	Dog	40	n.d.	yes	n.d.
	Rat	200-750	n.d.	yes	yes

n.d.= not done   ARF=acute renal failure
2,8,DHA= 2,8 dihydroxyadenine

Table I.

From the previous data (Table I) it appeared that the administration of 100 mg/kg (0.74 mmol/kg) of adenine intravenously led to conversion of adenine to 2,8–dihydroxyadenine in sufficient quantities to lead to crystallisation of this compound in the kidneys.

We had previously studied xanthine nephropathy in the pig (5) induced by feeding allopurinol and guanine and it appeared possible that the feeding of large doses of adenine might not only give another model of crystalline nephropathy, but could add to knowledge of adenine metabolism.

MATERIALS AND METHODS

Castrate littermate large white/landrace cross pigs weighing approximately 30 kg were used. Before the experiments, indwelling catheters were placed via the saphenous (femoral) artery into the abdominal aorta for injection and sampling. The animals were kept in metabolic cages throughout the study, but were unsedated and unrestrained. The diet was a skimmed milk/barley feed, low in purines. Spontaneously voided urines were collected, aliquoted and frozen; blood was sampled into heparin, spun immediately, the plasma separated and frozen. At slaughter, renal, gut and liver tissue were fixed in buffered formalin, absolute alcohol and snap-frozen in solid carbon dioxide/isopentane.

Creatinine and urea was measured in plasma and urine by standard AutoAnalyser techniuues (11) and creatinine clearances calculated as $U_{creat} \cdot V/P_{creat}$ expressed in ml/min.

Urinary and plasma purines were separated and quantitated by anion exchange chromatography followed by high voltage electro-phoresis on thin layer plates, as previously described (10,11).

Red cell nucleotides were analysed by high pressure liquid chromatography in eluates of TCA-precipitated red cells (Perrett, in press).

Alcohol fixed kidney tissue was sectioned, and examined unstained after paraffin embedding under polarised light; cryostat sections of untreated, snapfrozen tissue were also examined by polarised light. Formalin-fixed material and alcohol-fixed material was stained using haematoxylin-eosin, periodic acid-Schiff, and silver methenamine stains, and examined by conventional optical microscopy (11).

DESIGN OF EXPERIMENT

1.   Three pigs were given 100 mg/kg (0.74 mmol/kg) adenine in 500 ml saline over 30 mins intra-aortically. Samples were taken immediately before the injection, and 0.5, 1, 2, 4 and 24 hours following the injection. Urines were collected as voided. One pig was killed at 24 hours. In the second pig, a similar protocol was followed on two further consecutive days, so that by slaughter it had received 100 mg/kg/day of intravenous adenine for 3 consecutive days. In the third pig, oral allopurinol in a dose of 1.5 g/day (50 mg/kg/day) was begun the day after the first injection, and after a second day on allopurinol alone a further intravenous injection of adenine was given in the presence of oral allopurinol.

2.   Two pigs were given 100 mg/kg/day orally. The first pig was killed after 5 days on the regime, and only the kidneys were examined. In the second pig, the regime was continued for 7 days until slaughter, the urine being collected throughout with a daily blood sample.

RESULTS

The intravenous adenine was rapidly excreted, principally as allantoin, unchanged adenine, xanthine and hypoxanthine (Table II). Four infusions in three different animals all gave similar results. About one fifth of the injected adenine was recovered in unchanged form in the urine. Allowing for control excretion of allantoin (ca 650 mg/day: 4.5 mmol/day), about 42% of the adenine would appear to have been converted to allantoin, 5% to uric acid, together with 10% as xanthine and hypoxanthine, with a threefold increase in total purine excretion (Table II). Only one per cent was recovered as adenosine, and one half per cent or less as 2,8-dihydroxyadenine.

Following allopurinol administration, allantoin and uric acid excretion fell, while the excretion of xanthine and hypoxanthine increased in a ratio of 1.8:1 (Table III). During the adenine injection, however, the xanthine : hypoxanthine was reversed to 0.65:1. Before adenine injection, the allopurinol was excreted

EXCRETION OF PURINES FOLLOWING INTRA-AORTIC ADMINISTRATION OF 100mg/kg ADENINE IN THE PIG

(mean of 4 infusions)

	mg	(mmol)	%
Allantoin	1437	(9.4)	42.3
Adenine	674	(5.0)	22.1
Hypoxanthine	350	(2.6)	11.8
Xanthine	335	(2.2)	9.9
Uric acid	205	(1.1)	5.0
Adenosine	52	(0.2)	0.9
2,8, DHA	15	(0.09)	0.4
			92.4

(totals over 24 hours following infusion)

Table II.

EFFECT OF ORAL ALLOPURINOL UPON DISPOSAL OF INTRAVENOUS ADENINE IN THE PIG

	All	UA	Hx	X	Ad	Ads	DHA	Total
Diet alone	737 (4.66)	84 (0.5)	13 (0.1)	21 (0.14)	0	0	0	(5.4)
Diet + adenine	1375 (8.7)	307 (1.8)	303 (2.2)	389 (2.5)	797 (5.9)	36 (0.13)	7 (0.04)	(21.3)
Diet + Allopurinol	297 (1.9)	68 (0.4)	134 (1.0)	277 (1.8)	0	0	0	(5.1)
Diet + adenine + Allopurinol	240 (1.6)	117 (0.7)	852 (6.3)	619 (4.1)	334 (2.5)	0	0	(15.2)

(All figures in mg/24h, mmol/24h in brackets)   '0' = not detected

Table III.

almost equally as allopurinol riboside (46.9%), oxipurinol (45.8%) and 7.3% as unchanged drug (Table IV); during the adenine load, allopurinol riboside was undetectable in the urine.

In the pig treated with oral adenine, no adenine and only traces of adenosine and dihydroxyadenine were detectable in the urine (Table V). Allowing for control excretion of allantoin, probably 50% of the oral adenine was excreted as allantoin, and a further 10% as uric acid. Xanthine and hypoxanthine together accounted for 15%; three-quarters of the oral adenine thus being accounted for as urinary metabolites.

EFFECT OF PARENTERAL ADENINE ON THE FATE OF ORAL ALLOPURINOL IN THE PIG

	Allopurinol	Oxipurinol	Allopurinol Riboside	Total	% Dose
Diet alone	0	0	0		
Diet + Allopurinol (50 mg/kg/24h)	96 (0.7)	673 (4.4)	1216 (4.5)	(9.6)	87.0
Diet + Allopurinol (50 mg/kg/24h) + 100 mg/kg adenine i.v.	343 (2.5)	1081 (7.1)	Nil	(9.6)	87.0

(All figures in mg/24h, mmol/24h in brackets)   '0' = not detected

Table IV.

Creatinine Clearances

No alteration in plasma creatinine levels was noted and creatinine clearances were normal throughout all experimental regimes in all animals. No alteration in urine volume was noted in any experiments as has been found in previous studies of crystalline nephropathy in the pig (5).

Histology

There was no increase in kidney weight in treated animals or any macroscopic evidence of crystal deposition at post mortem (5, 11). Examination of either the snapfrozen, alcohol-fixed or formalin-fixed material by polarised light showed no evidence of crystal deposition; nor were there any microscopical changes which could be attributed to the different treatments in either the oral or intravenous adenine dosed pigs.

METABOLIC FATE OF ORAL ADENINE 100 mg/kg/day IN THE PIG

	All	UA	Hx	X	Ad	Ads	DHA
Diet alone	560 (3.2)	55 (0.3)	12 (0.1)	25 (0.2)	0	0	0
Diet + adenine Day 2	2115 (13.4)	286 (1.7)	149 (1.1)	229 (1.5)	0	T	T
Diet + adenine Day 7	2011 (13.2)	368 (2.2)	345 (2.5)	186 (1.2)	0	T	10 (0.06)

(All figures in mg/24h, mmol/24h in brackets)   '0' = not detected
                                                        T = trace

Table V.

DISCUSSION

It appears that parenteral or oral adenine is considerably less toxic in the pig than in other species, because of the low percentage of the injected adenine converted to 2,8-dihydroxy-adenine. In the other species studied, much larger percentages of administered adenine are converted to this compound at similar or smaller dosages. Although small numbers of crystals believed to be 2,8-dihydroxyadenine were observed in the urine during the first few hours after injection of adenine, no crystals were seen in the kidney under polarised light and renal function remained good.

Conversion of adenine to xanthine, hypoxanthine, uric acid and allantoin is very rapid in the pig, which is surprising since APRTase activity is low in this animal compared with human levels (11). The presence of traces of adenosine suggests that these compounds did arise via conversion of adenine to AMP. Obviously pig tissues must be investigated to exclude a direct route of oxidation.

The accumulation of hypoxanthine in excess of xanthine during adenine infusion in the pig treated with allopurinol is interesting

since it does not occur either in pigs fed guanine (11), or in
normal man (10) - in both xanthine excretion predominates over
hypoxanthine following allopurinol therapy. However, increase in
hypoxanthine excretion in excess of xanthine does occur with
allopurinol in the Lesch-Nyhan syndrome (15). It is difficult to
understand why HGPRTase should not be operative in the pig under
these circumstances. This finding may, however, be dose related;
alternatively it could indicate that significant reutilisation of
exogenously derived hypoxanthine does not normally occur.

The recovery of a high proportion of the oral adenine also in
the form of urinary metabolites (ca 75%) compared with approximately
50% for guanine at comparable or even higher dosages in a previous
experiment (11) is also of note. The percentage recovery of oral
adenine in urinary metabolites in other species would appear to be
dose related; at low doses a high proportion has been reported in
adenine nucleotides with little in the urine (1,9). With increasing
dosage the percentage recovered in the urine apparently increases
(1,9) and, in this study at least, at high doses this ratio has been
completely reversed. On this basis, the feeding of high doses of
adenine to Lesch-Nyhan subjects to promote feedback inhibition of
de novo synthesis by adenine nucleotides would appear unnecessary -
indeed it would tend to add to their urinary complication rather
than correct it.

The low rate of conversion of adenine to 2,8-dihydroxyadenine
in the pig by either route compared with other animal species
could be explained by several factors, (a) a rapid rate of
clearance of adenine coupled with (b) the presence of xanthine
oxidase solely in the liver and intestinal mucosa as in man (2);
the toxicity of adenine in the form of dihydroxyadenine to the
rabbit, dog and especially rat could be related to the wider tissue
distribution of xanthine oxidase in these species (2). The
latter species also excrete a much more concentrated urine containing
a higher level of purine end-product when expressed as mg/kg/ml
urine (2) - all of which would render precipitation of 2,8-dihydroxy-
adenine more likely. (c) A probable additional safety factor is
that the specific activity of human (and presumably pig) liver
xanthine oxidase is only 1/50th that in the rat.

The final interesting result of these studies concerns the
metabolism of allopurinol. On the control day, allopurinol
metabolism was similar to that which has previously been reported
in man and pig (10,11); the predominant metabolites at this dosage
being oxipurinol and allopurinol riboside. However, allopurinol
riboside was undetectable in the same pig on the following day
during adenine infusion. This may relate to the high hypoxanthine
levels during the combined adenine/allopurinol therapy mentioned
above. Lesch-Nyhan subjects on allopurinol also excrete no

allopurinol riboside (15) and this has been ascribed to their high
hypoxanthine levels; hypoxanthine being a much better substrate
for nucleoside phosphorylase would have a competitive advantage
over allopurinol in such circumstances (15).

SUMMARY

1.   Adenine administered either parenterally or orally is less
toxic to the pig than to other species; doses of 100 mg/kg are
rapidly catabolised and excreted largely as soluble purine end-
products in the urine.

2.   The low toxicity is explained by the excretion of less than 1%
of the dose as 2,8-dihydroxyadenine.

3.   These results suggest that adenine dosages which give rise to
kidney damage must be above a threshold-like level which varies
in the different mammalian species, and is higher in the pig than
in the rat, dog, rabbit or man.

REFERENCES

1.   Bendich, A., Bosworth Brown, G., Philips, F.S. and Thiersch,
     J.B. The direct oxidation of adenine *in vivo*. J. Biol.
     Chem., 183, 267-277 (1950).

2.   Cameron, J.S., Simmonds, H.A., Hatfield, P.J., Jones, A.S.
     and Cadenhead, A. The pig as an animal model for purine
     metabolic studies, in Purine Metabolism in Man (Ed. Sperling,
     P., de Vries, A. & Wyngaarden, J.B.), 41B, 691-   (Plenum
     Press, N.Y. 1974).

3.   Ceccarelli, M., Ciompi, M.L. and Pasero, G. Acute renal
     failure during adenine therapy in the Lesch-Nyhan syndrome, in
     Purine Metabolism in Man (Ed. Sperling, P., de Vries, A.
     & Wyngaarden, J.B.), 41B, 671-679 (Plenum Press, N.Y. 1974).

4.   Falk, J.S., Lindblad, G.T.O. and Westman, B.J.M. Histopatho-
     logical studies on kidneys from patients treated with large
     amounts of blood preserved with ACD-adenine. Transfusion,
     12, 376-381 (1972).

5.   Farebrother, D.A., Hatfield, P., Simmonds, H.A., Cameron, J.S.,
     Jones, A.S. and Cadenhead, A. Experimental crystal nephropathy
     (one year study in the pig). Clin. Nephrol., 4, 243-250 (1975).

6.   Hochdtadt-Ozer, J. The regulation of purine utilization in
     bacteria. J. Biol. Chem., 247, 2419-2426 (1972).

7. Minkowski, O. Untersuchen zur Physiologie und pathologie de Harnsaure bei Saugethieren. Arch. Exp. Pathol. u. Pharmacol., 41, 375-420 (1898).

8. Nicolaier, A. Ueber die Umwandlung des adenins im thierischen Organismus. Z. klin. Med., 45, 359-374 (1902).

9. Seegmiller, J.E., Klinenberg, J.R., Miller, J. and Watts, R.W.E. Suppression of glycine 15N incorporation into urinary uric acid by adenine 8^{14}C in normal and gouty subjects. J. Clin. Invest., 47, 1193-1203 (1968).

10. Simmonds, H.A. Urinary excretion of purines, pyrimidines and pyrazolopyrimidines in patients treated with allopurinol and oxipurinol. Clin. Chim. Acta, 23, 353-364 (1969).

11. Simmonds, H.A., Hatfield, P.J., Cameron, J.S., Jones, A.S. and Cadenhead, A. Metabolic studies of purine metabolism in the pig during the oral administration of guanine and allopurinol. Biochem. Pharmacol., 22, 2537-2551 (1973).

12. Simmonds, H.A., Van Acker, K.J., Cameron, J.S., McBurney, A. and Snedden, W. Purine excretion in complete adenine phosphoribosyltransferase deficiency: effect of diet and allopurinol therapy. (This Symposium).

13. Snyder, F.F. and Henderson, J.F. Alternative pathways of deoxyadenosine and adenosine metabolism. J. Biol. Chem., 248, 5899-5904 (1973).

14. Stern, I.J., Cosmos, F. and Garvin, P.J. The occurrence and binding of 2,8-dioxyadenine in plasma. Transfusion, 12, 382-388 (1972).

15. Sweetman, L. Urinary and CSF oxypurine levels and allopurinol metabolism in the Lesch-Nyhan syndrome. Fed. Proc., 27, 1055-1059 (1968).

16. Van Acker, K.J., Simmonds, H.A. and Cameron, J.S. Complete deficiency of adenine phosphoribosyltransferase (APRTase): report of a family. (This Symposium).

PATTERN OF PURINE-NUCLEOTIDE METABOLISM IN HEPATO-PANCREAS OF HELIX POMATIA (GASTROPODA)

M. M. Jeżewska and J. Barankiewicz

Institute of Biochemistry and Biophysics,

Polish Academy of Sciences

02-532 Warszawa, 36 Rakowiecka St., Poland

In 1970 Campbell and Bishop, describing the nitrogen metabolism in Moluscs, had written: "Terrestrial snails might thus serve as useful systems for investigating the specific derangements present in persons exhibiting the Lesch-Nyhan Syndrome" (1). For several years we have been concerned with the purine-nucleotide metabolism in land snail Helix pomatia (Pulmonata). This gastropod has been found to be purinotelic (11). During winter sleep, H. pomatia accumulates uric acid, which accounts for 75 % of the total purine content in the nephridium at the end of hibernation. During its active life, the nephridial excreta contain uric acid, xanthine and guanine, accounting for 34 - 49, 34 - 55 and 11 - 16 %, respectively of the total amount of purines excreted (12). An enzymic system allowing under normal, physiological conditions such diversity of the end-products of the nitrogen metabolism looked very promising. On the basis of investigations of the enzymes participating in the purine-nucleotide metabolism in the hepatopancreas, which is an organ homologous to the liver of Vertebrates, the scheme presented in Fig. 1. is suggested.

When the hepatopancreas has been injected in vivo with ^{14}C-bicarbonate, labelled AMP and GMP appeared in its ribonucleic acids (15); injected ^{14}C-glycine has been incorporated into uric acid, xanthine and guanine,

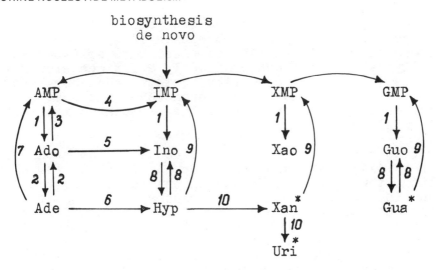

Fig. 1. Purine-nucleotide metabolism in H. pomatia
(Gastropoda) Enzymes found in hepatopancreas:
1. 5 -nucleotidase and non-specific phosphatase
2. Adenosine phosphorylase
3. Adenosine kinase
4. AMP aminohydrolase
5. Adenosine aminohydrolase
6. Adenine aminohydrolase
7. Adenine phosphoribosyltransferase
8. Inosine-guanosine phosphorylase
9. Hypoxanthine-guanine phosphoribosyltransferase
10. Dehydrogenase system hydroxylating hypoxanthine
    and xanthine

x - end-products of the protein nitrogen catabolism

with the proportion of the label amounting to 4.6, 3.3
and 1.2 % of the total dose, respectively (13). Therefore, the biosynthesis de novo as well as the catabolism of purine nucleotides occur in the hepatopancreas,
found later to be the main site of the latter process
in H. pomatia (2). The catabolic pathway of purine
nucleotides is initiated by their dephosphorylation to
the respective nucleosides by 5 -nucleotidase and a
non-specific phosphatase (2). In vitro the dephosphorylation of IMP and GMP is faster than that of AMP and
XMP, and it is inhibited by phosphate as well as by
pyrophosphate ions (2). The arising nucleosides, except

xanthosine, can undergo phosphorolysis to the respective
free purines and ribose-1-phosphate (4). In contrast to
the mammalian tissues (15), adenosine is splitted nearly
equally well as guanosine and only 2.5 times slower than
inosine (4). It is noteworthy that two phosphorylases
occur in the H. pomatia hapatopancreas: one specific
towards adenosine, and the another - towards inosine
and guanosine (6).

Free adenine which does not occur in the nephridial
excreta (11) must be salvaged or channeled to the catabolic pathway of IMP. Three aminohydrolases have been
found in the hepatopancrease, each specific towards a
different adenine compound (2). Among them AMP-aminohydrolase seems to play the most important role. Its
optimum pH is 8.5, approximating the physiological pH
8.0 of the hepatopancreatic tissue, and at the same
time its activity is rather high. The AMP deamination
is sensitive to inhibition by pyrophosphate and   resembles in this respect the dephosphorylation of nucleotides, but, in contrast, it is insensitive to phosphate (unpublished data). It can be mentioned that
AMP-aminohydrolase is postulated to take part in the
deamination of amino-acids by the purine-nucleotide
cycle in the mammalian liver (14). The role of the second aminohydrolase, acting on adenosine and 2´-deo-
xyadenosine, and inhibited by phosphate ions, is unclear because of its surprisingly low optimum pH 3.6 (10).
In turn, the activity of the third enzyme, adenine
aminohydrolase, with optimum pH 7.5, is low in comparison with the activity of AMP-aminohydrolase (2).
Therefore, it seems that adenine and adenosine from
exogenous sources as well as from the AMP catabolism
are rather returned to the pool of adenine nucleotides
by the enzymes of the salvage pathways present in the
hepatopancreas. Firstly, adenosine phosphorylase can
synthesize adenosine from adenine and ribose-1-phosphate,
and the initial rate of this reverse reaction is twice
that of the adenosine breakdown (6). The adenosine
synthesis is inhibited by ADP and ATP, but not by AMP
and pyrophosphate. Adenosine kinase present in the
hepatopancreas (unpublished data) may then transform
adenosine into AMP. The second possibility of the supplement of the adenine nucleotide pool at the expense
of free adenine is offered by the presence of very
active adenine phosphoribosyltransferase (5). The
synthesis of AMP by this enzyme is inhibited by AMP,
ADP and pyrophosphate.

Contrary to the adenine compounds, GMP, guanosine and guanine failed to be deaminated in the H. pomatia hepatopancreas (2). This explain the excretion of guanine as the end-product. Guanine can be converted into guanosine by inosine-guanosine phosphorylase (6). The initial rate of this reverse reaction is five times greater than that of the guanosine breakdown; however on account of the lack of guanosine kinase (unpublished data), the significance of the guanosine synthesis remains unclear. Guanine can be transformed directly into GMP by hypoxanthine-guanine phosphoribo-syltransferase present in hepatopancreas (5). The reaction is strongly inhibited by pyrophosphate and GMP, and less so - by GDP and GTP. Such reutilization of guanine seems to occur in vivo; namely, in the case of its absence, all snails would accumulate guanine during hibernation. In reality, however, in addition to guanine accumulating snails there are some which fail to do it (12); unfortunately, the underlying mechanism remains unclear. The lack of the enzymes deaminating the guanine compounds precludes the forma-tion of xanthine and indicates that the changes in the xanthine excretion cannot be related to changes in the intensity of the metabolism of guanine compounds.

It is probable that the cycles: AMP-adenosine-adenine-AMP and GMP-guanosine-guanine-GMP may operate in the H. pomatia hepatopancreas. The activities of the enzymes involved depend on the levels of phosphates, pyrophosphates and nucleotides and possibly these com-pounds could play some regulatory role in the courses of these cycles.

The conjectural operation of these cycles could provide some self-sufficiency and lack of dependence of the pools of the metabolically important adenine and guanine compounds on the main pathway of the pro-tein nitrogen catabolism: small precursors-IMP-inosine-hypoxanthine-xanthine-uric acid.

The intensity of this direct degradation of IMP must respond to the changes in the protein metabolism. During winter sleep the protein metabolism is five times less intense than during active life, when there are also lesser changes related to the alternate peri-ods of feeding and resting (7). Inosine phosphorolysis is activated by inosine itself (4), and this may pro-vide a quick response to the enhancement of the pro-tein metabolism. The resulting hypoxanthine is in turn

hydroxylated to xanthine and then to uric acid, both
these reactions depending on NAD (3). Both hypoxanthine
and xanthine can be converted into the respective nucleotides by hypoxanthine-guanine phosphoribosyltransferase
active also towards xanthine (5); however, the importance
of these reactions in the H. pomatia hepatopancreas, in
excess synthetizing IMP de novo, remains obscure. Experiments in vitro showed that if the formation of hypoxanthine in the reaction mixture proceeds very slow, then
only uric acid was formed and no xanthine accumulated.
In contrast, an increase in the rate of hypoxanthine
formation resulted in an accumulation of xanthine besides uric acid (2). This seemed to be in agreement
with the dependence of xanthine/uric acid ratio on
hypoxanthine concentration, found for the milk xanthine
exidase (8). However, in contrast to this enzyme, which
catalyzes the hydroxylation of hypoxanthine and xanthine
equally well, a dehydrogenase system in the H. pomatia
hepatopancreas hydroxylates these substrates at different
rates (10, unpublished data). During winter sleep, the
hydroxylating activity towards hypoxanthine is much
greater than towards xanthine, but the protein catabolism being reduced, it still provides nearly complete
transformation of slowly arising hypoxanthine to uric
acid. During the active life, the activity towards
xanthine increases, still being lower than that towards
hypoxanthine; it seems that this increase ensures the
formation of uric acid and xanthine at a rather constant ratio (1:1), while the protein metabolism becomes
several times more intense and the rate of hypoxanthine
formation increases considerably.

The presented peculiarities of the nucleotide metabolism in H. pomatia rule out this organism as a model
for the study of the Lesch-Nyhan mutation; nevertheless,
they are very interesting from the standpoint of evolu-
tion of enzymic systems in animals.

REFERENCES

1. Campbell, J. W., Bishop, S. H. (1970) in "Comparative
   Biochemistry of Nitrogen Metabolism", vol. I, pp. 103
   - 206, ed. J. W. Campbell, Academic Press, London and
   New York.
2. Barankiewicz J. (1973) Catabolism of IMP to uric acid
   in Helix pomatia (Gastropoda), Doctor's Dissertation
   (in Polish). Inst. Biochem. Biophys., Polish Academy
   of Sciences, pp. 1-75.

3. Barankiewicz, J. and Jeżewska, M. M. (1972) Bull.
    Acad. Pol. Sci. Ser. sci. biol., 20, 1 - 4.
4. Barankiewicz, J. and Jeżewska, M. M. (1973). Comp.
    Biochem. Physiol. 46B, 177 - 186.
5. Barankiewicz, J. and Jeżewska, M. M. (1975). Comp.
    Biochem. Physiol., 52B, 239 - 244.
6. Barankiewicz, J. and Jeżewska, M. M. (1976). Comp.
    Biochem. Physiol., 54B, 239 - 242.
7. Jeżewska, M. M. (1969). Acta Biochim. Polon. 16,
    313 - 320.
8. Jeżewska, M. M. (1973). Eur. J. Biochem. 36, 385 -
    390.
9. Jeżewska, M. M. and Barankiewicz, J. (1974).
    Abstracts 9th FEBS Meeting, s2e16, Budapest, Hungary.
10. Jeżewska, M. M. and Barankiewicz, J. (1976). Bull.
    Acad. Polon. Sci., Ser. sci. biol., in press.
11. Jeżewska, M. M., Gorzkowski, B. and Heller, J. (1963).
    Acta Biochim. Polon. 10, 55 - 65.
12. Jeżewska, M. M., Gorzkowski, B. and Heller, J. (1963).
    Acta Biochim. Polon. 10, 309 - 314.
13. Jeżewska, M. M., Gorzkowski, B. and Heller, J. (1964).
    Acta Biochim. Polon., 11, 135 - 138.
14. Lowenstein, J. M. (1972). Physiol. Rev., 52, 382-414.
15. Porembska, Z., Gorzkowski, B. and Jeżewska, M. M.
    (1964). Acta Biochim. Polon., 13, 107 - 111.
16. Zimmerman, T. P., Gersten, N. B., Ross, A. F. and
    Miech, R. P. (1971). Can. J. Biochem., 49, 1050-1054.

ISOZYMES OF AMP DEAMINASE

Nobuaki Ogasawara, Haruko Goto and Tomomasa Watanabe

Department of Biochemistry, Institute for Developmental

Research, Aichi Prefectural Colony, Aichi 480-03, Japan

AMP deaminase (AMP aminohydrolase EC, 3.5.4.6) catalyzes deamination of AMP to form IMP and ammonia. Although the physiological role of this enzyme remains obscure, it may be important to
stabilize the adenylate charge (1), in the conversion of adenine
nucleotide to inosine or guanine nucleotide (2-5) and furthermore
as a key enzyme in the purine nucleotide cycle (6-10). The deaminase exists in multiple molecular forms in different rat
tissues (11-15). On the basis of chromatographic, immunological,
and kinetic properties, three parental forms (types A, B and C)
have been detected in rat tissues. They could be resolved by
phosphocellulose column chromatography, and with respect to immunological properties, no crossreactivity occures between types A,
B and C. All these three parental isozymes have a sigmoidal velocity profile with respect to AMP in the absence of activators, and
are modified by alkali metals and nucleotides. Type A AMP deaminase is the only form found in skeletal muscle and is also the major
isozyme in diaphragma. Type B AMP deaminase is the major isozyme
of kidney, liver and testis. Type C AMP deaminase is the only form
found in heart. Other tissues such as brain, spleen, and lung,
where the multiple molecular forms exist, showed the B and C type
parental isozymes and presumably three B-C hybrids, while no A type
isozyme was observed in these tissues.

Previous studies demonstrated that there are five different
chromatographic forms of AMP deaminase in the adult rat brain,
these have been designated as isozymes I through V on the basis of
order of elution from the column (14). Isozymes I and V correspond in elution position to the only isozyme found in heart and
to the major component found in kidney or liver, respectively.
Kinetic, chromatographic, immunological, and other data showed

strongly that isozymes I and V are identical to the corresponding
isozymes in heart and kidney, respectively, and that the intermediate isozymes found in adult brain are tetrameric hybrids of
these two parental types (15).

In this paper, additional evidences are presented which
strongly support the existence of three parental isozymes. This
paper also reports isolation procedures and analysis of the subunit
structure of AMP deaminase from rat muscle (type A) and liver
(type B). We also analyzed a large number of rat tissues for multiple forms of AMP deaminase with the use of cellulose acetate
electrophoresis and by the techniques employing antisera prepared
against purified preparations of AMP deaminase from skeletal
muscle, kidney and heart.

MATERIALS AND METHODS

Materials

Adenosine monosulfate was purchased from Sigma Chemical Co.
AMP, ATP and beef liver glutamic dehydrogenase were obtained from
Boehringer Mannheim. Phosphocellulose was the product of Brown Co.
Sephadex G-200 and DE-52 were obtained from Pharmacia and Whatman,
respectively. Cellulose polyacetate membrane (Sepraphore III) was
purchased from Gelman Instrument Co. Other reagents were commer-
cial preparations of the highest purity available.

Preparation of Tissue Extracts

All tissues were obtained from male Wister rats decapitated
and exsanguinated. The tissues were removed and homogenized in a
Waring Blender in 5 volumes of Buffer A which contained 20 mM
potassium phosphate (pH 7.0), 50 mM NaCl and 0.1 % 2-mercapto-
ethanol. The homogemates were centrifuged at 20,000 g for 20 min
and the supernatant (30 ml each) was applied to the phosphocellulose column (0.9 x 7 cm) which was previously equilibrated with
Buffer A. After washing with 20 ml of the same buffer, the enzyme
was eluted in 3 ml of Buffer A contained 1.5 M NaCl. The concentrated enzyme solution was dialyzed against Buffer A and was used
for electrophoretic and immunological studies.

Enzyme Assay and Protein Determination

Enzyme activity was determined colorimetrically by estimating
production of ammonia. Standard assay mixture contained 30 mM AMP,

20 mM potassium phosphate (pH 7.0), 150 mM NaCl, 0.02 % 2-mercapto-
ethanol, and 0.05 % bovine serum albumin in a final volume of 0.25
ml. The amount of ammonia was determined by phenol-hypochlorite
reagents (16). To estimate apparent Km values for AMP in the absence and presence of activator, the reaction was carried out in
the assay mixture of 0.25 ml contained 20 mM Tris-HCl (pH 7.0),
0.05 % bovine serum albumin, various concentrations of AMP-Tris
and enzyme free from 2-mercaptoethanol. The amount of ammonia was
estimated by Nessler's reagent. The reaction was usually carried
out at 37° for 10 min. One unit of enzyme activity is defined as
the amount of enzyme that yields 1 μmole of ammonia per min under
the assay condition described above. Specific activity is defined
as units per mg protein. Protein concentrations were determine
by the method of Lowry et al (17) or by the fluorometric method of
Boehlen et al (18).

Preparation of Partially Purified Enzyme

AMP deaminases A, B and C were prepared from rat leg muscle,
kidney and heart respectively as described elsewhere (14). These
preparations gave a single activity peak when subjected to phosphocellulose column chromatography, indicating the presence of only
respective isozyme and no contamination of other isozyme. These
preparation were used to compare the immunological, kinetic,
electrophoretic and other properties (Table I).

Immunological Procedures

Antisera against three parental isozyme were prepared by
weekly injection into the rabbit foot pads with 1 ml of an emulsion
prepared equal volumes of enzyme and complete Freund's adjuvant,
followed by injecting the enzyme without adjuvant intraveously.
The antisera to AMP deaminase isozymes were the preparations used
in earlier studies (14). The studies on the effects of the antisera on the precipitation of the different AMP deaminase isozymes
were carried out in the following manner. The same amount of
isozyme, based on unit of activity (0.1 unit), was mixed with increasing volumes of antisera in a final volume of 0.4 ml. After
incubation for 18 hours at 4°, the samples were then centrifuged
for 15 min at 20,000 g and the supernatants were assayed for enzyme
activity. Non-immunized rabbit sera were used in place of the
antisera in control experiments.

Zone Electrophoresis

Cellulose acetate polyacetate strips (Sepraphor III) were
soaked for about 1 hour in electrophoresis buffer to which 1 mg

bovine serum albumin per ml had been added. Electrophoresis was
carried out for 2 hours at 250 V in buffer (pH 8.3) contained 0.05
M Tris, 0.384 M glycine and 0.1 % 2-mercaptoethanol in a cold
room, and the deaminase activity was detected by coupling with
glutamic dehydrogenase system. The region producing NADH oxidation was visually observed by the loss of fluorescence. After
electrophoresis the strip was put on a glassplate and upon it was
placed agar-reagent film (1 mm thick) containing 0.75 % Noble
agar, 100 mM NaCl, 4 mM MgCl$_2$, 2 mM ATP, 15 mM AMP, 5 mM α-Keto-
glutarate, 1 mM NADH, 20 mM cacodylate buffer (pH 7.0), and 5 units
of beef liver glutamic dehydrogenase. The agar was dissolved in
distilled water and then cooled to 50°. The remaining reagents at
37° were added to the agar solution, and the mixture was immediately placed on the cellulose strips. Development was allowed to
proceed at room temperature. AMP deaminase activity was seen as
dark spots on a blue fluorescent back ground when viewed under
ultraviolet light. After distinct spots had appeared, they were
recorded by contact printing on photographic enlarging paper according to the method developed for pyruvate kinase (19).

Other Methods

Sodium dodecyl sulfate polyacrylamide gel electrophoresis was
performed by the procedure of Weber and Osborn (20), using 7.5 %
gel.

The estimations of molecular weights by gel filtration on
Sephadex G-200 were similar to the method of Andrews (21). The
column (0.9 x 60 cm) was run at 4° and the folw rate of the eluant,
0.3 M potassium phosphate buffer (pH 7.0) contained 0.1 % 2-mer-
captoethanol, was 5 ml per hour. Phosphorylase a, catalase,
aldolase and bovine serum albumin were used as the molecular
weight standards.

RESULTS

Differences of Three Parental Isozymes

When the partially purified enzymes were individually subjected to chromatography on phosphocellulose and eluted with a
linear NaCl gradient from 0.05 M to 1.2 M, the heart, muscle and
kidney enzyme showed an activity peak at the NaCl concentration
of 0.25 M, 0.55 M and 0.75 M, respectively. Thus, these three
isozymes are chromatographically distinct.

Three isozymes have very different kinetic parameters both in
the absence and presence of activators. The apparent Km values

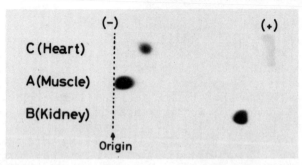

Fig. 1. Electrophoretic patterns of AMP deaminase activity of
three parental isozymes. Electrophoresis was performed as described under Materials and Methods. Activities were stained by
coupling with glutamic dehydrogenase system.

for AMP of 6, 24 and 37 mM were obtained for the heart, muscle and
kidney enzymes, respectively, in the absence of activator. In the
presence of 50 mM NaCl, all three isozymes were strongly activated
and Km values for AMP were lowered, but with apparent difference.

     The antisera against heart enzyme precipitated purified heart
AMP deaminase activity, but there was no effect on the deaminase
activity of muscle and kidney. The antisera to muscle enzyme removed the muscle enzyme, while no significant effect of the antisera on activity of the enzyme from heart and kidney. Similarly,
antisera against kidney enzyme precipitated the kidney enzyme
completely, but demonstrated no pronounced effect on the heart and
muscle enzyme. Thus, there was no crossreactivity between three
parental isozymes and they are immunologically distinct.

     An electrophoretic comparison of the three parental isozymes
was also carried out. In these experiments, AMP deaminase activity
was detected, by coupling with glutamic dehydrogenase systes, as a
dark spot on a blue fluorescent back ground of NADH. As shown in
Fig. 1, each isozyme exhibited a single activity spot, but their
mobilities were clearly different. Kidney enzyme was located at
the most anodic site and muscle enzyme at the most cathodic site.

     Further distinctions were observed on relative substrate spe-
cificity and potimal pH profile. The heart deaminase catalyzed
the deamination of adenosine monosulfate several times more efficiently than the enzyme from kidney. With heart enzyme the optimum pH was broad, while a sharp optimum at pH 6.0 was observed
with muscle and kidney enzyme. These differences along with chromatographic, kinetic, immunological, and electrophoretic differences between three isozymes are summarized in Table I.

TABLE I

DIFFERENCES OF THREE PARENTAL ISOZYMES

	Heart enzyme (C type)	Muscle enzyme (A type)	Kidney enzyme (B type)
Phosphocellulose column elution[1]	0.25 M NaCl	0.55 M NaCl	0.75 M NaCl
Km for AMP[2]			
- activator	6 mM	24 mM	37 mM
+ activator	1 mM	3 mM	6 mM
Antisera against[3]			
heart enzyme	+	-	-
muscle enzyme	-	+	-
kidney enzyme	-	-	+
Electrophoresis, distance to anode	1.5 cm	0.5 cm	6.5 cm
Adenosine mono sulfate/AMP[4]	0.27	0.16	0.04
Optimum pH	7, broad	6, sharp	6, sharp

1) Chromatography was carried out with a linear NaCl gradient from 0.05 M to 1.5 M in Brffer A.
2) Km values for AMP were determined as described under Materials and Methods in the absence and presence of 50 mM NaCl.
3) + and - indicate precipitation and no precipitation by antisera, respectively.
4) Activities were determined in the presence of 30 mM substrates.

Molecular Structures of Isozymes A and B

To obtain isozymes A and B to apparent homogeneity, frozen leg muscle and liver from adult male Wister rat were used, since the major isozyme in these tissues was isozyme A or B. 40 g of muscle were homogenized with Waring Blender for 2 min in 3 volumes of Buffer A contained 0.3 M NaCl and the supernatant was saved. The precipitate obtained was again homogenized and the mixture was centrifuged. The supernatants were combined and phosphocellulose (1.6 x 15 cm column volume) was added. Slurry was stirred for 30 min, and transferred to a Buchnner funnel, where it was washed with the extraction buffer. The cellulose phosphate was transferred to a column and the enzyme was eluted with a linear gradient between 0.3 M NaCl and 1.0 M NaCl in Buffer A. Fractions with peak enzyme activity were pooled and AMP deaminase was precipitated with ammonium sulfate between 30 % and 55 % saturation. The precipitate was dissoleved in 30 mM potassium phosphate (pH 7.0) containing 0.1

TABLE II

PURIFICATION OF MUSCLE ENZYME (TYPE A)

Fractions	Total volume	Total activity	Total protein	Specific activity
Supernatant	190	5,721	2,261	2.53
1st P-cellulose	124	2,164	9.92	218
DEAE cellulose	73	1,722	0.98	1,757
2nd P-cellulose	7.4	1,719	0.71	2,421

TABLE III

PURIFICATION OF LIVER ENZYME (TYPE B)

Fractions	Total volume	Total activity	Total protein	Specific activity
Supernatant	1,500	1,408	53,400	0.026
1st P-cellulose	83	745	324	2.30
AS-Dialysis	75	713	109	6.54
DEAE cellulose	27	461	4.46	103.4
2nd P-cellulose	7	223	0.21	1,061.9

% 2-mercaptoethanol and the solution was dialyzed against the same buffer overnight. The dialyzed solution was applied to DE-52 column (0.9 x 15 cm) and the enzyme was eluted with a gradient of 0.03 M to 0.2 M potassium phosphate (pH 7.0) containing 0.1 % 2-mercaptoethanol. The fractions with peak activity were pooled, applied to phosphocellulose column (0.9 x 5 cm) and the enzyme was eluted with 1.5 M NaCl in Buffer A. The results of representative purification are summarized in Table II.

For purification of liver enzyme, 300 g of frozen livers were homogenized in 5 volumes of Buffer A contained 0.3 M NaCl, the homogenate was centrifuged and the supernatant was saved. The supernatant was mixed for 1 hour with phosphocellulose, 5 x 30 cm column volume. The phosphocellulose was washed on a Buchnner funnel with extraction buffer, transferred to column and the enzyme was then eluted with Buffer A contained 1.5 M NaCl. Effluent fractions with enzyme activity were pooled and the enzyme was precipitated with ammonium sulfate between 30 % and 60 % saturation. The precipitate was dissolved in 30 mM potassium phosphate (pH 7.4) contained 0.1 % 2-mercaptoethanol and was dialyzed against the same buffer. The material was then applied to a DE-52 column (0.9 x 15 cm) and the enzyme was eluted with a linear gradient of 0.03 to 0.2 M potassium phosphate. Fractions with peak activity were pooled, applied to phosphocellulose column (0.9 x 10 cm) previously equilibrated with the extraction buffer, and the elution was carried out with a linear gradient of 0.3 to 1.0 M NaCl in Buffer

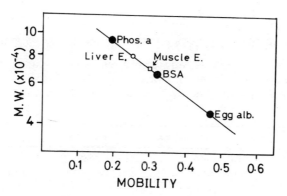

Fig. 2. Molecular weight determination of isozyme A and B subunits on 7.5 % polyacrylamide gel in sodium dodecyl sulfate according to Weber and Osborn (20). Protein standards used were egg albumin, bovine serum albumin and phosphorylase a. Isozymes A and B migrated with mobilities corresponding to molecular weights of 70,000 and 80,000, respectively.

A. Fractions with peak activity were pooled and concentrated by ultrafiltration. A representative purification is given in Table III.

Molecular weight determination of AMP deaminase A and B were carried out by gel filtration on Sephadex G-200. On the basis of the pattern obtained in gel filtration of the isozymes, it was likely that isozyme B was larger than isozyme A, by a molecular weight difference of about 40,000, with the molecular weight of isozyme A determined at 290,000. In disc gel electrophoresis in sodium dodecyl sulfate (Fig. 2), the isozyme A subunits migrated more rapidly than those from isozyme B, and the mobility difference corresponded to a difference in molecular weight of 10,000, with the molecular weight of the isozymes A and B subunits estimated at about 70,000 and 80,000, respectively. These results suggested that both isozymes A and B are tetramers composed of identical, or nearly identical, subunits.

Tissue Distribution of Isozymes

Characterization of AMP deaminase activity of various rat tissues was performed immunologically and also by electrophoretic analysis. The three antisera were tested for their ability to precipitate AMP deaminase activity from crude extracts of rat tissues and the results are summarized in Table IV. The antisera to purified heart enzyme precipitated most of the activity found in heart extracts and was also capable of precipitating some of the activities of brain, lung and spleen extract, however the effect of

TABLE IV

PRECIPITATION OF AMP DEAMINASE FROM CRUDE EXTRACTS

Crude extracts prepared and antisera precipitations carried out as described under Materials and Methods. Results indicate the maximum activity removed by the addition of serial dilution of antisera.

| Tissues | Maximum % removed by | | |
	Anti heart enzyme sera	Anti kidney enzyme sera	Anti muscle enzyme sera
Heart	90	2	0
Brain	44	50	0
Lung	30	62	0
Spleen	27	76	0
Kidney	8	98	0
Liver	7	91	0
Testis	1	92	0
Muscle	0	0	96
Diaphragm	0	0	92

the antisera was not evident in the extracts of other tissues. The anti kidney enzyme sera precipitated most of the activity of kidney, liver and testis extracts supporting a conclusion that the enzymes in these tissues are identical. The activities in brain, lung and spleen extracts were partially precipitated by the anti kidney enzyme sera and no activity was removed from heart, muscle and diaphragm. Thus, brain, lung and spleen gave precipitation data consistent with the presence of mixture of the two enzyme types in these tissues. The antisera to muscle enzyme precipitated most of the activity found in muscle and diaphragm extracts, but none of the activity in other tissue extracts.

The electrophoresis technique was also applied to analysis of AMP deaminase in crude extracts of rat tissues. Fig. 3. shows the activity pattern detected following electrophoresis of crude extracts. Extracts of heart, kidney, liver, testis, muscle and diaphragm showed a single activity spot. From electrophoretic mobility, it is reasonable to conclude that there are three main types of AMP deaminase; the type observed in heart extracts, the type in kidney, liver and testis, and the type in muscle and diaphragm. In extracts of brain, lung and spleen, no activity was detected that migrated to the anode more rapidly than the kidney enzyme or more slowly than the heart enzyme. With these three tissues, three spots of activity of intermediate mobility in addition to spots similar to heart and kidney enzyme were detected. As suggested by the antisera precipitation data, these pattern can be explained by hybrids of kidney and heart enzyme.

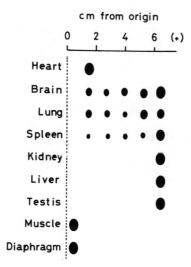

Fig. 3. Electrophoresis of AMP deaminase in extracts of various
rat tissues. Electrophoresis was carried out as described under
Materials and Methods.

DISCUSSION

     The data presented are consistent with the presence in rat
tissues of three parental isozymes of AMP deaminase. They were
different chromatographically, electrophoretically, kinetically,
immunologically and also different in other properties. These
three are AMP deaminase A (muscle type), B(kidney or liver type)
and C (heart type). Recently Coffee and Kofke (22) purified AMP
deaminase to apparent homogeneity from rat muscle and indicated
that native enzyme has a tetrameric structure consisting of four
polypeptide chains each having a molecular weight of 60,000.
However, from the data of gel electrophoresis in sodium dodecyl
sulfate, we obtained a molecular weight of 70,000 for subunits.
The reasons for this difference are not obvious, however the re-
peating experiments showed that the subunits of AMP deaminase in
muscle migrated apparently more solowly than bovine serum albumin
in electrophoresis. It is of interest to note that AMP deaminase
B has also a tetrameric structure consisting four identical subunits, however, having a larger molecular weight than those of
isozyme A. On the structure of isozyme C, the studies are in
progress, but the existence of five isozymes in brain, lung and
spleen, where both B and C subunits exist, strongly supports the
tetrameric structure of isozyme C. From the present immunological
and electrophoretic data, along with previous chromatographic data
(13), the distibution of isozyme in various rat tissues can be

summarized as following; heart C4; brain, lung and spleen, C4, C3B, C2B2, CB3 and B4; liver, kidney and testis, mainly B4; muscle and diaphragm, A4.

REFERENCES

1 Chapman, A. G. and Atkinson, D. E. (1973) J. Biol. Chem. 248, 8309-8312.
2 Cunningham, B. and Lowenstein, J. M. (1965) Biochim. Biophys. Acta 96, 535-537.
3 Setlow, B., Burger, R. and Lowenstein, J. M. (1966) J. Biol. Chem. 241, 1244-1245.
4 Setlow, B. and Lowenstein, J. M. (1967) J. Biol. Chem. 242, 607-615.
5 Askari, A. and Rao, S. N. (1968) Biochim. Biophys. Acta 151, 198-203.
6 Lowenstein, J. M. and Tornheim, K. (1971) Science 171, 397-400.
7 Tornheim, K. and Lowenstein, J. M. (1972) J. Biol. Chem. 247, 162-169.
8 Tornheim, K. and Lowenstein, J. M. (1973) J. Biol. Chem. 248, 2670-2677.
9 Tornheim, K. and Lowenstein, J. M. (1974) J. Biol. Chem. 249, 3241-3247.
10 Tornheim, K. and Lowenstein, J. M. (1975) J. Biol. Chem. 250, 6304-6314.
11 Ogasawara, N., Yoshino, M. and Kawamura, Y. (1972) Biochim. Biophys, Acta 258, 680-684.
12 Ogasawara, N., Goto, H., Watanabe, T., Kawamura, Y. and Yoshino, M. (1974) Biophys. Acta 364,353-364.
13 Ogasawara, N., Goto, H., Watanabe, T., Kawamura, Y. and Yoshino, M. (1974) FEBS Lett. 44, 63-66.
14 Ogasawara, N., Goto, H. and Watanabe, T. (1975) Biochim, Biophys. Acta 403, 530-537.
15 Ogasawara, N., Goto, H. and Watanabe, T. (1975) FEBS Lett. 58, 245-248.
16 Chaney, A. L. and Marbach, E. P. (1962) Clin. Chem. 8, 130-132.
17 Lowry, O. H., Rosebrough, N. J., Farr, A. L. and Randall, R. J. (1951) J. Biol. Chem. 193, 265-275.
18 Boehlen, p., Stein, S., Dairman, W. and Udenfriend, S. (1973) 155, 213-220.
19 Susor, W. A. and Rutter, W. J. (1971) Anal. Biochem. 43, 147-155.
20 Weber, K. and Osborn, M. (1969) J. Biol. Chem. 244, 4406-4412.
21 Andrews, P. (1964) Biochem. J. 91, 222-233.
22 Coffee, C. J. and Kofke, W. A. (1975) J. Biol. Chem. 250, 6653-6658.

Purification of Human Erythrocyte Adenosine Deaminase

Daddona, P.E. and Kelley, W.N.

Department of Internal Medicine and Biological
Chemistry. University of Michigan Medical School
Ann Arbor, Michigan, USA

Adenosine deaminase (ADA) catalyzes the irreversible
hydrolytic deamination of adenosine to produce inosine and
ammonia (1). Extremely reduced or absent ADA activity in man is
associated with one type of severe combined immunodeficiency
(SCID) (2). Previous studies of normal ADA from a variety of
species has provided evidence for substantial molecular
heterogeneity. Multiple forms of ADA varying in molecular
weight from 30,000 to 47,000 (designated small form) and 230,000
to 440,000 (designated large form) have been reported. (3-16).

Recent independent observations by Van der Weyden and
Kelley (17), Hirshhorn et al (18), and Chen et al (19) indicate
that an unusual molecular species of ADA is present in patients
with ADA deficient SCID disease. Knowledge of the structure of
human ADA would be important in relating the deficiency of ADA
to the immune dysfunction. For this reason we have undertaken
a purification of the normal small form of the enzyme from
human erythrocytes.

To date only partially purified preparations of human ADA
have been reported. Akedo et al (13) have purified the enzyme
to a specific activity of 47 µmol/min/mg from human stomach
tissue and Rossi et al (20) have achieved a specific activity
of 12 µmol/min/mg from human erythrocytes. In this communication,
we will describe the purification of human erythrocyte ADA to a
specific activity of 538 µmol/min/mg and apparent homogeneity.

Our initial approach to the purification of ADA from human
erythrocytes is shown in Table I.

TABLE I

Purification of Adenosine Deaminase

Step	Volume ml	Total Activity nmol/min x 10^5	Total Protein mg	Specific Activity nmole/min/mg	Purification	Recovery %
Crude Hemolysate Diluted 1:3	14,625	11.1	1.74×10^6	.636	1	100
1. pH Adjust 5.8	9,075	6.49	1.09×10^6	.592	.93	60
2. CM-Sephadex Batch Treatment	4,685	4.00	2.11×10^4	18.9	29.8	36
3. 40–60% $[NH_4]_2SO_4$	160	3.94	6.20×10^3	63.4	100	35
4. Heat Treatment 60° 15	15	3.94	3.22×10^3	122	192	35
5. CM-Sephadex Column, pH 5.8	15	3.48	824	422	664	31
6. DEAE-Sephadex Column, pH 6.0	0.875	1.11	85.7	1,300	2,045	10
7. Bio Gel A–0.5M 200–400 mesh	0.425	0.99	0.522	189,492	297,943	9

A

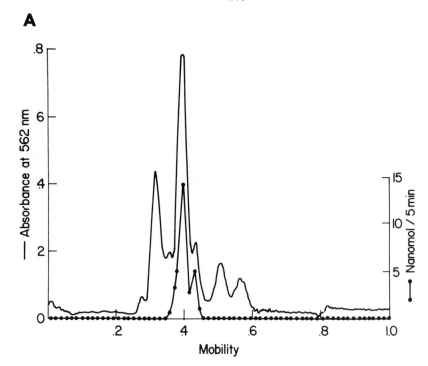

Fig 1A   Polyacrylamide gel electrophoresis of human ADA purified
298,00 fold. Polyacrylamide disc gel electrophoresis (5%) was
performed using a discontinuous Tris-glycine buffer, pH 8.3.
Gel was stained for protein with 0.3% Coomassie brilliant blue
and destained by heat facilitated diffusion. After destaining
the gel was scanned at 562nm. A companion polyacrylamide gel
was sliced into 1mm sections. Each slice was then placed in
10mM Tris-HCl buffer containing bovine serum albumin (1mg/ml),
pH 7.4 and eluted overnight. ADA activity in the supernatant
was then assessed by radiochemical assay. (21).

B

Fig 1B   SDS polyacrylamide gel electrophoresis of human ADA
purified 298,000 fold. The enzyme sample was dialyzed against
1000 volumes of 10 mM phosphate buffer, pH 7.0 for 24 hours and
then denatured in 1% SDS and 1% 2-mercaptoethanol at 100° for
5 min. SDS gel electrophoresis was then performed according to
Weber and Osborne (27).

We utilized a series of conventional purification steps including pH adjustment, negative batch adsorption, ammonium sulfate precipitation, heat treatment, CM-Sephadex, DEAE Sephadex and Bio Gel A - 0.5 M chromatography. The enzyme was purified approximately 298,000 fold achieving a specific activity of 189 μmol/min/mg at 37° with a 9% total recovery of enzyme activity. Polyacrylamide gel electrophoresis of this preparation revealed at least four protein bands two of which contained ADA activity by assay of a companion gel (Fig. 1A). Sodium dodecyl sulfate (SDS) polyacrylamide gel electrophoresis of the enzyme preparation showed two major bands of approximately equal intensity by Coomassie stain with apparent molecular weights of 56,000 and 42,000 (Fig. 1B).

Many attempts to improve this overall purification were unsucessful. We also tried preparative isoelectric focusing (23,15) and putative affinity columns to compliment this approach. Adenosine or inosine coupled to Sepharose 4 B via the ribose ring (2,3 position) with an iminobispropylamine spacer group (24,25) or adenosine coupled via the C-8 position with hexane spacer group (P-L Biochemicals, Inc) failed to bind ADA significantly over control columns. A series of homologous C-2 to C-10 n-alkane Sepharose columns (Miles-Yeda Ltd) as well as synthetically prepared phenyl and 4-phenylbutyl Sepharose columns (24,25) bound ADA activity although non-sepspecifically.

The failure of chemical affinity chromatography to provide a substantial improvement in the enzyme purification led us to develop an antibody affinity chromatography system. The small form of ADA from calf intestinal mucosa has been purified by Brady and O'Connell (7) to apparent homogeneity and a similar preparation can be obtained commerically. Since the physical and chemical properties of calf ADA appeared to us to be similar to human erythrocyte ADA, we questioned whether the calf enzyme might share common antigenic determinants with the human enzyme. Antisera produced against purified calf intestinal ADA was found to cross - react with human erythrocyte ADA (Fig. 2). Furthermore, the antisera could be highly purified by ammonium sulfate precipitation, DE-52 chromatography and specific immunoglobulin precipitation with an apparently homogeneous preparation of calf ADA. This resulted in a purified immunoglobulin fraction which was highly specific for

human erythrocyte ADA. An antibody affinity column was prepared
by coupling 27 mg of highly anti-calf ADA immunoglobulin to
2.5 ml of CNBr activiated Sepharose 4B (24).

As shown in Table II, crude hemolysate representing 30
units of outdated blood was purified approximately 100 fold
using steps similar to our initial method (Table I). Routinely,
all of the ADA activity present after Step 3 of the purification
procedure (Table II) could be bound to the 2.5 ml antibody
affinity column. The column was then washed successively with
50 mM phosphate buffer, pH 7.2 containing 0.5 and then 1.0 M
NaCl. The enzyme was eluted from the affinity column with a
5 - 8 M urea step gradient at pH 6.0. This urea gradient was
necessary to remove a small amount of weakly bound non-ADA
protein before the elution of enzyme activity. This technique
resulted in an overall 800,000 fold purification from crude
hemolysate to a specific activity of 538 μmol/min/mg at 37°
with a 35% total recovery of ADA activity (Table II).
The highly purified enzyme could be stored in a variety of
different buffers at 4° for over 1 month without significant
loss of enzyme activity.

The highly purified enzyme preparation was analyzed.
Polyacrylamide gel electrophoresis revealed 3 bands of protein
each of which were shown to contain ADA activity (Fig 3A).
SDS-polyacrylamide electrophoresis of this preparation revealed
one major band of protein by Coomassie stain, demonstrating the
human erythrocyte ADA to be a single polypeptide chain (Fig.
3B). Staining a companion SDS gel with periodic acid - Schiff
stain indicated the presence of carbohydrate on the protein.
The subunit molecular weight of ADA, estimated by SDS gel
analysis using the method of Weber and Osborn (27), was calculated
to be $41,700 \pm 700$. This estimated molecular weight apparently
was not influenced by the presence of carbohydrate on the
protein since variation of gel composition according to the
method of Segrest and Jackson (28) did not produce a lower
estimate of molecular weight. The Stokes radius was calculated
to be 24 A° by the method of Ackers (29) using gel filtration.
The sedimentation coefficient, determined by sucrose gradient
ultracentrifugation using the method of McCarty et al (30), was
estimated to be 3.8×10^{-13} sec.

TABLE II

Purification of Adenosine Deaminase

Step	Volume ml	Total Activity nmol/min x 10^5	Total Protein mg	Specific Activity nmol/min/mg	Purification	Recovery %
Crude Hemolysate Diluted 1:3	12,672	10.2	1.52×10^6	0.673	1	100
1. pH Adjust 5.8	13,526	10.2	1.16×10^6	0.881	1.3	100
2. CM-Sephadex Batch Treatment	12,626	8.78	3.53×10^4	24.9	37	86
3. 0-60% $[NH_4]_2SO_4$	362	8.22	1.21×10^4	67.5	100	80
4. Affinity Chromatography	0.5	3.63	.675	538,400	800,000	35

(From Ref 21)

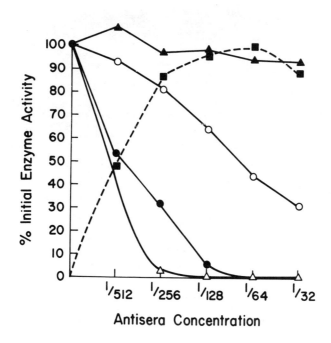

Fig 2 Precipitation of adenosine deaminase activity with anti-
calf adenosine deaminase antiserum. A constant amount of
adenosine deaminase was incubated with varying amounts of anti-
calf adenosine deaminase antiserum. This was normally followed
by a second incubation with goat anti-rabbit γ-globulin antiserum
(added at equivalence for rabbit γ-globulin). After centrifugation
the supernatant and precipitate were separated and the latter
washed once with 200 μl of PBS and resuspended in 50 μl of 50
mM Tris-HCl, pH 7.4. The percent of initial adenosine deaminase
activity remaining in the supernatant and that appearing in the
precipitate was determined. (Δ----Δ), calf adenosine deaminase
remaining in supernatant; (●----●), human erythrocyte adenosine
deaminase remaining in supernatant; (o----o), human erythrocyte
adenosine deaminase inhibition in the absence of goat anti-
rabbit antiserum; (■----■), human erythrocyte adenosine
deaminase activity appearing in the precipitate; (▲----▲),
adenine phosphoribosyltransferase activity in the presence of
goat anti-rabbit antiserum. (From Ref 21)

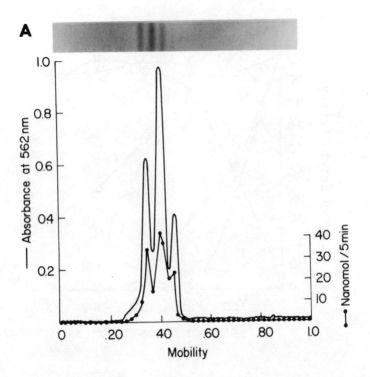

Fig. 3A. Polyacrylamide gel electrophoresis of human ADA purified 800,000 fold. Prepared as described in Fig. 1a. (From ref 21)

Fig. 3B. SDS polyacrylamide gel electrophoresis of human ADA purified 800,000 fold. Prepared as described in Fig. 1b. (From ref 21)

TABLE III

Amino Acid Composition of Human Erythrocyte
Adenosine Deaminase

Amino Acids	Residues/35,700
Lysine	20
Histidine	6
Arginine	10
Aspartic acid and asparagine	37
Threonine	17
Serine	21
Glutamic acid and glutamine	38
Proline	14
Glycine	30
Alanine	35
Valine	18
Methionine	8
Isoleucine	14
Leucine	28
Tyrosine	7
Phenylalanine	10
TOTAL	313

Detailed experimental methods previously reported (21)

Values for threonine and serine were increased by 5% and 10%
respectively to correct for destruction during hydrolysis.
Tryptophan and cystine were not determined by this method
nor was the yield of ammonia quantitated from acid amides.
(From Ref 21)

The amino acid composition of the highly purified human ADA
shown in Table III was consistent with a minimum molecular
weight of 35,700; this value was slightly low since tryptophan,
cystine and carbohydrate residues were not determined in the
analysis.   A partial specific volume of 0.729 cm^3 per g was
calculated from the amino acid composition according to Cohn and
Edsall (31).   From the Stokes radius, sedimentation coefficient
and partial specific volume, we calculated a molecular weight of
38,200 and a frictional ratio of 1.08 for the purified ADA.
remain to be established.

As shown by polyacrylamide gel electrophoresis (Fig 3A),
antibody affinity chromatography allowed the simultaneous
purification of all 3 electrophoretic forms of human erythrocyte
ADA.   These forms appeared to have a similar molecular weight
based on the observation that gel chromatography and sucrose

gradient ultracentrifugation of the mixture resulted in only one peak of enzyme activity. Furthermore, only a single protein band was observed under reducing conditions on SDS gel electrophoresis. The chemical and/or structural differences responsible for the electrophoretic heterogeneity of human ADA remain to be established.

The pH optimum, Km for adenosine, Ki for inosine, Stokes radius, sedimentation coefficient and apparent substrate specificity were all found to be essentially identical for both partially purified hemolysate and the highly purified enzyme (21). These data suggest that the denaturing effect of urea used in elution of the enzyme from the antibody did not alter or modify the highly purified ADA.

In conclusion the development of a highly specific antibody affinity column for the purification of human adenosine deaminase provides a technique for the purification of immunologically cross-reactive material from ADA deficient cells. In addition this technique will allow future study of the regulation of synthesis, degradation and post-translational modification of the human enzyme.

In summary we have shown the following:

1) Human erythrocyte ADA has been purified approximately 800,000 fold (sp. act. 538 μmol/min/mg at 37°) to apparent homogeneity using antibody affinity chromatography.

2) The enzyme was shown to be a single polypeptide chain with estimated molecular weight of approximately 38,000.

3) The 3 electrophoretic forms of erythrocyte ADA purified simultaneously by this technique were indistinquishable by SDS gel electrophoresis under reducing conditions.

4) Several properties of the highly purified ADA were identical to properties observed with an impure preparation of the enzyme.

REFERENCES

1. Conway, E.J., and Cooke, R. (1939) Biochem. J. 333, 479–492.

2. Giblett, E.R., Anderson, J.E., Cohen, F., Pollara, B., and Meuwissen, H.J. (1972) Lancet 2, 1067–1069.

3. Ma, P.F., and Fisher, J.R. (1968) Comp. Biochem. Physiol. 27, 105–112.

4. Ma, P.F., and Fisher, J.R. (1968) Comp. Biochem. Physiol. 27, 687–694.

5. Ma, P.F., and Fisher, J.R. (1968) Biochem. Biophys. Acta 159, 153–159.

6. Murphy, P.M., Noonan, M., Collins, P., Tully, E., and Brady, T.G. (1969) Biochim. Biophys. Acta 171, 157–166.

7. Brady, T.G., and O'Connell, W. (1962) Biochim. Biophys. Acta 62 216–229.

8. Pfrogner, N. (1967) Arch. Biochem. Biophys. 119, 141–146.

9. Cory, J.G., Weinbaum, G., and Suhadolnik, R.J. (1967) Arch. Biochem. Biophys. 118, 428–433.

10. Spencer, N., Hopkinson, D.A., and Harris, H. (1968) Ann. Hum. Genet. 32, 9–14.

11. Ressler, N. (1969) Clin. Chim. Acta 24, 247–251.

12. Ma, P.F., and Fisher, J.R. (1969) Comp. Biochem. Physiol. 31, 771–781.

13. Akedo, H., Nishihara, H., Shinkai, K., and Komatsu, K., and Ishikawa, S. (1972) Biochim. Biophys. Acta 276, 257–271.

14. Osborne, W.R.A., and Spencer, N. (1973) Biochem. J. 133, 117–123.

15. Van der Weyden, M.B., and Kelley, W.N. (1976) J. Biol. Chem., in press.

16. Piggott, C.O., and Brady, T.G. (1976) Biochim. Biophys.
     Acta 429, 600–607.

17. Van der Weyden, M.B., Buckley, R.H., and Kelley, W.N.
     (1974) Biochem Biophys. Res. Commun. 57, 590–595.

18. Hirschhorn, R., Beratis, N. and Rosen, R.S. (1975).
     Proc. Nat. Acad. Sci. USA. 73. 213–217.

19. Chen, S., Scott, C.R. and Swedberg, K.R., (1975). Am. J.
     Hum. Genet. 27, 46–52.

20. Rossi, C.A., Lucacchini, A., Montali, U., and Ronca, G.
     (1975) Int. J. Peptide Res. 7, 81–89.

21. Daddona, P.E., and Kelley, W.N. (1976) J. Biol. Chem., in
     press.

22. Davis, B.J. (1964) Ann. N.Y. Acad. Sci. 121, 404–427.

23. Vesterberg, O., and Svensson, H. (1966) Acta Chem. Scand.
     20, 820–834.

24. Cuatrecasas, P. (1970) J. Biol. Chem. 245, 3059–3065.

25. Gilham, P.T., (1970) Methods Enzymol. 21, 191–197.

26. Stevenson, K.J. and Landman, A. (1971) Can. J. Biochem.
     49, 119–126.

27. Weber, K., and Osborn, M. (1969) J. Biol. Chem. 244,
     4406–4412.

28. Segrest, J.P. and Jackson, R.L. (1972) Methods Enzymol
     28, 54–63.

29. Ackers, G.K. (1967) J. Biol. Chem. 242, 3237–3238.

30. McCarty, K.S., Stafford, D., and Brown, O. (1968) Anal.
     Biochem. 24, 314–329.

31. Cohn, E.J. and Edsall, J.T. (1943) in "Proteins, Amino
     Acids, and Peptides ", pp. 370, Reinhold Publishing, New
     York.

ADENOSINE DEAMINASE: CHARACTERIZATION OF THE MOLECULAR

HETEROGENEITY OF THE ENZYME IN HUMAN TISSUE

M. B. Van der Weyden and W. N. Kelley

Department of Medicine, Monash University, Australia and
Department of Medicine, University of Michigan, Ann Arbor,
Michigan

Adenosine deaminase (ADA) catalyzes the irreversible
hydrolytic deamination of adenosine to produce inosine and
ammonia. In human tissue this activity exhibits considerable
molecular heterogeneity in that "tissue specific isoenzymes"
have been demonstrated with either starch gel electrophoresis
(1-3) or gel filtration (3,4,7). The molecular weights of
the multiple forms apparent have varied from 30,000 to 47,000
(3-7) and 230,000 to 440,000 (3,4,7). Although the basis for this
heterogeneity has been obscure, with definition of the
characteristics of residual ADA activity in tissues of patients
with severe combined immunodeficiency and ADA deficiency (8,9),
it has been proposed that the molecular heterogeneity of ADA
in human tissue is the result of post-translational modification of a single locus product (9). For this hypothesis
to be valid, interconvertibility between the molecular forms
of ADA should be demonstrated. In this study we have examined
the nature of the molecular heterogeneity of ADA in human
tissue and demonstrate that the forms evident are interconvertible.

Molecular Heterogeneity

Human splenic tissue preparations when applied to an 8%
agarose gel filtration column yield an activity-elution profile
as shown in Fig.1. With filtration on either 8% agarose or
Sepharose 4B, a proportion of the ADA activity consistently elutes
in the void volum (V_o). The three soluble components of ADA
retarded by gel filtration have been arbitrarily designated as
the large, intermediate and small form of ADA (Table 1). Sucrose

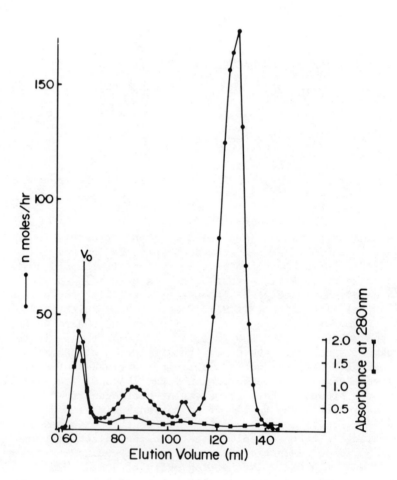

<u>Figure 1</u>.   Molecular heterogeneity of human ADA shown by
gel filtration. An aliquot (2 ml) of homogenate prepared
from human splenic tissue (see reference 10) was applied to
an 8% agarose column equilibrated with 10 mM Tris-HCl: Buffer
A (from Van der Weyden and Kelley, 1976).

TABLE I

Physical Properties of Soluble Human Adenosine Deaminase and Conversion Activity

	Sedimentation Coefficient[a]	Stokes Radius[a]	Molecular Weight	Frictional Ratio
	$\times 10^{13}$S	Å		
Small Form	3.8 ± 0.3 (51)	23.0 ± 0.9 (10)	36,000	1.06
Intermediate Form	7.3 ± 0.2 (11)	38.2 ± 1.9 (6)	114,000	1.20
Large Form	10.7 ± 0.4 (42)	67.2 ± 1.3 (14)	298,000	1.52
Conversion Activity	9.9 (1)	49 (1)	200,000	1.27

[a]The value given is the mean ± 1SD.

The figure in parenthesis indicates the number of determinations.

(From Van der Weyden and Kelley, 1976)

gradient ultracentrifugation of the same tissue preparation also
yields four components of ADA activity. With varying centrifugation
times, one component of activity consistently sediments to the
bottom of the gradient. The sedimentation coefficients of the
remaining three components of activity are listed in Table 1.
As determined by sucrose gradient ultracentrifugation, the
molecular form of ADA predominating in different tissues
appears to correlate with the specific activity of ADA in the
tissue extract (Table 2). The large form predominates in
those tissue extracts exhibiting lower enzyme activity, e.g.
lung or kidney, while the small form is the major species
in those tissue extracts exhibiting higher enzyme activity,
e.g. spleen or stomach.

Subcellular Distribution

Elucidation of the subcellular distribution of ADA in
human leukocyte preparations has disclosed that while the

TABLE II

Adenosine Deaminase in Human Tissue

Tissue	Specific Activity nmol/min/mg protein	Percentage of Total Activity Soluble Molecular Form		
		Large	Intermediate	Small
Spleen	128	10	2	86
Duodenum	127	9	0.7	89
Stomach	89	2	0.3	96
Jejunum	63	–	–	–
Pancreas	27	–	–	–
Cerebrum	27	–	–	–
Adrenal Gland	21	–	–	–
Lymph Node	23	–	–	–
Appendix	20	8	2	87
Testes	13	–	–	–
Cardiac Muscle	11	–	–	–
Liver	9	59	12	24
Cerebellum	9	–	–	–
Spinal Cord	9	–	–	–
Skeletal Muscle	8	–	–	–
Lung	7	74	–	–
Kidney	6	86	–	–
Thyroid	4	–	–	–

(From Van der Weyden and Kelley, 1976)

bulk of ADA activity is present in the 100,000 x g supernatent (cytosol), approximately 2% of the total activity present in the initial homogenate is associated with the 6000 x g and 100,000 x g pellets (Table 3). This particulate ADA activity does not appear to be the result of contamination with cytosol activity as no hypoxanthine-guanine phosphoribosyltransferase activity is detected in either fraction. Since the 6000 x g pellet exhibits substantial contamination with 5'nucleotidase activity, the subcellular organelle(s) to which this form of ADA is identified cannot be precisely defined. Sucrose gradient ultracentrifugation of the cytosol ADA activity reveals that while both the large and small forms of the enzyme are present, in contrast to tissue extracts no particulate ADA activity sedimenting to the bottom of the gradient is evident. Alternatively, examination of an extract of the 100,000 x g pellet in a similar fashion discloses that the major component of ADA activity is "particulate" with a minor component exhibiting the sedimentation velocity of the small form.

Following gel filtration of tissue extracts, 5'nucleotidase but not adenine phosphoribosyltransferase, hypoxanthine-guanine phosphoribosyltransferase or cytochrome C oxidase activity elutes with ADA activity in the void volume. This ADA activity when isolated, concentrated by ultrafiltration and reexamined by gel filtration reveals in addition to activity eluting in the void volume, the appearance of activity eluting as the small form (Fig. 2). Incubation of the ADA activity with 1% Triton X-100, results in a 3-fold increase in the total ADA activity eluting as the small form (Fig. 2). These observations provide further evidence that ADA activity sedimenting to the bottom of the sucrose gradients or eluting in the void volume with gel filtration is associated with subcellular particulate matter.

Interconversion of the Molecular Forms

Conversion of large to small: The large form of ADA isolated by gel filtration from tissue extracts in which the small form predominates and subjected after ultrafiltration to further gel filtration or sucrose density ultracentrifugation (Fig.3) yields three peaks of activities corresponding in Stokes radius and sedimentation coefficient to the large, intermediate or small species of ADA evident in the initial homogenate. The large form of ADA in tissue extracts in which large species predominates, when isolated as described above and subjected to sucrose density ultracentrifugation or repeat gel filtration, yields a single peak of activity corresponding to the native large form. This form dissociates to the small form under conditions of acidic pH (Fig.4).

TABLE III

Adenosine Deaminase Activities in Subcellular Fractions of Human Leukocytes

Fraction	Volume (ml)	Protein (mg)	Adenosine Deaminase	Hypoxanthine-Guanine Phosphoribosyl-Transferase	Total Activity (nmol/min) 5' Nucleotidase		Cytochrome C Oxidase
					AMP as Substrate	IMP as Substrate	
Homogenate +2% Triton	2	20 ND	134 136	34.8 33.8	12.3 ND	21.4 ND	0.026 ND
6000 x g pellet +2% Triton	1	0.86 ND	1.9 2.3	<0.001 <0.001	2.9 3.0	4.0 3.3	0.014 ND
100,000 x g pellet +2% Triton	1	1.10 ND	1.7 2.7	<0.001 <0.001	6.0 4.4	7.7 6.9	0.003 ND
100,000 x g Supernatant +2% Triton	11	9.86 ND	69.0 73.5	23.6 24.5	<0.001 <0.001	3.5 3.1	-- ND

ND = Not Determined.

The technique employed in the subcellular fractionation was as reported by Van der Weyden and Kelley (1976)

(From Van der Weyden and Kelley, 1976)

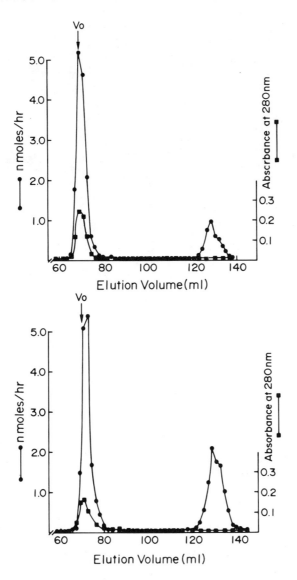

<u>Figure 2.</u>   Repeat gel chromatography of particulate ADA
from human spleen with and without treatment with Triton X-100.
The enzyme activity that eluted between 62 and 68 ml from the
8% agarose column (see Fig.1) was pooled and concentrated by
ultrafiltration. Aliquots of this preparation were incubated
at 4° for 45 min with either Buffer A or 1% Triton and applied
to identical 8% agarose columns equilibrated with Buffer A.
A. Preparation incubated with Buffer A.   B. Preparation
incubated with 1% Triton. (From Van der Weyden and Kelley, 1976).

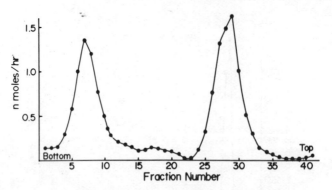

Figure 3. Sucrose gradient ultracentrifugation of the large molecular form of ADA from human splenic tissue. The enzyme activity that eluted between 83–94 ml from the 8% agarose column (see Fig.1) was pooled and concentrated by ultrafiltration. An aliquot of this was applied to a sucrose gradient and centrifuged at 4° as outlined in reference 10. (From Van der Weyden and Kelley, 1976).

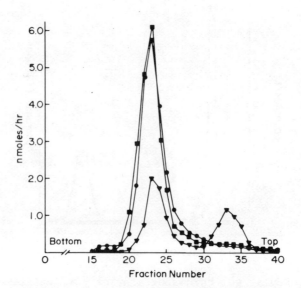

Figure 4. Dissociation of the large form of ADA to the small form at pH 3.4 shown by sucrose gradient ultracentrifugation. The large form of ADA was isolated by gel filtration from human kidney tissue homogenate. A portion of this enzyme preparation was incubated at 4° for 30 min either in 50 mM Tris HCl pH 7.4 (■——■), 50 mM sodium succinate pH 7.4 (●——●) or 50 mM sodium succinate pH 3.4 (▲——▲). An Aliquot of each reaction mixture was layered on to sucrose gradients and centrifuged at 4° as outlined in reference 10. (From Van der Weyden and Kelley, 1976).

Conversion of intermediate to small: The intermediate molecular species when isolated by either gel filtration or by sucrose density ultracentrifugation and reexamined by sucrose gradient ultracentrifugation yields a major peak of activity corresponding to the small form and minor peaks of activity with sedimentation coefficients corresponding to the intermediate and large forms.

Conversion of small to large: Incubation of the isolated small molecular species with tissue extracts in which the large species of ADA predominates (see Table 2) leads to the aggregation of the small to the intermediate and large molecular forms (Fig.5). This conversion of the small to the large species is independent of the source of the small form whether isolated from human erythrocytes, spleen, stomach or small intestine or by dissociation of the large form. The distribution of this "conversion activity" in different tissue extracts as shown in Table 4 exhibits a direct relationship to the presence of the large molecular species in the respective extract, being absent in those extracts in which the small predominates and present in those tissue extracts in which the large species of ADA predominates. The aggregation phenomenon produced by conversion activity occurs at both 4° and 37° (Fig. 6) and exhibits a broad pH optimum of 5.0– 8.0, but does not exhibit the characteristics of a catalytic process as the reaction is associated with loss of conversion activity. The calculated molecular weight and frictional ratio of conversion activity is listed in Table 1.

TABLE IV

Tissue Distribution of Conversion Activity

Tissue	Percentage of Total Adenosine Deaminase Activity Soluble Molecular Form			Conversion Activity units/mg protein
	Large	Intermediate	Small	
Kidney	90	–	–	3.9
Lung I	74	–	–	0.73
Liver I	80	–	–	0.33
Liver II	59	12	24	0.03
Lung II	25	3	63	Not Detectable
Appendix	8	2	87	Not Detectable
Stomach	2	0.3	96	Not Detectable
Spleen	10	2	86	Not Detectable

(From Van der Weyden and Kelley, 1976)

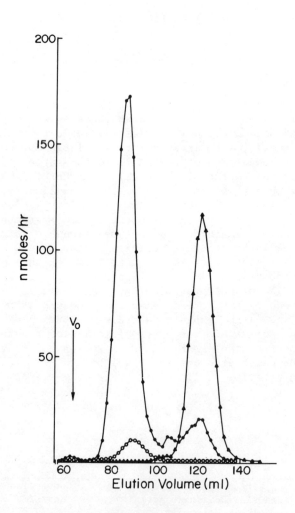

<u>Figure 5.</u>   Conversion of the small form of adenosine deaminase
to the large and intermediate species by human kidney extract
shown by gel filtration.   The isolated small form was incubated
at 37° for 60 min with 5 mg bovine serum albumin (▲—▲) or 6.5
mg human kidney extract (●—●).   The samples were sequentially
applied and eluted from the 8% agarose column.   Kidney extract
(6.5 mg ○—○) was also equilibrated alone at 37° for 60 min and
applied to the same column. (From Van der Weyden and Kelley, 1976).

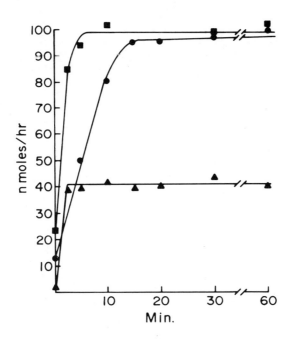

<u>Figure 6</u>.   Effect of temperature and protein on the rate
of conversion of the small molecular form of ADA to the large
molecular form by conversion activity partially purified from
kidney tissue (11). A quantity of the partially purified small
form (10 μg, specific activity 6.3 units/mg) was incubated with
0.88 units of conversion activity at 37° (▲——▲) or 2.2 units
of conversion activity at 4° (●——●) or 37° (■——■). The
formation of the large form under these conditions was quantitated
as outlined in reference 10. (From Van der Weyden and Kelley,
1976).

TABLE V

Electrophoretic Variants of Adenosine Deaminase from Human Tissue

Tissue	Molecular Form	Donor	Ampholyte Range	Isoelectric Points	Relative Activity
Small Intestine	Small	A	4-6	4.64	0.17
				4.75	1.00
				4.87	0.07
Spleen	Small	B	4-6	4.69	0.17
				4.83	1.00
Spleen	Small	C	4-6	4.72	1.00
				4.84	2.86
				4.96	3.10
Kidney	Large	A	4-6	4.64	0.57
				4.74	1.00
				4.80	1.22
				5.01	0.95
				5.09	0.95
				5.16	2.32
Kidney	Large	D	4-6	4.76	1.00
				4.98	0.21
				5.02	0.27
				5.14	0.59
				5.24	0.74
Kidney	Large	E	3-6	4.65	0.85
				4.72	1.00
				4.83	0.65
				4.98	0.30
				5.04	0.47
				5.13	0.93
Lung	Large	A	4-6	4.65	0.27
				4.75	1.00
				5.06	0.32
				5.14	0.15
				5.24	0.07
Liver	Large	A	4-6	4.75	1.00
				4.98	0.18
				5.02	0.10
				5.09	0.26
				5.24	0.06
Produced in vitro	Large	A	4-6	4.62	0.63
				4.71	1.00
				4.80	0.33
				4.85	0.28
				4.88	0.25
				4.91	0.32
				5.02	0.28
				5.20	0.13

The maximum activity in each peak is expressed relative to the peak of activity appearing at pH 4.71 to 4.76 which in most cases was found to be the major peak of activity. Spleen from donor B did not exhibit a peak of activity between pH 4.71 and 4.76, thus activity is expressed relative to the peak appearing at pH 4.83.

(From Van der Weyden and Kelley, 1976)

Electrophoretic Heterogeneity

The small form of ADA isolated from either small intestine or splenic tissue, exhibits two or three different electrophoretic forms with isoelectric focusing (Table 5). Sucrose gradient ultracentrifugation of each electrophoretic variant yields a single peak of activity with a sedimentation coefficient ranging from 3.6-3.8 corresponding to that of the native small form. Isoelectric focusing of the large molecular species of ADA obtained by gel filtration from one of several different tissues reveals five to six different electrophoretic forms (Table 5). The large form elaborated by incubation of the isolated small form with partially purified conversion activity yielded a pattern similar to that observed with the native large form. Sucrose gradient ultracentrifugation of each electrophoretic variant from kidney or the large form produced in vitro yields a single peak of activity with sedimentation coefficients of 10.6-10.8 which correspond to that of the native large form. In contrast, gel filtration of the electrophoretic variants with pI(s) of 4.65 and 4.75 from lung, indicates a mixture of the large and small molecular forms, while the electrophoretic variants with pI(s) of 5.06, 5.14 and 5.24 remain large. The results with the large form from liver are essentially the same as observed with the large form from lung.

In summary, these studies have demonstrated that ADA exists in multiple forms in human tissue. One form of the enzyme appears to be particulate and the three soluble forms are interconvertible. The small form of ADA is convertible to the large form only in the presence of a protein which has an apparent molecular weight of 200,000. This conversion of the small form of the enzyme to the large occurs at 4°, exhibits a pH optimum of 5.0-8.0 and is associated with loss of conversion activity. The large form of ADA predominates in tissue extracts exhibiting the lower enzyme specific activities and abundant conversion activity. The small form of ADA predominates in tissue preparations exhibiting the higher enzyme specific activities and no detectable conversion activity. The small form of ADA shows several electrophoretic variants by isoelectric focusing. The electrophoretic heterogeneity observed with the large form is similar to that observed with the small form with the exception that several additional electrophoretic variants are uniformly identified. No organ specificity is demonstrable for the different electrophoretic forms. These observations support the hypothesis based initially on the study of ADA in tissues from patients with severe combined immunodeficiency (8,9) that the molecular heterogeneity of ADA in human tissue is the result of post-translational modification of the protein (9).

REFERENCES

1. Spencer, N., Hopkinson, D.A., and Harris, H. 1968.
    Adenosine deaminase polymorphism in man. Ann.Hum.Genet.
    32:9-14.

2. Ressler, N. 1969. Tissue-characteristic forms of adenosine
    deaminase. Clin.Chim.Acta. 24:242-251.

3. Edwards, Y.H., Hopkinson, D.A., and Harris, H. 1971.
    Adenosine deaminase isoenzymes in human tissue. Ann.Hum.Genet.
    35:207-219.

4. Akedo, H., Nishihara, H., Shinkai, K., Komatsu, K., and
    Ishikawa, S. 1972. Multiple forms of human adenosine
    deaminase. I. Purification and characterization of two
    molecular species. Biochem.Biophys.Acta. 276:257-271.

5. Osborne, W.R.A. and Spencer, N. 1973. Partial purification
    and properties of the commonly inherited forms of adenosine
    deaminase from human erythrocytes. Biochem.J. 133:117-123.

6. Agarwal, R.P., Sagar, S.M., and Parks, R.E.,Jr. 1975.
    Adenosine deaminase from human erythrocytes. Purification
    and effects of adenosine analogs. Biochem.Pharmac. 24:
    693-701.

7. Ma, P.F. and Mager, T.A. 1975. Comparative studies of human
    adenosine deaminase. Int.J. Biochem. 6:281-286.

8. Hirschhorn, R., Levytska, V., Pollara, B., and Meuwissen,
    H.J. 1973. Evidence for control of several different
    tissue-specific isoenzymes of adenosine deaminase by
    a single genetic locus. Nature (New Biology) 246:200-202.

9. Van der Weyden, M.B., Buckley, R.H., and Kelley, W.N. 1974.
    Molecular form of adenosine deaminase in severe combined
    immunodeficiency. Biochem.Biophys.Res.Commun. 57:590-595.

10. Van der Weyden, M.B. and Kelley, W.N. 1976. Human
    adenosine deaminase: distribution and properties. J. Biol.
    Chem. (in press).

11. Nishihara, H., Ishikawa, S., Shinkai, K., and Akedo, H.
    1973. Multiple forms of human adenosine deaminase. II.
    Isolation and properties of a conversion factor from
    human lung. Biochim.Biophys.Acta. 302:429-442.

HUMAN 5'-NUCLEOTIDASE: MULTIPLE MOLECULAR FORMS AND REGULATION

Irving H. Fox and P. Marchant

Purine Research Laboratory, University of Toronto

Rheumatic Disease Unit, Wellesley Hospital, Toronto

The ability to rapidly degrade purine nucleotides to uric acid in man under experimental conditions suggests that there may be important regulatory mechanisms to maintain a constant intracellular nucleotide environment. This has been illustrated by the rapid infusion of 50 grams of fructose in man which leads to a 35% increase in the serum uric acid within 30 minutes (1). The basis for this abrupt rise in the serum uric acid is a decrease in hepatic ATP and inorganic phosphate following the phosphorylation of fructose to fructose-1-P. There is a resultant increased synthesis of AMP and IMP and a release from inhibition of the catabolic pathways leading from these nucleotides to uric acid. The purine ribonucleotide catabolic pathways to uric acid include a dephosphorylation step in which purine nucleoside monophosphates are hydrolyzed to the corresponding nucleosides and inorganic phosphate. One important enzyme catalyzing this reaction is 5'-nucleotidase. The regulation of the human enzyme has not previously been evaluated in detail (for review see reference 2). We have prepared this enzyme from human placental microsomes and studied its properties (2,3).

Preparative isoelectric focusing was performed with a gradient from pH 5.0 to 7.0 (Figure 1). Multiple electrophoretic forms of 5'-nucleotidase were evident. A total of 7 electrophoretic variants were isolated during the preparation of 6 placentas. Only 3 to 6 variants were found in a single placenta. To assess the physical basis of these variants, four of the electrophoretic peaks were studied on column with biogel A 0.5 (Figure 2). Three molecular weight species were evident, a heavy, medium and a light form. Peak II had a medium form only, peak III medium and light forms, peak IV and VII had medium and heavy forms. Further studies indicated that

249

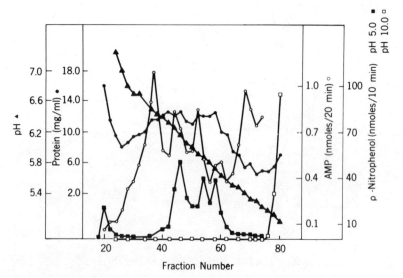

Fig. 1. Preparative isoelectric focusing of human 5'-nucleotidase.
The horizontal axis indicates the fraction numbers. In this
gradient 5 electrophoretic variants of 5'-nucleotidase were evident.
Acid phosphatase focused as 3 peaks. Alkaline phosphatase was at
the end of the gradient.

at least the medium and light forms had a monomer-dimer relationship and were interconvertable. The apparent molecular weight of
the medium and light forms of the enzyme was estimated to be 86,500
and 43,500 respectively. This suggests that variable states of
enzyme aggregation accounted for the electrophoretic variants. No
difference between these 4 peaks could be discerned when specific
activity, Michaelis constants and inhibitory properties were compared.

Other characteristics of the human 5'-nucleotidase were
evaluated. The substrate preference was specific for 5'-nucleoside
monophosphates. The pH optimum was 9.8, but became a plateau from
7.4 to 9.8 with 5 mM Mg. Although no divalent cation was required,
Mg stimulated the reaction by 134%. The Km for Mg was 3 μM. Since
the intracellular Mg is 1 mM under normal conditions, Mg usually
stimulates 5'-nucleotidase. The enzyme was stabilized at -20°C by
ampholytes and sucrose. The Km for AMP ranged from 12 to 18 μM, for
GMP from 33 to 67 μM and for CMP 170 to 250 μM.

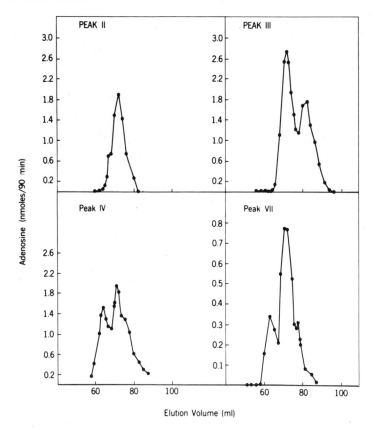

Fig. 2. Gel filtration studies of 5'-nucleotidase.

The regulation by nucleotides was then studied using inhibition kinetics (Table 1). ADP was a competative inhibitor with respect to AMP. The Ki was 20 μM. This is well below the known intracellular concentration of ADP which is 10 times higher. Inhibition studies using ATP also showed competative inhibition. The Ki was 54 μM. The intracellular concentration of ATP is about 20 times greater than this. Thus it would appear that 5'-nucleotidase is inhibited by these nucleotides under normal conditions. Allopurinol ribonucleotide, also a substrate, was a competative inhibitor with respect to AMP. The Ki was 10 μM. The apparent Ki of other nucleotide inhibitors was estimated from plots of inhibitor concentration against the reciprocal of the velocity. Nucleoside diphosphates were relatively effective inhibitors with the exception of CDP. Nucleoside triphosphates except XTP and TTP were also effective inhibitors. In contrast some reaction products adenosine, cytidine, inosine and inorganic phosphate were ineffective inhibitors.

TABLE 1

INHIBITION OF 5'-NUCLEOTIDASE

Compound	Apparent Ki mM	Ki Slope mM
GMP	0.1	
Allopurinol Ribonucleotide		0.01
ADP		0.02
GDP	0.2	
UDP	0.2	
CDP	0.7	
NAD	0.2	
ATP		0.05
GTP		0.12
TTP	3.0	
UTP	0.2	
CTP	0.5	
XTP	0.8	
Adenosine	0.4	
Inosine	3.0	
Cytidine	5.0	
NaF	6.0	
P_i	42.0	

TABLE 2

A COMPARISON OF NUCLEOTIDE 5'-PHOSPHOMONOESTERASE
ACTIVITY OF TWO PLACENTAL ENZYMES

	5'-Nucleotidase	Alkaline Phosphatase
K_m AMP	14 μM	400 μM
Mg 5 mM % Stimulation	134	19
% Maximum Activity at pH 7.4	90	18
Substrate Specificity	5'-Nucleoside monophosphate	Any phosphorylated compound
Ki (Apparent) Pi	42 mM	0.6 mM
Inhibition by	Nucleotides	Any phosphorylated compound

Since alkaline phosphatase is also capable of 5'-phosphomono-esterase activity, we have studied this enzyme for comparison with 5'-nucleotidase (Table 2). AMP may be preferentially hydrolyzed by 5'-nucleotidase since the Km is 30 times higher for alkaline phosphatase. This latter enzyme is only minimally stimulated by 5 mM MgCl$_2$ and has only 18% of its maximum activity at pH 7.4. A relatively greater specificity for nucleoside monophosphates was evident for 5-nucleotidase as compared to the more general ability of alkaline phosphatase to hydrolyze almost any phosphorylated compound. Normal intracellular concentrations of inorganic phosphate, which are 1 mM, will substantially inhibit alkaline phosphatase but not 5'-nucleotidase.

To summarize, we have evaluated some important properties of human placental 5'-nucleotidase. Seven electrophoretic variants appeared to be pseudoisozymes based variable aggregation of a monomer form. 5'-nucleotidase and alkaline phosphatase are probably inhibited under normal conditions. Hydrolysis of nucleoside mono-phosphates may preferentially occur by 5'-nucleotidase rather than alkaline phosphatase. Stimulation of ribonucleotide catabolism may occur by decreased inhibition of 5'-nucleotidase and alkaline phosphatase with a diminution of the intracellular concentration of ATP and Pi respectively. It is hoped that these observations will extend our understanding of purine catabolism in man.

REFERENCES

1. Fox, I.H. and Kelley, W.N. 1972. Studies on the mechanism of fructose induced hyperuricemia. Metabolism 21:713-721.

2. Fox, I.H. and Marchant, P.J. 1976. Purine catabolism in man: characterization of placental microsomal 5'-nucleotidase. Can. J. Biochem. 54:462-469.

3. Fox, I.H. and Marchant, P.J. Purine catabolism in man: inhibition of 5'-phosphomonoesterase activities from placental microsomes. (in preparation)

Subcellular distribution of purine degrading enzymes

in the liver of the carp (Cyprinus carpio l.)

Hans Goldenberg

1st Department of Medical Chemistry at the
University of Vienna, Austria

Microbodies (peroxisomes[1]) have been established as
being the intracellular seat of urate oxidase in the
livers of various animals and in plants (2,3,4,). So
far, two possibilities of intraparticulate localization
of this enzyme within microbodies have been found.
While in some plant tissues, like in castor bean endo-
sperm, there is a soluble uricase in the peroxisome
matrix (5), in most animals it is firmly bound to a crystalline core (nucleoid) within the microbody (2,6,7,8)
from which it can not be solubilized (9).

Peroxisomes of carp liver, detectable histochemi-
cally by their catalase content (fig 1), have no crystalline inclusions, but do, however, contain urate oxidase according to cell fractionation experiments (10).
The same is probably true for other teleost species (11).
The site of uricase within the particles has not been
examined so far.

The fact that carp degrades purines to urea and
glyoxylic acid (12) and the finding of xanthine dehydrogenase and allantoinase in peroxisomes (3) give rise to
the speculation, that these organelles might have a
major part in purine catabolism.

These two problems were investigated by differential and density gradient centrifugation experiments
on carp liver homogenates and by attempts to solubilize
urate oxidase from isolated peroxisomes.

254

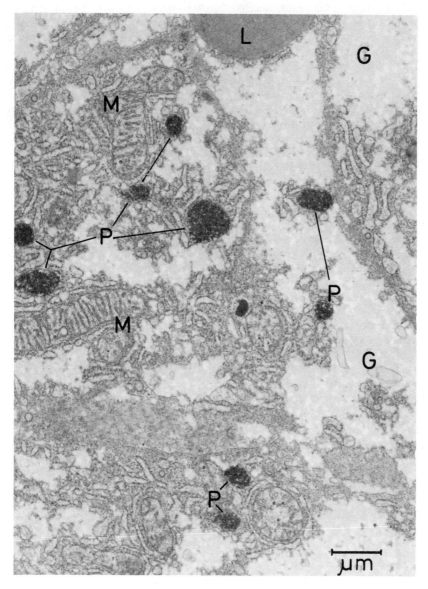

Figure 1   Carp liver, tissue section.
Identification of peroxisomes (P) by DAB-staining for
peroxidatic catalase activity. Fixation with glutaric
aldehyde. M:Mitochondria, G:Glycogen, L:Lipid droplets

METHODS

Freshly excised carp liver was freed from adhering fat and homogenized with one stroke in three volumes of 0.25M sucrose containing 0.1% ethanol, 2% dextrane 10 and 3mM imidazol-HCl, p_H= 7.2. The homogenate was spun in the MSE HS 18 centrifuge at 600xg for 10 min. The residue, containing nuclei, unbroken cells and cell debris, was washed twice, suspended in sucrose and designated N (nuclear fraction). The combined supernatants were further fractionated according to DeDuve's method for rat liver (2). Four cytoplasmic fractions, namely a "heavy mitochondrial"(M), a "light mitochondrial"(L), a microsomal and asoluble fraction (Mc and S) were obtained (fig 2). Protein, marker enzymes and purine catabolic enzymes were assayed in all fractions.

Peroxisomes were isolated by use of their high buoyant density in sucrose by centrifugation in a linear sucrose density gradient (1.2-1.8M)in a B14 aluminium zonal rotor in the MSE SS50 centrifuge. Peroxisomes band at a density of 1.225-1.23g/cm^3. The gradient fraction with the highest uricase activity was treated with two volumes of 50mM Tris-HCl buffer, p_H=7.6 containing 0.6M(19% w/w) sucrose and KCl in the concentrations given in table 1. The mixtures were incubated at 25°C for 30min and the membranes from the particles osmotically ruptured were pelleted at 100000xg for 30min and resuspended in buffer (13). All pellet and supernatant fractions were assayed for catalase and urate oxidase activities.

Enzyme Assays

Cytochrome oxidase: The method of Cooperstein and Lazarow was used for the mitochondrial marker.
A
$$\Delta \log O.D._{550} = 1 \ /min$$
corresponds to 1 unit of activity (14).

Catalase: Assay as described (15). 1 unit is a first order rate constant
$$k = \frac{1}{\Delta t} \ln \frac{c_1}{c_2} = 1$$
for degradation of H_2O_2. T=25°C.

Acid phosphatase: Assay of Gianetto and DeDuve (16) with ß-glycerophosphate as substrate. 1 unit = 1 µmol of inorganic phosphate liberated per minute. T=37°C.

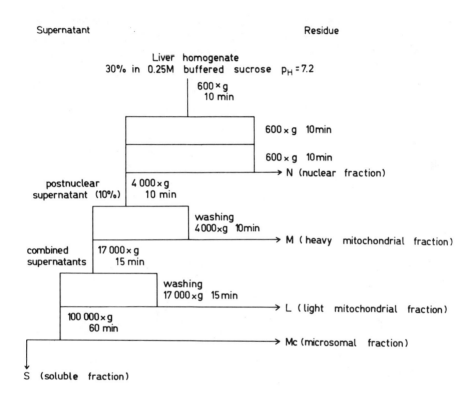

Figure 2

Differential centrifugation scheme:
Centrifugation is done in the 8x50ml angle rotor of
the MSE HS18 highspeed centrifuge. Each particulate
fraction is suspended in 2ml of 0.25M sucrose per
gram original tissue.

Arylesterase: Optical measurement of release of
α-naphthol from 0.4mM α-naphthylacetate by coupling to
an azo dye with Fast Red TR salt at 540nm. p_H=8.6,T=30°.
1 unit = 1μmol of α-naphthol formed per minute.
Urate oxidase: Registration of uric acid cleavage at
293nm (15). p_H=9.0, T=37°. 1 unit = 1 μmol of uric acid
oxidized per minute.
Xanthine oxidase: Assayed as sum of oxidase and dehydrogenase activities (17), with addition of NAD^+(0.5mM),
lactate dehydrogenase (from rabbit skeletal muscle, diluted 1:3000) and pyruvate(1.7mM) at 293 nm. $p_H \doteq 8.0$,
T=37°. 1 unit = 1 μmol of uric acid formed per minute.
Allantoinase: Assay of Brown (18). Allantoic acid formed
during incubation is hydrolyzed to urea and glyoxylic
acid for 15 min at 60°C. Glyoxylate is determined as
dinitrophenylhydrazone (19).
Allantoicase: Optical assay system of Brown et al (20)
with lactate dehydrogenase from skeletal muscle which
readily reduces glyoxylate formed during allantoate
hydrolysis at p_H=7.4 and 25°C, in the presence of NADH,
whose consumption is followed at 340nm.
Protein is determined by the Lowry method.

Morphology

Catalase positive particles in carp liver were identified by the method of Novikoff (21) with 3,3'-di-
aminobenzidine in glutaric aldehyde fixed tissue.
Details are described in ref. 10. The isolated peroxisomal fraction was fixed with 2.5% glutaric aldehyde,
diluted with water and pelleted by spinning for 1h at
100000xg. Cytochemistry was performed with small pieces
of the pellet in the same manner as with the original
tissue.

RESULTS

Fig. 3 shows the enzyme distrbution in carp liver after
differential centrifugation. There is a clear distinction
into a pure mitochondrial (cytochrome oxidase), a mito-
chondrial-peroxisomal (cytochrome oxidase-catalase)
and a microsomal (arylesterase) fraction.Lysosomes
(indicated by acid phosphatase) obviously are destroyed
to a great part by homogenization. Urate oxidase quite
clearly follows catalase in subcellular distribution
with the exception, that there is no uricase activity
in the soluble fraction. Partial solubilization of
urate oxidase is possible by osmotic breakage of iso-

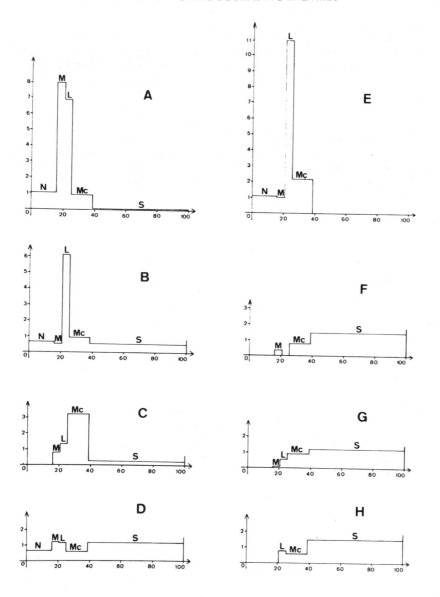

Figure 3

Distribution of markers and purine degrading enzymes
after differential centrifugation. Abscissa: percentage
of total protein. Ordinate: relative specific activity.
N: Nuclear -, M: heavy mitochondrial -, L: light mitochondrial,
Mc: microsomal and S: soluble fraction.
A: cytochrome oxidase, B: catalase, C: aryl esterase, D: acid
phosphatase, E: urate oxidase, F: xanthine oxidase, G: allantoinase, H: allantoicase

lated peroxisomes. Small amounts of KCl in the dilution
medium completely solubilize the enzyme (table 1).
The other purine degrading enzymes assayed are for
their largest part found in the soluble fraction.
Small amounts present in the particulate fractions are
probably due to protein adsorption or incomplete separation of pellets and supernatants after centrifugation.

Omission of NAD$^+$ and NADH-consuming system in the
xanthine oxidase assay gives lower values for total
activity, but the same subcellular distribution.

DISCUSSION

In all tissues containing urate oxidase so far investigated by cell fractionation this enzyme was found
to be localized in peroxisomes (microbodies). In the
livers of mammals these organelles contain crystalline
cores which have been established to contain the urate
oxidase. In fact, this enzyme is the only protein con-
stituent of the core identified so far. There are, however, tissues containing nucleoids in the peroxisomes
but being deficient of urate oxidase, e.g., rat kidney.
(22). The reverse case has not yet been detected. In
carp liver the distribution of urate oxidase between
subcellular fractions is much the same as in rat liver (2
There is no activity in the soluble fraction, which
shows that the enzyme must be quite firmly bound to the
organelle. On the other hand, no nucleoid can be detected, neither in tissue blocks, nor in isolated
organelles (fig. 1 and 4). This suggests a binding of
the enzyme to the particle membrane, which is confirmed
by the results of the osmotic experiments. While some
catalase is found soluble in fractionation and is easily released by osmotic breakage of the organelles, a
complete solubilization of urate oxidase is possible
only by addition of salt to the osmotic medium.

A general role of peroxisomes in purine catabolism
has been suggested by the results of Scott et al (3),
who found xanthine dehydrogenase in peroxisomes of
chicken liver and allantoinase in those of frog liver.
A localization of all four purine degrading enzymes
(xanthine oxidase, urate oxidase, allantoinase and allantoicase) would be the first important metabolic pathway to be detected organized in peroxisomes (fig 5).
Moreover, xanthine oxidase might be able to produce

Table 1

Treatment	Enzyme release %	
	Catalase	Urate Oxidase
50mM Tris–HCL pH 7.6 + 19% Saccharose	98 %	40%
Same medium + 0.15M KCL	100%	98 %
Same medium + 1M KCL	100%	100 %

The peroxisomal gradient fraction from the density
gradient was diluted 1+2 with buffered sucrose containing KCl in the given concentrations and incubated
at 25°C for 30min. Membranes were collected by centrifugation at 100000xg for 30min and suspended in buffer.
The sum of enzyme activities in pellet and supernatant is taken as 100%. The table shows the percentage
of activity found in the supernatant.

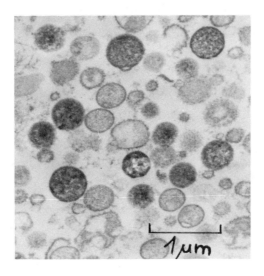

Figure 4

Peroxisomal fraction from a sucrose density gradient.
The suspension was fixed in 2.5% glutaric aldehyde,
diluted with water and centrifuged for 1h at 100000xg
The pellet was cut in narrow stripes and processed in
the same manner as tissue blocks (10).

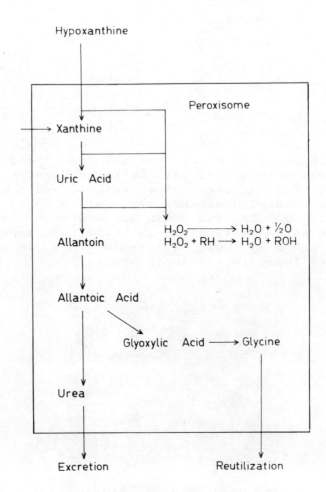

Figure 5

Function of peroxisomes in purine degradation,
hypothesis.

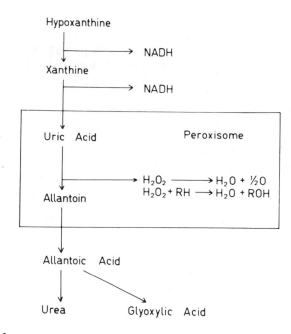

Figure 6

Function of peroxisomes in purine degradation in carp liver according to cell fractionation results.

H_2O_2 which is to be destroyed by peroxisomal catalase, a reaction being probably the most important function of these organelles (1). More likely, however, the enzyme in vivo is working as NAD-dependent dehydrogenase, and is converted to oxidase in vitro (23,24,25).

Our results show, that apart from urate oxidase no purine catabolizing enzymes are housed in peroxisomes. Xanthine oxidase, allantoinase and allantoicase are constituents of the hyaloplasma. The passage of catabolism products through an organelle for one single metabolic step is made by purine metabolites for protection of the cell from hydrogen peroxide (fig 6), whose most powerful producer in metabolism probably is urate oxidase (26).

This work was supported by "Fonds zur Förderung der wissenschaftlichen Forschung in Österreich",Projekt Nr.2368.

REFERENCES

1)DeDuve,C.,Baudhuin,P.,(1966),Physiol.Rev.46,323-357
2)DeDuve,C.,Beaufay,H.Jacques,P.,Rahman-Li,Y.,Sellinger,
   O.Z.,Wattiaux,R.DeConinck,S.,(1960),Biochim.Biophys.
   Acta 40,186-187
3)Scott.P.,Visentin,L.,Allen,J.M.,(1969),Ann.N.Y.Ac.Sci.
   168,244-264
4)Huang,A.,Beevers,H.,(1971)Plant Physiology 48,637-641
5)Huang,A.,Beevers,H.,(1973)J.Cell Biol.58,379-389
6)Baudhuin,P.,Beaufay,H.,Deduve,C.,(1965)J.Cell Biol.26,
                                                 219-243
7)Afzelius,B.A.,(1965)J.Cell Biol.26,835-843
8)Hruban,Z.,Swift,H.,(1966)Science 146,1316-1317
9)Hayashi,H.,Suga,T.,Niinobe,S.,(1973)Biochim.Biophys.
   Acta 297,110-119
10)Kramar,R.,Goldenberg,H.,Böck,P.,Klobučar,N.,(1974)
    Histochemistry 40,137-154
11)Heusequin,E.,(1973)Arch.Biol.(Liège)84,243-279
12)Brunel,A.,(1937)Bull.Soc.Chim.Biol.19,1027-1036
13)Brown,R.,Lord,J.,Merrett,M.,(1974)Biochem.J.144,559-
                                                 -566
14)Cooperstein,S.,Lazarow,A.,(1951)J.Biol.Chem.189,665-
                                                 -670
15)Böck,P.,Goldenberg,H.,Hüttinger,M.,Kolar,M.,Kramar,R.
    (1975)Exp.Cell Res. 90,15-19
16)Gianetto,R.,DeDuve,C.,(1955)Biochem.J. 59,433-438
17)Stirpe,F.,DellaCorte,E.,(1970)Biochem.J.117,97-100
18)Brown,G.,Jr.,(1964)Am.Zool.4,310
19)Tonhazy,N.,White,N.,Umbreit,W.,(1950)Arch.Biochem.
                        Biophys.28,36-42
20)Brown,G.,Jr.,James,J.,Henderson,R.,Thomas,W.,
    Robinson,R.,Thompson,A.,Brown,E.,Brown,S.,(1966)
    Science 153,1653-1654
21)Novikoff,P.,Novikoff,A.,(1972)J.Cell Biol.53,532-560
22)Ericsson,J.,(1964)Acta Pathol.Microbiol.Scand.Suppl.
                        168,1-121
23)Stirpe,F.,DellaCorte,E.,(1970)J.Biol.Chem.244,3855-
                                                 -3863
24)Waud,W.,Ragajopalan,K.,(1976)Arch.Biochem.Biophys.
                        172,354-379
25)Krenitzky,T.,Tuttle,J.,Cattan,E.,Wang,P.,(1974),
    Comp.Biochem.Physiol.49B,687-703
26)Boveris,A.,Oshino,N.,Chance,B.,(1972)Biochem.J.128,
                                                 617-630

PURINE ENZYME ABNORMALITIES: A FOUR YEAR EXPERIENCE

Irving H. Fox

Purine Research Laboratory, University of Toronto

Rheumatic Disease Unit, Wellesley Hospital, Toronto

Specific inherited purine enzyme abnormalities are a rare occurrence. During the past 4 year period in Toronto, Canada, we have screened a population of patients for abnormalities of erythrocyte purine enzymes. Many of these patients were hyperuricemic although a substantial proportion were only assayed because their erythrocytes were used for other experimental studies. Hemolysates from certain patients were specifically referred because an enzyme abnormality was suspected. In the majority of these cases, none was usually found.

Measurement of erythrocyte purine enzymes has been performed in up to 648 patients (Table 1). Seven patients or 1.1% had a partial deficiency of adenine phosphoribosyltransferase (APRT). This agrees with the incidence of APRT deficiency previously reported (1). Six patients or 0.9% had a deficiency of hypoxanthine-guanine phosphoribosyltransferase (HGPRT). Other enzymes were assayed in smaller numbers of patients but no abnormalities were found. A few children with congenital immunodeficiency disease were screened for abnormalities of adenosine deaminase and purine nucleoside phosphorylase and were normal. In addition one patient presented with hypouricemia, hypouricosuria, and elevated urinary oxypurines. These findings were compatible with hereditary xanthine oxidase deficiency.

Our 6 patients with HGPRT form a spectrum conforming to the known features of this X-linked disorder (Table 2). They were male ages 12 to 35 with a history of renal calculi. Only 3 to 6 patients had gout. Two patients were severely defective neurologically, but only one of these had self-mutilation. This patient died at age 12 of acute aspiration. One patient had only mild central nervous system dysfunction.

TABLE 1

SCREENING FOR ERYTHROCYTE PURINE ENZYME ABNORMALITIES

Enzyme	Patients Screened	Number Deficient	% of Total	Number of Increased	% of Total
Adenine Phos-phoribosyltransferase	648	7	1.1	–	–
Hypoxanthine-guanine Phosphoribosyl-transferase	648	6	0.9	–	–
Phosphoribosyl-pyrophosphate Synthetase	612	0	0	0	0
Adenosine Kinase	237	0	0	–	–
Adenosine Deaminase-Total	153	0	0	–	–
Immunodeficiency	13	0	0	–	–
Purine Nucleoside Phosphorylase-Total	87	0	0	–	–
Immunodeficiency	2	0	0	–	–

TABLE 2

CLINICAL FEATURES OF 6 CASES OF

HYPOXANTHINE-GUANINE PHOSPHORIBOSYLTRANSFERASE DEFICIENCY

Patient	Sex	Age	Renal Calculi	Gout	CNS Abnormality
1	M	14	+	–	Retarded, choreo-athetosis
*2	M	12	+	–	Retarded, choreo-athetosis, self-mutilation
3	M	35	+	+	–
4	M	25	+	+	–
5	M	21	+	+	–
6	M	16	+	–	Slow learner, slight spasticity

* Died of aspiration pneumonia.

All patients were hyperuricemic and were overexcretors of uric acid (Table 3). The uric acid to creatinine ratio ranged from 0.53 to 1.95. Erythrocyte HGPRT ranged from $<$0.001 to 36.1 nanomoles/hr/mg. The latter value was 40% of the normal mean. Three patients had an elevation of erythrocyte APRT. PP-ribose-P synthetase in hemolysate was normal.

We have evaluated 7 patients with partial APRT deficiency. There were 3 females and 4 males, 5 patients were hyperuricemic, 1 patient had a renal calculus and 3 patients had gout. One was an overexcretor of uric acid. There were a number of other disorders found in these patients. One subject (case 2) was completely healthy and had no other abnormalities. Another subject had rheumatoid arthritis. Our total experience and the 4 previously published cases (2-5) are listed in Table 4. There were 7 male and 4 females, 9 of 11 were hyperuricemia, 6 of 11 were gouty and 2 of 11 had renal calculi. The question arises whether hyperuricemia and gout are associated with APRT deficiency or are merely a bias introduced by the investigator's interest. We have analyzed previously published pedigrees (3-5) and eliminated the selective bias by excluding the propositus. Twenty-four of 26 family members with APRT deficiency were normouricemic and none was gouty. Thus no definite clinical

TABLE 3

LABORATORY DATA ON 6 CASES OF

HYPOXANTHINE-GUANINE PHOSPHORIBOSYLTRANSFERASE DEFICIENCY

Patient	Serum Uric acid (mg/dl)	Urine Uric acid (mg/24 hr)	Urine Uric acid Creatinine	HGPRT APRT PRPP Synthetase nmoles/hr/mg		
1	8.3	1209	1.95	$<$.001	57.0	41.5
2	-	-	-	.003	52.5	53.5
3	11.3	1279	0.67	34.2	36.5	46.5
4	11.1	1135	0.53	36.1	32.0	34.4
5	9.1	2289	-	4.0	29.1	29.3
6	10.6	1051	1.4	0.03	64.7	50.7
Normal	$<$ 7.0	$<$ 590	$<$ 0.75	87.3 ±16.6	24.4 ±6.8	43.8 ±11.3

abnormalities can be associated with APRT deficiency. We have analyzed the male to female ratio in a similar manner. Fourteen females and 12 males were found among the family members with APRT partial deficiency. This is similar to the expected 1:1 ratio of an autosomally inherited abnormality. The gene coding for APRT has been localized to chromosome 16.

Laboratory determinations in APRT partial deficiency showed that serum uric acid ranged from 3.1 to 13.6 mg/dl. Urine uric acid ranged from 376 to 1100 mg/24 hr. The urine uric acid to creatinine ratio was within the normal range in all patients. Excessive synthesis of uric acid was evident only in patient 7. Values for erythrocyte APRT ranged from 25 to 42% of the normal mean value. Four of 7 patients had deficient leukocytes with 30 to 45 of the normal mean value. No previous deficiency has been found in leukocytes in the 1 case in which it was estimated (3).

Finally we have observed a 49 year old lady with hypouricemia and recurrent renal calculi (Table 5). The serum uric acid was 0.2 mg/dl, the urine uric acid was 15 mg/24 hr and urine oxypurines 2300 micromoles/24 hr. These observations were compatible with xanthine oxidase deficiency, although direct biopsy of deficient tissue was not permitted. Urine xanthine-hypoxanthine ratio was 17.6. The clearance of uric acid and the ratio of the uric acid clearance to the creatinine clearance were normal.

Our studies of purine enzyme defects of man suggest that although patients screened were a selected population, abnormalities may be more common than previously considered. Adenine and hypox-anthine-guanine phosphoribosyltransferase partial deficiencies are the most frequent abnormalities detected. The other enzyme abnor-malities may be rare since no others were detected in our screening program.

TABLE 4

CLINICAL FEATURES OF PROPOSITI FOR

ADENINE PHOSPHORIBOSYLTRANSFERASE DEFICIENCY

	Male	Female	Hyperuricemia	Gout	Renal Calculi
Current 7 cases	4	3	5	3	1
Previous 4 cases (2-5)	3	1	4	3	1
Total 11 cases	7	4	9	6	2

TABLE 5

LABORATORY CHARACTERISTICS SUGGESTIVE OF

OF XANTHINURIA IN 1 PATIENT

Test	Value
Serum uric acid (mg/dl)	0.2
Urine uric acid (mg/24 hr)	15
Urine oxypurines (micromole/24 hr)	2300
Urine xanthine/hypoxanthine	17.9
Curate (ml/min)	5.2
Curate/Ccreatinine (%)	5.2

REFERENCES

1. Srivastava, S.K., Villacorte, D. and Beutler, E. 1972. Correlation
    between adenylate metabolizing enzymes and adenine nucleo-
    tide levels of erythrocytes during blood storage in various
    media. Transfusion 12:190-197.

2. Kelley, W.N., Levy, R.I., Rosenbloom, F.M., Henderson, J.F. and
    Seegmiller, J.E. 1968. Adenine phosphoribosyltransferase
    deficiency: a previously undescribed genetic defect in man.
    J. Clin. Invest. 47:2281-2289.

3. Fox, I.H., Meade, J.C. and Kelley, W.N. 1973. Adenine phos-
    phoribosyltransferase deficiency in man. Am. J. Med. 55:
    614-620.

4. Delbarre, F., Auscher, C., Amor, B., de Gery, A., Cartier, P.
    and Hamet, M. 1974. Gout with adenine phosphoribosyl-
    transferase deficiency. Biomedicine 21:82-85.

5. Emmerson, B.T., Gordon, R.B. and Thompson, L. 1975. Adenine
    phosphoribosyltransferase deficiency: its inheritance
    and occurrence in a female with gout and renal disease.
    Aut. N.Z. J. Med. 5:440-446.

FIBROBLAST PHOSPHORIBOSYLPYROPHOSPHATE AND RIBOSE-5-PHOSPHATE

CONCENTRATION AND GENERATION IN GOUT WITH PURINE OVERPRODUCTION

Michael A. Becker

Department of Medicine, University of California,
San Diego and San Diego Veterans Administration
Hospital, La Jolla, California 92161 U.S.A.

Despite the identification of several hereditary enzyme abnormalities which result in excessive uric acid synthesis (1-4), the underlying metabolic defects in the great majority of individuals with gout and purine overproduction have not been defined by routine screening of enzyme activities (5,6). For this reason, a different approach to the identification of additional enzyme abnormalities or of subtle variants of known enzyme abnormalities has been developed (7). This approach is based on our present understanding of the regulation of the rate of purine synthesis *de novo* schematically summarized in Figure 1.

Study of certain enzyme abnormalities, notably deficiencies of hypoxanthine-guanine phosphoribosyltransferase (HGPRT)[1] and excessive PP-ribose-P synthetase activity, have highlighted the role of the associated increase in intracellular PP-ribose-P concentration in determining excessive rates of purine nucleotide synthesis. This effect appears to be mediated through a weighting of the interaction of PP-ribose-P and inhibitory purine ribonucleotides toward net activation of the first enzyme of purine synthesis *de novo*, PP-ribose-P amidotransferase (PAT) (9,10). Of particular interest for the present study is the observation that increased PP-ribose-P concentration can result either from diminished utilization of this compound (as in HGPRT deficiency) (8,11) or from increased production of PP-ribose-P (as is the case with excessive PP-ribose-P synthetase activity) (4,8). Two further suggestions

----

[1]Abbreviations: HGPRT, hypoxanthine-guanine phosphoribosyltransferase; PP-ribose-P, 5-phosphoribosyl 1-pyrophosphate; PAT, PP-ribose-P amidotransferase; PRT, phosphoribosyltransferase; FGAR, formylglycinamide ribotide.

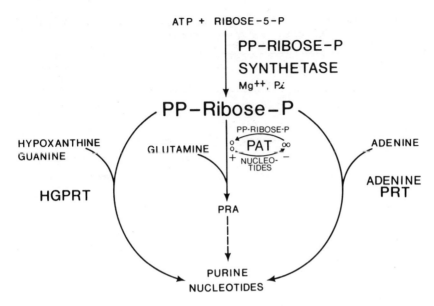

Figure 1.   Schematic representation of purine nucleotide synthesis and its regulation.   See text for discussion.   PRA, phosphoribosyl-amine.   Figure modified from Becker, Meyer, and Seegmiller (8).

arise from consideration of this scheme. First, any additional stimulus to increased PP-ribose-P production and concentration is potentially a stimulus to purine overproduction which should be reflected by increased concentration of ribose-5-phosphate, an immediate precursor of PP-ribose-P and itself the product of a number of metabolic pathways. Second, a potential mechanism for purine overproduction without increased PP-ribose-P concentration is provided by the antagonistic interaction of PP-ribose-P and purine nucleotides on PAT (9,10). Thus, alterations in nucleotide concentrations or in the responsiveness of PAT to purine nucleotides (12) are potential causes of excessive uric acid production not requiring increased PP-ribose-P concentration.

These considerations have led to the prediction that measurement of PP-ribose-P concentration and rate of generation and ribose-5-phosphate concentration would provide a functional assessment of purine metabolism which in the cells of uric acid overproducers would allow tentative subclassification of any such patient into one of the four subgroups shown in Table 1. The value of this subclassification would then lie in establishing the association of each subgroup with abnormality in a particular enzyme reaction or group of reactions as tentatively proposed in Table 1.

TABLE 1

Proposed Classification of Abnormalities Associated With
Excessive Purine Production

	PP-Ribose-P		Ribose-5-Phosphate	Example of Associated
Subgroup	Concentration	Generation	Concentration	Enzyme Abnormality
I	Increased	Increased	Increased	Glucose-6-phosphatase deficiency*
II	Increased	Increased	Normal or Decreased	Increased PP-ribose-P synthetase activity
III	Increased	Normal	Normal	Hypoxanthine-guanine phosphoribosyltrans- ferase deficiency
IV	Normal or decreased	Increased	Normal	Feedback resistant PP-ribose-P amido- transferase*

*Association of enzyme abnormality with pattern of PP-ribose-P and ribose-5-P determinations
proposed but not yet demonstrated.

Table 1.   Reprinted from Becker (7).

In this study, these predictions have been tested by determination of PP-ribose-P concentration and generation and ribose-5-
phosphate concentration in fibroblasts cultured from controls and
from patients with uric acid overproduction. The patterns predicted
for HGPRT deficiency and excessive PP-ribose-P synthetase activity
have been directly validated by determinations made in cells from
patients with these enzyme aberrations; conversely, the usefulness
of the subclassification procedure in directing further study has
been indicated by identification of subtle forms of alteration in
each of these two enzymes among patients with uric acid overproduction of previously undetermined cause.

Fibroblasts were cultured (11) from skin biopsy specimens of
24 individuals who were assigned to groups according to clinical
criteria, daily urinary uric acid excretion and routine measurement at saturating substrate concentrations of HGPRT, adenine PRT
and PP-ribose-P synthetase in erythrocyte lysates. Group 1 was
composed of five normal male skin biopsy donors; Group 2 of five
male patients with gout and normal daily urinary uric acid excretion (that is, excretion of less than 600 mg/day while receiving
a purine-free diet)(13); Group 3 contained a total of seven patients
with excessive uric acid excretion and gross enzyme abnormalities:
four patients had deficiencies of HGPRT activity ranging from

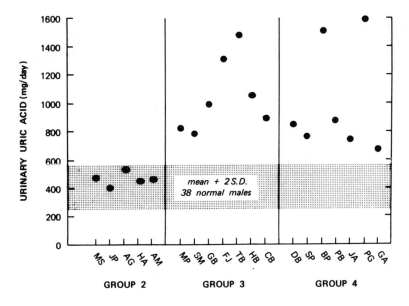

Figure 2. Daily urinary uric acid excretion of patients in groups 2, 3 and 4 while receiving a purine-free diet.

virtually complete (1)   to less severe   (3)   with 3% residual activity; three patients, members of a single family (8), had excessive PP-ribose-P synthetase activity; Group 4 consisted of seven male patients with excessive uric acid excretion and normal (six patients) or nearly normal (one patient) enzyme activities.

Figure 2 depicts the daily urinary uric acid excretion of each of the patients in groups 2, 3 and 4 compared with the values for a group of normal individuals. Uric acid excretion of all patients in groups 3 and 4 was markedly increased. In Figure 3, red cell lysate activities of HGPRT, adenine PRT (3) and PP-ribose-P synthetase (8) are presented. In contrast to the profound HGPRT deficiencies and associated increased adenine PRT activities of the first four patients in Group 3 and the markedly increased PP-ribose-P synthetase activities of the three other patients in this group, enzyme activities in Group 4 patients were normal or minimally abnormal.

Measurement of HGPRT and PP-ribose-P synthetase in each of the 24 fibroblast cultures grown to confluence confirmed the findings in erythrocyte lysates. In contrast to previous studies in HGPRT-deficient fibroblasts (14) and lymphoblasts (15), PP-ribose-P synthetase activity was normal in each of the HGPRT-deficient fibroblast strains. An excessive rate of purine synthesis *de novo* in

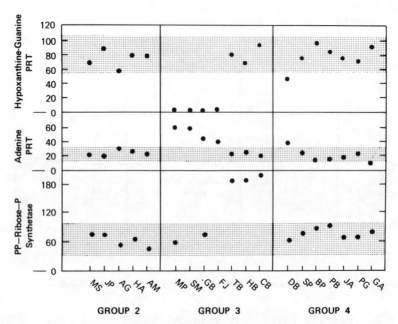

Figure 3. Erythrocyte lysate enzyme activities (nmoles/hr/mg protein) of patients in groups 2, 3 and 4. Dotted areas represent mean ± 2 S.D. derived from 28 normal controls.

fibroblasts from each patient in groups 3 and 4 was shown (Figure 4) by measurement of the rate of accumulation of formylglycinamide ribotide (FGAR) in cells incubated with the glutamine antagonist azaserine (11,16,17). The magnitude of excessive purine synthesis in fibroblasts from these patients ranged from 1.9- to 5.6-fold the values for cells from the controls of groups 1 and 2.

The PP-ribose-P concentration of each fibroblast strain is shown in Figure 5, the brackets denoting ± 1 standard deviation. Fibroblasts from each of the 14 patients with purine overproduction showed increased PP-ribose-P concentration. The uniformity of increased PP-ribose-P concentration in fibroblasts from Group 4 patients contrasts with previous reports (8, 18) of a low incidence of increased erythrocyte PP-ribose-P concentration in patients with uric acid overproduction of unknown etiology. Therefore, PP-ribose-P concentrations were measured in erythrocytes from six of the seven Group 4 patients. In two patients, erythrocyte PP-ribose-P concentrations were normal while the concentrations were increased in the other four patients tested. These findings and earlier reports of normal erythrocyte PP-ribose-P concentrations in some individuals with partial HGPRT deficiencies (19, 20) suggest that fibroblast

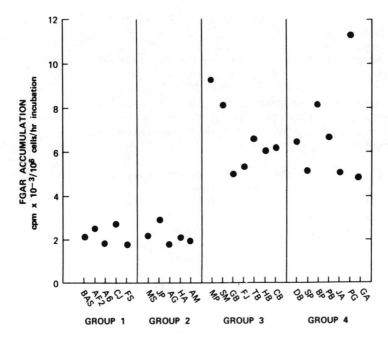

Figure 4.   Purine synthesis *de novo* in fibroblast strains.   Incorporation of {^{14}C}formate into FGAR was determined as previously described (11, 17).

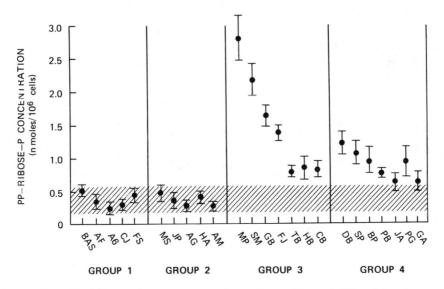

Figure 5.   PP-Ribose-P concentrations in cultured fibroblasts. Solid circle indicates mean and brackets ± 1 S.D. from mean for each cell strain.   Shaded bar indicates ± 2 S.D. from mean for strains from Group 1 patients.   Reprinted from Becker (7).

PP-ribose-P determinations may be more reliable than those performed in erythrocytes.

Generation of PP-ribose-P by intact fibroblasts from each of the 24 individuals studied is shown in Figure 6. Fibroblasts from all four patients with known HGPRT deficiency showed normal PP-ribose-P generation in contrast to the increased generation seen in cells from patients with excessive PP-ribose-P synthetase activity. Among Group 4 patients, PP-ribose-P generation was normal in two cell strains and increased in the other five.

Figure 7 shows the intracellular ribose-5-phosphate concentrations of these fibroblast strains. The ribose-5-phosphate concentrations in cell strains from individuals in groups 1 to 3 were quite similar. Both decreased and increased concentrations of this compound were, however, found among strains from Group 4 patients. In patient B.P., a decreased concentration of ribose-5-phosphate was found while two patients, P.G. and G.A., showed increased concentrations. All three of these cell lines showed increased concentration and generation of PP-ribose-P.

These studies show that among fibroblasts from Group 3 patients, the pattern of findings in patients with HGPRT deficiency is

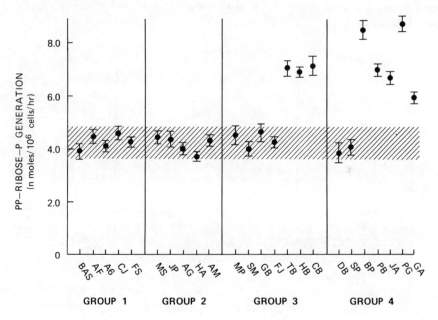

Figure 6. Generation of PP-ribose-P in cultured fibroblasts. Symbols are those described for Figure 5. Reprinted from Becker (7).

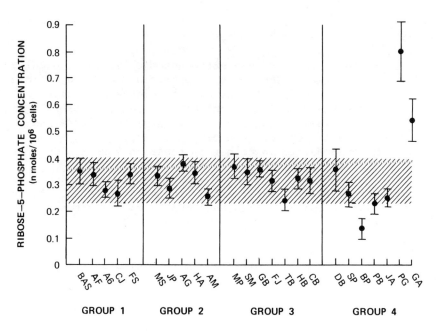

Figure 7. Ribose-5-phosphate concentrations in cultured fibroblasts. Symbols are those described for Figure 5. Reprinted from
Becker (7).

distinguishable from that of patients with excessive PP-ribose-P
synthetase activity. In HGPRT deficiency PP-ribose-P concentration
is increased but PP-ribose-P generation and ribose-5-phosphate
concentration are normal. Excessive PP-ribose-P synthetase activity is associated with increased concentration and generation of
PP-ribose-P, while ribose-5-phosphate concentration is normal.

     In order to assess the usefulness of this method of evaluation
in directing further study of specific enzymes, Group 4 patients
were subclassified (Table 1) according to the results shown in
Figures 5, 6 and 7. Two patients (P.G. and G.A.) were assigned to
subgroup 1; three others (B.P., P.B. and J.A.) to subgroup 2; and
two (D.B. and S.P.) to subgroup 3. Since all cell lines had increased PP-ribose-P concentration, no likely subgroup 4 abnormality
(such as a feedback insensitive PAT) was encountered among this
group of uric acid overproducers.

     One patient with previously undefined enzyme abnormality was
selected from each of the three subgroups and extensive analyses
of HGPRT and PP-ribose-P synthetase were carried out on erythrocyte and fibroblast preparations from each. These studies (7),
which included kinetic, thermal, and electrophoretic evaluation of

partially purified preparations of these enzymes, showed the following:

> Patient D.B. had an electrophoretically and thermally altered HGPRT with a four-fold decreased affinity for the substrates hypoxanthine and guanine.
> Patient B.P. had an electrophoretically altered PP-ribose-P synthetase with a three-fold increased affinity for ribose-5-phosphate.
> No abnormality in either of these enzymes was detected in material from patient P.G.

In summary, this initial experience with the measurement of fibroblast PP-ribose-P concentration and generation and ribose-5-phosphate concentration:

1. has demonstrated a high frequency of increased PP-ribose-P concentration in fibroblasts from purine overproducers

2. has directed attention to specific enzyme reactions resulting in identification of subtle enzyme defects

3. has suggested that excessive ribose-5-phosphate concentration may underlie purine overproduction in some individuals.

ACKNOWLEDGEMENTS

This work was supported in part by: grant AM-19187 from the National Institutes of Health, a Veterans Administration Clinical Investigatorship and a Veterans Administration Grant (MRIS 0865) to the author; grants AM-13622, AM-05646, and GM-17702 from the National Institutes of Health to Dr. J. Edwin Seegmiller, for whose support and advice the author is grateful.

REFERENCES

1. Seegmiller, J. E., Rosenbloom, F. M. and Kelley, W. N.: *Science (Wash, D.C.)* 155: 1682-1684 (1967).
2. Alepa, F. P., Howell, R. R., Klinenberg, J. R. and Seegmiller, J. E.: *Am. J. Med.* 42: 58-66 (1967).
3. Kelley, W. N., Rosenbloom, F. M., Henderson, J. F. and Seegmiller, J. E.: *Proc. Natl. Acad. Sci. U.S.A.* 57: 1735-1739 (1967).
4. Sperling, O., Boer, P., Persky-Brosh, S., Kanarek, E. and De Vries, A.: *Rev. Eur. Etud. Clin. Biol.* 17: 703-706 (1972).
5. Yü, T.-F., Balis, M. E., Krenitsky, T. A., Dancis, J., Silvers, D. N., Elion, G. B. and Gutman, A. B.: *Ann. Intern. Med.* 76: 255-265 (1972).

6.   Becker, M. A. and Seegmiller, J. E.: *Annu. Rev. Med.* 25: 15-28 (1974).
7.   Becker, M. A.: *J. Clin. Invest.* 57: 308-318 (1976).
8.   Becker, M. A., Meyer, L. J. and Seegmiller, J. E.: *Am. J. Med.* 55: 232-242 (1973).
9.   Holmes, E. W., McDonald, J. A., McCord, J. M., Wyngaarden, J. B. and Kelley, W. N.: *J. Biol. Chem.* 248: 144-150 (1973).
10. Holmes, E. W., Wyngaarden, J. B. and Kelley, W. N.: *J. Biol. Chem.* 248: 6035-6040 (1973).
11. Rosenbloom, F. M., Henderson, J. F., Caldwell, I. C., Kelley, W. N. and Seegmiller, J. E.: *J. Biol. Chem.* 243: 1166-1173 (1968).
12. Henderson, J. F., Rosenbloom, F. M., Kelley, W. N. and Seegmiller, J. E.: *J. Clin. Invest.* 47: 1511-1516 (1968).
13. Seegmiller, J. E., Grayzel, A. I., Laster, L. and Liddle, L.: *J. Clin. Invest.* 40: 1304-1314 (1961).
14. Martin, D. W., Jr., Graf, L. H., Jr., McRoberts, J. A. and Harrison, T. M.: *Clin. Res.* 23: 263a (1975).
15. Reem, G. H.: *Science (Wash., D.C.)* 170: 1098-1099 (1975).
16. Henderson, J. F.: *J. Biol. Chem.* 237: 2631-2635 (1962).
17. Boyle, J. A., Raivio, K. O., Becker, M. A. and Seegmiller, J. E.: *Biochim. Biophys. Acta* 269: 179-183 (1972).
18. Fox, I. H. and Kelley, W. N.: *Ann. Intern. Med.* 74: 424-433 (1971).
19. Greene, M. L. and Seegmiller, J. E.: *J. Clin. Invest.* 48: 32a (1971).
20. Sperling, O., Eilam, G., Persky-Brosh, S. and De Vries, A.: *J. Lab. Clin. Med.* 79: 1021-1026 (1972).

FAMILIAL DISTRIBUTION OF INCREASED ERYTHROCYTE PP-RIBOSE-P LEVELS

Roberto Marcolongo, Giuseppe Pompucci, Vanna Micheli

Service of Rheumatology and Institute of Biochemistry

University of Siena, Siena, Italy

Studies in vitro and in vivo have suggested the critical role
played by the intracellular concentration of PP-ribose-P (PRPP) in
the regulation of purine biosynthesis de novo (5,6,14,15). Because
abnormalities of PRPP synthesis may contribute to the pathogenesis
of some cases of hyperuricemia and gout, we studied the concentration of this compound in a number of gouty patients known to produce
excessive quantity of uric acid. Twenty-two male patients, aged
35-72, affected by primary gout with purine overproduction were
examined. Control subjects were 11 males, aged 25-45, who were
not affected by gout or any other metabolic disease. Fluid intake
was not restricted and no subject was taking drugs that might
affect urate levels in blood and urine. Serum and urinary uric
acid determinations were made by an enzymatic-spectrophotometric
method (12). PRPP concentration was determined in erythrocytes
and leukocytes according to the method of Micheli et al. (10).
HGPRTase activity was assayed in erythrocyte hemolysates by the
radiochemical method of Cartier and Hamet (4). The activity of
PRPP synthetase was measured in erythrocyte lysates according
to the method of Becker et al. (2). PP-ribose-P amidotransferase
activity was assayed in leukocyte lysates according to the method
of Wyngaarden and Ashton (16). Glutathione reductase activity
was measured in erythrocyte lysates by the method of Beutler (3).
Intracellular generation was assayed by the method of Hershko
et al. (5,6) after incubating arythrocytes in the presence of 40
mM glucose.

TABLE I

Serum and urinary uric acid and erythrocyte PP-ribose-P content in normal controls and gouty patients

subjects	serum uric acid mg%	urinary uric acid mg/24 hrs	erythrocyte PRPP nmol/g Hb	leukocyte PRPP nmol/10^6 cells
normal controls	4.9 ± 1.2 *	530 ± 115	16.3 ± 9.3	0.124 ± 0.08
gouty patients	8.5 ± 2.1	875 ± 180	14.3 ± 7.2	0.117 ± 0.05
Family B				
B.E.	7.5	845	9.0	0.058
B.N.	8.4	950	27.1	0.202
Family F				
F.F.	7.8	795	188.3	0.627
F.P.	9.2	1100	80.9	0.712
Family G				
G.I.	3.4	375	6.1	0.104
G.N.	8.3	930	63.2	0.373
G.O.	8.5	820	197.1	0.367

* mean ± standard deviation

TABLE II

HGPRTase, PP-ribose-P synthetase and PP-ribose-P amidotransferase in normal controls and gouty patients

subjects	erythrocyte HGPRTase nmol/mg protein/hr	erythrocyte PP-ribose-P synthetase nmol/mg protein/hr	leukocyte PP-ribose-P amidotransferase Units/10^6 cells
normal controls	98 ± 11 *	56.81 ± 16.76	1.34 ± 0.58
gouty patients	102 ± 19	60.30 ± 14.35	1.20 ± 0.44
gouty patients with increased intracellular PRPP content			
B.N.	97	94.98	3.12
G.N.	88	38.13	3.20
G.O.	105	54.39	3.25
F.F.	107	58.41	0.97
F.P.	102	73.23	1.50

* mean ± standard deviation

The results obtained are summarized in Tables I and II. Five gouty patients showed both erythrocyte and leukocyte PRPP concentrations significantly higher than those observed in other gouty patients and in controls (patients B.N., F.F., F.P., G.N., G.O.). These subjects, two of which were brothers (G.N., G.O.) and other two uncle and granson (F.P.,F.F.), had no particular clinical featurea in respect to other gouty patients. The elevated intracellular PRPP content did not appear to result from a decreased utilization of PRPP because of the normal erythrocyte HGPRTase activity, nor is due to an increased PRPP synthetase activity which resulted within the normal range. Patients B.N.,G.N. and G.O. showed, however, an increased activity of leukocyte PP–ribose–P amidotransferase (9). A familial survey has been carried out on serum uric acid levels and erithrocyte PRPP content among the relatives of the gouty patients F.F. and F.P. This study disclosed an increased erythrocyte PRPP content or hyperuricemia in some male and female relatives. The determinable genealogy of this family is shown in Fig. 1. The first and second generation about which we had information, although only by history, comprise subjects I-1 and I-2, and subjects from II-2 to II-10 who are believed to have been free of arthritis and nephrolithiasis, and subject II-1 who is said to have had a severe tophaceous gout. The study of the oldest generation available to us revealed that two of the siblings (III-1 and the first probandus III-5) of the gouty subject II-1 had gouty arthritis and the third (III-3) had asymptomatic hyperuricemia. The son (the second probandus IV-3) of the gouty subject III-1 had also gouty arthritis. All other members of the third, fourth and fifth generations tested had no clinical indications of gouty arthritis or nephrolithiasis. Increased serum uric acid levels were present in subjects III-9, III-10, IV-20, V-15, V-37, V-47 and V-48. An increased erythrocyte PRPP content has been observed in subjects III-29, IV-5, IV-10, IV-12, IV-23, IV-29, IV-36, IV-43, IV-44, IV-51, IV-52 and V-5. The pattern of occurence of increased PRPP concentration did not show any definite modality of genetic transmission, thus the mode of inheritance of the observed PRPP abnormality remains to be defined. From this point of view, it must be emphasized that there are several suggestions of the limitations of PRPP determinations only in erythricytes, as showed by the normal erythrocyte PRPP content in some subjects with partial HGPRTase activity deficiency in whom an increased PRPP concentration would be expected (1).

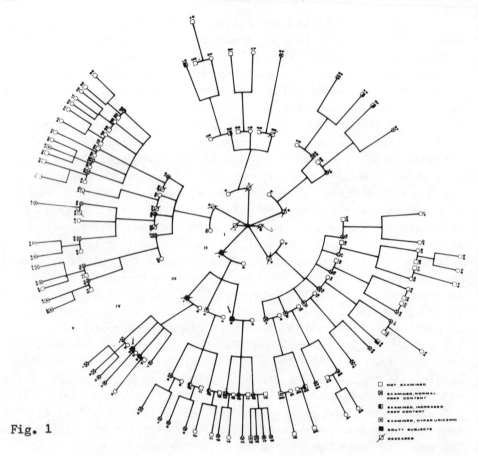

Fig. 1

Several mechanisms could account for the increased erythrocyte
PRPP content in the gouty patients of the Family F and in some of
their relatives. The problem is whether such an alteration is primary or secondary to other metabolic disturbances. Of particular interest is the finding that the increased PRPP levels are not related
to an increase in the activity of PRPP synthetase that could provide
the more likely mechanism for the abnormally high intracellular PRPP
content in the absence of HGPRTase deficiency. It must be emphasized
that inorganic phosphate (Pi) is an essential cofactor for the activity of PRPP synthetase. According to Becker et al. (2) our assays*
were carried out at Pi concentration of 32.0 mM. However, an increased PRPP synthetase activity was demonstrable in the family
reported by Sperling et al. (13) only at Pi concentrations below
2.0 mM, therefore our cases will require a more careful analysis
of enzyme activity at different Pi concentrations.

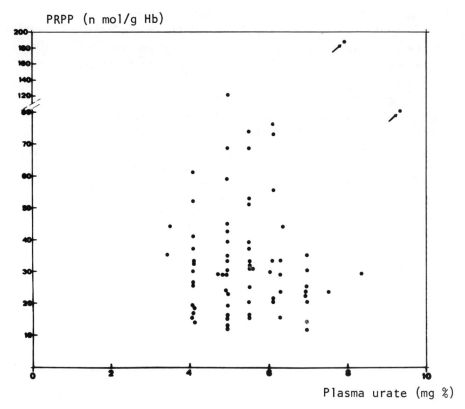

Fig. 2

The lack of correlation between PRPP levels and PRPP synthetase
activity could suggest that additional causes of increased intracellular PRPP concentration other than increased PRPP synthetase
activity and HGPRTase deficiency must exist, as suggested by Becker
et al. (2). A significant role of ribose-5-P has been suggested in
determining an accelerated PRPP production and purine overproduction
in patients with glucose-6-phosphatase deficiency (type 1, glycogen
storage disease) and in gouty patients in whom normal activities of
PRPP synthetase were observed (1,7). Our experiments showed that
generation of PRPP from glucose in erythrocytes of the gouty patients examined was comparable to that from the controls. An
increased activity of glutathione reductase would presumably increase
cellular PRPP levels through an accelerated metabolic rate of
the exose monophosphate shunt (7,8,15), however, our patients showed
an enzyme activity 4.80 ± 1.60 U/g Hb) comparable to that from normal controls (4.12 ± 1.10 U/g Hb).

The basic abnormality underlying an increased PRPP concentra—
tion in gouty patients of Family F and in some of their relatives
and its possible significance and correlation with purine and uric
acid production remain obscure. Of particular interest is the
finding of lack of any correlation between increased intracellular
PRPP concentration and serum uric acid values among the relatives
examined (Fig. 2, the arrows indicate gouty patients F.F. and F.P.).
On the other hand, an increased PRPP availability itself seems not
to be sufficient to increase serum uric acid, as demonstrated in
normal subjects after glucose and ribose loads that determine a
significant erythrocyte PRPP increase but no modifications of serum
and urinary uric acid (11). It remains to be defined whether alterations of the intracellular PRPP concentration are associated in
all circumstances with corresponding alterations in the rate of
purine synthesis de novo.

REFERENCES

1. Becker, M.A. (1976). J.Clin.Invest., 57, 308.
2. Becker, M.A., Meyer, L.J., Seegmiller, J.E. (1973). Amer.J.Med.,
   55, 232.
3. Beutler, E. (1971). In "Red Cell Metabolism", Grune & Stratton,
   New York, 64.
4. Cartier, P., Hamet, M. (1968). Clin.Chim.Acta, 20, 205.
5. Hershko, A., Hershko, C., Mager, J. (1968). Israel J.Med.Sci.,
   4, 939.
6. Hershko, A., Razin, A., Mager, J. (1969). Bioch.Biophys.Acta,
   184, 64.
7. Kelley, W.N., Holmes, E.W., Van der Weyden, M.B. (1975). Arthr.
   Rheum., 18, 673, suppl.
8. Long, W.K. (1967). Science, 155, 712.
9. Marcolongo, R., Pompucci, G., Micheli, V., Marinello, E. (1974).
   Reumatismo, 26, 223.
10. Micheli, V., Pompucci, G., Marcolongo, R. (1975). Clin.Chim.Acta,
    65, 181.
11. Pompucci, G., Micheli, V., Marcolongo, R. (1976). In press.
12. Praetorius, E., Poulsen, H. (1953). Scand.J.Clin.Lab.Invest.,
    5, 273.
13. Sperling, O., Boer, P., Persky-Broch, S., Kanarek, E., De Vries,
    A. (1972). Europ.J.Clin.Biol.Res., 17, 703.
14. Sperling, O., Ophir, R., De Vries, A. (1971). Rev.Europ.Et.Clin.
    Biol., 16, 147.
15. Wyngaarden, J.B. (1974). Amer.J.Med., 56, 651.
16. Wyngaarden, J.B., Ashton, D.M. (1959). J.Biol.Chem., 234, 1492.

X-LINKED PATTERN OF INHERITANCE OF GOUT DUE TO MUTANT FEEDBACK-RESISTANT PHOSPHORIBOSYLPYROPHOSPHATE SYNTHETASE

E. Zoref, A. de Vries and O. Sperling

Tel-Aviv University Medical School, Department of Chemical Pathology, Tel-Hashomer, and the Rogoff-Wellcome Medical Research Institute, Beilinson Medical Center, Petah Tikva, Israel

We have recently reported a new familial enzyme abnormality associated with excessive purine production, gout and uric acid lithiasis (1-3). Phosphoribosylpyrophosphate (PRPP) synthetase in the erythrocytes and cultured skin fibroblasts from the propositus (O.G.) exhibited feedback-resistance to inhibition by several cellular compounds such as GDP and ADP. As a result the enzyme was superactive in the normal physiological milieu, and consequently PRPP content and availability for nucleotide synthesis were increased. In cultured fibroblasts the increased PRPP availability was manifest in accelerated de novo synthesis of purine nucleotides (3).

The father (I.G.), one brother (A.G.) and all four sons of the propositus were found normal by the clinical and biochemical criteria studied (Tables 1 and 2). On the other hand, the propositus' other brother (H.G.) and his mother (D.G.) were found to be affected. Whereas the biochemical and clinical manifestations of the defect in this brother (H.G.) were similar to that in the propositus (O.G.), the manifestations in the mother (D.G.) were attenuated. She exhibited hyperuricosuria but was normouricemic and never had suffered from gouty arthritis or kidney stones. No abnormality could be detected in her erythrocytes but her fibroblasts exhibited increased PRPP content and availability and accelerated synthesis de novo of purine nucleotides. In preliminary experiments, the increase above normal in these parameters in the mother's (D.G.) fibroblast cultures, amounted to 56-72% of the increase exhibited by the fibroblasts from her affected son (O.G.) (3). However, in further studies on the rate of purine synthesis de novo in several cultures from additional biopsies of the mother's skin, a broad range of values was found, ranging from slightly above the upper normal limit to

TABLE 1

SERUM LEVEL AND URINARY EXCRETION OF URIC ACID IN AFFECTED FAMILY

Subject		Age (yr)	Wt (kg)	Uric acid [a]	
				Serum (mg/100ml)	Urine (mg/24h)
Control subjects [b]		20–45		< 7.0	< 800
Affected family					
Propositus	O.G.	38	95	13.5	2,400
Father	I.G.	68	78	6.7	880
Mother	D.G.	68	70	5.3	1,100
Healthy brother	A.G.	23	80	6.0	
Affected brother	H.G.	35	70	13.6	2,250
Sons	Al.G.	16	61	5.2	361
	Am.G.	9	26	3.7	330
	Ra.G.	10	26	3.7	300
	T.G.	4	16		220

[a] Values in the patient O.G. were obtained after 10 days cessation of treatment on low purine diet. Uric acid was determined by an anzymatic spectrophotometric method. The uric acid values given for the control subjects are the upper normal limit for normal subjects on normal home diet in our laboratory (colorimetric autoanalyzer method).

[b] Three females and two males.

approximately 80% of the increase found with her son's fibroblasts (Table 2). On the basis of these data, the mother (D.G.) could be heterozygous either for an X-linked recessive trait or for an autosomal dominant trait, the latter with variable expression. An autosomal dominant trait has been suggested for a family with mutant PRPP synthetase reported by Becker et al (4).

If the mother (D.G.) is a heterozygote for X-linked recessive trait, then her fibroblast cultures should contain two cell populations, normal and mutant (5). On the other hand, if the mother is heterozygous for autosomal dominant trait, her fibroblast cultures should contain only the mutant cell population. The finding of the broad range of rates of activity of purine synthesis de novo in different cell cultures from the mother suggested the presence in them of two cell populations in varying proportion. Studies were therefore carried in our laboratory to directly demonstrate the

TABLE 2

PRPP CONTENT AND AVAILABILITY AND DE NOVO SYNTHESIS OF PURINE NUCLEOTIDES IN CULTURED FIBROBLASTS[a]

Subjects		PRPP Content	[8-^{14}C]Adenine incorporation into cellular nucleotides	Excretion into medium of labeled purines synthetized from [^{14}C]formate
		(nmol/mg protein)	(pmol/mg protein/min)	(cpm/mg protein/h)
Control subjects[b]		0.126±0.0768	106.81±24.51 (12)	722±354.55 (9)
Affected family				
Porpositus	O.G.	0.466 (5)	188.0 (10)	±1,187 (7)
Father	I.G.	0.062 (2)	91.9 (2)	548 (2)
Mother	D.G.	0.329 (2)	157.8 (4)	830-9,000 (20)
Healthy brother	A.G.		128.3 (5)	1,201 (3)
Sons	Al.G.		88.3 (1)	857 (1)
	Am.G.		76.6 (1)	1,383 (1)
	Ra.G.		109.1 (1)	690 (1)

a Values represent for control subjects mean ± 1 SD, for mother (D.G.) the range of values obtained in 20 experiments on different cultures derived from 3 biopsies, and for all other subjects means only. Number of experiments is indicated in parenthesis.

b Five subjects for PRPP content and purine syntehsis de novo and six for adenine incorporation.

presence of two cell populations in the mother's cell cultures. This was done by selecting the mutant cells from mixed cultures containing mutant and normal cells. The method utilized was essentially that described by Green and Martin (6). The fibroblast growth medium (Eagle's Minimal Essential Medium, containing Earle's balanced salt solution with 15% fetal calf serum) was modified to allow survival of only the mutant cells, i.e. cells possessing increased capacity to produce purine nucleotides. 6-Methyl-mercaptopurine-riboside (0.2 mM) was added to block purine synthesis de novo, and hypoxanthine (0.2 mM) and uridine (0.5 mM) were added to allow salvage synthesis of inosinic acid and uridylic acid, respectively. Normal cells, under the selective conditions being unable to produce a sufficient amount of purine nucleotides to sustain life, died within one week. In contrast, under the same selective conditions and during the same time period, the mutant cells, containing the feedback-resistant superactive PRPP synthetase, survived the selection and multiplied normally following transfer to the normal growth medium (Table 3). In a subsequent experiment, cultures from normal subjects, from the propositus and from his mother, as well as normal:mutant cell mixtures (1:1 ratio) representing artificial mosaicism for X-linkage of mutant PRPP synthetase, were exposed to the selective condition for 7-10 days. Special care was taken to ensure that at the onset of selection all cultures contained less than 250 cells/cm^2 growth surface. This was required since at higher cell density a contact-dependent metabolic cooperation between mutant and normal cells prevented selection (7). Following selection, the cell cultures were allowed to grow in fresh regular growth media until confluency, and then trypsinized and propagated for one more generation. As controls, samples of each culture were grown in parallel in the normal medium. Subsequently, the rate of purine synthesis de novo was measured (3,7) in all cell cultures (Table 3).

The measurement of the rate of purine synthesis de novo in cell cultures allows reliable identification of normal or mutant cell type in homogeneous cultures, as well as the determination of the proportion between normal and mutant cells in mixed cultures. As can be seen from Table 3, exposure to the selective condition resulted in death of normal cells, whether in pure cultures or in mixed mutant:normal cultures. No survivors were found in exposed pure normal cultures, while the survivors of the mixed mutant:normal cultures exhibited excessive purine synthesis, typical for the mutant cells. All cultures from the mother whether originally with low, intermediate or high rates of purine synthesis, exhibited following exposure augmentation in the rate of purine synthesis, the increase being the greater, the lower the rate in the control unexposed cultures.

These results provide definite evidence for the presence of two cell types in the fibroblast cultures from the mother, one with

TABLE 3

PURINE SYNTHESIS DE NOVO IN CULTURED FIBROBLASTS FOLLOWING
EXPOSURE TO SELECTIVE CONDITIONS

Source of cells	Excretion into medium of labeled purines synthetized from [^{14}C]formate (cpm/mg protein/h)[a]	
	Not exposed to selection	Exposed to selection
Normal subjects		
1	830	All cells killed
2	1,350	All cells killed
Propositus O.G.	15,000	16,000
	(14,000–17,000)	(14,000–18,000)
Mixture of normal and mutant cells 1:1		
O.G. + 1	9,100	15,000
O.G. + 2	9,800	15,760
O.G.'s mother (D.G.)	1,850	10,560
	4,810	11,830
	5,900	18,330
	10,830	15,260
	13,300	19,600

[a] Values were obtained in representative experiments.Only for propositus O.G. values represent means and the range is given in parenthesis.

normal and the other with mutant PRPP synthetase. The death of the normal cells in the mother's cultures, similar to the death of the normal cells in the mixed normal:mutant cultures, resulted in the augmentation in the rate of purine synthesis in these cultures.The rate of purine synthesis increased from the low value in the control unexposed culture, reflecting the average of the activities of the normal and the mutant cells, to the high rate characteristic for the mutant cells.

On the basis of the available clinical and biochemical data we conclude that in this gouty family with mutant feedback-resistant PRPP synthetase the abnormality is transmitted as an X-linked recessive trait.

ACKNOWLEDGMENT

This study was supported in part by a research grant (No. 78) from the United-States-Israel Binational Science Foundation (BSF), Jerusalem, Israel.

REFERENCES

1. Sperling, O., Boer, P., Persky-Brosh, S., Kanarek, E. and de Vries, A. Rev. Europ. Etud. Clin. Biol., 17:703-706, 1972.
2. Sperling, O., Persky-Brosh, S., Boer, P. and de Vries, A. Biochem. Med., 7:389-395, 1973.
3. Zoref, E., de Vries, A. and Sperling, O. J. Clin. Invest., 56:1093-1099, 1975.
4. Becker, M.A., Meyer, L.J., Wood, A.W. and Seegmiller, J.E. Science, 179:1123-1126, 1973.
5. Lyon, M.F., Nature, 190:372-373, 1961.
6. Green, E.D. and Martin, D.W. Jr. Proc. Natl. Acad. Sci. U.S.A., 70:3698-3702, 1973.
7. Zoref, E., de Vries, A. and Sperling, O. Nature, 260:786-788, 1976.

INCIDENCE OF APRT DEFICIENCY

B.T. EMMERSON, L.A. JOHNSON and R.B. GORDON

University of Queensland Department of Medicine

Princess Alexandra Hospital, Brisbane, Australia

A partial deficiency of adenine phosphoribosyltransferase (APRT) has been reported in a number of subjects with gout, as well as in asymptomatic individuals (1-4). The association with gout has been attributed to the chance association of two common conditions and no mechanism has been demonstrated to relate APRT deficiency to a disorder of purine metabolism. The incidence of APRT deficiency in 700 normal subjects was assessed to test this hypothesis. A spectrophotometric assay was developed which was ideally suited for large scale surveys. The assay was based on the change of UV-absorbance which resulted on the conversion of Ad to its nucleotide and was dependent upon the simultaneous and quantitative removal of hemoglobin and AMP from the incubation solution. No significant difference was found between the values obtained by this method and the standard radiochemical method (5).

APRT activities showed a distribution ranging continuously between 10 and 40 nmoles/mg of Hb/hr, with a mean value of 25 $\pm$ 5 nmoles/mg of Hb/hr. Three individuals were found with APRT activities less than 3 SD below the mean and these were considered to exhibit a partial APRT deficiency. This incidence was not significantly different from that found in subjects with gout and suggested that the association of APRT deficiency and gout may be explicable by the incidence of APRT deficiency in the normal population.

These studies are reported in more detail in the following paper which has been submitted for publication: Johnson, L.A., Gordon, R.B. and Emmerson, B.T. "Adenine phosphoribosyltransferase A simple spectrophotometric assay and the incidence of mutation in the normal population".

REFERENCES

1. KELLEY, W.N., LEVY, R.I., ROSENBLOOM, F.M., HENDERSON, J.F.
    and SEEGMILLER, J.R. (1968). J. Clin. Invest. 47:2281.

2. FOX, I.H., MEADE, J.C. and KELLEY, W.N. (1973). Am. J. Med.
    55:614.

3. DELBARRE, F., AUSCHER, C., AMOR, B. and DeGERY, A. (1974).
    In Sperling, O., DeVries, A. and Wyngaarden, J.B. (Eds)
    "Purine Metabolism in Man". Advances in Experimental
    Medicine and Biology, vol. 41A, Plenum Press, New York,
    p. 333.

4. EMMERSON, B.T., GORDON, R.B. and THOMPSON, L. (1975). Aust.
    N.Z. J. Med. 5:440.

5. EMMERSON, B.T., THOMPSON, C.J. and WALLACE, D.C. (1972).
    Ann. Intern. Med. 76:285.

The authors are grateful to the National Health and Medical
Research Council of Australia for financial support.

COMPLETE DEFICIENCY OF ADENINE PHOSPHORIBOSYLTRANSFERASE: RE-

PORT OF A FAMILY

K.J. Van Acker, H.A. Simmonds and J.S. Cameron

Department of Pediatrics, University of Antwerp, Belgium, and Department of Medecine, Guy's Hospital Medical School, London, U.K.

The study of inborn errors of purine metabolism in man has
led to a better understanding of the latter. Deficiencies of
enzymes involved in the purine reutilization pathways mainly con-
cern the hypoxanthine-guanine phosphoribosyltransferase (HGPRT),
the clinical expression being the Lesch-Nyhan syndrome or eventually
gout. Partial deficiency of adenine phosphoribosyltransferase
(APRT), the enzyme catalyzing the transport of the ribosylphosphate
moiety of phosphoribosylpyrophosphate (PRPP) to adenine with formation of AMP, has only rarely been described (2,3,5-9). Complete
deficiency of this enzyme was reported in only one patient (2).
We report here on two siblings with complete APRT deficiency: in
the youngest child the deficiency was detected during an investigation for urolithiasis, in the older brother it was demonstrated
on the occasion of the family study.

The propositus, a boy, was admitted to the Pediatric Department for the first time at the age of 20 months for investigation
of urolithiasis with abdominal colics, dysuria and pain in the
urethra while eliminating small stones. Since birth he had passed
gravel and since the age of about 9 months small stones and crystals appeared in the urine. He was subsequently treated during 6
months with a low protein diet, a treatment which resulted in a
diminished rate of formation and even a temporary disappearance of
the stones. Analysis of the stones with conventional techniques
had revealed pure uric acid; plasma and urine uric acid were reported normal on this occasion. He was subsequently treated with
alkali and allopurinol without much result.

Clinical examination at the age of 20 months revealed a
boy with a normal length and weight, a slightly increased head
circumference (51 cm) but no abnormal clinical findings except
for a marked muscular hypotonia and an hyperlaxity of the
joints.   Due to the hypotonia the motor developmental milestones
had been retarded, but intelligence was normal. Self-mutilation
was never observed. Routine laboratory investigations were not
contributory; the glomerular filtration rate was normal (creatinine clearance 86 ml/min./1.73 m^2). On a normal diet the plasma
uric acid was slightly increased (6,3 mg %), urine uric acid was
normal (8,6 mg/kg/24 hrs) (colorimetric method). The urinary
sediment contained numerous crystals. No calcifications were
seen on an X-ray film of the abdomen. Analysis of the stones
again revealed pure uric acid. Treatment with allopurinol, 100
mg/day, and alkali was continued but had no appreciable effect
on the stone formation.

The patient was reinvestigated one year later, at the age
of 2 ^{7}/12 years. The clinical situation essentially remained
unchanged except for an appreciable diminution of the hypotonia:
the child was now able to walk normally. Routine laboratory investigations were normal again, as were the glomerular filtration
rate (creatinine clearance 105 ml/min./1.73 m^2), the renal concentrating capacity and the renal acidification capacity. Aminoacids
in plasma and urine were also normal. The urinary sediment still
contained a large amount of crystals. EMG and EEG gave normal
results.

During this period purine metabolism was studied more thoroughly. Plasma and urine acid were normal (enzymatic method) but appreciable amounts of adenine and its oxydation products 8.OH-adenine
and 2.8.Di-OH-adenine were detected in the urine (TABLE 1). Re-
examination of the stones with UV-spectrophotometry, mass spectro-
metry, infrared spectrophotometry, high voltage electrophoresis
and thin layer chromatography showed that the principal constitu-
ent was not uric acid but 2.8.Di-OH-adenine (10).

APRT activity was then determined in erythrocyte lysate: it
was less than 1% of normal. Mixing of the hemolysate from the
patient with hemolysate from normal controls yielded intermediate
levels, thus excluding inhibition or absence of activation of the
APRT in the erythrocytes. HGPRT activity was within normal limits.
Hematological investigations were performed on several occasions:
megaloblasts were never observed; G$_6$PD, Pyruvate Kinase and Glutathione Reductase activity in the erythrocytes was normal.

Treatment with low purine diet and allopurinol (125 mg/day)
was started and the alkali supplement was withdrawn. Within days
stones were no longer eliminated and the crystals disappeared from

TABLE 1

Urinary excretion of purines in the propositus (low purine diet).

urinary excretion (mmol/24 hrs)

Uric acid	0.71
Xanthine	0.03
Hypoxanthine	0.025
	0.765
Adenine	0.112
8.OH-Adenine	0.016
2.8.Di-OH-Adenine	0.141
	0.269

the urinary sediment. No more colics were noted. The child was reinvestigated after 3 months: the beneficial effect of the treatment was still present.

The patient's family was studied extensively. The pedigree is shown in the FIGURE 1. It should be noted that the mother (II_2) had 4 spontaneous abortions before giving birth to her 2 living children. The father (II_2) was operated on for urolithiasis in 1974 but no data are available. There was no history of urolithiasis or gout in any other family member. The paternal grandmother (I_3) underwent nephrectomy probably for cystic disease of the kidney, while the maternal grandmother (I_1) suffered from proteinuria for some years. An unusually high number of individuals (8, not indicated in the pedigree) from 2 generations are said to have died from cancer.

APRT activity was measured in erythrocyte lysate from all the living family members shown on the pedigree with the exception of the father. Complete APRT deficiency (less than 0.8 % of normal) was demonstrated in the older brother of the propositus. This is a 7-year-old boy without any complaint, who shows no abnormalities on clinical examination except perhaps for a slight hyperlaxity of the joints, and who has a normal intelligence. This patient is being investigated at present: although the urine contains increased amounts of adenine and its oxydation products, no crystals are seen in the sediment and there is no stone formation. Five individuals, all from the mother's side, were found to be heterozygotes with APRT activities in the erythrocyte lysate ranging from 20 % to 47 % of normal. HGPRT activity was normal in all. Plasma uric acid in the 13 individuals in whom it was measured, was normal (TABLE 2). As in the younger brother, inhibition or lack of

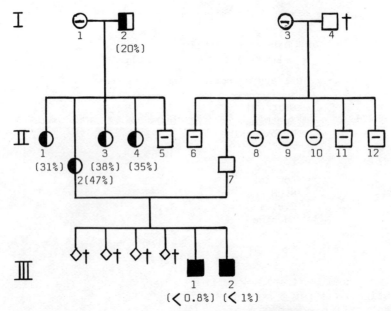

FIGURE 1.   Pedigree of the D'H. family. ■ ● complete APRT deficiency; ◨ ◖ partial APRT deficiency; ⊟ ⊖ normal APRT
activity; ☐ not examined.

activation of APRT was excluded in the 7-year-old boy by mixing
hemolysate from this patient with hemolysate from normal controls.

DISCUSSION

APRT deficiency has been described only rarely: partial deficiency was observed in 41 individuals from 6 families and only one
patient with total deficiency is mentioned in the literature
(TABLE 3). It is not clear from these observations to what extent and in what way purine metabolism is influenced by APRT
deficiency: abnormal urate production, gout and urolithiasis
occurred in family members with normal APRT activity as well as
in those with partial deficiency. The patient with complete
deficiency could not be fully investigated: urolithiasis, very
comparable to that seen in our patient, was the only clinical
symptom. As opposed to the X-linked Lesch-Nyhan syndrome, the
transmission of the defect in the APRT deficient families suggests
an autosomal inheritance.

The clinical symptoms in the propositus described here, were
limited mainly to abdominal colics and urethral pain due to the

TABLE 2

Uric acid levels, APRT and HGPRT activity in
the different members of the D'H. family.

	PLASMA URIC ACID (mmoles/l)	APRT[x]	HGPRT[x]
I1	0.23	28.7	133.1
I2	0.12	6.08	133.1
I3	0.23	20.8	123.0
II1	0.15	7.4	115.3
II2	0.12	6.08	133.1
II3	0.20	9.3	124.1
II4	0.09	8.5	121.4
II5	0.13	23.2	138.4
II6	0.21	33.4	122.0
II8	-	28.8	127.0
II9	0.17	25.5	114.4
II10	0.16	29.4	125.4
II11	0.25	21.6	117.0
II12	0.19	29.7	129.0
III1	0.20	<0.25	123.2
III2	0.13	<0.3	125.1
control adult female		24.1	117.4
control adult male		29.5	134.1

x expressed as nanomoles of base converted per mg of hemoglobin per hour.

elimination of crystals and small stones. The muscular hypotonia
was transient and can probably be ascribed to the longstanding
low protein diet before admission. Hyperlaxity of the joints
was prominent and may be of importance as it was also present to
a lesser degree in the older brother as the only clinical symptom.
No clinical symptoms were observed in the heterozygotes.

Considering the role of APRT in purine metabolism, it can
be predicted that total deficiency of this enzyme will lead to
accumulation of adenine and increased excretion of adenine and its
oxydation products 8.OH-adenine and 2.8.Di-OH-adenine. The latter
is unsoluble over a wide range of pH. A comparable situation has
been obtained in different animal species and in normal and disea-
sed man (1,4) by administration of adenine: the problem is discussed in detail elsewhere in this symposium (1). Why only the youngest of the two brothers described here, formed 2.8.Di-OH-adenine
stones and crystals is not clear at present.

TABLE 3

FAMILIES WITH APRT DEFICIENCY DESCRIBED IN THE LITERATURE.

Examined Family Members	APRT Activity	Hyper		Gout	Urolithiasis	Hyperlipopro-teinemia	Reference
		Uricemia	Uricosuria				
15 (3 generations)	Partial deficiency: 4 (21-31% activity)	none		none		1 (Type II)	(9)
	Normal: 11						
13 (3 generations)	Partial deficiency: 8 (8-44% activity)	1		1	none	1 (Type IV)	(8)
	Normal: 5	2		none		normal in propos.	
23 (4 generations)	Partial deficiency: 8 (12-41% activity)	2		1	1	normal in propos.	(5)
	Normal: 15	3		1	3		
24 (3 generations)	Partial deficiency: 13 (32-49 %activity)	1		1	none	normal	(6,7)
	Normal: 11	1		1	1		
number not mentioned (3 generations)	Partial deficiency: 5 (25-54% activity)	2 (slight)	(slight)	none		1 (Type IV)	(3)
	Normal: ?						
6 (3 generations)	Total deficiency: 1	1 (slight)		none	1	1	(2)
	Partial deficiency: 3						
	Normal: 2						
	Total deficiency: 1	(1)		-	1	1	
	Partial deficiency: 41	6		3	1	1	
	Normal: 44	6		2	3	2	

Allopurinol blocks the conversion of adenine to its oxydation
products: the beneficial effect of this treatment on the stone
formation in the propositus is therefore not unexpected. Lack
of this effect during prior treatment of the child with allopurinol
may be due to the simultaneous treatment with alkali and to the
lower dose given. The latter may also be responsable for the poor
results in patients with Lesch-Nyhan syndrome treated with adenine
and allopurinol (4).

The influence of APRT deficiency on other pathways of purine
metabolism is at present being investigated: part of this study
is discussed elsewhere in this symposium (11): uric acid in
plasma and urine was normal in the individuals from this family in
whom it was studied.

Although APRT values are not known yet in the father and
cannot be obtained in the paternal grandfather, autosomal recessive
transmission of the APRT deficiency seems the most likely possibility. This would be in accordance with the data from the literature. As in the other families, APRT activity was always less
than the expected 50 % of normal in the heterozygotes. Although
other causes cannot be excluded with certainty, the possibility
that the 4 abortions in the family represent lethal forms of the
APRT deficiency must be taken into account.

REFERENCES

1. Cameron, J.S., Simmonds, H.A., Cadenhead, A. and Farebrother, D.:
   Metabolism of intravenous adenine in the pig.
   This symposium.

2. Cartier, P. et Hamet, M.: Une nouvelle maladie métabolique: le
   déficit complet en adénine phosphoribosyltransférase avec
   lithiase de 2.8.-di-hydroxyadénine. C.R.Acad. Sc. Paris 279 :
   883, 1974.

3. Cartier, P.: Les déficits enzymatiques du métabolisme des purines. Symposium Interdisciplinaire sur l'hyperuricémie, Paris,
   Mai 1975.

4. Ceccarelli, M., Ciompi, M.L. and Pasero, G.: Acute renal failure
   during adenine therapy in Lesch-Nyhan Syndrome.
   In Purine Metabolism in Man (Edit. Sperling, O, De Vries, A. and
   Wyngaarden, J.B.) Plenum Press, New York, 41 B : 671, 1974.

5. Delbarre, F., Auscher, C., Amor, B. and de Gery, A.: Gout with
   adenine phosphoribosyltransferase deficiency. Ibid. 41 A : 333,
   1974.

6. Emmerson, B.T., Gordon, R.B. and Thompson, L.: Adenine phosphoribosyltransferase deficiency in a female with gout. Ibid 41 A :

327, 1974.

7. Emmerson, B.T., Gordon, R.B. and Thompson, L.: Adenine phospho-
    ribosyltransferase deficiency: its inheritance and occurrence
    in a female with gout and renal disease. Aust.N.Z.J.Med. 5 :
    440, 1975.

8. Fox, I.H. and Kelley, W.N.: Adenine phosphoribosylferase defi-
    ciency: report of a second family. In Purine Metabolism in
    Man (Edit. Sperling, O., De Vries, A. and Wyngaarden, J.B.)
    Plenum Press, New York 41 A : 319, 1974.

9. Kelley, W.N., Levy, R.I., Rosenbloom, F.M., Henderson, J.F.
    and Seegmiller, J.E.: Adenine Phosphoribosyltransferase defi-
    ciency: a previously undescribed genetic defect in man.
    J.Clin.Invest. 47 : 2281, 1968.

10. Simmonds, H.A., Van Acker, K.J., Cameron, J.S. and Snedden, W.:
    The identification of 2.8.-Dihydroxyadenine: a new component
    of urinary stones. Biochem. J. 1976 (in press).

11. Simmonds, H.A., Van Acker, K.J., Cameron, J.S., Mc Burney, A.
    and Snedden, W.: Purine excretion in complete adenine phosphori-
    bosyltransferase deficiency: effect of diet and allopurinol
    therapy.
    This symposium.

PURINE SYNTHESIS AND EXCRETION IN MUTANTS OF THE WI-L2 HUMAN LYMPHOBLASTOID LINE DEFICIENT IN ADENOSINE KINASE (AK) AND ADENINE PHOSPHORIBOSYLTRANSFERASE (APRT)

Michael S. Hershfield,* Elaine B. Spector,† and J. Edwin Seegmiller
Department of Medicine, University of California, San Diego, La Jolla, California 92093

Human lymphoblasts offer several advantages over fibroblasts for the study of mutations which affect purine metabolism in human cells in tissue culture. Because fibroblasts become senescent after a more or less finite number of doublings, a mutant once selected and cloned offers limited material for detailed biochemical analysis. Lymphoblasts divide indefinitely so that one may repeatedly obtain sufficient quantities of cloned mutants for biochemical study. An additional advantage is that although these lines maintain their karyotype in culture (1,2), they can with continued passage accumulate a high frequency of spontaneous mutations in genes which are not essential for survival. We have found the frequency with which mutations in some autosomal genes occur in the WI-L2 line of human splenic lymphoblasts (3) in some cases approaches that for X-linked genes, such as hypoxanthine-guanine phosphoribosyltransferase (HPRT)[1] (4). Thus, we have been able to select mutants from WI-L2 which are virtually completely deficient in adenosine kinase (AK) and adenine phosphoribosyltransferase (APRT). The gene for APRT has been mapped to human autosomal chromosome

- - - - - - - - - -

[1] Abbreviations: HPRT, hypoxanthine-guanine phosphoribosyltransferase; AK, adenosine kinase; APRT, adenine phosphoribosyltransferase; 6-MMPR, 6-methylmercaptopurine ribonucleoside; PP-ribose-P, phosphoribosylpyrophosphate; FGAR, formylglycinamide ribonucleotide; EHNA, erythro-9-(2-hydroxy-3-nonyl)adenine.

* Present address: Duke University Medical Center, Durham, North Carolina, 27710.

† Present address: Mental Retardation Unit, The Center for the Health Sciences, University of California, Los Angeles, Los Angeles, California, 90024.

16 (5) and that for AK tentatively to chromosome 10. Such mutants
selected *in vitro* are essentially isogenic with their parental line
and, in the case of the mutants to be described, have identical rates
of growth. Since in cultured cells, the rate of *de novo* purine biosynthesis is likely to be closely tied to the rate of cell division,
the ability to eliminate this variable is important, though often
overlooked in studies comparing purine metabolism in cells derived
from patients with "normal" cells derived from different individuals with widely different genetic backgrounds.

SELECTION OF MUTANTS

AK and APRT-deficient mutants of WI-L2 were selected in mass
culture for resistance to the analogs 6-methylmercaptopurine ribonucleoside (6-MMPR) and 2,6-diaminopurine, respectively, and were
then cloned in soft agarose. Whereas growth of the parent strain
was completely inhibited in 0.5 μM 6-MMPR, AK$^-$ mutants were not
inhibited even at 1 mM (6). The APRT deficient lines to be discussed grew normally in at least 200 μM diaminopurine, whereas growth
of the parent strain was inhibited at less than 10 μM. Sensitivity
of the adenosine kinase mutants to adenosine and of the diaminopurine resistant lines to adenine are the subject of another paper
in this symposium (Snyder, Hershfield, and Seegmiller, #5).

Thirty-three 6-MMPR resistant clones derived from three
separate selections with and without treatment with ethylmethane
sulfonate were tested for the ability to phosphorylate {^{14}C}6-MMPR,
both in intact cells and in extracts in the presence of magnesium
and ATP, and all had <1% of parental activity. These were all thus
considered to have arisen from the same mutation in the original
population and one representative, MTI^{r}107a, was selected for
further study. However, the differences between this mutant and
the parent strain which we observed were subsequently confirmed in
MTI^{r}U3 (Table I), an independently selected spontaneous 6-MMPR
resistant mutant derived from WI-L2. In this presentation we shall
deal primarily with diaminopurine resistant strains which were found
to be most deficient (i. e., <1%) in APRT activity, although we
have also obtained and cloned diaminopurine as well as 8-azaadenine
resistant mutants (EBS, in preparation) which contain 10-30% residual APRT activity. All mutants have retained their biochemical
characteristics after 6 months to 2½ years growth in the absence of
selective agents.

ENZYME ACTIVITIES IN MUTANT CELLS

Although 6-MMPR is converted to intracellular 6-MMPR-5'-phos-
phate by adenosine kinase (7), 6-MMPR resistant Ehrlich ascites
tumor lines have been described which are much more deficient in

TABLE 1

SPECIFIC ACTIVITIES OF ADENOSINE KINASE, HGPRT, AND APRT IN EXTRACTS
OF WI-L2 AND MUTANT LYMPHOBLASTS[A]

CELL LINE	ADENOSINE KINASE	HGPRT	APRT
		NMOL/MIN/MG PROTEIN	
WI L2	2.63	7.99	8.89
MTI[R]107A	< 0.02	5.67	9.43
MTIRU-3	< 0.02	6.14	9.22
MTI-TG	0.03	0.07	11.12
DAP[R]	----	5.17	< 0.05

[A]ASSAYED IN 100,000 X G SUPERNATANTS

their ability to phosphorylate the analog than adenosine itself (8). When extracts of WI-L2 are subjected to glycerol gradient sedimentation, the ability to phosphorylate $\{^{14}C\}$-labeled 6-MMPR and adenosine cosediments with APRT activity. Both AK and 6-MMPR kinase activities were undetectable in gradients of MTI[r]107a extracts while the HPRT and APRT activities were found in amount and position comparable to gradients containing WI-L2 extracts. We have been unable to detect the ability to phosphorylate adenosine in extracts of the mutant cells over a wide range of pH and substrate concentrations in the presence of an inhibitor of adenosine deaminase, EHNA (6), Table 1. Similarly, the diaminopurine resistant lines we shall discuss incorporated less than 1% of the $\{^{14}C\}$-adenine into intracellular nucleotide as did the parent strain at an extracellular concentration of $\{^{14}C\}$adenine of 0.5 mM, and in extracts retained less than 1% of parental APRT activity. All of these mutants had essentially the same specific activity for HPRT as the parent strain. We shall also present experiments performed with a derivative of MTI[r]107a which was selected for resistance to thioguanine, called MTI-TG, which has less than 2% of parental HPRT activity (Table I). In some studies we used the HPRT deficient derivatives of WI-L2, AG[r]9 clone 3-1 and AG[r]9 clone 35-1 characterized by Drs. George Nuki and Julia Lever (1). The residual HPRT activity in these strains was the same as in MTI-TG.

TECHNICAL POINTS REGARDING STUDIES OF PURINE SYNTHESIS
IN CULTURED CELLS

We have compared rates of purine synthesis *de novo* in mutants and WI-L2 using the assay described earlier (this symposium #4). In some experiments the medium in which cells were labeled was subjected to acid hydrolysis and labeled purines isolated by Dowex 50 and two-dimensional thin-layer chromatography (9). Our experiments were performed on cells under actual growth conditions in medium containing 10% dialyzed calf serum which had also been heated at 62° for 5 hours to inactivate adenosine deaminase activity. All cell lines grew with the same doubling time of 16-20 hours in this medium. The use of extensively dialyzed calf serum was essential because of the presence in undialyzed calf serum of amounts of hypoxanthine sufficient to inhibit *de novo* purine synthesis to greater than 90% in the lines with normal HPRT activity without affecting HPRT$^+$ lines. In such medium the rates of *de novo* synthesis in HPRT$^+$ lines were initially 5-7-fold less than in the mutant but as medium hypoxanthine was used up the rate of *de novo* purine synthesis rose in the normal line, approaching that in the HPRT$^-$ line after about 6 hr, after which similar rates were found in both (9). Thus, the use of medium containing undialyzed calf serum may well account for the nearly 10-fold faster apparent rates of purine synthesis *de novo* reported for HPRT$^-$ fibroblasts (10) compared to the very much smaller increase we report here.

Many studies of the regulation of purine synthesis in cells in culture have been performed in balanced salt solutions or in culture medium lacking serum. We have not used either of these conditions, first to avoid nutritional step-down conditions which might alter intermediary metabolism in an uncontrolled manner. Second, lymphoblasts and fibroblasts do not divide in the absence of serum and we wished to compare cells under growth conditions. Third, and perhaps more important, lymphoblasts resuspended in medium lacking serum are subject to almost immediate damage. Simply resuspending pelleted lymphoblasts in serum-free medium resulted in 20-35% lysis of cells with each resuspension, documented both by Coulter counter and direct hemocytometer counts, which did not occur in the presence of serum or 4 mg/ml bovine serum albumin (9).

IMP BRANCH POINT REGULATION IN AK$^-$ LYMPHOBLASTS
IN PURINE FREE MEDIUM AND IN THE PRESENCE OF ADENOSINE

To assess the effect of loss of AK activity on regulation of purine synthesis, we compared control of both proximal and distal steps in the pathway in normal and mutant lymphoblasts. A representative study is shown in Table 2-A, in which WI-L2 and MTI 107a (AK$^-$) were compared after 20 hr growth to various cell densities

TABLE 2

A. PURINE SYNTHESIS AT VARIOUS CELL DENSITIES IN PURINE-FREE
   MEDIUM

Cell density (cells/ml x 10^{-5})	Rate of de novo synthesis (relative to WI-L2)	G/A ratio WI-L2	G/A ratio MTI^{r}107a
2.6	0.81	1.09	0.71
5.4	0.75	1.11	0.73
8.4	0.77	1.10	0.56
11.9	0.86	0.92	0.50

B. EFFECT OF ADENOSINE ON DE NOVO PURINE SYNTHESIS

	Adenosine μM	Rate of de novo synthesis (% of control)	G/A ratio
HPRT$^-$	0	100	0.94
	0.8	83	1.31
	8.0	23	3.21
AK$^-$	0	100	0.56
	0.8	61	0.48
	8.0	19	0.51
AK$^-$-HPRT$^-$	0	100	0.59
	0.8	115	0.59
	8.0	102	0.57

A. Four parallel subcultures of WI-L2 and MTI^{r}107a at various
initial cell densities were grown for 18 hr (one doubling) in
medium containing 10% dialyzed, heated fetal calf serum, after
which they were labeled for 60 min with {^{14}C}formate.

B. Cultures in the medium described in A. were preincubated with
or without adenosine for 15 min and then labeled for 30 min with
{^{14}C}formate.

in purine free medium.   Whereas the guanine to adenine (G/A) labeling
ratio was close to one in WI-L2 during most of log phase, the ratio
was about 0.7 in the mutant even at low density and remained much
lower at higher cell densities.   HPRT deficiency did not significantly alter the G/A ratio,and at this point insufficient experiments have been performed with the APRT mutants to permit comment.
Table 2-B shows that when adenosine enters a cell only via the AK
pathway, as in an HPRT$^-$ cell, it resembles adenine, inhibiting
de novo synthesis and diverting residual IMP from adenine to
guanine nucleotide synthesis, i. e., it increases the G/A labeling
ratio.   However, the inhibition of *de novo* synthesis in the AK$^-$
strain is not accompanied by change in this ratio since adenosine
must be deaminated before it can give rise to intracellular nucleotides via the HPRT reaction.   In the AK$^-$-HPRT$^-$ double mutant, adenosine has no effect on *de novo* synthesis since both routes of
nucleotide synthesis are absent.

TABLE 3

DE NOVO PURINE SYNTHESIS AND EXCRETION AND PP-RIBOSE-P CONCENTRATION
IN WI-L2 AND MUTANT HUMAN LYMPHOBLASTS

CELL LINE	WI-L2	AK$^-$	APRT$^-$	HPRT$^-$	AK$^-$-HPRT$^-$
PP-RIBOSE-P (PICOMOLS/10^6CELLS)	173.4	119.6	76.3	610.4	662.4
FGAR LABELLING	CPM/10^6 CELLS/60 MIN (RELATIVE TO WI-L2)				
SODIUM ^{14}C FORMATE mM					
0.17	98070 = 1.0	0.89	0.88	1.27	0.92
2.67	18780 = 1.0	0.88	0.98	1.23	0.86
TOTAL PURINE SYNTHESIS CELLS + MEDIUM					
0.17	161935 = 1.0	1.02	0.82	1.54	1.32
2.67	44715 = 1.0	0.93	0.83	1.12	1.08
MEDIUM					
0.17	10555 = 1.0	0.93	0.79	4.40	4.16
2.67	1812 = 1.0	1.0	1.44	4.85	4.55

Parallel cultures of the indicated strains were grown for 30 hr in
medium containing 10% heated, dialyzed fetal calf serum to mid-log
phase and the indicated parameters measured in duplicate.

PP-RIBOSE-P CONCENTRATIONS AND RATES OF PURINE SYNTHESIS AND EXCRETION IN NORMAL AND MUTANT LYMPHOBLASTS

Table 3 presents the results of an experiment in which rates of purine synthesis and excretion were compared at two different formate concentrations in parallel cultures of WI-L2, the adenosine kinase mutant, the APRT⁻ mutant, an HPRT⁻ line and a mutant lacking both HPRT and AK activities. In this experiment we also measured the intracellular concentration of PP-ribose-P in all five cultures as well as the rate of labeling of formylglycinamide ribonucleotide (FGAR) in the presence of azaserine at both formate concentrations. The important points are first, neither AK nor APRT deficiency results in elevation of PP-ribose-P concentration whereas, as has previously been shown (11), cell lines deficient in HPRT activity have considerable elevation in intracellular PP-ribose-P concentration. Second, all lines labeled FGAR in the presence of azaserine at roughly similar rates with no more than a 25% increase noted in the HPRT⁻ lines. Third, in the absence of azaserine, the overall rates of purine synthesis *de novo*, that is, the sum of labeling intracellular as well as medium purines, were similar in all lines and the maximum increase noted in any of the experimental points for one of the HPRT⁻ mutants was less than 50%. Fourth, the only striking difference relating to rates of purine synthesis *de novo* in any of these mutants is over a four-fold increase in excretion of labeled purines into the medium by both the HPRT⁻ and HPRT⁻-AK⁻ double mutant. This double mutant produces and excretes purines at the same rate as the mutant lacking only HPRT activity. The excreted purines were ∿70% hypoxanthine, 15-20% guanine and the remainder adenine. In these experiments, the newly synthesized purine excreted by HPRT⁺ lines in one hour represented about 5% of total purines synthesized *de novo*, and during 4 hr this decreased to about 1.5% of total synthesis. In the HPRT⁻ mutants the excreted purines represented approximately 15% of total synthesis after 1 and 4 hr of labeling. Thus, after 4 hr the HPRT⁻ mutants excreted 7-10 times as much purine as did HPRT⁺ cells but, in terms of overall *de novo* synthesis "overproduced" by <25%. This important information regarding overall rates of *de novo* synthesis is not available from previous studies (10,12) in which labeling of medium purines was measured but without simultaneous determination of labeling of all intracellular purines. In addition, we pointed out that the studies of Zoref *et al*. (10) were conducted in medium containing 15% undialyzed calf serum. The study by Chan *et al*. (12) reported increased excretion of purines by mutants lacking AK, as well as by HPRT⁻ and APRT⁻ cells from a variety of sources. In that study, cells in "late exponential phase" were incubated for 48 hr with a very low concentration of sodium {¹⁴C}formate--0.018 mM--before labeling of excreted hypoxanthine and xanthine were determined. Differences in growth rate during the labeling period were not evaluated and no measure of overall rates of purine synthesis were undertaken.

It should be noted that the intracellular concentrations of PP-ribose-P shown in Table 3 are as much as 5-20 times higher than previously have been reported for both normal and HPRT⁻ human lymphoblast lines (11,13,14). In these earlier studies, PP-ribose-P concentration was measured in lymphoblasts which had been incubated or washed several times in isotonic solutions devoid of serum which we have shown severely traumatizes these cells. Under these conditions, PP-ribose-P concentration is lowered in both normal and HPRT⁻ lines but the absolute concentration in the normal line may be lowered to such a degree that it becomes limiting for *de novo* purine synthesis (9). This might account for earlier reports (1,11) of 2-4-fold higher relative rates of labeling of FGAR in HPRT mutants of the WI-L2 line in experiments conducted in serum free medium.

Our finding (Table 3) that under growth conditions in purine free medium normal cells synthesize purines at a rate that approaches that of HPRT⁻ lines despite a 3- to 4-fold increased PP-ribose-P concentration in the mutant line is not consistent with the concept that PP-ribose-P concentration limits the rate of purine synthesis *de novo* in normal lymphoblasts. This was demonstrated earlier in this symposium (MSH and JES, #4) where we found no increase in rate of *de novo* synthesis despite about a 5-fold increase in PP-ribose-P concentration during logarithmic growth of normal, AK⁻ and AK⁻-HPRT⁻ lymphoblasts.

SOURCES OF PURINES EXCRETED BY HPRT⁻ LYMPHOBLASTS

The excess newly synthesized hypoxanthine excreted by HPRT⁻ cells could be derived from dephosphorylation of IMP or via an hypothetical "adenosine cycle" (15): AMP → adenosine → inosine → hypoxanthine → IMP → AMP. The rate of the adenosine cycle is thought to be controlled by relative rates of dephosphorylation of AMP and reconversion of adenosine to AMP catalyzed by AK. A cell lacking AK and HPRT activities should excrete purines at a significantly greater rate than normal or HPRT⁻ cells. This was not observed with our mutant MTI-TG. We suggest that the pathway IMP → inosine → hypoxanthine accounts for the excess hypoxanthine excreted by HPRT⁻ lymphoblasts. Furthermore, the ribose-1-PO$_4$ formed in amounts equimolar to the hypoxanthine could be converted to PP-ribose-P. In normal cells the PP-ribose-P and hypoxanthine formed would be converted back to IMP whereas in HPRT⁻ cells the base would be lost and the intracellular PP-ribose-P concentration increased (Fig. 1). The ability of inosine to increase PP-ribose-P concentration, without affecting rate of *de novo* synthesis, is shown in Table 4.

We propose that the series of reactions by which newly

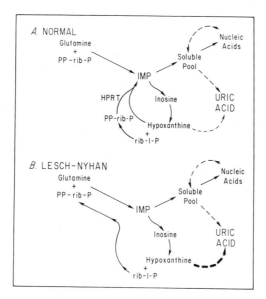

Figure 1. Proposed source of newly synthesized hypoxanthine excreted by HPRT⁻ cells: the "shunt" pathway (16).

TABLE 4

EFFECT OF INOSINE ON PP-RIBOSE-P CONCENTRATION

IN HPRT DEFICIENT LYMPHOBLASTS

Inosine	PP-ribose-P	Rate of de novo purine synthesis
µM	(relative to control)	
None	1.00	1.00
100	1.81	0.99

MTI-TG (HPRT⁻-AK⁻) lymphoblasts were incubated for 10 min with or without 100 µM inosine and then labeled with {¹⁴C}formate for 50 min. PP-ribose-P concentrations were determined at the beginning and end of formate pulse and since the values differed by less than 5%, they were averaged.

synthesized IMP is converted to hypoxanthine represents the "shunt" pathway originally suggested by Benedict *et al.* (16) to account for the extremely rapid labeling of urinary uric acid in some uric acid overexcretors given a dose of isotopically labeled glycine (Fig. 1). In children with the Lesch-Nyhan syndrome dephosphroylation of newly synthesized IMP combined with a defect in purine base reutilization could contribute to both the overproduction of uric acid and the increased intracellular PP-ribose-P concentration characteristic of the disease.

Our results indicate that virtually complete deficiency of neither AK nor APRT activities result in overproduction or overexcretion of newly synthesized purines in cultured human lymphoblasts. However, AK and possibly APRT deficiency appear to alter regulation of the IMP branch point, apparently without detriment to growth in purine free medium. We do not yet know how these effects are mediated or what their consequences might be to man. Individuals with AK deficiency have not yet been reported. It is known that a child with virtually complete APRT deficiency excreted excessive amounts of adenine (17) but whether the adenine was derived from endogenous synthesis or diet or cell turnover is not clear. It is hoped that further study of mutants such as those described will help answer these questions and they will undoubtedly pose new ones.

ACKNOWLEDGEMENTS

This work was supported in part by United States Public Health Service grants AM-13622, AM-05646, and GM-17702, and by grants from the National Foundation and the Kroc Foundation. MH is the recipient of National Institues of Health Research Fellowship AM-00710-01.

REFERENCES

1.   Lever, J. E., Nuki, G., and Seegmiller, J. E.: *Proc. Nat. Acad. Sci. U.S.A.* 71: 2679-2683, 1974.
2.   Miller, O. J., Miller, D. A., Allderice, P. W., Dev, V. G., and Grewal, M. S.: *Cytogenetics* 10: 338-346, 1971.
3.   Levy, J., Virolainen, M., and Defendi, V.: *Cancer* 22: 517-524, 1968.
4.   Hershfield, M. S., Spector, E. B., and Seegmiller, J. E.: In preparation.
5.   Tischfield, J. A. and Ruddle, F. H.: *Proc. Nat. Acad. Sci. U.S.A.* 71: 45-49, 1974.
6.   Hershfield, M. S., Trafzer, R. T., and Seegmiller, J. E.: In preparation.
7.   Bennett, L. L., Jr., Schnebli, H. P., Vail, M. H., Allan, P. W., and Montgomery, J. A.: *Mol. Pharmacol.* 2: 432, 1966.

8. Lomax, C. A., and Henderson, J. F.: *Can. J. Biochem.* 50: 423-427, 1972.
9. Hershfield, M. S., and Seegmiller, J. E.: In preparation.
10. Zoref, E., de Vries, A., and Sperling, O.: *J. Clin. Invest.* 56: 1093-1099, 1975.
11. Nuki, G., Lever, J. E., and Seegmiller, J. E.: *Adv. Exp. Med. Biol.* 41A: 255-267, 1974.
12. Chan, T.-S., Ishii, K., Long, C., and Green, H.: *J. Cell. Physiol.* 81: 315-322, 1973.
13. Wood, A. W., Becker, M. A., and Seegmiller, J. E.: *Biochem. Genet.* 9: 261-274, 1973.
14. Reem, G. H.: *Science* 190: 1098-1099, 1975.
15. Balis, M. E.: *Fed. Proc.* 27: 1067-1074, 1968.
16. Benedict, J. D., Roche, M., Yü, T.-F., Bien, E. J., Gutman, A. B., and Stetten, DeW., Jr.: *Metabolism* 1: 3-12, 1952.
17. Cartier, M. P., and Hamet, M.: *C.R. Acad. Sci. Paris* 279: 883-886, 1974.

PARTIAL HPRT DEFICIENCY : HETEROZYGOTES EXHIBIT ONE CELL

POPULATION IN INTACT CELL ASSAYS

R.B. GORDON, L. THOMPSON and B.T. EMMERSON

University of Queensland Department of Medicine

Princess Alexandra Hospital, Brisbane, Australia

Heterozygotes for the partial deficiency of HPRT exhibit
levels of enzyme activity in cell lysates which are intermediate
between the activity values obtained for normal individuals and
those for severely deficient patients (1). The likely interpretation of this finding is that these intermediate enzyme
activities reflect varying proportions of HPRT-normal and HPRT-
deficient cells. The present report summarises studies on
heterozygotes from several families with partial HPRT deficiency.

To ascertain whether a correlation existed between the
proportion of normal cells and the enzyme activity in cell-free
lysates, the proportion of mutant cells was estimated using an
autoradiographic procedure (2). In one family the expected high
proportion of mutant cells was not detected and this family was
studied in more detail.

The HPRT activities in lysates are shown in Table 1. The
mutation for partial deficiency extended to the three cell types
studied. Hemizygotes from families L and C had very little
activity. However, the hemizygote from family B had approximately
15% of normal activity in the three cell lysates. Heterozygotes
from this family exhibited a range of activities. With heterozygotes from families in which the hemizygote had very low activity
(L and C), there was reasonable agreement between the number of
mutant cells detected by autoradiography and that expected on the
basis of enzyme activity (Table 2). Heterozygotes from a L-N
family (family W) also showed the expected proportion of mutant
cells (Table 2). No difference was observed between heterozygotes
from family B and controls. With our quantitative autoradio-

graphic procedure, 5 - 10% of control cells have low grain counts
similar to mutant cells.

TABLE 1

HPRT activity in cell lysates

$nm \ h^{-1} \ mg^{-1}$

Cell	Erythrocyte	Lymphocytes	Fibroblast
Family L			
* G.L.	.02	.25	7
R.L.	10	45	103
Family C			
* C.C.	.02	6	-
* A.C.	.02	-	27
M.C.	44	72	46
W.C.	59	138	152
Family B			
* F.B.	13	37	30
V.B.	18	45	60
M.B.	28	84	70
L.B.	71	131	-
Normal	101 ± 12 (50)	214 ± 21 (8)	157 ± 32 (6)

* Hemizygote

TABLE 2

Percentage of mutant cells in HPRT heterozygotes

Family	Lymphocytes		Fibroblasts	
	Expected from lysate activity	Observed by autoradiography	Expected from lysate activity	Observed by autoradiography
Family W				
E.W.			55	21
J.W.			83	72
G.W.			82	68
Family L				
R.L.	79	60		
Family C				
W.C.	38	40	4	< 5
Family B				
V.B.	95	5-10	76	5-10
M.B.	73	5-10	68	5-10
L.B.	47	5-10		

In addition, autoradiography of fibroblast cells from the
hemizygote from family B revealed a cell population similar to
controls (Figure 1). Studies of the mutant enzyme in erythrocytes
from family B have revealed no evidence of increased enzyme
lability or decreased *in vivo* half-life. Therefore, the 15% of
normal activity measurable in cell lysates from this hemizygote
probably represents the catalytic capability of the enzyme and not
residual activity of a highly unstable enzyme.

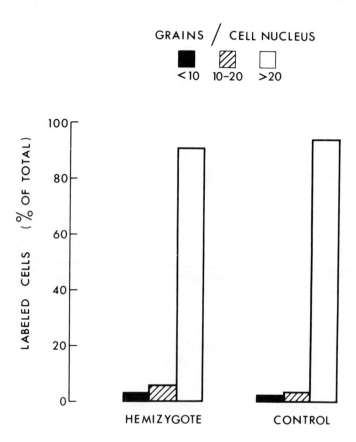

Fig. 1. Autoradiography of fibroblasts of the hemizygote from
family B and a normal control following incubation with ^{3}H-Hx.

Because of difficulties associated with quantitative auto-radiography it was decided to look at other procedures for the determination of the proportion of mutant cells in family B. One such procedure proposed by Albertini and DeMars (3) measures the incorporation of ^{3}H-thymidine into nucleic acid material of PHA-stimulated lymphocytes and studies the inhibition of this incorporation by various concentrations of the purine analogue 8-azaguanine. The rationale of this procedure is that 8-azaguanine is incorporated into nucleotide material via the HPRT reaction and subsequently causes damage to the cell. Thus cells which have taken up sufficient drug fail to incorporate ^{3}H-thymidine into nucleic acids. The number of mutant cells is therefore proportion-al to the extent of ^{3}H-incorporation. Studies using this technique showed no significant difference between the response to 8-azaguanine of control lymphocytes and lymphocytes from both the hemizygote and a heterozygote from family B.

The uptake of hypoxanthine by fibroblasts from Lesch-Nyhan patients has been found to be neglibible in comparison with uptake by control cells (4, 5). The hypoxanthine uptake by fibro-blasts from the hemizygote and heterozygotes from family B was measured with the expectation that differences in the uptake rates would allow an estimation of the proportion of mutant cells. The results of these uptake studies on fibroblasts in monolayer culture again revealed no significant differences between hemi-zygous, heterozygous and control cells.

Studies on uptake of purine bases by whole cells suggest that base transport across the cell membrane and subsequent conversion to phosphorylated intermediates by phosphoribosyltransferase reactions are separate steps (6, 7, 8). Further, studies with Novikoff rat hepatoma cells in culture have suggested that purine-base transport across the cell membrane was the rate-limiting step in the conversion of extracellular substrate to intracellular phosphorylated intermediates (8).

In the procedures utilised, it has not been possible to detect a mutant cell population in either the hemizygote or heterozygotes from one family with partial HPRT deficiency (family B, Table 1 and 2). A possible explanation for this may be that, with experiments using intact cells (i.e. base-uptake studies and autoradiography), a hypoxanthine transport rate is observed and that this transport may be the rate-limiting step in the incorporation of hypoxanthine into cells. It is suggested that in members of the HPRT deficient family B, mutant cells have sufficient enzyme activity to be undistinguished from normal cells. It is only when the rate of the HPRT reaction is very low, as with L-N cells (family W) or with families in which the hemizygote has very low activity (families L and C), that the HPRT step becomes

the rate-limiting one and differences between control and mutant
cells are observed in intact assays.

REFERENCES

1. EMMERSON, B.T., THOMPSON, C.J. and WALLACE, D.C. (1972). Ann.
        Intern Med. 76:285-287.

2. FUGIMOTO, W.Y. and SEEGMILLER, J.E. (1970). Proc. Nat. Acad.
        Sci. 65:577-584.

3. ALBERTINI, R.J. and DeMARS, R. (1974). Biochem. Genet. 11:
        397-411.

4. BENKE, P.J., HERRICK, N. and HERBERT, A. (1973). Biochem. Med.
        8:309-323.

5. RAIVIO, K.O. and SEEGMILLER, J.E. (1973). Biochim. Biophys.
        Acta 299:273-282.

6. HAWKINS, R.A. and BERLIN, R.D. (1969). Biochim. Biophys. Acta
        173:324-337.

7. BERLIN, R.D. (1970). Science 168:1539-1545.

8. ZYLKA, J.M. and PLAGEMANN, P.G.W. (1975). J. Biol. Chem.
        250:5756-5767.

The authors are grateful to the National Health and Medical
Research Council of Australia for financial support.

DIMINISHED AFFINITY FOR PURINE SUBSTRATES AS A BASIS FOR GOUT WITH MILD DEFICIENCY OF HYPOXANTHINE-GUANINE PHOSPHORIBOSYLTRANSFERASE

L. Sweetman, M. Borden, P. Lesh, B. Bakay and M. A.

Becker*, Departments of Pediatrics and Medicine*, University of California, San Diego, La Jolla, California

Deficiency of HGPRT activity is a well-established cause of excessive purine nucleotide and uric acid production. Nevertheless, only a very small proportion of patients with gout or uric-acid stone formation are deficient in this enzyme and, in these patients residual enzyme activity is less than 30% of normal activity (1,2,3). Since routine screening of HGPRT activity has usually been restricted to measurement of the maximal rate of the enzyme reaction determined at saturating substrate concentrations, the possibility remains that significant numbers of patients with functional deficiencies in HGPRT activity exist in whom the enzyme abnormality is minimally, or not at all, expressed as a diminished maximal reaction velocity. The present study, in which HGPRT activity was measured at sub-saturating substrate concentrations, was undertaken in order to assess the frequency of occurrence of such variants of HGPRT deficiency. These studies confirm the occurrence of functionally important HGPRT deficiencies which are poorly reflected in the routine screening procedures, and indicate that mutation in substrate binding to the enzyme, though rare, is a definite mechanism of HGPRT deficiency.

Heparinized whole blood from 42 patients with excessive uric acid production was studied. The assay of HGPRT was based on the lanthanum precipitation of the nucleotide product (4) following a 15 min incubation at 60°C of 1 μl of heparinized blood in 1 ml 60 mM Tris-HCl pH 7.4 with 0.01 mM ^{14}C-hypoxanthine (H)(1.25 μCi/μmole), 0.25 mM 5-phosphoribosyl-1-pyrophosphate (PRPP), and 6 mM $MgCl_2$. As seen in Table 1, normal enzyme activity at subsaturating substrate concentrations was 1000-1300 nmoles/min/ml packed cells. All samples from patients fell within this range with the exception of patient D.B. whose enzyme activity was 197 nmoles/min/ml packed cells.

TABLE I

ERYTHROCYTE HGPRT ACTIVITY AT SUBSATURATING CONCENTRATIONS OF SUBSTRATES

		nmole/min/ml packed cells
NORMAL		1000 - 1300
Patient	D.B.	197
Nephew	P.M.	188
Mother	P.B.	669
Sister	S.M.	571

When assayed at saturating substrate concentrations, samples from this patient showed 60% of normal activity. The patient was a 29-year-old hyperuricemic male (11 mg%) with gout since age 22. Uric acid excretion was 850-900 mg/day. Erythrocyte adenine phosphoribosyltransferase activity and PRPP concentrations were twice normal while PRPP synthetase activity was normal.

The pedigree of the B. family, shown in Fig. 1, is consistent with X-linked inheritance.

Polyacrylamide gel electrophoresis (5) of HGPRT from erythrocytes of D.B. is shown in Fig. 2. There was reduced activity of the faster migrating bands in contrast to other partial deficient variants which migrate 12-15% faster than normal (6).

Kinetic analysis (7) of HGPRT in erythrocytes of the proband and family are shown in Table 2. Vm values were determined at saturating concentrations of substrate and apparent Km values were determined with the alternate substrate at saturating concentrations.

The proband had a VM of 60% of normal with four-fold elevations of the Km's for H and G but with a normal Km for PRPP. A nephew (P.M.) showed a similar abnormality in substrate affinity while the mother (P.B.) and sister (S.M.) had intermediate kinetic properties.

The properties of fibroblasts of the patient are given in Table III. In the cells of D.B., the mild deficiency of HGPRT at saturating substrate concentrations was confirmed as was the increased PRPP concentration and rate of purine synthesis. However, ribose-5-phosphate concentration and rate of PRPP generation were normal.

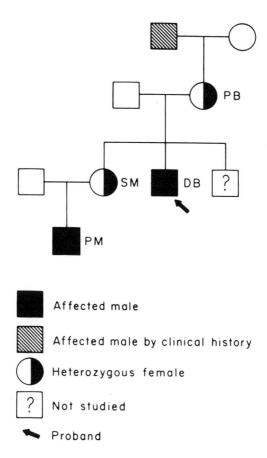

Fig. 1   Pedigree of the D.B. family with variant HGPRT.

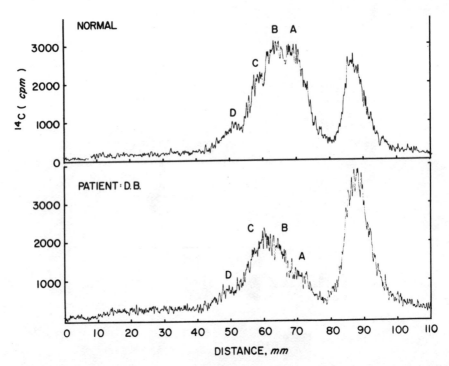

Fig. 2   Electrophoretic profile of HGPRT (left) and APRT (right) from erythrocytes.

TABLE II

KINETIC PROPERTIES OF ERYTHROCYTE HGPRT

	D.B.	P.B.	S.M.	P.M.	CONTROL
APPARENT Vm (nmoles/hr/mg protein)					
HGPRT	44	52	54	45	81 ± 13 (S.D.)
APRT	41	32	28	40	21 ± 05 (S.D.)
APPARENT Km (μM)					
H	19	10	10	44	4 - 6
G	75	29	28	83	18 - 22
PRPP	17	–	–	–	17 - 39

TABLE III

PROPERTIES OF FIBROBLASTS

	D.B.	NORMAL
APPARENT Vm (nmoles/hr/mg/protein)		
HGPRT	88	117 – 168
APRT	232	196 – 261
CONCENTRATIONS (nmoles/10^6cells)		
PRPP	1.15 $\pm$ 0.17	0.20 – 0.55
Ribose-5-phosphate	0.37 $\pm$ 0.07	0.23 – 0.40
PRPP GENERATION (nmoles/hr/10^6cells)	2.73	2.53 – 4.13
PURINE SYNTHESIS (cpm/hr/10^6cells)	4900	2100 – 2700

SURVIVAL OF FIBROBLASTS IN AZAGUANINE

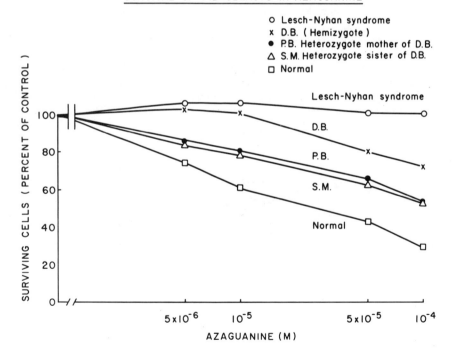

Fig. 3   8-Azaguanine resistance of fibroblasts with variant HGPRT.

The observed Vm of 60% with four-fold increased Km's for purines would suggest only a moderate impairment of HGPRT activity. However, the degree of purine overproduction in the patient indicated a more severe functional impairment. This was evaluated by the resistance of fibroblasts to 8-azaguanine in culture. Fibroblasts (1.6 x 10^5 cells per 60 mm plate) were grown in triplicate in Eagle's Minimal Essential Medium with 10% fetal calf serum and various concentrations of 8-azaguanine. After 72 hours, cells were harvested and counted. The results are expressed as percent of cells surviving in 8-azaguanine relative to control cultures grown without 8-azaguanine.

Normal cells were susceptible to 8-azaguanine while those of a patient with the Lesch-Nyhan syndrome with severe deficiency of HGPRT were resistant, as shown in Fig. 3. Cells from patient D.B. were as resistant as the severely deficient cells except at the highest 8-azaguanine concentrations. The heterozygotes P.B. and S.M. were intermediate in resistance, indicating mosaicism. These results indicate a severe functional deficiency of HGPRT in spite of only moderate kinetic deficiencies.

Some of the previously described kinetic variants of HGPRT have been associated with more readily identifiable deficits in enzyme activity than is shown by this patient. McDonald and Kelley (8) described a patient with classical features of the Lesch-Nyhan syndrome with approximately ten-fold increases in Km's for H, G, and PRPP but with a normal Vm at saturating substrates. Fox and colleagues (9) described a gouty patient with normal Km values but a Vm of about 30% of normal and abnormal responsiveness to product inhibitors. These variant enzymes would give deficient values at the normal saturating substrate levels. On the other hand, a hyperuricemic patient reported by Benke and colleagues (10) had a normal Vm, normal Km's for H and G but a 10-fold higher Km for PRPP. This variant was not detected in the initial assays at saturating concentrations of substrates.

All of the variants described would be readily detectable as having below normal activity in our screening assay at low substrate concentrations. These findings suggest that despite a rather low frequency of kinetic mutation in HGPRT, screening of uric acid overproducers using substrates at the concentrations of normal Km's is a warranted procedure which may reveal subtle kinetic variants of HGPRT which are potentially overlooked by the conventional enzyme assays.

ACKNOWLEDGEMENTS

We thank Laurence Meyer for his expert technical assistance. This work was supported in part by the National Foundation March of Dimes Grant No. 1-377, U.S. National Institutes of Health Grant No. AM-18197, and a U.S. Veteran's Administration Research Grant No. MR1S 0865.

REFERENCES

1. Kelley, W. N., et al.: Ann. Intern. Med. 70: 155-206 (1969).

2. Yü, T. F., et al.: Ann. Intern. Med. 76: 255-264 (1972).

3. Emmerson, B. T., and Thompson, L.: Quart. J. Med. XLII: 423-424 (1973).

4. Bakay, B., Telfer, M. A., and Nyhan, W.L.: Biochem. Med. 3: 230-243 (1969).

5. Bakay, B.: Anal. Biochem. 40: 429-439 (1971).

6. Bakay, B., et al.: Biochem. Genet. 7: 73-85 (1972).

7. Becker, M. A.: J. Clin. Invest. 57: 308-318 (1976).

8. McDonald, J. A. and Kelley, W. N. In Adv. in Exp. Med. and Biol. (O. Sperling, A. DeVries and J. B. Wyngaarder, Eds.), 41A: 167-175 (1974).

9. Fox, I. H., et al.: J. Clin. Invest. 56: 1239-1249 (1975).

10. Benke, P. J., Herrick, N., and Hebert, A.: J. Clin. Invest. 52: 2234-2240 (1973).

PURINE AND PYRIMIDINE NUCLEOTIDE CONCENTRATIONS IN CELLS WITH
DECREASED HYPOXANTHINE-GUANINE-PHOSPHORIBOSYLTRANSFERASE (HGPRT)
ACTIVITY

G. Nuki, K. Astrin, D. Brenton, M. Cruikshank, J. Lever
and J. E. Seegmiller
Departments of Medicine
Welsh National School of Medicine, Cardiff and the
University of California at San Diego, La Jolla, Californi

Severe and partial deficiency of the enzyme Hypoxanthine-
guanine-phosphoribosyl-transferase (HGPRT) (EC.2.4.2.8.) is associated with accelerated rates of purine biosynthesis de novo,
both in vivo (1,2,3) and in cultured skin fibroblasts (4,5) and
lymphoblasts (6) from patients with these diseases. Although the
biochemical basis for this accelerated de novo purine biosynthesis
has been attributed to elevated intracellular levels of phosphoribosylpyrophosphate (PP-ribose-P) both in fibroblasts (4) and
lymphoblasts (6,7) recent in vitro experiments with the enzyme
amidophosphoribosyltransferase (EC.2.4.2.14) have re-emphasised
the potential regulatory role of purine ribonucleotides as feedback inhibitors at this first, and presumed rate limiting step in
the de novo pathway ; the activity and physical properties of the
enzyme being controlled by a critical interaction of PP-ribose-P
and purine ribonucleotides (8). In an effort to investigate the
role of purine ribonucleotides in the regulation of de novo purine
biosynthesis in living human cells deficient in HGPRT, intracellular
ribonucleotide concentrations have been measured in HGPRT⁻ human
lymphoblasts, fibroblasts and erythrocytes and in appropriate HGPRT⁺
controls.

In previous studies we have suggested (7,9) that cloned human
lymphoblast mutants selected by resistance to 8-azaguanine or 8-
azahypoxanthine provide a model in vitro system where phenotypic
abnormalities associated with HGPRT deficiency can be studied in a
more uniform genetic background than that provided by cells cultured
from affected patients. Biochemical characterisation of a number
of clones of drug resistant lymphoblast mutants selected with (9)
or without (7,9) prior mutagenesis suggested that the degree of
HGPRT deficiency correlated well with intracellular concentrations

326

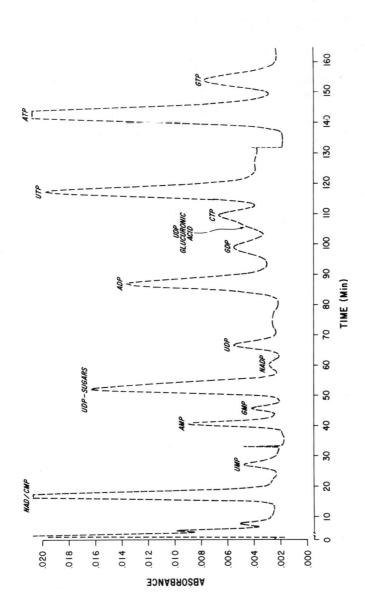

Fig. 1. Nucleotide profile of extract of 2×10^{6} Wl-L$_2$ lymphoblasts. Machine sensitivity attenuated to give full scale absorbance of 0.16 at 132 minutes.

of PP-ribose-P and with rates of <u>de novo</u> purine biosynthesis.
Some results of these studies are summarised in Table I.

INTRACELLULAR NUCLEOTIDE CONCENTRATIONS

Intracellular nucleotide concentrations were measured by
high pressure liquid chromatography using a Varion LCS 1000 (Varion)
Aerograph) machine linked to a Varion A-25 10 m volt double pen
recorder. Nucleotide separations on 20-25 μL extracts were
effected by a 300 cm x 1 mm internal diameter stainless steel
column packed with a pellicular strong anion cross linked
polystyrene resin with groups of the trimethyl benzyl ammonium
type following elution with a concave gradient (dilute buffer –
0.015 M Potassium dihydrogen phosphate P.H. 3.85 concentrated
buffer – 0.25 M potassium chloride P.H. 4.60). Details of cell
culture (9) extraction and chromatography methods (10) are reported
elsewhere. All purine and pyrimidine di and triphosphates were
well separated, but AMP and IMP, GMP and XMP and CMP and NAD were
not. Using simultaneous traces with full scale absorbance ranges
of 0.01 and 0.02 absorbance units (U.V. detector 254 mμ), as little
as 20 picomoles of the sharply eluted mononucleotides and 100
picomoles of the more broadly eluted triphosphate peaks could be
detected after loading an extract of approximately 2 x 10^6 lymphoblasts (Fig.1). The identity of peaks was verified by measurements
of retention time, cochromatography with standards and by measurement of absorbance ratios of individual peaks at different wavelengths.

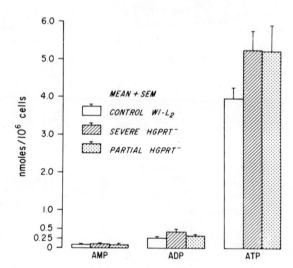

Fig.2. Intracellular concentrations of adenine
nucleotides in human lymphoblasts. Mean
+ S.E.M. of six separate determinations.

Table II

SUMMARY OF CHARACTERISTICS OF HGPRT⁻ LYMPHOBLAST MUTANTS

CLASS	GROWTH IN*			RADIO-† AUTOGRAPHY	HGPRT† (% WILD TYPE)	PP-RIBOSE-P (P moles/10^6 cells)	RATE DE NOVO ++ PURINE SYNTHESIS FGAR (CPM/10^6 cells/hour)
	8-AG	TG	HAT				
WILD TYPE NORMAL CONTROL	0	0	+	++++	100	10	36000
PARTIAL HGPRT⁻	+	-	+	++	57-63	26-37	44000 -100000
INTERMEDIATE HGPRT⁻	+	+	+	0	10-15	40-54	82000 -90000
SEVERE HGPRT⁻	+	+	0	0	<1	49-71	82000 -150000

* 8-AG − 20 μM 8-Azaguanine
  TG   −20 μM Thioguanine
  HAT −0.16 μM Aminopterin, 100 μM hypoxanthine, 20 μM Thymidine
† Radioautography with (3H) Hypoxanthine
† 100 % HGPRT = 327 n moles/hr/mg protein Hypoxanthine converstion to IMP
+ Rate of de novo purine synthesis estimated by incorporation of (^{14}C) Formate into formylglycinamide Ribonucleotide (FGAR) in the presence of 0.3 mM Azaserine

All methods described in Lever et al (1974) and Nuki et al (1973)

PURINE RIBONUCLEOTIDES

Intracellular concentrations of adenine and guanine nucleotides in wild type (Wl-L$_2$) lymphoblasts, in two clones severely deficient in HGPRT (AG^{r}9 Cl$_2$ SC$_1$, AG^{r}9 Cl$_{35}$ SC$_1$) and in two partially deficient clones (AG^{r}9 Cl 16 SC$_3$, AG^{r}9 Cl$_{20}$ SC$_1$) are shown in Figures 2 and 3. The cells were all grown in parallel in autoclavable Eagles minimal essential medium (Augto-Pow Flow Laboratories) supplemented with 2mM glutamine, sodium pyruvate, sodium bicarbonate, non essential aminoacids and 10% foetal calf serum (9), to a density of 8 − 12 x 10^5 cells/ml prior to extraction as previously described (10). The HGPRT deficient cells did not show any decrease in intracellular purine ribonucleotide concentrations when compared with the wild type controls. An apparent increase in intracellular adenine nucleotides was not statistically significant. An adenine to guanine nucleotide ratio of approximately 4:1 is observed in

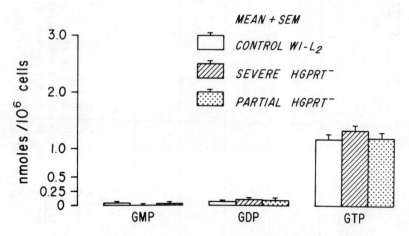

Fig.3. Intracellular concentrations of guanine nucleotides in human lymphoblasts. Mean + S.E.M. of six separate determinations.

both HGPRT$^+$ and HGPRT$^-$ cells and the high ATP/ADP and ATP/AMP ratios throughout are consistent with a high energy status ; attesting to the fact that no serious degradation has occurred during the preparation of the extracts. Intracellular adenine and guanine nucleotide concentrations were also similar when a long term lymphoblast cell line from the peripheral blood of a patient with Lesch-Nyhan syndrome was compared with similar lines from persons with normal purine metabolism (10, 11).

Adenine nucleotide pools were found not to be diminished when intracellular nucleotides in the erythrocytes of three boys with Lesch-Nyhan syndrome were compared with erythrocyte nucleotides in three control subjects (Table II).

TABLE II

INTRACELLULAR CONCENTRATIONS OF PURINE AND
PYRIMIDINE NUCLEOTIDES IN HUMAN ERYTHROCYTES
(n moles/10^9 RBC)

Mean (+ range)

Nucleotide	Controls (HGPRT$^+$)	Lesch-Nyhan (HGPRT$^-$)
UMP	–	17.3 (11.6-22.2)
AMP/IMP	1.4(1.1-1.8)	1.3 (1.1-1.5)
UDPG	5.2(4.1-5.9)	21.7 (14.0-26.8)
ADP	15.5(14.4-16.6)	21.1 (16.8-24.5)
UTP	0.6(0-1.8)	3.3 (1.7-5.1)
ATP	75.4(71.0-81.0)	128.9 (105.0-168.0)
Total Adenine Nucleotides	92.3(86.5-98.4)	151.3(122.9-194-0)
Total pyrimidine Nucleotides	5.8(4.1-7.7	42.3 (27.3-54.1)
ATP/ADP	4.9	6.1
ATP/AMP	54.0	99.2

As noted previously by others (12) erythrocytes are characterised by very low concentrations of guanine nucleotides.

Attempts at measuring intracellular nucleotides in lymphocytes from the same persons were unsuccessful as lymphocyte separation by a variety of procedures was always accompanied by considerable degradation of ATP.

Measurement of intracellular nucleotide concentrations in extracts of confluent skin fibroblasts of comparable passage number from a patient with Lesch-Nyhan syndrome, a patient with gout and partial HGPRT deficiency and from a control showed no significant changes in nucleotide pools (Fig.4). The ATP/ADP ratio was remarkably high (12.6 - 17.1) despite harvesting by trypsinisation prior to extraction with perchloric acid.

These measurements suggest that the increase in de novo purine biosynthesis observed in HGPRT deficient cells is not the result of altered feedback regulation of amidophosphoribosyl-transferase due to lowering of purine ribonucleotide concentrations. Similar results were reported by Rosenbloom et al (4) in a single previous study, but it was not possible to measure GMP or GDP at the time.

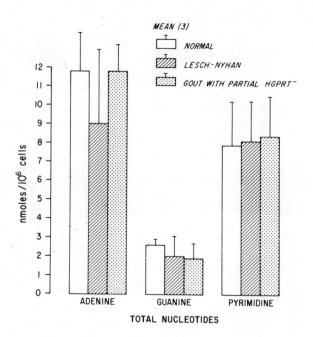

Fig. 4. Intracellular concentrations of purine
and pyrimidine nucleotides in human fibroblasts.
Each bar represents mean + range of three
separate determinations.

There is evidence to suggest that rates of de novo purine synthesis in vivo are well below their potential maximal velocity, so that availability of PP-ribose-P rather than enzyme amount is the major regulatory factor (13).

PYRIMIDINE NUCLEOTIDES

Superimposition of the nucleotide profile of a severe HGPRT⁻ lymphoblast mutant over a control profile extracted from an equivalent number of cells (Fig.5) reveals considerable increases in the UDP sugars, UTP and CTP/UDP glucuronic acid. The latter hybrid peak has been quantitatively separated by spectrophotometric analysis at several wavelengths, and the results confirmed by shifting the UDP glucuronate with a UDP glucuronate decarboxylase purified from the yeast cryptococcus laurentii, which was kindly donated by Dr. Feingold of the University of Pittsburg. Analysis of the intracellular pyrimidine nucleotides and sugar nucleotides shows a 1.5-2 fold increase in UDP sugars UDP and UTP in the severe HGPRT⁻ cells ; a 3-8 fold increase in UDP glucuronic acid and a 2-3 fold increase in CTP (Fig.6). Total pyrimidine nucleotides are increased 2-3 fold, so that their intracellular concentration approaches that of adenine nucleotides in HGPRT⁻ mutants with less than 1% residual activity. The differences are statistically significant and are essentially the same when calculated as n moles/mg. protein, or when expressed as mM concentrations (11). The increases in pyrimidine nucleotides appear to be related to both the degree of HGPRT deficiency and to intracellular PP-ribose-P concentrations.

Similar differences in pyrimidine nucleotide concentrations were observed when long term lymphoblasts derived from the peripheral blood of a boy with Lesch-Nyhan syndrome were compared with peripheral blood lines from normal subjects (10,11). Increases in pyrimidine nucleotides were also detected in erythrocytes from Lesch-Nyhan boys (Fig.7 and Table II), but not in HGPRT⁻ fibroblasts (Fig.4).

Astrin and others (10) have shown that growth of HGPRT⁺ (W1-L2) and HGPRT⁻ (UM-11) lymphoblasts is unimpaired in the presence of 5×10^{-5} M adenine, adenosine or inosine, but is accompanied by a marked fall in intracellular uridine and cytidine concentrations in the HGPRT⁻ cells alone, suggesting that availability of PP-ribose-P may be regulating synthesis of pyrimidine nucleotides.

PP-ribose-P is a common substrate for a number of purine, pyridine and pyrimidine enzymes. (Fig. 8). Orotate phosphoribosyl-transferase (EC 2.4.2.10) and orotidylic acid decarboxylase (EC 4.1.1.23) activity are increased in red blood cells from children with Lesch-Nyhan syndrome (14), but this is not the case in leuco-cytes (14) or in HGPRT⁻ lymphoblasts in culture (Argubright and Becker, Personal communication).

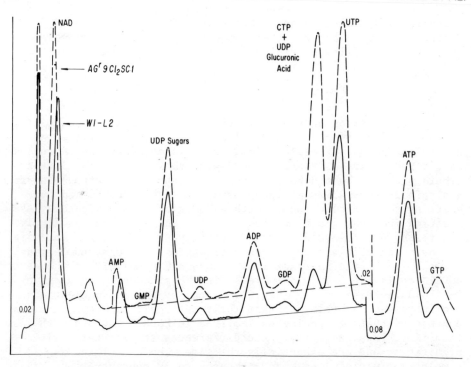

Fig.5. Nucleotide profiles of extracts of equivalent
numbers of W1-L$_2$ HGPRT$^+$ and AG^{r}9 CL$_2$ SC$_1$ (HGPRT$^-$)
human lymphoblasts superimposed.

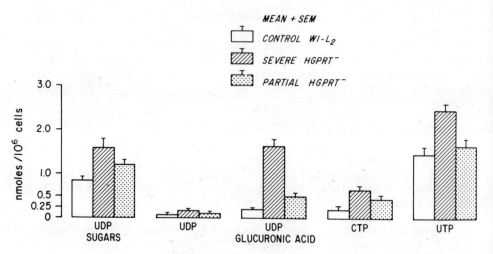

Fig. 6. Intracellular concentrations of pyrimidine
nucleotides in human lymphoblasts. Mean + S.E.M.
of six separate determinations.

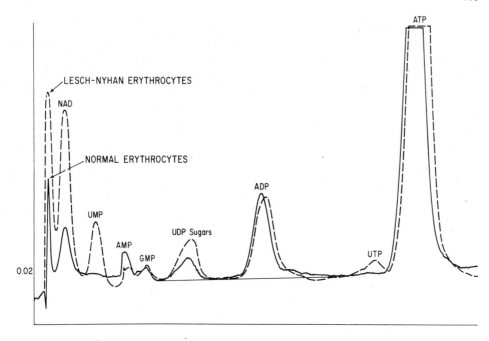

Fig. 7. Nucleotide profiles of extracts of equivalent
numbers of HGPRT$^+$ and HGPRT$^-$ human erythrocytes
superimposed.

Tatibana and Shigesada (15,16) have demonstrated that PP-ribose-
P in concentrations as low as 4-9 μM is an effective allosteric
activator of the glutamine dependent carbamyl phosphate synthetase
II (EC 2.7.2.5), the first enzyme in the pathway of de novo
pyrimidine synthesis (Fig. 9).

It is tempting to speculate that under certain conditions the
initial steps of purine and pyrimidine biosynthesis de novo might be
coordinately regulated by PP-ribose-P.

Preliminary experiments in which rates of de novo pyrimidine
synthesis have been estimated in HGPRT$^-$ and control lymphoblasts by
measurement of the incorporation of (^{14}C)Na HCO$_3$ into acid soluble
pyrimidine nucleotides and RNA have however failed to show any
evidence of accelerated pyrimidine synthesis in HGPRT$^-$ cells (Fig.
10). Moreover, elevation of intracellular PP-ribose-P concentrations
in Wl-L$_2$ lymphoblasts following incubation with increasing concentrations of inorganic phosphate was not accompanied by changes in
intracellular nucleotide concentrations (Fig.11).

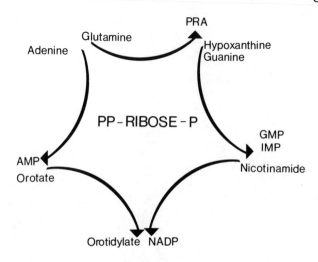

Fig. 8.

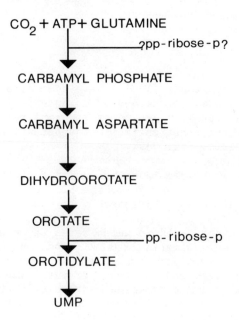

Fig. 9. Effects of PP–ribose–P on the pathway of _de novo_
         pyrimidine biosynthesis.

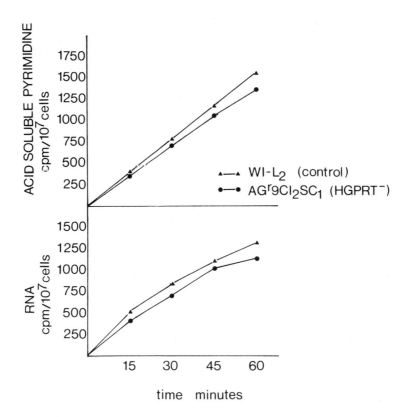

Fig. 10. Incorporation of (14_C) Na HCO_3 into acid soluble pyrimidine
nucleotides and RNA. Wl-L2 ($HGPRT^+$) and AG^r9 Cl_2 SC_1 (HGPRT⁻) lymphoblasts grown in parallel to density of 12 x 10^5 cells/ml. 30 x 10^6
cell aliquots re-suspended in 25 ml fresh Autopow MEM, with 10%
dialysed foetal calf serum and 2 mM glutamine. 50 µci (14_C) Na HCO_3
(4.7 µci/µmole) added after preincubation in capped tubes at 37°C for
60 minutes. Cells washed x 2 ice cold PBS with 4mg/mlBSA. Cell pellet
re-suspended in 2 ml normal saline, freeze thawed x 3 liquid nitrogen
and equal volume 4N. PCA added. Acid soluble fraction hydrolysed
100°C 60 minutes, placed on activated charcoal column (Norit A MCB)
and eluted x 3 with 5 ml ethanol - H20 - NH_3 (2:2:1). Eluate evaporated to dryness, redissolved in 250 µL counted on cellulose TLC
sheet in Beckman liquid scintillation counter using Toluene/PPO/
POPOP phosphor. Recoveries (≈ 80%) checked with (14_C) UTP. Purine
and pyrimidine nucleotides separated by 2 dimensional TLC. Acid
precipitate hydrolysed 1N. KOH at 37°C 15 hours. 2 ml 5% TCA and
400 µL 6N HCl added to precipitate DNA. 1 ml aliquot RNA supernatant
counted in 10 ml Aquasol (New England Nuclear).

Although the concept of coordinate regulation of the rates of de novo purine and pyrimidine synthesis by availability of PP-ribose-P is an attractive one, no evidence has been found to suggest that it accounts for the striking elevations of pyrimidine nucleotides found in HGPRT⁻ lymphoblasts.

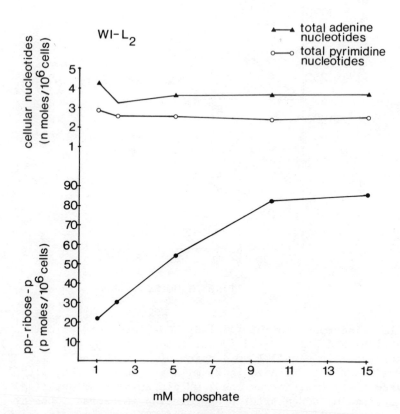

Fig. 11.   Effects of inorganic phosphate concentration on intracellular PP-ribose-P and nucleotides. Wl-L2 lymphoblasts grown in Autopow-MEM with 2 mM glutamine and 10% dialysed foetal calf serum to cell to cell density of 6.5 x 10⁵ cells/ml.   60 ml. Aliquots re-suspended in fresh medium with phosphate buffer P.H. 7.4 at same cell density and incubated with shaking for 45 minutes at 37°C in capped flasks gassed with 5% CO_2 in air.   Extraction procedures for PP-ribose-P (6) and nucleotides (10) modified by the addition of BSA (4 mg/ml) to PBS washes to reduce the lymphoblast lysis which occurs in protein free media (Hershfield-personal communication).

SUMMARY

1. In order to investigate the role of purine ribonucleotides
    in the regulation of de novo purine synthesis in living
    human cells deficient in HGPRT, intracellular ribonucleotide
    concentrations have been measured in HGPRT$^-$ lymphoblasts,
    fibroblasts and erythrocytes and in appropriate HGPRT$^+$
    controls by high pressure liquid chromatography.

2. Purine ribonucleotide concentrations were not reduced in
    HGPRT$^-$ cells, supporting the hypothesis that accelerated
    purine biosynthesis de novo results from increased avail-
    ability of PP-ribose-P and not from altered feedback
    regulation by purine ribonucleotides in HGPRT deficient cells.

3. Striking increases in intracellular concentrations of some
    pyrimidine nucleotides and nucleotide sugars were detected
    in HGPRT$^-$ lymphoblasts and erythrocytes, but not in fibro-
    blasts.

4. The possibility that this abnormality of pyrimidine metab-
    olism might result from coordinate regulation of purine and
    pyrimidine synthesis de novo by PP-ribose-P was not sub-
    stantiated by measurements of rates of pyrimidine synthesis
    and experimental elevation of intracellular concentrations
    of PP-ribose-P following incubation of cells with inorganic
    phosphate.

ACKNOWLEDGEMENTS

W1-L$_2$ and UM-11 cells were generously supplied by Dr. Richard
Lerner, Scripps Clinic and Research Foundation and Dr. Arthur Bloom,
University of Michigan at Ann Arbor respectively. The work was
supported in part by grants from the National Institutes of Health
AM 13622, AM 05646 and GM 17702. G.N. was in receipt of a Merck
International fellowship in Clinical Pharmacology. J.L. was a post
doctoral fellow of the Arthritis Foundation.

REFERENCES

1.   Seegmiller, J.E., Rosenbloom, F.M. and Kelley, W.N. (1967).
      Science 155, 1682-1684.

2.   Kelley, W.N., Rosenbloom, F.M., Henderson, J.F. and
      Seegmiller, J.E. (1967).
      Proc.Nat.Acad.Sci. U.S.A. 57, 1735-1739.

3.   Kelley, W.N., Greene, M.L., Rosenbloom, F.M., Henderson, J.F.
      and Seegmiller, J.E. (1969).
      Ann.Int.Med. 70, 155-206.

4.   Rosenbloom, F.M., Henderson, J.F., Caldwell, I.C., Kelley,
      W.N. and Seegmiller, J.E. (1968).
      J.Biol.Chem. 243, 1166-1173.

5.   Rosenbloom, F.M., Henderson, J.F., Kelley, W.N. and Seegmiller,
      J.E.
      Biochim.Biophys.Acta. 166, 258-260.

6.   Wood,A.W., Becker, M.A. and Seegmiller, J.E. (1973).
      Biochem.Genet. 9, 261-274.

7.   Nuki,G., Lever, J.E. and Seegmiller, J.E. (1974).
      Adv.Exp.Med.Biol. 41A, 255-267.

8.   Holmes, E.W., Wyngaarden, J.B. and Kelley, W.N. (1973).
      J.Biol.Chem. 248, 6035-6070.

9.   Lever, J.E., Nuki, G. and Seegmiller, J.E. (1974).
      Proc.Nat.Acad.Sci. U.S.A. 71, 2679-2683.

10.   Astrin, K.H., Brenton, D.P., Cruikshank, M.C. and Seegmiller,
      J.E. (1976).
      Biochem.Med. (in press).

11.   Nuki, G., Astrin, K.H., Brenton, D.P., Cruikshank, M.C.,
      Lever, J. and Seegmiller, J.E. (1976).
      in Purine and Pyrimidine Metabolism (Ciba Found.Symp. 48).
      Elsevier/Exerpta Medica/North Holland, Amsterdam (in press).

12.   Scholar, E.M., Brown, P.R., Parks, R.E.jnr. and Calabresi, P.
      (1973).
      Blood 41, 927-936.

13.   Henderson, J.F. (1975).
      Biochem.Soc.Trans. 3, 1195-1198.

14.   Beardmore, T.D., Meade, J.C. and Kelley, W.N. (1973).
      J.Lab.Clin.Med. 81, 43-52.

15.   Tatibana, M. and Shigesada, K. (1972).
      Biochem.Biophys.Res.Commun. 46, 491-497.

16.   Tatibana, M. and Shigesada, K. (1972).
      J.Biochem. 72, 549-559.

ELECTROPHORETIC VARIATION IN PARTIAL DEFICIENCIES OF

HYPOXANTHINE-GUANINE PHOSPHORIBOSYLTRANSFERASE

Irving H. Fox and Sheila Lacroix

Puring Research Laboratory, University of Toronto

Rheumatic Disease Unit, Wellesley Hospital, Toronto

In the study of inherited abnormalities of man the demonstration of variable properties of the assayable gene products have provided evidence for structural gene mutations. Different variants from the normal have suggested heterogeneity of these mutations. Structural alterations of hypoxanthine-guanine phosphoribosyltransferase (HGPRT) have been expressed by abnormalities of enzyme kinetic properties and physical properties in circulating erythrocytes and cultured cells.

Recently we have evaluated the electrical charge properties of the mutant erythrocyte HGPRT enzyme from 4 patients with a partial deficiency. The activity of erythrocyte HGPRT in the 4 patients with an enzyme partial deficiency is shown in Table 1. The values ranged from 5 to 44% of normal activity. The mother of patient 2 had 59% of normal activity.

Isoelectric focusing was used to evaluate the electrical charge properties of normal and mutant erythrocyte HGPRT. All isofocusing was performed in a 110 ml sucrose gradient using an LKB ampholine 8100 column (1). The results of isoelectric focusing from a normal male patient is in Figure 1. Adenine phosphoribosyltransferase had an isoelectric pH of 4.5. HGPRT was resolved into 3 major isoenzymes at pH 5.76, 5.82 and 6.02.

Table 2 shows the isoelectric pH for hypoxanthine-guanine phosphoribosyltransferase hemolysates from 2 normal males and 4 male patients with a partial deficiency of HGPRT. The data from our study and others suggest that the normal major isoenzymes have isoelectric pH values from 5.6 to 6.1 (1-5). The isoenzymes in the

TABLE 1

SPECIFIC ACTIVITY OF ERYTHROCYTE HYPOXANTHINE–GUANINE
PHOSPHORIBOSYLTRANSFERASE

Patient	HGPRT (nanomoles/hr/mg)	Percent of Normal
Normal	87.3 ± 16.6	
1	4.0	5
2	34.2	39
3	34.7	40
4[a]	38.4	44
Mother of 2	51.4	59

a. Hemolysate was generously supplied by Dr. J.F. Henderson.

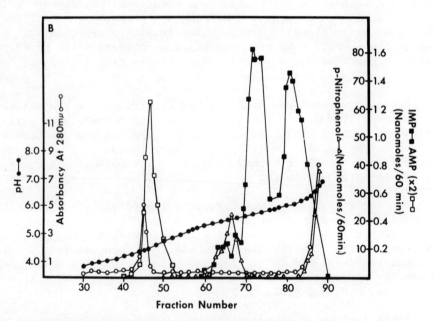

Fig. 1. Isoelectric focusing of normal hemolysate (male). The pH gradient (●—●) ran from 3.5 to 7.0. Hypoxanthine-guanine phosphoribosyltransferase ■—■ , adenine phosphoribosyltransferase □—□.

TABLE 2

ISOELECTRIC pH VALUES OF ERYTHROCYTE HYPOXANTHINE-GUANINE
PHOSPHORIBOSYLTRANSFERASE

Patient			Isoelectric pH's			
Normal Male			5.76	5.82		6.02
Normal Male			5.70	5.82		6.10
Patient 1	5.30	5.40				
Patient 2			5.50	5.70		
Patient 3			5.55	5.76	5.85	
Patient 4			5.50	5.75	5.80	5.85

hemolysate from patient 1 were the most abnormal found. The 4
patients shown have HGPRT activity peaks below pH 5.60 and none
have a major peak of activity at pH 6.0. Although HGPRT isoenzymes
appear to be based upon a non-genetic alteration of the enzyme
protein and usually can vary considerably with the normal pH range,
the mutant HGPRT demonstrates variability in a different pH range.
This indicates that the enzyme protein has a different electrical
charge and supports the concept of structural alterations of HGPRT.
The differences noted among the mutant HGPRT enzymes themselves
suggest that there is considerable heterogeneity of these mutations.
Our results indicate that electrophoretic variation is a common
occurrence in HGPRT partial deficiency.

The existence of an abnormal electrical charge of mutant HGPRT
had some other useful applications. The leukocyte HGPRT from patient
2 was found within the range of normal. The question arose whether
this was mutant HGPRT or normal HGPRT. Isoelectric focusing of
leukocyte extract demonstrated peaks of enzyme activity at pH 5.40,
5.59 and 5.67. These values were very similar to the hemolysate
values and indicated mutant HGPRT with normal activity. Normal
leukocyte HGPRT and erythrocyte HGPRT have isoenzymes in the same
range.

Another application for isofocusing arose when we considered
whether the erythrocytes from the mother of patient 2, with 59% of
normal HGPRT activity, contained mutant enzyme activity. In this
situation the abnormal electrical charge of the mutant enzyme could
be used as a marker to detect mutant enzyme (6). Isoelectric
focusing of hemolysate from the mother of patient 2 is shown in
Figure 2. There were a series of peaks of enzyme activity below and

above pH 5.6 and encompassing the range of mutant and normal isoenzymes. Isoelectric focusing of normal female hemolysate demonstrated peaks of HGPRT activity in the normal range. Thus the mother of patient 2 has two populations of red cells, one with mutant enzyme and one with normal enzyme. These observations imply random inactivation of the X chromosome and are in accordance with the Lyon hypothesis.

In conclusion, our studies of human partially deficient HGPRT by isoelectric focusing has provided evidence for: a) structural gene mutations and genetic heterogeneity, b) mutant enzyme with a normal specific activity and c) mosaicism in erythrocytes from a heterozygote for HGPRT partial deficiency.

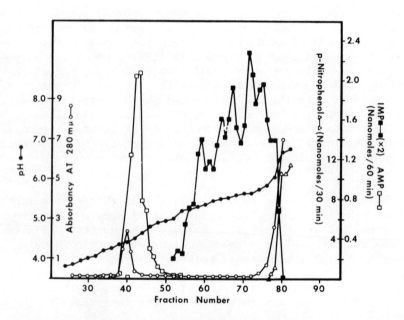

Fig. 2 Isoelectric focusing of hypoxanthine-guanine phosphoribosyl-transferase of a female carrier. The format is the same as Fig. 1.

REFERENCES

1. Fox, I.H., Dwosh, I.L., Marchant, P.J., Lacroix, S., Moore, M.R. Omura, S. and Wyhofsky, V. 1975. Hypoxanthine-guanine phosphoribosyltransferase: characterization of a mutation in a patient with gout. J. Clin. Invest. 56:1239-1249.

2. Arnold, W.J. and Kelley, W.N. 1971. Human hypoxanthine-guanine phosphoribosyltransferase purification and subunit structure. J. Biol. Chem. 246:7398-7404.

3. Kelley, W.N. and Arnold, W.J. 1973. Human hypoxanthine-guanine phosphoribosyltransferase: studies on the normal and mutant forms of the enzyme. Fed. Proc. 32:1656-1659.

4. Der Kaloustian, V.M., Awdeh, Z.L., Haddal, R.T. and Wakid, N.W. 1973. Analysis of human hypoxanthine-guanine phosphoribosyltransferase isoenzymes by isoelectric focusing in polyacrylamide gel. Biochem. Genet. 9:91-95.

5. Bakay, B. and Nyhan, W.L. 1975. Heterogeneity of hypoxanthine-guanine phosphoribosyltransferase from human erythrocytes. Archiv. Biochem. Biophy. 168:26-34.

6. Fox, I.H., Marchant, P.J. and Lacroix, S. 1976. Hypoxanthine-guanine phosphoribosyltransferase: mosaicism in the peripheral erythrocytes of a heterozygote for a normal and a mutant enzyme. Biochem. Genet. 14:587-593.

XANTHINE OXIDASE ACTIVITY IN A GOUTY PATIENT WITH PARTIAL
DEFICIENCY OF HGPRT

A. CARCASSI

Medical Clinic

Piazza Selva 78 – 53100   SIENA   ITALY

Xanthine oxidase is an enzyme which is directly uricogenetic,
as it catalyzes the aerobic dehydrogenation of hypoxanthine to
xanthine and of xanthine to uric acid, these being the final steps
in purine metabolism in man. In man, the enzyme is present mainly
in the liver and in the jejunal mucosa.

Increased hepatic xanthine oxidase was detected previously in
7 out of 8 subjects with primary gout, who were over-excretors.
The mean increase was fourfold (1). No rise was observed in levels
of enzymatic activity in the jejunal mucosa of gouty patients (2).
No relationship was observed in the same subjects between hepatic
xanthine oxidase and levels of serum uric acid, while a close relationship was observed with urinary uric acid excretion (3).

Xanthine oxidase is a readily induced enzyme, the activity of
which is increased in the liver of protein-depleted mice by repeated injections of xanthine (4), and in chicks by the administration of inosine (5), or by starvation (6) or other conditions (7).

In normal man, hepatic xanthine oxidase activity is increased
by feeding with hypoxanthine, ribonucleic acid, EA-TDA (8), or by
rapid infusion of fructose (9). The data indicate that increased
xanthine oxidase observed in gouty patients is very probably related to increased production of enzyme, and not genetically determined.

346

One of the subjects studied in our previous work (1) had a
twelve-fold increase in enzymatic activity in liver xanthine oxidase, while jejunal enzymatic activity was within the normal range
(2). He was a 52-year old man who had a history of "juvenile gout"
since the age of 30 years, and of renal lithiasis from the age of
25 years. The acute attacks of gouty arthritis were very frequent,
occurring every week before he arrived in our Clinic.

MATERIAS AND METHODS

The methods used in the study have been described in our previous works (1,2). HGPRT activity was assayed by the radiochemical
method of Cartier and Hamet (10). According to Cartier and Hamet
the mean values (+ 2 SE) are 111 + 13 nm/mg proteins/h.

RESULTS

Studies on purine metabolism in our patient revealed hyperuricemia (10 mg %) and increased urinary uric acid excretion (1150 mg/
/24 hr) on a purine-free diet. HGPRT activity was later studied in
red cells haemolysates in the same subject, and was found to be
29 nm/mg proteins/h which is 25 % of the normal range.

DISCUSSION

Partial deficiency of HGPRT activity was found in gouty patients by Kelley et al. in 1967 (11), and subsequently by many others
authors. The partial deficiency of this X-linked enzymatic activity
may vary from zero to a low percentage of the normal, without any
correlation between severity of HGPRT deficiency and the clinical
picture (12). It represents a very rare cause of gout (13), and
all the subjects had an early onset of the disease.

Partial deficiency of HGPRT, together with the increased PRPP
synthetase activity found by Sperling in 1972 (14) and subsequently
by Becker (15), represents a possible, rational explanation of the
pathogenesis of "juvenile gout", that has been indicated for many
years as a particular, rare, strictly inherited type of gout (16).

The mechanism of accelerated purine synthesis in subjects with
deficiency of HGPRT incudes:

a)   decreased concentration of inosinic and guanilic acid, leading to reduced feed-back inhibition of PRPP amido-transferase,

b)   decreased utilization of PRPP for direct synthesis of gua-
nine-monophosphate (GMP) and inosine-monophosphate (IMP), making
this substrate potentially available for purine biosynthesis "de
novo",

c)   increased concentration of hypoxanthine or guanine and
activation of PRPP amidotransferase (17).

Sorenson has found that a subject with a partial deficiency of
HGPRT equivalent to 12,5 % of normal enzymatic activity, had a ten-
fold increase in urinary hypoxanthine in comparison with the normal
values, during treatment with allopurinol (18).

These findings imply that hypoxanthine is the major source of
the excessive uric acid production and that increased availability
of hypoxanthine is probably the main factor producing increased
xanthine oxidase observed in our patient.

SUMMARY

A patient with juvenile gout and partial deficiency of HGPRT
is presented. In This subject, hepatic xanthine oxidase activity
showed a twelve-fold increase.   Xanthine oxidase is a readily
induced enzyme and this increased activity is probably correlated
with the increased availability of hypoxanthine observed in such
patients.

REFERENCES

1 – Carcassi A., Marcolongo R., Marinello E., Riario-Sforza G.,
    Boggiano C.
    Liver Xanthine Oxidase in Gouty Patients
    Arthr. Rheum. 12,17, 1969
2 – Riario-Sforza G., Carcassi A., Bayeli P.F., Marcolongo R.,
    Marinello E., Montagnani M.
    Attività xantina-ossidasica nella mucosa digiunale di sog-
    getti gottosi.
    B.S.I.B.S. 45,785, 1969

3 - Ciccoli L., Riario–Sforza G., Marinello E., Marcolongo R.,
Carcassi A.
Relazione tra aumento della xantina ossidasi epatica, iper-
uricemia e uricuria nei soggetti gottosi.
Reumatismo 21, 277, 1969

4 - Mangoni A., Pennetti V., Spadoni M.A.
Aumento adattativo di xantina ossidasi in topini alimentati
con diete a diverso contenuto proteico.
B.S.I.B.S. 31, 1397, 1955

5 - Della Corte E., Stirpe F.
Regulation of xanthine dehydrogenase in cick liver.
Biochem. J. 102, 520, 1967

6 - Stirpe F., Della Corte E.
Regulation of xanthine dehydrogenase in cick liver. Effect
of starvation and of administration of purine and purine
nucleosides.
Biochem. J. 94, 309, 1965

7 - Bray R.C.
Xanthine Oxidase, The Enzymes Vol. 7 Ed. by P. Bayer, Lardy
H. Myrback K. New York , Academic Press, pp. 533–556

8 - Marinello E. Carcassi A., Riario–Sforza G., Ciccoli L.,
Marcolongo R.
Effetto di varie sostanze iperuricemizzanti sulla xantina
ossidasi epatica di soggetti normali.
Reumatismo 21, 310, 1969

9 - Riario–Sforza G., Ciccoli L., Marinello E., Marcolongo R.,
Carcassi A.
Comportamento della xantina ossidasi nel fegato di soggetti
iperuricemici in seguito a carico di fruttoso.
Reumatismo 21, 322, 1969

10 - Cartier P., Hamet M.
Les activités purine phosphoribosyltransférasiques dans les
globules rouges humains. Technique de dosage.
Clin. Chim. Acta 20, 205, 1968

11 - Kelley W.N., Rosenbloom F.M. Henderson J.F. Seegmiller J.E.
A specific enzyme defect in gout associated with overpro-
duction of uric acid.
Proc. Natl. Acad. Sci. U.S. 57, 1735, 1967

12 - De Bruyn C.H.M.M., Oei T.L.
Heterogeneity in X-linked gout.
Scand. J. Rheum. suppl. 8, abst. 31–08 , 1975

13 - Yü T.F., Balis M.E., Kretinsky T.A.,   Dancis J. Silvers D.N.
     Elion G.B., Gutman A.B.
     Rarity of X-linked partial hypoxanthine-guanine phosphoribo-
     syltransferase deficiency in a large gouty population.
     Ann. Intern. Med. 76, 255, 1972
14 - Sperling O., Boer P., Persky-Broch S., Kanarek E. De Vries A.
     Altered kinetic property of erythrocyte phosphoribosylpyro-
     phosphate synthetase in excessive purine production.
     Europ. J. Clin. Biol. Res. 17, 103, 1972
15 - Becker M.A. Meyer L.J.Seegmiller J.E.
     Gout with purine overproduction due to increased phosphori-
     bosylpyrophosphate synthetase activity.
     Amer. J. Med. 55, 232, 1973
16 - Izar G. Lenzi F.
     La Gotta. Relazione 55° Congr. Soc. Ital. Med. Int. ed.
     Pozzi, Roma 1954
17 - Kelley W.N., Greene M.L., Rosenbloom F.M., Henderson J.F.,
     Seegmiller J.E.
     Hypoxanthine-guanine phosphoribosyltransferase deficiency
     in gout.
     Ann. Intern. Med. 70, 155, 1969
18 - Sorenson L.B.
     Mechanism of excessive purine biosynthesis in hypoxanthine-
     guanine phosphoribosyltransferase deficiency.
     J. Clin. Invest. 49, 968, 1970

EXPERIENCE WITH DETECTION OF HETEROZYGOUS CARRIERS AND PRENATAL

DIAGNOSIS OF LESCH-NYHAN DISEASE

B. Bakay, U. Francke, W. L. Nyhan and J. E. Seegmiller[*]

Departments of Pediatrics and Medicine[*], University of

California, San Diego, La Jolla, California   92093

For the purpose of genetic counselling, considerable efforts have been expended on the development of reliable methods necessary to identify the carriers of the Lesch-Nyhan syndrome, as well as to test in utero the sex of the fetus and its hypoxanthine phosphoribosyl transferase (HPRT, E.C. 2.4.2.8) activity.

It is now well established that the molecular abnormality responsible for the Lesch-Nyhan syndrome is virtually total absence of HPRT activity (1, 2). Since the structural gene for this enzyme is located on the X-chromosome , the deficiency of HPRT is inherited in X-linked fashion. The males who inherit an X-chromosome with the abnormal gene are totally deficient in HPRT. In contrast, female carriers have shown mosaicism of deficient and normal cells as predicted by the X-inactivation hypothesis of Lyon. Consequently, some of the cells of heterozygous females have normal level of HPRT activity and the others have none. The two cell populations seem to prevail in all tissues of adult heterozygotes except of bone marrow, which presumably, due to selection against HPRT⁻ cells, produce blood cells with normal level of HPRT only (4, 5). Therefore, identification of carriers of total HPRT deficiency is not possible by measurement of HPRT in blood cells.

Carrier identification was achieved by cloning of skin fibroblasts or growth in selective media which separate populations of HPRT⁺ and HPRT⁻ cells (3, 6 - 8). However, these techniques are laborious and time consuming.

Several years ago, Gartler and his associates had shown that mosaicism can be demonstrated rapidly by analysis of individual hair roots, plucked from the scalp of suspected heterozygotes (9).

We have studied this system and we have found that we could assay
simultaneously HPRT and adenine phosphoribosyl transferase (APRT) in
the hair roots using a polyacrylamide gel electrophoretic method
(10). In this assay, individual hair roots are placed on top of the
stacking gel, covered with buffer and freeze-thawed four times with
a stream of freon. Following the separation of enzymes by electrophoresis, the gels are incubated in substrate solution containing
PRPP and ^{14}C-labelled hypoxanthine and adenine. Next, the gels are
placed in 0.1 $\underline{M}$ lanthanum chloride and washed in water (11). This
results in removal of soluble purine bases, while precipitated
nucleotides remain fixed in the gel. The gels are then fractionated
by a mechanical fractionator and analyzed for ^{14}C in a liquid scintillation counter using a flow cell (12).

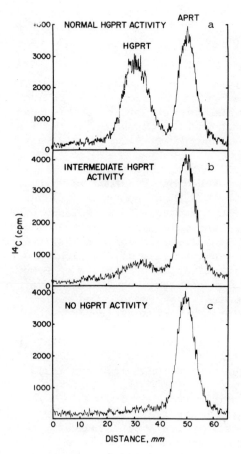

Fig. 1. Electrophoretic analysis of hair roots of a heterozygous
carrier of the Lesch-Nyhan syndrome for HPRT and APRT activity.
a. root containing normal cells; b. root with mixed cell population; c. root containing HPRT-deficient cells.

Normal roots produced tracings in which faster migrating APRT formed a symmetrical peak of activity on the right, while slower migrating HPRT produced a broad zone of activity in the middle of the profile (Fig. 1a). In this assay, APRT which is not X-linked but coded for by a gene on chromosome 16, served as internal reference for HPRT activity and viability of the hair root.

Roots originating from HPRT$^-$ cells produced profiles containing only the APRT peak (Fig. 1c). Roots containing a mixed cell population yielded an intermediate level of HPRT activity (Fig. 1b). In order to achieve the necessary statistical margin of confidence, we usually analyzed 30 hair roots (10).

The results of screening of 70 female relatives of patients with the Lesch-Nyhan syndrome from 29 different kindreds, are compiled in Table 1. Of 25 mothers, 19 were heterozygous carriers and six were normal homozygotes.

Interestingly, among 29 probands there were six who had normal homozygous mothers. Furthermore, there were five probands whose mothers were heterozygotes but whose grandmothers were normal homozygotes. Those probands and the four mothers appear to represent new mutations. The occurrence of new mutations in 47 Lesch-Nyhan families from different medical centers has been discussed in an article which appeared earlier this year (13).

For in utero detection of fetuses with the Lesch-Nyhan syndrome, fetal cells obtained by amniocentesis were propagated in tissue culture. They were karyotyped for sex determination and were analyzed for HPRT and APRT activity using the electrophoretic method and incorporation of ^{3}H-labelled hypoxanthine by autoradiography.

For enzyme analysis, amniotic cells were washed and lysed by freezing and thawing, and centrifuged 20 min at 24,000 x g in a refrigerated centrifuge. Fibroblasts with normal HPRT content, as well as cells containing no HPRT activity were used as controls. The normal amniotic cells produced a profile which resembled profiles of normal skin fibroblasts (Fig. 2a). Skin fibroblasts from patients with the Lesch-Nyhan syndrome produced profiles containing a normal amount of APRT activity but no HPRT activity (Fig. 2b).

Amniotic cells from the third pregnancy of heterozygote M.R., who has two boys with the Lesch-Nyhan syndrome, yielded profiles indicating that her fetus lacked HPRT activity (Fig. 2c). Absence of HPRT activity in her amniotic cells was confirmed by autoradiography. According to chromosomal analysis, M.R. was carrying a male fetus. On the basis of these tests, she decided to terminate her pregnancy. The aborted fetus proved to be a male. However,

TABLE 1

RESULTS OF DETECTION OF HETEROZYGOTES BY
ELECTROPHORETIC ANALYSIS OF HAIR ROOTS

MOTHERS		MATERNAL GRANDMOTHERS		SISTERS		MATERNAL AUNTS	
+/-	+/+	+/-	+/+	+/-	+/+	+/-	+/+
DB	JD	DE	JA	BM	BM	SB	JA
DE	HU	JG	MA	BM	BM	–	JA
JA*	MC	LG	LJ	DA	HU	–	LJ
JG	OK	SB	SE	DB	HU	–	MD
JM	PB	–	WI	LG	JG	–	MD
LG	SM	–	–	LG	LJ	–	NB
LJ*	–	–	–	LJ	LJ	–	RO
MD	–	–	–	LJ	MA	–	SE
MS	–	–	–	LJ	MC	–	–
MY	–	–	–	MY	RW	–	–
PR	–	–	–	TC	RW	–	–
RO	–	–	–	WI	TC	–	–
RW	–	–	–	MS	TC	–	–
SB	–	–	–	MS	SM	–	–
SCH	–	–	–	–	SM	–	–
SE*	–	–	–	–	–	–	–
SW	–	–	–	–	–	–	–
TC	–	–	–	–	–	–	–
XG	–	–	–	–	–	–	–
19	6	4	5	14	15	1	8

* MATERNAL GRANDMOTHERS WERE NORMAL HOMOZYGOTES

due to saline effect, most of the enzymes of the fetal tissues
were inactive.

Amniotic cells from a concurrent pregnancy of L.C., a younger
sister of M.R. whose carrier status was unknown, yielded a profile
with normal HPRT activity (Fig. 2d). Autoradiography demonstrated
that amniotic cells of L.C. were incorporating ^{3}H-hypoxanthine.
According to chromosomal analysis, L.C. was carrying a male fetus.

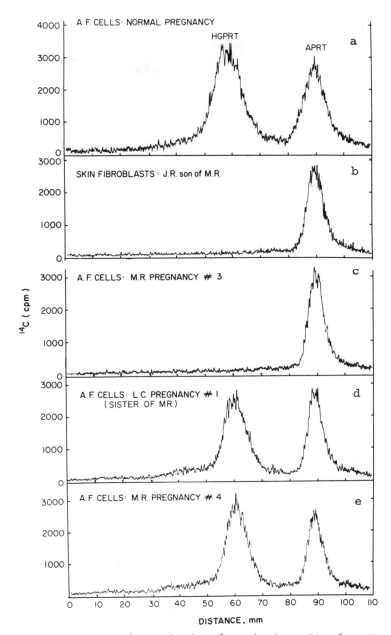

Fig. 2. Electrophoretic analysis of amniotic cells for HPRT and APRT activity. (a) cells with normal HPRT activity. (b) skin fibroblasts with no HPRT activity. (c) cells from third pregnancy of M.R. (d) cells from first pregnancy of L.C., a sister of M.R. (e) cells from fourth pregnancy of M.R.

TABLE 2

RESULTS OF PRENATAL DIAGNOSIS OF THE LESCH-NYHAN SYNDROME

DATE	MOTHER	GENO-TYPE OF MOTHER BY HAIR ROOT ANA-LYSIS	KARYO-TYPE OF FETUS	HPRT OF A.F.CELLS BY ELECTRO-PHORETIC ASSAY	INCORP. OF ^{3}H-HX BY A.F.CELLS	GENO-TYPE of FETUS	ACTION
MAY 1973	M.R.	+/-	XY*	-	-	-/Y	TERMIN.
MAY 1973	L.C.	+/+	XY*	+	+	+/Y	- - -
NOV 1973	A.S.	+/-	XY	-	N.A.	-/Y	TERMIN.
FEB 1974	M.R.	+/-	XX*	+	+	+/+	- - -
MARCH 1974	D.G.	+/-	XY	-	-	-/Y*	TERMIN.
JUNE 1974	J.S.	+/+	XY*	+	+	+/Y	- - -
AUG 1975	M.G.	+/-	XY*	+	+	+/Y	- - -
NOV 1975	M.R.	+/-	XX*	+	+	N.A.	- - -
MAY 1976	O.N.	N.A.	XY	+	+	+/Y	- - -

A.F. = AMNIOTIC FLUID   HX = HYPOXANTHINE

N.A. = NOT AVAILABLE   * = CONFIRMED

Subsequent analysis of hair follicles revealed that L.C. was a normal homozygote. Later, she delivered a normal boy.

We also had had the opportunity to examine amniotic fluid cells from the fourth and fifth pregnancy of M.R.. In her fourth pregnancy, she was carrying a homozygous normal female fetus (Fig. 2e). No HPRT⁻ amniotic cells were detected by autoradiography and 8-azaguanine selection. The fifth time, she was carrying another female which had normal HPRT activity,but its genotype was not established. Just a few weeks ago, M.R. delivered a normal girl.

To date, we have monitored 9 pregnancies at risk (Table 2). In each case, amniotic cells obtained by amniocentesis were grown in tissue culture, karyotyped and analyzed for HPRT activity by electrophoresis, and for incorporation of ^{3}H-hypoxanthine by autoradiography. There were four male and two female fetuses with a normal HPRT level. Mothers, L.C. and J.S., who had normal male fetuses were normal homozygotes. Mother M.G. was a heterozygote. The genotype of mother O.N. was not known.

There were three HPRT-deficient male fetuses whose mothers were heterozygotes. These pregnancies were terminated. In two of them the absence of HPRT was confirmed by concurrent autoradiographic analysis of incorporation of ^{3}H-hypoxanthine and by analysis of fetal tissues.

Use of the electrophoretic assay of HPRT and APRT for the detection of heterozygous carriers and for the prenatal diagnosis of the Lesch-Nyhan syndrome has several advantages. First, it is thoroughly tested, reliable and expedient system. Second, in this system, each sample is assayed simultaneously for HPRT and APRT. This permits the use of APRT as an internal reference. The third and most important advantage of this system, is that HPRT is assayed after it has been separated from enzymes which may interfere with the assay. This is especially important when analyzing fibroblastoid cells in which highly active 5'-nucleotidase converts the formed IMP to inosine and thus complicates measurement of HPRT. In general, this system can be used without restriction on a variety of samples.

ACKNOWLEDGMENTS

Chromosome studies on amniotic cells were done in Dr. O. W. Jones' laboratory. Most of the radioautographic studies were performed in Dr. J. E. Seegmiller's laboratory. The competent technical assistance of Carol Wagner, Marsha Graf, Sharen Carey and Norma Busby is gratefully acknowledged. This work was supported by grants from U.S. Public Health Service MCT-000274 and GM-17702,and The National Foundation, March of Dimes, NF-1377.

REFERENCES

1. Lesch, M., and Nyhan, W.L., Am. J. Med. 36: 561 (1964).

2. Seegmiller, J.E., Rosenbloom, F.M., and Kelley, W.N., Science 155: 1682 (1967).

3. Migeon, B.R., Der Kaloustian, V.M., Nyhan, W.L., Young, W.J., and Childs, B., Science 160: 425 (1968).

4. Nyhan, W.L., Bakay, B., Connor, J.D., Marks, J.F., and Keele, D.K., Proc. Nat. Acad. Sci. USA 65: 214 (1970).

5. Albertini,R.J., and DeMars, R., Biochem. Genet. 11: 397 (1974).

6. Salzman, J., DeMars, R., and Bencke, P., Proc. Nat. Acad. Sci. USA 60: 545 (1968).

7. Migeon, B.R., Biochem. Genet. 4: 377 (1970).

8. Felix, J.S., and DeMars, R., J. Lab. Clin. Med. 27: 596 (1971).

9. Gartler, S.M., Scott, R.C., Goldstein, J.L., Campbell, B., Science 172: 572 (1971).

10. Francke, U., Bakay, B., and Nyhan, W.L., J. Ped. 82: 472 (1973).

11. Bakay, B., and Nyhan, W.L., Biochem. Genet. 5: 81 (1971).

12. Bakay, B., Anal. Biochem. 40: 429 (1971).

13. Francke, U., Felsenstein, J., Gartler, S.M., Migeon, B.R., Dancis, J., Seegmiller, J.E., Bakay, B., and Nyhan, W.L., Amer. J. Hum. Genet. 28: 123 (1976).

HGPRT-POSITIVE AND HGPRT-NEGATIVE ERYTHROCYTES IN HETEROZYGOTES

FOR HGPRT DEFICIENCY

B.T. EMMERSON, L.A. JOHNSON and R.B. GORDON

University of Queensland Department of Medicine

Princess Alexandra Hospital, Brisbane 4102, Australia

A method for autoradiography of tritium labelled mononucleo-tides in red blood cells has been developed and used to examine heterozygotes for HGPRT deficiency. Although HGPRT-positive and HGPRT-negative cell populations have been demonstrated by auto-radiography in cultured fibroblasts from heterozygotes for the severe HGPRT deficiency found in the Lesch-Nyhan syndrome, erythrocyte lysates have always shown normal HGPRT activity. In heterozygotes for the partial HGPRT deficiency, on the other hand, erythrocyte lysates have shown a range of activities between 20% of normal and completely normal values. Although erythrocytes do not synthesize nucleic acids, they are capable of accumulating significant quantities of labelled mononucleotides when incubated with labelled purines. However, these soluble mononucleotides are lost when the selective permeability of the cell membrane is destroyed by fixing or drying. Thus the problem of autoradio-graphically demonstrating tritiated hypoxanthine uptake by red cells involves fixing the labelled nucleotide. This has been achieved by precipitating the mononucleotides with lanthanum simultaneously with fixation of the red cells with osmium tetroxide.

Washed erythrocytes were incubated on a microscope slide within a small chamber prepared by cementing a glass ring to a cover slip. Tritiated hypoxanthine in a phosphate-glucose buffer was mixed with erythrocytes in the chamber, covered with a microscope slide, inverted and incubated at 37° for 30 minutes, during which time the red cells become attached to the slide. The slide was then washed with saline to remove labelled hypoxanthine, fixed in a mixture of osmium tetroxide and lanthanum chloride, washed again, dried and fixed with methanol. After dipping in photographic emulsion

359

(Kodak NTB 2) the slides were exposed for 3 days, developed and
mounted.

The method was validated with artificial mixtures of normal
and HGPRT-deficient (Lesch-Nyhan) erythrocytes. Erythrocytes from
hemizygotes for the Lesch-Nyhan syndrome all showed less than 5
grains/cell (unlabelled), whereas in normal individuals, over 90%
of red cells had more grains than this. In mixtures of 75% normal,
25% severely deficient cells, 50% normal and 50% severely deficient
cells and 25% normal and 75% severely deficient cells, grain
counting closely approximated the expected ratios of normal :
deficient cells.

One family with the Lesch-Nyhan syndrome and two families
with a partial HGPRT deficiency were studied. Erythrocytes from
two heterozygotes for the Lesch-Nyhan syndrome showed normal
labelling. In two other families in which the hemizygotes had a
partial deficiency of HGPRT activity in an erythrocyte lysate,
erythrocytes were unlabelled on autoradiography. A heterozygote
from each family, each of whom demonstrated an intermediate level
of HGPRT activity in an erythrocyte lysate, showed both normally
labelled and unlabelled erythrocytes on autoradiography. The
proportion of HGPRT deficient erythrocytes observed on autoradiography agreed satisfactorily with that calculated from the HGPRT
activity in the erythrocyte lysate of the heterozygote and the
residual HGPRT activity of the particular mutation.

Thus, in partial HGPRT deficiency, the autoradiographic
technique has demonstrated two cell populations of erythrocytes
in heterozygotes. In the severe HGPRT deficiency of the Lesch-
Nyhan syndrome, on the other hand, although HGPRT deficiency in
erythrocytes could be identified by the technique, no circulating
HGPRT deficient erythrocytes were detected in two heterozygotes
whose heterozygosity had been established by autoradiography of
fibroblasts. This was in accordance with the finding of normal
HGPRT activity in erythrocyte lysates from such heterozygotes.
This lends direct support to the suggestion that the mosaicism in
the heterozygote for the partial HGPRT deficiency state extends to
the erythrocyte precursors of the haemopoietic system, whereas the
erythrocyte precursors in heterozygotes for the Lesch-Nyhan
syndrome are all normal.

These results will be reported more extensively in a paper
to be published separately.

STUDY OF IMMUNOREACTIVE MATERIAL IN PATIENTS WITH DEFICIENT HPRT

ACTIVITY

B. Bakay, M. Graf, S. Carey, and W. L. Nyhan

Department of Pediatrics, University of California

San Diego, La Jolla, California   92093

Patients with severe deficiency of hypoxanthine phosphoribosyl transferase (HPRT, E.C.: 2.4.2.8) activity, who usually have
the Lesch-Nyhan syndrome are phenotypically distinct from the
patients with partially active enzyme who have renal stone disease
or gout (1,2,3). There is a number of possible genetic mechanisms
which could lead to the deficiency of HPRT in various patients. In
most cases the molecular abnormality appears to be a structural
gene mutation which results in the synthesis of a catalytically inefficient enzyme protein with 0.1% to 50% of normal HPRT activity.
However, when enzymatic activity is undetectable, it is not known
whether a completely inactive enzyme protein is synthesized,or no
protein is synthesized. To distinguish between these two possibilities, the general approach has been to use an antibody produced
against normal enzyme and test the cell extracts of enzyme deficient individuals for cross-reacting material (CRM). Earlier investigators concluded that erythrocytes from patients with the
Lesch-Nyhan syndrome contained normal amounts of catalytically in-
competent HPRT protein which crossreacted with antibodies prepared
against HPRT from hemolysates of normal individuals (4 - 6). However, we were unable to confirm these findings (7) and therefore
studied this problem further.

Three New Zealand white rabbits were injected intradermally
with a 300 fold purified Crude Enzyme prepared from lysates of
normal human red cells; by removal of hemoglobin and pH 5 insoluble proteins and filtration through Sephadex G-100 (8). Three
other rabbits were injected with a 4200 fold purified Native Enzyme
which was prepared from the Crude Enzyme by isotachophoresis and
gel filtration (8). Another three rabbits were immunized with
2900 fold purified Heated Enzyme prepared by heating hemolysate

fractions at 85°C in the presence of PRPP, as described by Olsen and Milman (9). The specific activities of the Crude, Native and Heated Enzyme preparations were 250, 3700 and 2200 nmoles IMP/min/ mg protein, respectively.

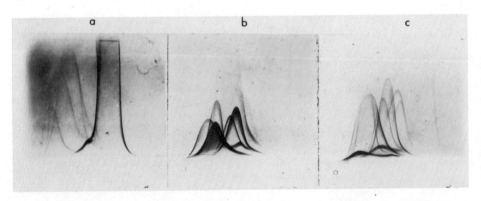

Fig. 1   Crossed immunoelectrophoretic analysis of immunoglobulins produced against (a) Crude, (b) Native, and (c) Heated Enzyme, using Crude Enzyme as antigen.

The reaction with hemolysate from normal individuals revealed that immunoglobulin prepared against Heated Enzyme precipitated 4 and 55 times more HPRT activity than immunoglobulin prepared against Crude or Native Enzyme, respectively. When tested against Crude Enzyme by crossed immunoelectrophoresis, these immunoglobulins produced several rockets (Fig. 1). According to autoradiographic analysis, some rockets had enzymatic activity and were apparently representing isozymes of HPRT.

The reaction of HPRT with immunoglobulin yielded complexes which were sedimented by centrifugation (Fig. 2). Assay of HPRT activity before and after centrifugation revealed that sedimented insoluble HPRT-IgG complexes retained up to 60% of the initial activity. Furthermore, the residual activity of these complexes was not affected by further increase of antibody concentration.

The effect of enzymatically deficient but immunochemically normal enzyme on the immunoprecipitation reaction was examined using an aged purified normal enzyme which had lost about 80% of its activity due to prolonged storage at 4°C. This preparation neutralized about five times more antibody per unit of enzyme activity than normal Native enzyme, demonstrating that the enzymatically inactive protein was fully immunoreactive. As shown in Fig. 2, the immunoprecipitation of the normal enzyme was greatly reduced in the presence of aged enzyme, proving that competitive assay of CRM was working.

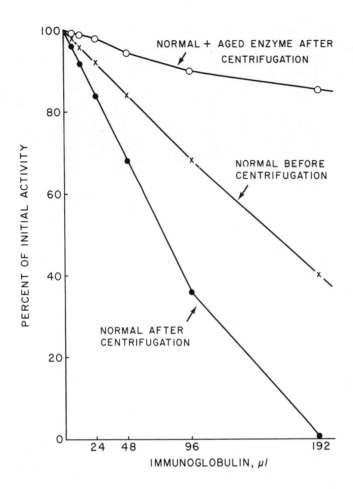

Fig. 2   Precipitation of normal HPRT by immunoglobulin.

The same competitive immunoprecipitation reaction was used for the detection of CRM in the hemolysates of 13 patients with total and 5 patients with partial deficiency of HPRT. Mixtures containing one part of hemolysate from normal individual and four parts of hemolysate from HPRT-deficient patients were reacted with different amounts of immunoglobulin, centrifuged and assayed for HPRT activity (Fig. 3). Without exception, mixtures containing hemolysates from patients with the Lesch-Nyhan syndrome lost the same amount of activity as normal hemolysates alone. Thus, hemolysates of these patients contained no CRM which was capable of blocking the immunoprecipitation of normal HPRT.

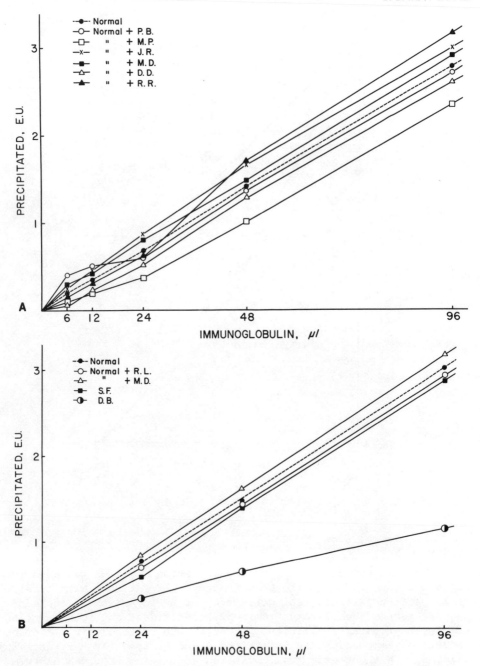

Fig. 3   Precipitation of normal HPRT in the presence of hemolysates from patients with (A) the Lesch–Nyhan syndrome, and (B) with partial deficiency of HPRT.

Hemolysates from patients M.D. and R.L., with 2.5 and 4.6% of normal HPRT activity, respectively, also failed to block the precipitation of normal HPRT (Fig. 3b). However, hemolysate from D.B. with 50% of normal activity, contained CRM which considerably reduced the precipitation of normal HPRT.

In addition, to sedimentable insoluble complexes, reaction of HPRT with antibody also produced soluble complexes. Following immunoprecipitation reaction and centrifugation, supernatants were separated by polyacrylamide gel electrophoresis (10). The gels were incubated first with PRPP and ^{14}C-labeled hypoxanthine; then, with 0.1$\underline{M}$ LaCl$_3$ and washed in water (10). This resulted in removal of soluble purine bases, while precipitated nucleotides remained fixed in the gel. The gels were then fractionated by a mechanical fractionator and analyzed for ^{14}C in a liquid scintillation counter using a hollow flow cell (10).

In this system, free normal HPRT produced Peak A in the middle of the profile (Fig. 4a). After incubation with immunoglobulin, part of the HPRT activity migrated a notably shorter distance and formed Peak B (Fig. 4b-c). Peak B contained the soluble HPRT-antibody complexes, which displayed the same amount of activity as the free enzyme. With increasing immunoglobulin concentration, the activity in Peak A decreased and that in Peak B first increased and then decreased until no activity was detected in the gel (Fig. 4d).

This system proved to be especially valuable for the detection of CRM in patients with partial deficiency of HPRT (Fig. 5). After reacting with immunoglobulin, hemolysates of patients D.B.; a heterozygous female S.F.; and patient T.L., who diplayed 50%, 34% and 5.5% of normal activity, respectively, produced profiles in which there was activity in Peak A and Peak B. This indicated that HPRT of these individuals reacted with immunoglobulin and formed slow migrating soluble HPRT-antibody complexes. Hemolysates of R.L., who displayed 4.6% of normal activity and who is a half brother of S.F., produced a profile which contained Peak A but contained no Peak B (Fig. 5). This was in agreement with the competitive immunoprecipitation assay and confirmed that R.L. contained no CRM.

In the profile of S.F., there was a considerable amount of activity in the faster migrating part of Peak A, which represented the product of her abnormal cells (11). This confirmed our previous observations and indicated that during electrophoresis, HPRT produced by her abnormal cells reacted with HPRT produced by her normal cells, and thus, caused the activation of abnormal enzyme (11). Also, the profile of S.F. shows that immunoglobulin reacted with her normal enzyme, but not with her deficient enzyme. This is in agreement

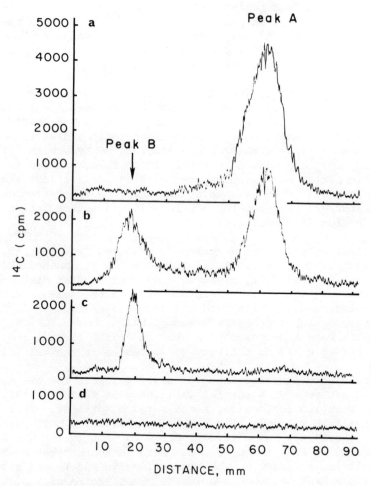

Fig. 4   Electrophoretic analysis of soluble HPRT-antibody complexes in presence of free normal HPRT.

with the results obtained with hemolysate of her hemizygous half
brother R.L. shown in Fig. 5 and with results obtained by immunoprecipitation reaction shown in Fig. 3.   These findings prove con-
clusively that in this kindred, the mutant gene produces an enzymatically inefficient product which does not react with antibody
against the normal enzyme.

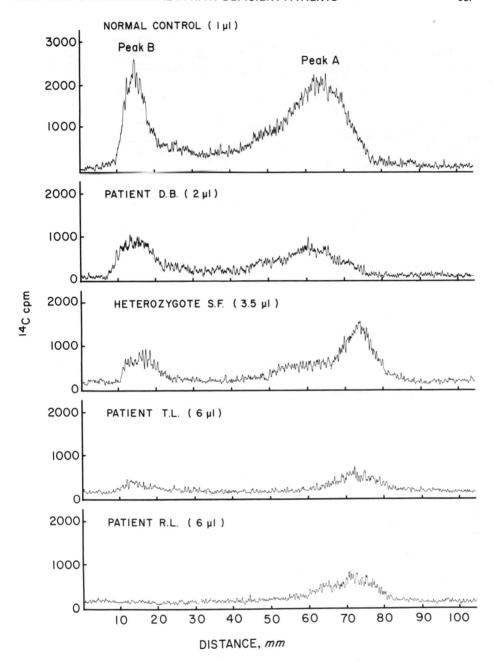

Fig. 5. Electrophoretic analysis of soluble HPRT-antibody complexes in presence of free enzyme from patients with partially deficient enzyme.

In conclusion, these studies have shown that hemolysates from patients with the Lesch-Nyhan syndrome as well as some patients with partially active HPRT, contain no material which could react with antibodies produced in rabbits against normal human red cell HPRT enzyme. In contrast, hemolysates of two patients containing partially active HPRT, contained CRM which yielded both the insoluble and soluble complexes. These results support our earlier observations (7) as well as those of other investigators (12,13). Finally, lack of CRM in HPRT-deficient patients indicates considerable heterogeneity in the HPRT locus which results in production of variant enzymes with altered recognition sites.

ACKNOWLEDGMENTS

We wish to thank Drs. Michael A. Becker, Department of Medicine, Nancy Fawcett, University of Miami, Florida, and Melvin A. Hoch, The Permanente Medical Group, Haywood, California, for making available their patients for this study. This work was supported by grants from the National Science Foundation BMS-74-21424, the National Foundation, March of Dimes NF-1377, and U.S. Public Health Service MCT-000274 and GM-17702.

REFERENCES

1. Lesch, M. and Nyhan, W.L. (1964) Amer. J. Med. 36, 561-570.

2. Seegmiller, J.E., Rosenbloom, F.M., and Kelley, W.N. (1967) Science 155, 1682-1684.

3. Kogut, M.D., Donnell, G.N., Nyhan, W.L. and Sweetman, L. (1970) Amer. J. Med. 48, 148-161.

4. Rubin, C.S., Dancis, J., Yip, L.C., Nowinsky, R.C., and Balis, M.E. (1971) Proc. Nat. Acad. Sci. USA 68, 1461-1464.

5. Arnold, W.J., Meade, J.C., and Kelley, W.N. (1972) J. Clin. Invest. 51, 1805-1812.

6. Müller,M.M., Dobrovits, H.and Stemberger, H. (1972) J. Klin. Chem.Klin. Biochem. 10, 535-538.

7. Bakay, B., Bazaral, M., and Nyhan, W.L. (1975) Ped. Res. 9, 311.

8. Bakay, B., and Nyhan, W.L. (1975) Arch. Biochem. Biophys. 168, 26-34.

9. Olsen, A.S. and Milman, G. (1974) J. Biol. Chem. 249, 4030-4037.

10. Bakay, B. (1971) Anal. Biochem. 40, 429-439.

11. Bakay, B., Nyhan, W.L., Fawcett, N., and Kogut, M.D. (1972) Biochem. Genet. 7, 73-85.

12. Changas, G.S., and Milman, G. (1975) Proc. Nat. Acad. Sci. USA 72, 4147-4150.

13. Upchurch, K.S., Leyva, A., Arnold, W.J., Holmes, E.W., and Kelley, W.N. (1975) Proc. Nat. Acad. Sci. USA 72, 4142-4146.

EFFECTS OF INOSINE ON PURINE SYNTHESIS IN NORMAL AND HGPRT-DEFICIENT HUMAN FIBROBLASTS

Michael A. Becker

Department of Medicine, University of California, San Diego and San Diego Veterans Administration Hospital, La Jolla, California 92161 U.S.A.

Antagonistic roles for 5-phosphoribosyl 1-pyrophosphate (PP-ribose-P[1]) and purine nucleotides in effecting regulation of the rate of purine nucleotide synthesis *de novo* have been proposed from the kinetic and structural effects of these compounds on the allosteric enzyme PP-ribose-P amidotransferase described in the elegant studies of Holmes, Wyngaarden and Kelley (1, 2). From experiments in intact cells, Bagnara, Letter and Henderson (3) have pointed out additional or alternative mechanisms for the control of purine production such as the utilization of PP-ribose-P in competing reactions and purine nucleotide inhibition of PP-ribose-P synthesis.

A corollary of these contemporary concepts of the regulation of this pathway is that conditions in which increased PP-ribose-P concentrations occur without increased purine nucleotide concentrations should be accompanied by accelerated rates of purine synthesis *de novo*. Indeed, a number of studies have demonstrated parallel changes in the rates of purine synthesis corresponding to increases or decreases in PP-ribose-P concentrations (Reviews, references 4, 5); however, as reported here, when normal or HGPRT-deficient human fibroblasts are incubated with inosine, increases in PP-ribose-P concentrations are attended by decreased rates of purine synthesis. In the case of normal fibroblasts, the decrement

---

[1]Abbreviations: PP-ribose-P, 5-phosphoribosyl 1-pyrophosphate; HGPRT, hypoxanthine-guanine phosphoribosyltransferase; ribose-5-P, ribose-5-phosphate; FGAR, formylglycinamide ribotide.

in purine synthetic rate in the face of increased PP-ribose-P concentration appears best explained by simultaneous increases in nucleotide feedback inhibitor concentrations. In HGPRT-deficient cells, however, failure of increased PP-ribose-P concentration to stimulate purine synthesis is presently unexplained.

Fibroblasts were cultured from skin biopsy specimens obtained from 4 normal individuals and 4 children with severe HGPRT deficiency (6) and the Lesch-Nyhan syndrome (7). As seen in Table 1, residual HGPRT activity in the mutant fibroblasts ranged from 1.1 to 2.8% of normal enzyme activity and the mutant cells were markedly impaired in their ability to incorporate labeled inosine or hypoxanthine (11, 12) into purine nucleotides, showing less than 3% of the normal rates of incorporation. As previously reported (11), HGPRT-deficient fibroblasts had baseline PP-ribose-P concentrations from 2- to 5-fold greater than normal cells. Ribose-5-phosphate (ribose-5-P) concentrations were, however, comparable in the two cell types.

Addition of 5 mM inosine to suspensions of normal and mutant cells in Krebs-Ringer phosphate buffer with 1 mM glucose resulted in increases in both PP-ribose-P and ribose-5-P concentrations compared to control cells incubated in buffer with glucose alone (Table 2). The maximal increments in PP-ribose-P concentrations occurred after 20 to 30 minutes while the more dramatic 18- to 33-fold increments in ribose-5-P concentrations were observed during the first minute of incubation with inosine. The response of PP-ribose-P concentration was dependent on the concentration of added inosine and was maximal at between 5 and 10 mM inosine. More modest increases in fibroblast PP-ribose-P and ribose-5-P concentrations resulted from addition of the artificial proton acceptor

TABLE 1

BIOCHEMICAL CHARACTERIZATION OF FIBROBLAST STRAINS

Source of fibroblast strain	HGPRT activity*	Incorporation of labeled isotope into purine nucleotides[+]		Baseline concentration[§]	
		8-[^{14}C] inosine	8-[^{14}C] hypoxanthine	PP-ribose-P	Ribose-5-P
	nmoles/h/mg protein	nmoles/h/10^6 cells		nmoles/10^6 cells	
Normal:					
C. H.	133.4	2.46	1.48	0.32	0.26
D. Bu.	141.1	2.32	1.61	0.32	0.36
C. J.	118.7	2.19	1.39	0.21	0.22
Aug.	132.3	2.24	1.09	0.17	0.18
Lesch-Nyhan syndrome:					
P. M.	2.1	0.03	0.02	0.66	0.36
S. M.	3.1	0.03	0.03	0.90	0.20
M. P.	1.4	0.02	0.01	0.78	0.22
M. J.	2.2	0.04	0.03	0.68	0.38

*Activity determined by method of Kelley and Meade (8) with [8-^{14}C] hypoxanthine as substrate.

[+]Final concentration of each isotope was 2.0 mM. Incorporation of label into purine nucleotide determined by method of Crabtree and Henderson (9).

[§]Concentrations determined as previously described (10).

methylene blue, the responses to which were maximal at 50 to 100
µM (Table 2).

In contrast to the case with methylene blue which stimulates
the pentose phosphate pathway by accelerating the rate of NADPH
oxidation (13), the increment in PP-ribose-P concentration during
incubation with inosine does not appear to involve stimulation of
this pathway. In studies in which the specific radioactivity of
the ribose moiety of labeled inosine was compared with that of
the PP-ribose-P extracted from mutant cells during incubation with
this labeled nucleoside, direct donation of the ribose moiety of
inosine to the ribosylphosphate moiety of PP-ribose-P was demonstrated. Thus the effect of inosine on PP-ribose-P concentration
most likely results from phosphorylation of inosine with production
of equimolar quantities of hypoxanthine and ribose-1-phosphate;
isomerization to ribose-5-P; and condensation of ribose-5-P with
ATP to form PP-ribose-P. While direct stimulation of PP-ribose-P
synthetase by inosine is not excluded by these studies, purified
human PP-ribose-P synthetase is neither stimulated nor relieved
of the effects of inhibitors by inosine.

Despite increased PP-ribose-P concentrations in fibroblasts
incubated with inosine, rates of purine synthesis *de novo* estimated
by the incorporation of $\{^{14}C\}$formate into formylglycinamide ribotide (FGAR) in the presence of azaserine (14), were markedly
diminished in normal fibroblasts and more modestly diminished in
mutant cells incubated with 5 mM inosine (Table 3). The effect of
inosine on FGAR accumulation was relatively constant throughout the
period of incubation and in experiments in which the specific
activity of $\{^{14}C\}$formate was varied over a 10-fold range, did not
appear to result from alteration in formate pool size by inosine.
Since in normal cells nucleotide formation from the hypoxanthine
generated during the phosphorylation of inosine was possible, it
seemed likely that increased purine nucleotide concentrations might
explain the diminished purine synthetic rate observed in the face
of increased PP-ribose-P concentrations. In view of the even
higher PP-ribose-P concentrations in the mutant cells and their
impaired ability to incorporate hypoxanthine or inosine into purine
nucleotides, the lack of a stimulatory effect of inosine on purine
synthesis in these cells was unexpected.

Stimulation of FGAR accumulation in both normal and mutant
cells during incubation with methylene blue (Table 3) made it
unlikely that saturation of PP-ribose-P amidotransferase by baseline concentrations of PP-ribose-P explained the lack of stimulation
of purine synthesis in the mutant cells treated with inosine.
Another possibility, that residual HGPRT activity in mutant cells
was sufficient to allow generation of purine nucleotide inhibitors,
was directly tested by measurement of intracellular purine

TABLE 2

RELATIVE INCREMENTS IN FIBROBLAST PP-RIBOSE-P AND RIBOSE-5-P CONCEN-
TRATIONS DURING INCUBATION WITH 5 mM INOSINE AND 50 µM METHYLENE BLUE

Fibroblast group	PP-Ribose-P Concentration		Ribose-5-P Concentration	
	Inosine	Methylene blue	Inosine	Methylene blue
	Values relative to control (1.0)		Values relative to control (1.0)	
Normal	2.4 - 3.8	1.7 - 2.2	20.4 - 33.1	6.8 - 10.6
Mutant	2.7 - 5.3	1.9 - 3.4	18.5 - 33.7	6.7 - 10.0

Cell suspensions were incubated for 30 minutes in buffer alone (Control)
or in buffer with stated addition prior to the above measurements.

TABLE 3

EFFECTS OF INOSINE AND METHYLENE BLUE ON RATES OF PURINE SYNTHESIS
DE NOVO IN NORMAL AND MUTANT FIBROBLASTS

Fibroblast strain	Sodium [^{14}C]-formate incorporation into FGAR[*]		
	no addition	5 mM inosine	50 µM methylene blue
	cpm/10^6 cells	% inhibition	% stimulation
Normal:			
C. H.	2490	76	78
D. Bu.	2720	64	85
C. J.	2260	80	71
Aug.	2615	74	58
Mutant:			
P. M.	8455	8	49
S. M.	8120	19	37
M. P.	9240	2	46
M. J.	7875	15	51

[*]Incubations were carried out for 60 minutes at 37° as previously
described (11); FGAR was extracted, separated by thin layer
chromatography (15), and counted.

TABLE 4

PURINE NUCLEOTIDE CONCENTRATIONS IN NORMAL AND
HGPRT - DEFICIENT FIBROBLASTS

Fibroblast strain	Inosine 5 mM	Nucleotide Concentrations		
		Total adenine Nucleotides	Total guanine Nucleotides	IMP
Normal		nmoles/10^6 cells		
C.H.	-	4.50	1.84	0.03
	+	5.74	2.13	0.05
C.J.	-	5.29	1.65	--
	+	6.96	2.26	0.02
Mutant				
S.M.	-	4.48	1.42	--
	+	4.50	1.43	--
M.P.	-	4.52	1.87	0.02
	+	4.51	1.77	0.02

nucleotide pools in the presence and absence of inosine (Table 4).

Purine nucleotide concentrations in two normal and two mutant cell strains were measured by high pressure liquid chromatography (16). Baseline purine nucleotide concentrations in normal and mutant cells were comparable as previously noted (11). Incubation of normal fibroblasts with 5 mM inosine resulted in increased adenine and guanine nucleotide concentrations (Table 4). No significant alteration in nucleotide concentration occurred, however, during exposure of HGPRT-deficient fibroblasts to the same concentration of inosine.

In mutant cells, then, inhibition of purine synthesis during incubation with inosine occurs without detectable increases in purine nucleotide pool sizes suggesting that in HGPRT-deficient cells nucleotide feedback inhibition is not responsible for counteracting the stimulatory effect of increased PP-ribose-P concentration. Since, however, the more marked inhibition of purine synthesis in normal cells is accompanied by increased purine nucleotide concentrations, purine nucleotide feedback inhibition of PP-ribose-P amidotransferase activity may contribute at least in part to the inhibition of purine synthesis in these cells during incubation with inosine. Thus while a stimulatory role for PP-ribose-P seems well established by many previous studies (4, 5) and while an inhibitory role for purine nucleotides best explains the inhibition of purine synthesis observed here in normal cells, the failure of increased PP-ribose-P concentrations resulting from incubation with inosine to accelerate purine synthesis in HGPRT-deficient fibroblasts is not explained in full by consideration of the concentrations of purine nucleotides and PP-ribose-P.

In summary,

1. Incubation of normal and HGPRT-deficient fibroblasts with inosine results in *increased* PP-ribose-P concentrations

2. The increased PP-ribose-P concentrations are accompanied by *decreased* rates of purine synthesis *de novo*, more marked in normal cells

3. Increased purine nucleotide concentrations during incubation with inosine provide a likely explanation for the inhibition of purine synthesis in normal cells

4. The lack of accelerated purine synthesis in mutant cells under these conditions is not fully explained by consideration of PP-ribose-P and purine nucleotide concentrations.

ACKNOWLEDGEMENTS

This study was supported by: a Veterans Administration Clinical Investigatorship, a Veterans Administration grant (MRIS 0865) and grant AM-18197 from the National Institutes of Health to the author; and grants AM-13622, AM-05616 and GM-17702 from the National Institutes of Health to Dr. J. Edwin Seegmiller to whom the author is grateful for support and advice.

REFERENCES

1. Holmes, E.W., McDonald, J.A., McCord, J.M., Wyngaarden, J.B. and Kelley, W.N.: *J. Biol. Chem.* 248: 144-150 (1973).
2. Holmes, E.W., Wyngaarden, J.B. and Kelley, W.N.: *J. Biol. Chem.* 248: 6035-6040 (1973).
3. Bagnara, A.S., Letter, A.A., and Henderson, J.F.: *Biochim. Biophys. Acta* 374: 259-270 (1974).
4. Fox, I.H., and Kelley, W.N.: *Ann. Intern. Med.* 74: 424-433 (1971).
5. Becker, M.A. and Seegmiller, J.E.: *Annu. Rev. Med.* 25: 15-28 (1974).
6. Seegmiller, J.E., Rosenbloom, F.M. and Kelley, W. N.: *Science (Wash. D.C.)* 155: 1682-1684 (1967).
7. Lesch, M. and Nyhan, W.L.: *Am. J. Med.* 36: 561-570 (1964).
8. Kelley, W.N. and Meade, J.C.: *J. Biol. Chem.* 246: 2953-2958 (1971).
9. Crabtree, G.W. and Henderson, J.F.: *Cancer Res.* 31: 985-991 (1971).
10. Becker, M.A.: *Biochim. Biophys. Acta* in press.
11. Rosenbloom, F.M., Henderson, J.F., Caldwell, I.C., Kelley, W. N. and Seegmiller, J.E.: *J. Biol. Chem.* 243: 1166-1173 (1968).
12. Raivio, K.O. and Seegmiller, J.E.: *Biochim. Biophys. Acta* 299: 273-282 (1973).
13. Brin, M. and Yonemoto, R.H.: *J. Biol. Chem.* 230: 307-317 (1958).
14. Henderson, J.F.: *J. Biol. Chem.* 237: 2631-2638 (1962).
15. Boyle, J.A., Raivio, K.O., Becker, M.A. and Seegmiller, J.E.: *Biochim. Biophys. Acta* 269: 179-183 (1972).
16. Brown, P.R.: *J. Chromatogr.* 52: 257-262 (1970).

AN ALTERNATE METABOLIC ROUTE FOR THE SYNTHESIS OF INOSINE

5'-PHOSPHATE (IMP) IN THE LESCH–NYHAN ERYTHROCYTE

Bertram A. Lowy and Marjorie K. Williams

Department of Biochemistry
Albert Einstein College of Medicine
Bronx, New York 10461

The cells of the Lesch–Nyhan individual, which are deficient in hypoxanthine–guanine phosphoribosyltransferase (HGPRT) activity, lack the ability to convert the 6-keto purines, hypoxanthine and guanine, to the corresponding nucleotides (1). Mammalian cells, in general, lack appreciable kinase activity for the 6-keto purine nucleosides, inosine and guanosine (2,3). Thus, the major route to IMP and GMP in the Lesch–Nyhan cell is via the overall pathway of de novo purine nucleotide synthesis in which the purine ring of IMP is built up on carbon 1 of a molecule of ribose 5-phosphate (4). The IMP is readily converted to GMP in normal human and Lesch–Nyhan cells.

Normal human and Lesch–Nyhan cells possess the enzyme adenine phosphoribosyltransferase (APRT) and thus can convert the 6-amino purine, adenine, to AMP (1). However, there appears to be no real function for mammalian APRT, since free adenine is virtually non-existent in mammalian cells. Human cells do possess the enzyme adenosine kinase and can convert adenosine to AMP (5).

The Lesch–Nyhan syndrome is characterized by elevated levels of uric acid (6) and cellular PRPP (7) as well as a markedly elevated rate of purine nucleotide synthesis by the de novo pathway. Among the postulated metabolic consequences of the HGPRT deficiency are a poor regulation of the de novo pathway (1), an inadequate level of cellular folates (8,9), and a reduced level of brain or serum glutamine (10) or glutamine metabolites. However, the relationship between the enzymic deficiency and the mental, physical and behavioral manifestations remains unresolved.

Since the known metabolic aberration in the disease is the failure to form IMP and GMP from the purine bases, it was of interest to seek alternate metabolic routes to these compounds thus by-passing the deficiency without involving the overall de novo pathway.

A number of years ago, we reported that the intact normal human erythrocyte lacked the capacity for the overall synthesis of purine nucleotides by the de novo pathway (11), probably a consequence of the erythroid cell maturation process (12). The cell does, however, possess the ability to carry out the final reactions of the pathway (11). Thus, if 4-amino-5-imidazolecarboxamide (AICA) or ribosyl-AICA (rAICA) are provided to the cell the imidazole compounds enter the cell and are converted to the corresponding nucleotide by the reactions summarized in Figure 1.

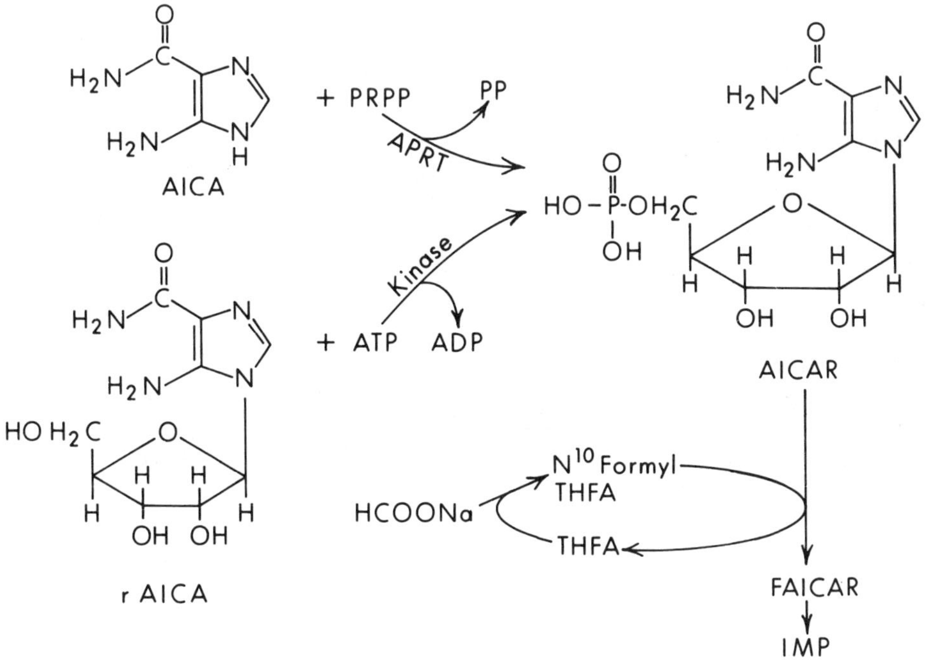

Figure 1.
Enzymic capacity of human erythrocyte for IMP synthesis from 4-amino-5-imidazolecarboxamide (AICA) and its ribosyl derivative (rAICA)

The first reaction is catalyzed by APRT which can utilize AICA as substrate even though it is not actually a 6-amino purine (13). The ribosyl-AICA is phosphorylated by a kinase, possibly the adenosine kinase of the erythrocyte. The ribotide, AICAR, is one of the final intermediates of the de novo pathway. In the presence of N^{10}-formyl tetrahydrofolic acid, formed from the THFA of the erythrocyte and added sodium formate, the AICAR is then formylated and the resultant formyl-AICAR (FAICAR) undergoes ring closure to IMP. The IMP may be converted to GMP via xanthosine 5'-phosphate (XMP). The human erythrocyte is unable to convert IMP to AMP (14). Although the normal human erythrocyte possesses both HGPRT and APRT, the reactions cited have, in effect, led to the formation of IMP and GMP without utilizing HGPRT.

We have now extended these findings to the Lesch-Nyhan erythrocyte and have demonstrated two by-pass mechanisms for IMP synthesis in the HGPRT deficient cell.

Nucleotide synthesis in erythrocytes obtained from a normal adult male and from a Lesch-Nyhan patient was compared by incubating aliquots of red blood cells with adenine-8-C^{14} and hypoxanthine-8-C^{14} (Table I). Incubation with labeled adenine resulted in about equivalent labeling of the AMP of the normal and the Lesch-Nyhan cell and similar extents of deamination to IMP. In contrast, labeled hypoxanthine led to extensive labeling of IMP in the normal erythrocyte, but only a very small amount of label was detected in the IMP of the Lesch-Nyhan cell. The results are in accord with the known deficiency of HGPRT in the Lesch-Nyhan cell and the presence of APRT and adenylate deaminase in normal and in Lesch-Nyhan erythrocytes. When 4-amino-5 imidazolecarboxamide and sodium formate-C^{14} were the substrates, extensive labeling of IMP occurred in both the normal and Lesch-Nyhan erythrocytes. Thus AICA, a substrate for APRT, was converted to the ribonucleotide, which after formylation and ring closure, gave rise to labeled IMP. The presence of both APRT and HGPRT in the normal cell did not result in a greater extent of labeling in that cell. AICA is a known inhibitor of HGPRT (15).

The utilization of AICA and sodium formate-C^{14} for the synthesis of IMP in the Lesch-Nyhan erythrocyte was extended to two additional Lesch-Nyhan patients (Table II). Although a variability exists in the extent of IMP labeling among the patients, the utilization of the substrates was similar to that of the normal and did not occur from labeled sodium formate alone. In the absence of AICA, the radioactivity in the total IMP fraction was very small and could be due to the small number of reticulocytes in the erythrocyte preparation.

TABLE I

Utilization of adenine, hypoxanthine and 4-amino-5-imidazolecarboxamide (AICA) for synthesis of purine nucleotides by erythrocytes from a patient with Lesch-Nyhan syndrome and a normal adult male.

	Incubated with			
	Adenine-8-C^{14} 5.0 µCi, 1.6 µmoles		Hypoxanthine-8-C^{14} 5.0 µCi, 1.6 µmoles	Sodium formate-C^{14} 5.0 µCi, 3.1 µmoles AICA 1.5 µmoles
	Isolated			
	AMP	IMP	IMP	IMP
Subject	Total cpm per ml cells			
Normal (BL)	7040	2750	268900	43440
Patient (JA)	7290	2300	2960	45020

Aliquots of washed erythrocytes (0.85 ml) were incubated for 2 hours at 37° in equal volumes of isotonic sodium phosphate buffer (pH 7.4) containing glucose (15 µmoles). Carrier IMP was added. Nucleotides were isolated, purified and assayed for radioactivity.

TABLE II

Utilization of 4-amino-5-imidazolecarboxamide (AICA) and its
ribosyl derivative (rAICA) by erythrocytes from patients with Lesch-Nyhan
syndrome and a normal adult male.

	Incubated with		
	Sodium formate-C^{14} 10 µCi, 3.0 µmoles	Sodium formate-C^{14} 10 µCi, 3.0 µmoles AICA 1.5 µmoles	Sodium formate-C^{14} 10 µCi, 3.0 µmoles rAICA 1.5 µmoles
	Isolated		
	IMP	IMP	IMP
Subject	Total cpm per ml cells		
Normal (BL)	2170	24910	60480
Patient (CW)	3250	32240	24990
(JA)	830	9080	14410
(PB)	500	18240	21660

Aliquots of washed erythrocytes (1.0 ml) were incubated for 2 hrs at 37° in
equal volumes of isotonic sodium phosphate buffer (pH 7.4) containing
glucose (15 µmoles). Carrier IMP was added. Total IMP was isolated,
purified and assayed for radioactivity.

TABLE III

Utilization of 4-amino-5-imidazolecarboxamide (AICA) and its ribonucleotide (AICAR) by lysates of erythrocytes from patients with Lesch-Nyhan syndrome.

	Incubated with	
	Sodium formate-C^{14} 10 µCi, 3.0 µmoles AICA 1.5 µmoles PRPP 3.0 µmoles	Sodium formate-C^{14} 10 µCi, 3.0 µmoles AICAR 1.5 µmoles
	Isolated	
	IMP	IMP
Subject	Total cpm per ml cells	
Patient (JA)	14500	—
(PB)	—	6500

Lysates (1.0 ml) prepared from washed erythrocytes were incubated for 2 hrs at 37° in equal volumes of isotonic sodium phosphate buffer (pH 7.4) and glucose (15 µmoles). ATP (3 µmoles) was added initially and at 30 minute intervals. Carrier IMP was added. Total IMP was isolated, purified and assayed for radioactivity.

When ribosyl-AICA was used as substrate, along with sodium formate-C^{14}, considerable labeling of the IMP also occurred in the patients' erythrocytes and in the normal. Since ribosyl-AICA is not cleaved by human erythrocyte purine nucleoside phosphorylase (16), the conversion of the ribosyl compound to the ribotide probably occurred by direct phosphorylation, prior to formylation and ring closure to form the IMP.

The utilization of imidazole compounds also occurred in lysates prepared from Lesch-Nyhan erythrocytes (Table III). When AICA, PRPP and labeled sodium formate were incubated with a lysate, extensive labeling of IMP occurred. When the ribotide, AICAR, was used along with sodium formate-C^{14}, it was also well utilized for IMP synthesis.

In summary, we have described two by-pass mechanisms for the synthesis of IMP in the HGPRT deficient Lesch-Nyhan erythrocyte, in the absence of a functioning overall de novo pathway. One mechanism utilizes AICA plus PRPP and the APRT of the cell, and the other mechanism utilizes ribosyl-AICA plus ATP and a kinase. The ribotide, AICAR, produced by either mechanism is formylated by N^{10}-formyl THFA and converted to IMP (Figure 1).

Extension of the present study to other Lesch-Nyhan cell types could lead to a clarification of the role of the purine nucleotides, as feedback inhibitors, and of PRPP, as a stimulator, of the initial enzyme, the glutamine PRPP amidotransferase. Although both ribosyl-AICA and AICA are converted to IMP, only AICA requires PRPP. Thus it should be possible to form IMP with or without a pronounced effect on the characteristically elevated PRPP concentration of the Lesch-Nyhan cell.

The reaction mechanisms described may also provide a biochemical basis for a new approach to the early treatment of the disease, and a possible therapeutic use of AICA and rAICA.

This investigation was supported by a grant from the National Institutes of Health (AM 17622).

REFERENCES

1. Seegmiller, J.E., Rosenbloom, F.M., and Kelley, W.N., Science 155, 1682 (1967).

2. Friedmann, T., Seegmiller, J.E. and Subak-Sharpe, J.H., Exptl. Cell Res. 56, 425 (1969).

3. Payne, M.R., Dancis, J., Berman, P.H. and Balis, M.E., Exptl. Cell Res. 59, 489 (1970).

4. Buchanan, J.M., and Hartman, S.C., Adv. Enzymol. 21, 200 (1959).

5. Lowy, B.A. and Williams, M.K., Blood 27, 623 (1966).

6. Lesch, M. and Nyhan, W.L., Am. J. Med. 36, 561 (1964).

7. Rosenbloom, F.M., Henderson, J.F., Caldwell, I.C., Kelley, W.N. and Seegmiller, J.E., J. Biol. Chem. 243, 1166 (1968).

8. Kelley, W.N., Greene, M.L., Rosenbloom, F.M., Henderson, J.F. and Seegmiller, J.E., Ann. Intern. Med. 70, 155 (1969).

9. Newcombe, D.S., Pediatrics, 46, 508 (1970).

10. Ghadimi, H., Bhalla, C.K. and Kirchenbaum, D.M., Acta Paediat. Scand. 59, 233 (1970).

11. Lowy, B.A., Williams, M.K. and London, I.M., J. Biol. Chem. 237, 1622 (1962).

12. Lowy, B.A., Cook, J.L. and London, I.M., J. Biol. Chem. 236, 1442 (1961).

13. Kelley, W.N., Fed. Proc. 27, 1047 (1968).

14. Lowy, B.A. and Dorfman, B., J. Biol. Chem. 245, 3043 (1970).

15. Krenitsky, T.A., Papaioannou, R. and Elion, G.B., J. Biol. Chem. 244, 1263 (1969).

16. Lewis, A.S. and Lowy, B.A., unpublished observation.

ADENOSINE METABOLISM IN PERMANENT LYMPHOCYTE LINES AND IN ERYTHROCYTES OF PATIENTS WITH THE LESCH-NYHAN SYNDROME

Gabrielle H. Reem

New York University Medical Center

New York, N.Y. 10016

Adenosine metabolism of mammalian lymphocytes is of particular interest since the discovery of Giblett that hereditary severe combined immunodeficiency is linked with adenosine deaminase deficiency (1). In mammalian cells adenosine is phosphorylated to adenylate (AMP) or deaminated to inosine (Fig. 1 reaction 2), which is converted to hypoxanthine. Cells derived from patients with the Lesch Nyhan (L.N.) syndrome are deficient in hypoxanthine guanine phosphoribosyltransferase (HGPRT, EC 2.4.2.8) and therefore cannot convert hypoxanthine to inosinate (IMP). Consequently these cells provide a convenient model system to assess the importance of adenosine as a source of purine ribonucleotides in mammalian cells. The phosphorylation of adenosine to adenylate (AMP) is catalysed by adenosine kinase (AK, EC 2.7.1.20, Fig. 1 reaction 1); this pathway could serve as an additional means of purine salvage. Since adenosine is toxic to mammalian cells in culture (2), and plays a role in suppressing the proliferative response of lymphocytes to PHA (3), the present study of the metabolic fate of adenosine was carried out in cultured lymphocytes and in erythrocytes obtained from normal subjects and patients with the L.N. syndrome. The purpose of this study is to determine whether in HGPRT deficient cells, adenosine could indeed serve as an alternative source of purine ribonucleotides.

Lymphocyte cultures were incubated with varying amounts of tritium labeled adenosine. The culture medium was separated from the cells by centrifugation, purine ribonucleotides extracted from the cells; ribonucleotides and the metabolites of adenosine in the medium were identified by thin layer chromatography (5). It was found that the amounts of purine ribonucleotides increased when increasing amounts of adenosine (43 to 604 µM) were incubated,

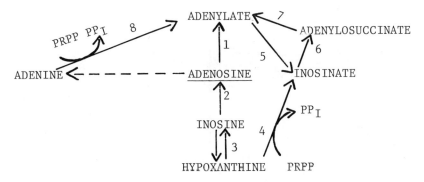

Fig. 1. Metabolic pathways of adenosine.

while the percentage of labeled material recovered in ribonucleotides decreased, indicating that at higher concentrations of adenosine, the major portion of adenosine was deaminated (Table 1).
Control and L.N. lymphoblasts incorporated similar amounts of
adenosine. At high concentrations of adenosine (above 300 μM),
adenosine was almost entirely deaminated. The metabolites of
adenosine found in the medium were mostly inosine and hypoxanthine
(Fig. 2). No more than 5% of unmetabolized adenosine was found
in the medium after 15 minutes of incubation. At higher concentra-

TABLE 1.   INCORPORATION OF [^{3}H] ADENOSINE INTO LESCH-NYHAN AND
CONTROL LYMPHOBLASTS IN CULTURE

	ISOTOPICALLY LABELLED RIBONUCLEOTIDES FORMED			
ADDITIONS	CONTROL		LESCH NYHAN	
Adenosine				
M	nmoles*	%	nmoles*	%
4.3	1.9	44	0.9	21
64	10	16	8	13
304	22	7	11	4
604	44	7	24	4

* nmoles per 5 x 10^6 cells per 15 minutes

Lymphoblasts were grown in RPMI medium with 20% fetal calf serum.
Cells were harvested by centrifugation, washed 3 times in phosphate buffered saline and incubated 15 minutes at 37°C in Hamk's
balanced solution (pH 7.4) buffered with Tris-HCl and Hepes with
100 μCi of ^{3}H adenosine and varying amounts of adenosine as noted.

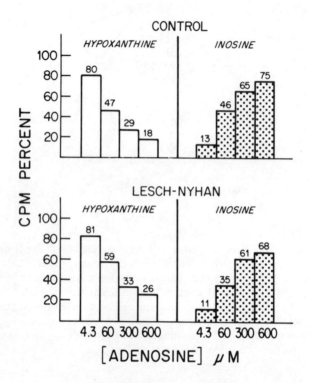

Fig. 2.   Products of [³H] adenosine in media of cultured lymphoblasts.

tions of adenosine, inosine was the major metabolite, while at
lower concentrations hypoxanthine predominated (Fig. 2).

Since L.N. and control lymphoblasts incorporated similar
amounts of adenosine into purine ribonucleotides, the incorporation
of adenosine, inosine and hypoxanthine in L.N. lymphoblasts was
compared with that of the control cell line (Table 2). The incorporation of [8-¹⁴C] adenosine exceeded that of inosine and
hypoxanthine two fold in the control cells. This finding could
reflect a higher rate of incorporation of adenosine, or a smaller
size of the adenosine pool. While the control cell line incorporated approximately equal amounts of inosine and hypoxanthine into
ribonucleotides, the L.N. cells failed to incorporate either inosine
or hypoxanthine. Clearly, the ribonucleotides synthesized from
adenosine in the L.N. cells were formed by a route other than that
dependent upon deamination of adenosine. The most likely pathway

TABLE 2. INCORPORATION OF [8-^{14}C] ADENOSINE [8-^{14}C] INOSINE AND
[8-^{14}C] HYPOXANTHINE INTO PURINE RIBONUCLEOTIDES OF
LESCH-NYHAN AND CONTROL LYMPHOBLASTS

PRECURSOR	[^{14}C] INCORPORATION INTO PURINE RIBONUCLEOTIDES	
	CONTROL	LESCH-NYHAN
	CPM/10^9 cells	
[8-^{14}C] Adenosine	142	130
[8-^{14}C] Inosine	43	1.7
[8-^{14}C] Hypoxanthine	50	1.7

Lymphoblasts were incubated with 1 μCi of [8-^{14}C] adenosine,
[8-^{14}C] inosine or [8-^{14}C] hypoxanthine as noted above.

would be by the conversion of adenosine to adenylate catalysed by
AK (reaction 1 Fig. 1). However, the possibility of conversion to
adenine and salvage by adenine phosphoribosyltransferase (APRT,
EC 2.4.2.7) needs to be reevaluated, particularly since it was
found very recently that APRT deficient children form renal calculi consisting of dihydroxyadenine (6,7).

The purine ribonucleotides labeled following incubation with
[^{3}H] adenosine were determined and it was found that adenine
nucleotides predominated (Fig. 3). Approximately two thirds or
more of ribonucleotides formed were adenine nucleotides. No
difference between control and L.N. lymphoblasts was apparent,
suggesting that in L.N. cells AMP was deaminated to IMP. Adenosine, therefore, can serve as a source of AMP and IMP in L.N. cells
and could indeed be a biologically important source of purine ribonucleotides in HGPRT deficient cells.

To test this hypothesis further analogous studies were carried
out in erythrocytes. Incubation of freshly obtained erythrocytes of
a child with the L.N. syndrome with [8-^{14}C] adenosine, [8-^{14}C] inosine and [8-^{14}C] hypoxanthine resulted only in the incorporation
of adenosine into ribonucleotides (Table 3). The metabolites isolated from the incubation medium were hypoxanthine and inosine
(Table 4). Less than 1% of the labeled compounds in the medium was
adenosine. The distribution of labeled adenosine into ribonucleotides was determined in fresh erythrocytes incubated with [^{3}H]
adenosine in Hank's solution (Fig. 4). Adenine nucleotides represented 55% to 65% of total nucleotides.

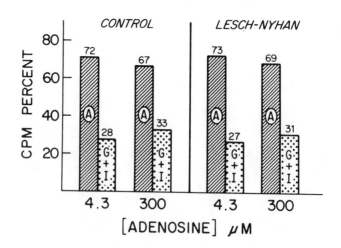

Fig. 3. Distribution of [³H] adenosine into purine ribonucleotides of Lesch-Nyhan and control lymphoblasts in culture.

Cells were incubated as described for Table 1.

TABLE 3.   INCORPORATION OF PURINE RIBONUCLEOSIDES AND HYPOXANTHINE
INTO LESCH-NYHAN ERYTHROCYTES

PRECURSOR	INCORPORATION INTO PURINE RIBONUCLEOTIDES
	CPM*
[8-¹⁴C] Adenosine	19400
[8-¹⁴C] Inosine	200
[8-¹⁴C] Hypoxanthine	200

* CPM per 10 µl packed RBC

Freshly harvested erythrocytes of a child with the Lesch-Nyhan syndrome were incubated (90 min) in Hank's balanced solution with 1 µCi [8-¹⁴C] adenosine, [8-¹⁴C] inosine or [8-¹⁴C] hypoxanthine as noted.

TABLE 4.   METABOLITES OF ADENOSINE AND HYPOXANTHINE IN MEDIA OF
LESCH–NYHAN ERYTHROCYTES

	ISOTOPICALLY LABELLED METABOLITES			
PRECURSORS	AR	I	Hx	Λ
	%			
[8-14C] Adenosine	<1	38	62	<1
[8-14C] Inosine	<1	15	85	<1
[8-14C] Hypoxanthine	<1	4	96	<1

Freshly harvested erythrocytes of a child with the Lesch–Nyhan
syndrome were incubated (90 min) in Hank's balanced solution with
1 µCi[8-14C] adenosine, [8-14C] inosine or [8-14C] hypoxanthine as
noted.

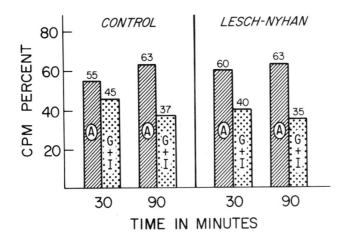

Fig. 4.   Distribution of [3H] adenosine in purine ribonucleotides
of erythrocytes of a child with Lesch–Nyhan syndrome and a control
subject

Erythrocytes cannot form purines de novo and therefore are dependent upon preformed purines; they cannot convert IMP to AMP. Since L.N. erythrocytes cannot salvage circulating hypoxanthine for IMP synthesis, they are largely dependent upon AMP synthesis as a source of IMP and guanine ribonucleotides. This study shows that adenosine is readily taken up by control and L.N. erythrocytes and therefore this pathway may contribute significantly to ribonucleotide synthesis in L.N. erythrocytes. The deamination of AMP appears to be an important pathway for the synthesis of IMP in L.N. erythrocytes.

SUMMARY

Adenosine metabolism was studied in permanent lymphocyte cell lines and in erythrocytes of L.N. patients and compared with that of normal subjects. It was found that L.N. lymphocytes incorporated adenosine into purine ribonucleotides in spite of the fact that they cannot incorporate inosine or hypoxanthine, the major metabolites of adenosine. These findings suggest that adenosine may be a potentially important salvage pathway for purines and a source of IMP and guanine nucleotides in L.N. cells.

REFERENCES

1. Giblett, E.R., Anderson, J.E., Cohen, F., Pollara, B. and Meuwissen, H.J., 1972, Lancet 2, 1067-1069.

2. Ishii, K., and Green, H., 1973, J. Cell Science 13, 429-439.

3. Green, H., and Chan, T., 1973, Science 182, 836-837.

4. Snyder, F.F., and Henderson, J.F., 1973, J. Biol. Chem. 248, 5899-5904.

5. Crabtree, G.W., and Henderson, J.F., 1971, Cancer Research, 31, 985-991.

6. Debray, H., Cartier, P., Temstet, A., and Cendron, J., 1976, J. Ped. Research 10, 762-766.

7. Acker, van K.J., Simmonds, A.H., and Cameron, J.S., 1976, J. Clin. Chem. Clin. Biochem. 277.

PURINE-NUCLEOSIDE PHOSPHORYLASE AND ADENOSINE AMINO-
HYDROLASE ACTIVITIES IN FIBROBLASTS WITH THE LESCH-
NYHAN MUTATION

J. Barankiewicz and M. M. Jeżewska

Institute of Biochemistry and Biophysics

Polish Academy of Sciences

02-532 Warszawa, 36 Rakowiecka St., Poland

The Lesch-Nyhan syndrome is an X-linked disease
(Lesch, Nyhan, 1964) characterized by a nearly total
deficiency of the hypoxanthine-guanine phosphoribosyltransferase activity in several tissues including the
fibroblasts (Seegmiller et al., 1967; Rosenbloom et al.,
1967, 1968). The de novo synthesis of purine nucleotides is greatly enhanced, and the levels of hypoxanthine and xanthine are considerably increased (Rosenbloom et al., 1967; Balis et al., 1967). It has been
postulated (Lee et al., 1973) that there is some coordinate control of the activities of the enzymes
participating in the purine metabolic pathway; this
control is suggested to be related to the levels of
the intermediate metabolites. Barankiewicz et al.
(1975) have found that in the mammary glands of mice
the direction of changes in hypoxanthine-guanine
phosphoribosyltransferase and purine-nucleoside phosphorylase activities was the same for both enzymes,
i.e. they both either increased or dropped at the same
time. Therefore, it was of interest to determine
whether the activities of purine-nucleoside phosphorylase and adenosine aminohydrolase in the fibroblasts
lacking hypoxanthine-guanine phosphoribosyltransferase
are altered.

MATERIALS AND METHODS

The following reagents were used: $[8 - {}^{14}C]$-labelled adenine, guanine and hypoxanthine – from Radiochemical Center (Amersham, England); adenosine, 2'-deoxyadenosine, guanosine, 8-azaguanine, inosine and α-D-ribose-1-phosphate dicyclohexylammonium salt $\cdot H_2O$ – from Calbiochem (Los Angeles, USA); Tris(hydroxymethyl)methylamine – from Koch-Light Lab. Ltd. (Colnbrook, Bucks, England); 1,4-bis(2-phenyloxazyl)-benzene (POPOP) and 2,5-diphenyloxazole (POP) – from Fluka AG (Switzerland); all other reagents, pA grade, were purchased from Polskie Odczynniki Chemiczne (Gliwice, Polska).

The materials examined, comprising blood and cultures of fibroblasts, were obtained from the Genetics Department of the Psychoneurological Institute in Warsaw. Fibroblasts cultured under standard conditions (Harnden, 1960) were developed from skin biopsies of several normal children, a patient with the Lesch-Nyhan syndrome (described in an original report by Pawlus et al., 1973), his mother and his sister. The two last persons were examined for the L-N heterozygosity (according to the method of deMars, 1971), and 8-azaguanine-resistant populations of fibroblasts were obtained; it was confirmed by the method of Kelley and Meade (1971) that while lacking the hypoxanthine-guanine phosphoribosyltransferase activity, they showed a normal level of adenine phosphoribosyltransferase. Thus, both the mother and sister of the L-N boy are L-N heterozygotes. Cultures of normal and hypoxanthine-guanine phosphoribosyltransferase-lacking fibroblasts of L-N mice were also provided by the Genetics Department of the Psychoneurological Institute in Warsaw.

The purine-nucleoside phosphorylase activity was determined in the fibroblast extracts and in the erythrocyte hemolysates prepared according to methods of Kelley and Meade (1971) and Kelley et al. (1967), respectively. The standard incubation mixture contained in a total volume of 0.2 ml: 0.1 mM $[8-{}^{14}C]$-purine (guanine, hypoxanthine or adenine, spec. act. 5 Ci/mmol), 1 mM ribose-1-phosphate, 0.1 M Tris-HCl buffer, pH 7.4, and 0.01 mg protein. In blanks R-1-P was omitted. Incubation was carried out for 15 min at 37°C. The reaction was stopped by an addition of 0.03 ml of 3 N $HClO_4$. The chromatographic resolution of the substrates and products as well as the determination of their radio-

activity were performed as described previously (Barankiewicz and Jeżewska, 1976).

The adenosine aminohydrolase activity was determined by the method described by Kalckar (1947). The sample contained in 3-ml volume: 75 mM adenosine or 2 -
deoxyadenosine, 50 - 100 μg of fibroblast protein in
0.1 M Tris-HCl buffer, pH 7.4. The decrease in the absorbance was followed at 265 nm for 15 min with a Unicam
SP-500 spectrophotometer. Light path 1 cm, temperature
23 - 25°C.

The enzymic activities were expressed in nmoles of
purine nucleoside formed/1 mg protein per hour. Total
protein content in the sample was assayed by the method
of Lowry et al. (1950), with bovine plasma albumin as
standard.

RESULTS AND DISCUSSION

The purine-nucleoside phosphorylase activity levels
in the fibroblasts from the investigated persons   remained within a very wide range, which in Table 1   is
arbitrarily divided into three subranges. The activities
of the fibroblast cultures obtained from normal children
were contained within the highest and middle subranges.
In contrast, the enzymic activities in the fibroblast
cultures derived from two skin biopsies obtained from
the Lesch-Nyhan boy, D. Rz., in most cases remained
within the lowest subrange. Similarly, in the 8-aza-
guanine-resistant fibroblasts of mouse, lacking hypoxanthine-guanine phosphoribosyltransferase, the activity of purine-nucleoside phosphorylase was two times
lower, as compared with normal murine fibroblasts. This
could suggest that in the fibroblasts there is a positive correlation between the levels of these two
enzymes.

However, the fibroblasts derived from heterozygotes:
mother (T. Rz.) of the L-N boy and his sister (A. Rz.),
when cultured in a non-selective medium, displayed the
purine-nucleoside activity remaining within all three
subranges (Table 1). High activities of purine-nucleoside phosphorylase predominated in most cultures of the
T. Rz. fibroblasts, whereas most cultures of the A. Rz.
fobroblasts showed a lower activity, though in all
these cultures the levels of hypoxanthine-guanine

Table 1. Purine-nucleoside phosphorylase activity of normal and L-N mutant fibroblasts

| | Medium | Enzymic activity (nmoles of guanosine formed/1mg fibroblast protein per hour) | | |
		1 - 500	500-1000	>1000
Normal adult B.C.	NS			1478
Normal children				
T.Cz.	NS		618	1258
L.J.	NS		715	1103
A.W.	NS			1022
R.K.	NS		702	1472
	HAT			1977
	AzaG	degeneration		
L-N boy D.Rz.	NS	109[a], 190[a], 353[b], 372[a], 383[a], 403[b], 450[a]	700[b]	
	HAT	degeneration		
	AzaG	277[b]	702[b]	
His mother T.Rz. (heterozygote)	NS	232[a]	574[b], 654[a], 884[a]	1720[b], 1236[b], 2020[b]
	HAT			1652[b]
	AzaG			1160[a], 2020[b]
His sister A.Rz. (heterozygote)	NS	115[a], 167[a], 205[b], 358[b]		2351
	HAT		537[b]	
	AzaG	434[b]	758[b]	

Media: NS - nonselective; HAT - supplemented with hypoxanthine (3×10^{-5} M), aminopterine (10^{-7} M) and thymidine (3×10^{-5} M); AzaG - supplemented with 8-azaguanine (2×10^{-5})
a - cultures obtained from the first skin biopsies
b - cultures obtained from the second skin biopsies

phosphoribosyltransferase did not differ from those in normal fibroblasts.

There was an analogous difference between the 8-azaguanine-resistant fibroblasts of T. Rz. and A. Rz., both showing a low level of hypoxanthine-guanine phosphoribosyltransferase. In the two cultures of the L-N fibroblasts from T. Rz. high activity of purine-nucleoside phosphorylase was found, whereas both L-N fibroblast cultures derived from A. Rz. showed a significantly lower activity. Therefore, it seems that there is no correlation between the levels of hypoxanthine-guanine phosphoribosyltransferase and purine-nucleoside phosphorylase in the fibroblasts with the Lesch-Nyhan mutation.

The level of purine-nucleoside phosphorylase remained unchanged during two weeks of storage at $-18^{\circ}C$, in extracts from the fibroblasts of a normal boy, L-N boy and two heterozygotes, when cultured in a non-selective medium (Table 2). In contrast, a decrease in the enzymic activity during storage was observed in the fibroblasts of T. Rz. and A. Rz., growing in medium containing hypoxanthine, aminopterine and thymidine (HAT medium), whereas the activity in normal fibroblasts cultured in the HAT-medium remained unchanged. Even greater decreases in the phosphorylase activity were observed during the storage of the extracts of the 8-azaguanine-resistant fibroblasts from the heterozygotes and the L-N boy. The activities initially remaining within the highest and middle subranges passed to the lowest subrange after two weeks at $-18^{\circ}C$. In the 8-azaguanine-resistant fibroblasts exhibiting low activity of purine-nucleoside phosphorylase no changes occurred. The observed decreases in the investigated enzymic activity in extracts from the fibroblasts of persons with the L-N mutation, growing in HAT and aza-G media, remains unexplained.

No differences in the level of purine-nucleoside phosphorylase were found, when the hemolysates of the erythrocytes of normal children, L-N boy D. Rz., his mother R. Rz. and his sister A. Rz. were examined. In all cases the activity was 524-546 nmoles of guanosize formed/1 mg of hemolysate protein per hour; thus, it remained within the same range as that reported by Fujimoto and Seegmiller (1970).

Substrate specificity of purine-nucleoside phosphorylase in fibroblasts from the L-N boy was the same as in those from normal children. The activity towards hypoxanthine was as high as that towards guanine, and - similarly as in other tissues (Zimmerman et al., 1971) - only traces of activity towards adenine were present in normal and mutant fibroblasts.

Table 2. Purine-nucleoside phosphorylase activity
during storage at -18°C

| | Medium | Enzymic activity (nmoles of guanosine formed/1 mg fibroblast protein per hour) | | |
		after extraction	after storage for 1 week	1 weeks
	NS			
Normal boy R.K.		702	661	-
		1470	1501	1847
L-N boy D.Rz.		702	834	645
His mother T.Rz.		574	607	533
(heterozygote)		1720	1676	1620
His sister A.Rz.		205	210	288
(heterozygote)		2351	2437	2303
	HAT			
Normal boy R.K.		1977	1872	1835
His mother T.Rz.		1652	1246	981
His sister A.Rz.		537	415	257
	AzaG			
L-N boy D.Rz.		277	303	-
		758	305	273
His mother T.Rz.		1160	700	384
		2020	821	547
His sister A.Rz.		434	457	309
		758	305	273

Media as in Table 1

Adenosine aminohydrolase activity levels in the
fibroblasts from a normal boy, L-N boy and both heterozygous females did not differ significantly. They remained within the range of 354 - 615 and 519 - 757
nmoles of product formed/1 mg fibroblasts protein per
hour, with 2'-deoxyadenosine and adenosine as substrates,
respectively. The stability of the enzyme was not investigated.

The above results showed that the lack of hypoxanthine-guanine phosphoribosyltransferase did not
change the levels of purine-nucleoside phosphorylase
and adenosine aminohydrolase in the fibroblasts.

Acknowledgements. We are greatly indebted to Dr Irena Głogowska and Dr Teresa Abramowicz from the Genetics Department of the Psychoneurological Institute in Warsaw for supplying all fibroblast cultures. The technical assistance of Miss Ewa Winter is gratefully acknowledged.

REFERENCES

Balis, M. E., Krakoff, J. H., Berman, P. H., and
    Dancis, J. (1967) Science, 156, 1122 - 1123.
Barankiewicz, J. and Jeżewska, M. M. (1976) Comp.
    Biochem. Physiol., 54B, 239 - 242.
Barankiewicz, J., Jeżewska, M. M. and Chomczyński, P.
    (1975) FEBS Letters, 60, 384 - 387.
Fujimoto, W. Y. and Seegmiller, J. E. (1970) Proc. Nat.
    Acad. Sci., 65, 577 - 583.
Harnden, D. G. (1960) Brit. J. Exptl. Pathol., 40,
    31 - 37.
Kalckar, H. M. (1947) J. Biol. Chem., 167, 461.
Kelley, W. N. and Meade, J. C. (1971) J. Biol. Chem.,
    246, 2953 - 2958.
Kelley, W. N., Rosenbloom, F. M., Henderson, J. F. and
    Seegmiller, J. E. (1967) Proc. Nat. Acad. Sci., 57,
    1735 - 1739.
Krenitsky, T. A., Neil, S. M. and Miller, R. L. (1970)
    J. Biol. Chem., 245, 2605 - 2611.
Lee, P. C., Nickels, J. S. and Fisher, J. R. (1973)
    Arch. Biochem. Biophys., 158, 677 - 680.
Lesch, M. and Nyhan, W. L. (1964) Am. J. Med., 36, 561
    - 570.
Lowry, O. H., Rosebrough, N. J., Farr, A. L. and
    Randall, R. J. (1951) J. Biol. Chem., 193, 265-275.
deMars, R. (1971) Fed. Proc., 30, No 3, part I, 944 -
    955.
Pawlus, M., Zaremba, J. S., Czartoryska, B., Barankie-
    wicz, J., Dymecki, J. and Zaremba, J. M. (1973)
    Proc. 3rd Congress of Intern. Assoc. Scientific
    Study of Mental Retardation, Haga, PZWL ed.,
    Poland, 220 - 228.
Rosenbloom, F. M., Henderson, J. F., Caldwell, J. C.,
    Kelley, W. N. and Seegmiller, J. E. (1968) J. Biol.
    Chem., 243, 1166 - 1173.
Rosenbloom, F. M., Kelley, W. N., Miller, J., Henderson,
    J. F. and Seegmiller, J. E. (1967) J. Am. Med.
    Assoc., 202, 175 - 177.
Seegmiller, J. R., Rosenbloom, F. M. and Kelley, W. N.
    (1967) Science, 155, 1682 - 1684.

ALTERED EXCRETION OF 5-HYDROXYINDOLEACETIC ACID AND GLYCINE IN PATIENTS WITH THE LESCH-NYHAN DISEASE

L. Sweetman, M. Borden, S. Kulovich, I. Kaufman, and

W. L. Nyhan, Department of Pediatrics, University of

California, San Diego, La Jolla, California

Mizuno and Yugari (1,2) reported that 5-hydroxytryptophan (5-HTP) given orally at 1-8 mg/kg/day was effective in preventing self-mutilation in four patients with the Lesch-Nyhan syndrome. The proposed mechanism of action was through alteration of seretonin (5-hydroxytryptamine, 5-HT) levels in the brain.

We undertook a program to assess the usefullness of 5-HTP in modifying the self-mutilative behavior of a larger series of patients with the Lesch-Nyhan syndrome. In addition, we studied the effect of 5-HTP and Carbidopa (MK-486), an inhibitor of the peripheral metabolism of 5-HTP.

The metabolism of 5-HTP is shown in Fig. 1. The compound is decarboxylated by aromatic L-amino acid decarboxylase to 5-HT in the brain and in the periphery (3). Unlike 5-HTP, 5-HT does not cross the blood-brain barrier. 5-HT is oxidized by monoamine oxidase to 5-hydroxyindoleacetic acid (5-HIAA) which is excreted (4). Carbidopa is an inhibitor of aromatic L-amino acid decarboxylase, but acts only in the periphery because it does not cross the blood brain barrier (5). By preventing peripheral metabolism of 5-HTP, Carbidopa potentiates the elevation of 5-HT in the central nervous system (6).

Urinary excretion of 5-HIAA was measured twice in nine patients with the Lesch-Nyhan syndrome prior to treatment using the colorimetric method of Goldenberg (7). The mean excretion was $8.6^{\pm}2.23$ (S.D.) µg/mg creatinine which was significantly greater (P < 0.005) than the mean of $5.1^{\pm}1.26$ found for 5 normal males of similar age (Fig. 2). Mizuno and Yugari (2) had reported normal 5-HIAA excretion in their four patients and the excretion was normal in three of

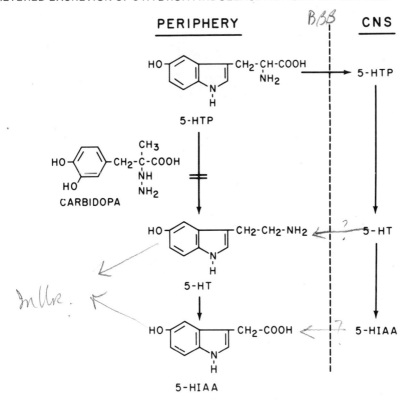

Fig. 1   Metabolism of 5-Hydroxytryptophan.

our patients. Our results suggest an increased catabolism of sero-
tonin in some of the patients with the Lesch-Nyhan syndrome.

When 5-HTP was administered orally to patients with the
Lesch-Nyhan syndrome, there was a large increase in excretion of
5-HIAA which was approximately proportional to the amount of 5-HTP
(Fig. 3.). Simultaneous administration of Carbidopa greatly reduced
this excretion, indicating that peripheral decarboxylation and oxidation of 5-HTP was the major source of urinary 5-HIAA. Table I
shows that increasing Carbidopa from 7 mg/kg/day to 16 mg/kg/day
caused relatively little furthur decrease in 5-HIAA excretion.

The effect of 5-HTP therapy on the self-mutilation by the
patients with the Lesch-Nyhan syndrome was variable. Some patients
showed no response, while most responded to various low doses. All
developed tolerance to 5-HTP within a three-month period.

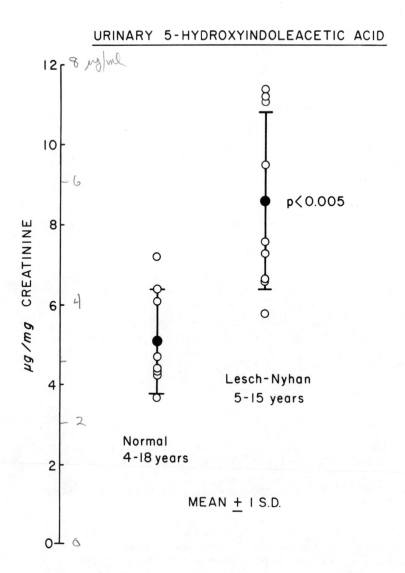

Fig. 2   Excretion of 5-HIAA by normals and Lesch-Nyhan patients
           prior to 5-HTP treatment.

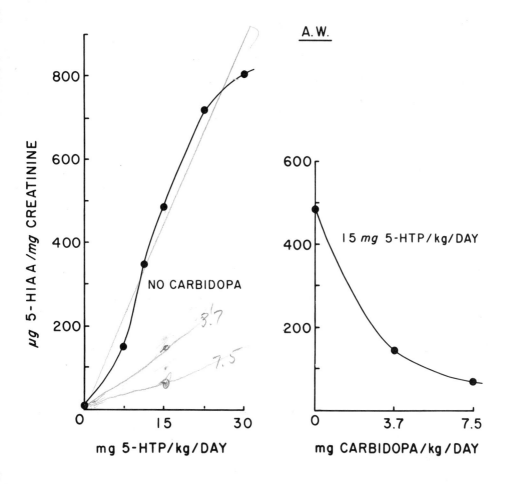

Fig. 3   Excretion of 5-HIAA by a Lesch-Nyhan patient receiving
          5-HTP and Carbidopa.

TABLE I

URINARY 5-HYDROXYINDOLEACETIC ACID

LESCH-NYHAN PATIENT	5-HTP mg/kg/day	CARBIDOPA mg/kg/day	5-HIAA µg/mg CREATININE
D.D.	0.0	0.0	6.7
D.D.	4.0	8.1	44.8
D.D.	4.0	12.1	42.3
D.D.	4.0	16.2	31.8
R.S.	0.0	0.0	9.5
R.S.	0.0	7.2	· 7.0
R.S.	3.6	7.2	59.7
R.S.	3.6	14.4	28.2
P.B.	0.0	0.0	4.5
P.B.	7.1	7.1	52.9
P.B.	7.1	14.3	55.3
A.W.	0.0	0.0	11.2
A.W.	14.9	0.0	487.0
A.W.	14.9	3.7	144.0
A.W.	14.9	7.5	69.5

5-HTP = 5-HYDROXYTRYPTOPHAN
5-HIAA = 5-HYDROXYINDOLEACETIC ACID

Routine electrophoresis for urinary amino acids gave the surprising result that all patients receiving Carbidopa had a 2-5 fold increase in glycine. 5-HTP alone did not affect glycine excretion. Quantitative chromatographic analysis of urinary amino acids of five patients receiving Carbidopa showed a mean increase of 4-fold in glycine excretion (Table II). Two patients had above normal control excretions of glycine but all patients showed at least a 2-fold elevation on Carbidopa, and all were above the normal range. There were no consistent changes in other amino acids with Carbidopa treatment. There was no increase in total urinary oxypurines in two patients studied.

TABLE II   GLYCINE EXCRETION

PATIENT	5-HTP mg/kg/day	CARBIDOPA mg/kg/day	GLYCINE mg/mg CREATININE	GLYCINE % of CONTROL
Normal	0.0	0.0	0.025 - 0.254	-
A.W.	0.0	0.0	0.127	-
A.W.	29.8	0.0	0.083	65
A.W.	14.9	7.5	0.580	457
O.K.	0.0	0.0	0.317	-
O.K.	31.7	10.6	0.745	235
P.B.	0.0	0.0	0.317	-
P.B.	14.3	14.3	0.510	395
D.D.	0.0	0.0	0.305	-
D.D.	4.0	16.2	0.853	280
R.S.	0.0	0.0	0.082	-
R.S.	0.0	7.2	0.520	634
R.S.	3.6	14.4	0.458	559
			MEAN	427%

TABLE III   GLYCINE LEVELS

PATIENT	5-HTP mg/kg/day	CARBIDOPA mg/kg/day	GLYCINE mg/dL	GLYCINE % of CONTROL
		P L A S M A		
Normal	0.0	0.0	0.78 - 2.32	-
P.B.	0.0	0.0	1.42	-
P.B.	14.3	14.3	2.10	148
D.D.	0.0	0.0	1.07	-
D.D.	0.0	8.1	1.54	144
R.S.	0.0	0.0	1.59	-
R.S.	0.0	7.2	2.14	135
R.S.	3.6	14.4	2.05	129
			MEAN	139%
	C E R E B R O S P I N A L	F L U I D		
D.D.	0.0	0.0	0.150	-
D.D.	4.0	8.1	0.166	111
R.S.	0.0	0.0	0.194	-
R.S.	1.8	10.8	0.442	228
			MEAN	169%

To determine whether Carbidopa was affecting renal tubular absorption to cause the increased urinary glycine, levels were measured in plasma (Table III). Three patients all showed an increase in plasma glycine on treatment with Carbidopa with a mean increase of 39%. However, the values did not exceed normal plasma levels. One patient showed no increase in cerebrospinal fluid levels of glycine while another had a 2-fold increase while receiving Carbidopa.

The finding of elevated glycine in plasma and urine suggests a metabolic effect of Carbidopa. The major catabolic pathway for glycine is believed to be decarboxylation by the glycine cleavage system (8). It appears that high concentation of Carbidopa may inhibit glycine decarboxylation as well as aromatic L-amino acid decarboxylase.

ACKNOWLEDGEMENTS

We thank Jack Leslie for his expert technical assistance in the analysis of amino acids. This work was supported in part by grants from The National Foundation March of Dimes (#1-377), and by U.S. Public Health Service Grants #RR-00827 and #HD 04608 from the National Institute of Child Health and Development.

REFERENCES

1. Mizuno, T., and Yugari, Y.: Lancet p.761 (1974).

2. Mizuno, T., and Yugari, Y.: Neuropädiatric 6: 13-23 (1975).

3. Sims, K.L.: In Adv. Biochem. Psychopharm. "Serotonin-New Vistas", (E. Costa, G.L. Gessa and M. Sandler, Eds.) 11: 43-50, Raven Press, New York (1974).

4. Neff, N.H. et al.: In Adv. Biochem. Psychopharm. "Serotonin-New Vistas", (E. Costa, G.L. Gessa and M. Sandler, Eds.) 11: 51-58, Raven Press, New York (1974).

5. Annotated Bibliography, MK-486 and Levodopa, Merck, Sharp and Dohme (1974).

6. Modigh, K.: In Adv. Biochem. Psychopharm. "Serotonin-New Vistas", (E. Costa, G.L. Gessa and M. Sandler, Eds.) 10: 213-217, Raven Press, New York (1974).

7. Goldenberg, H.: Clin. Chem. 19: 38-44 (1973).

8. Kikuchi, G.: Mole. Cell. Biol. 1: 169-187 (1973).

XANTHINURIA, LITHIASIS AND GOUT IN THE SAME FAMILY

C.Auscher, A de Gèry, C.Pasquier and F.Delbarre

Institut de Rhumatologie. Centre de Recherches sur les
Maladies ostéo-articulaires. U.5 INSERM; ERA 337 CNRS
27 rue du Fg St Jacques 75014 Paris, France

Xanthinuria is a rare hereditary disorder characterized by
a deficiency of xanthine oxidase activity and by the resultant
excretion of xanthine and hypoxanthine as the chief end products
of purine metabolism.

The mode of inheritance of xanthinuria is not clearly
determined. Most of the known data are consistent with the
interpretation that xanthinuria is transmitted as an autosomal
recessive disorder (1). However, it has been reported relatives
having increased urinary excretion of oxypurines together with
normal uric acid one (2,3,4,).

The disease is often associated with urolithiasis. But
it is not known whether renal lithiasis is related to the low
solubility of xanthine or to a genetic abnormality.

Delbarre et al.(5,6,) have previously described the
case of a xanthinuric man who, at first, was thought to be gouty.
The study of his large family, in which xanthinuria, lithiasis
and gout coexist, is presented in this report.

BIOCHEMICAL INVESTIGATION

To locate abnormality of oxypurine excretion, the ratios of oxypurines to creatinine (Ox/C) and of uric acid to creatinine (UA/C) were measured in morning samples of urine in different age-groups of a control population. The individuals were chosen among the personnel or their relatives. They were aged from 2 to 84 years. They were in good health and did not take any drug for a least a week.

As Shown (fig.1) means of Ox/C values are significantly related to age untill 8 years, as it is for UA/C (fig.1) (7,8). Furthermore, whatever the age between 2 and 84 years is, there is a linear regression and a significant correlation (p < 0.001) between both values (fig.2).

Therefore, in order to avoid misinterpreting of slightly increased levels of oxypurines, age and uric acid excretion were taken into account to locate abnormality in oxypurine excretion.

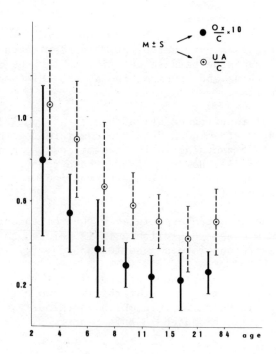

Fig.1 Ratios of oxypurines to creatinine (Ox/C) and uric acid to creatinine (UA/C) of morning samples of purine in different age-groups of a control population. The mean (M ± s) of each group : 2-4 ; 4-6 and 6-8 years is significantly different from the following (Wilcoxon test).

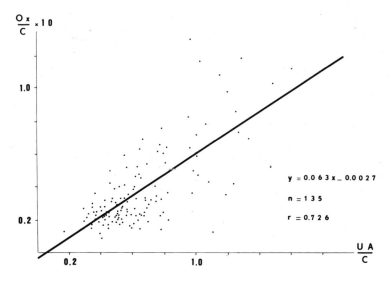

<u>Fig.2</u> Relationship between Ox/C and UA/C values in a control
population aged 2 to 84 years.

RESULTS

The propositus (VI.15) has four sibs (Tab.1). Three of them
have xanthinuria (VI.13 - VI.14 - VI.16). The fourth (VI.12) has
increased level of plasma and urinary uric acid and gout. Three
out of the sibs have suffered from nephretic colics (VI.12 - VI.13
and VI.14). The gouty eldest brother (VI.12) had excreted an uric
acid stone. It may be seen on table 1 that the two xanthinuric sibs
who had suffered from renal lithiasis excrete the same amount of
xanthine as the two others.

The pedigree of the family is shown on the figure 3. In order
to obtain as much information as possible concerning the possibility
of a familial factor of gout or lithiasis, offspring of the
second marriage of each member III.3 and III.4 were reported.
In the fifth generation the mother (V.16) of affected sibs got
married to a third cousin (V.5).
In the seventh generation it should be pointed that all members
are under the age of twenty at the time of our investigation.

N°	SEX	PLASMA[1]		URINE[2]				LITH.	GOUT
		UA	H+X	UA	H+X	X	Cr.		
VI.12	M	9.80	0.230	833	33	—	1380	U	+
VI.13	M	1.29	0.620	9*	248*	164*	980*	X	—
VI.14	F	0.96	0.420	8	441	322	1030	X	—
VI.15	M	1.20	0.580	17	464	330	1410	—	—
VI.16	F	1.20	0.360	12	226	183	770	—	—

/ propositus   (1) mg%
* mg/l   (2) mg/24h

Table 1 : Data of the siblings (UA and U = uric acid; H = hypoxanthine; X = xanthine and Cr = creatinine)

According to the biochemical test we described in the first
part of this report, no abnormality of oxypurine excretion was
found in the 62 relatives studied (fig.3). But eight relatives
have got either uric acid lithiasis, increased plasma uric acid
or gout (see fig.3).

CONCLUSION

This family is the largest that, so far, has been reported
concerning xanthinuria and the data presented in this report let
us assume the following conclusions :
First : xanthinuria is inherited as an autosomal recessive
trait indeed :
a) the four affected sibs (of the two sexes) were born from con-
sanguineous parents. Their coefficient of inbreeding is F = 1/256
b) no abnormality of oxypurine excretion was found either in the
presumed obligate heterozygotes or in the relatives studied.

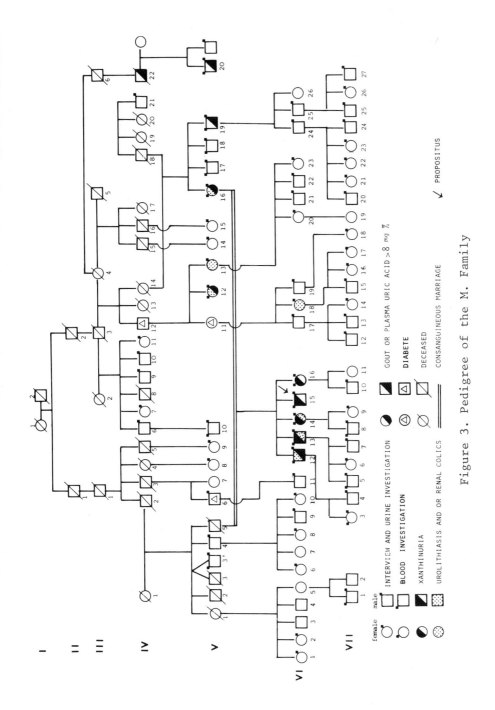

Figure 3. Pedigree of the M. Family

Secondly : an hereditary factor could be the cause of
xanthine urolithiasis of the two affected sibs indeed :
a) xanthine lithiasis does not appear to be related to the low
solubility of xanthine in urine (see tab.1),
b) eight relatives have got uric acid lithiasis or increased plasma
uric acid or gout. Among them, there are the mother (V.16) and
the brother (VI.12) of the xanthinuric sibs. Probabilities of
these relatives to be heterozygotes for xanthinuria were calculated
on table 2. It should only be pointed out that they were quite high.
By elsewhere none of these members were offspring of the couples
III.2-III.3 and III.4-III.5

Of course the coexistence of xanthinuria and gout in the
same family may be a coincidence. Nevertheless, it is of interest
because both these inherited disorder share purine metabolism.

N°	plasma uric ac. mg%	lithiasis nature	gout	probability " p "
IV.22	?	–	+	1/16
V.12	12,00	–	+	1/8
V.13	?	U.A	–	1/8
V.16	8,16	U.A	–	1
V.19	8,24	–	–	1/2
V.20	9,00	–	–	1/32
VI.12	9,20	U.A	+	2/3
VI.18	?	?	–	1/32

Tab. 2 Probabilities " p " of the relatives having increased
plasma uric ac., gout or lithiasis to be heterozygote
for xanthinuria.

AKNOLEDGEMENTS

We wish to thank Dr. R. Berger for his helpful discussion about genetics and to Mrs. F. Loyer and R. Reibaud for their excellent technical assistance.

REFERENCES

(1) - Wyngaarden, J.B. : Xanthinuria. In metabolic basis of inherited disease. Stanbury, J.B., Wyngaarden, J.B., Fredrickson D.S. ed. Third Ed. Mc Graw pub. New-York 1972 pp 992-1002.

(2) - Cifuentes-Delatte, L. and Castro-Mendoza, H. : Xanthinuria familiar. Rev. Clin. Esp. 1967 107, 244-256.

(3) - Rapado, A., Castro-Mendoza, H.J. : Genetics of xanthinuria. Urinary calculi Proc. : Int. Symp. Renal Stones Res. Madrid 1972 pp 80-83. Karger, Basel 1973

(4) - Wilson, D.M.and Tapia, H.R. : Xanthinuria in a large kindred. In : Purine metabolism in man. Sperling O., Vries A. de, Wyngaarden, J.B. (ed) in Advances in Experimental Medicine. New York Plenum Pub. 1974 41 B pp 343-349.

(5) - Delbarre, F., Weissenbach, R., Auscher, C. and Gery A.de : Accès de goutte chez un xanthinurique. Nouv. Press. Med. 1973 37 (2) 2465-2466.

(6) - Auscher, C., Pasquier, C., Mercier, N. and Delbarre F. : Urinary excretion of 6-hydroxylated metabolite and oxypurines in a xanthinuric man given allopurinol or thiopurinol In : Purine metabolism in man. Sperling O., Vries, A. de, Wyngaarden, J.B. (ed). In advances in Experimental Medicine. New York Plenum pub. 1974 41 B pp 663-667.

(7) - Danchot, J., Auscher, C., Le Gô, A. and Delbarre, F. : Etude du rapport uraturie-créatininurie dans des groupes d'enfants normaux et d'enfants inadaptés mentaux. Rev. Epid. Med. Soc. et Santé Publique 1972 20, 2, 177-183.

(8) - Kaufman, J.M., Greene, M.L. and Seegmiller, J.E. : Urine uric acid ratio to creatinine ratio. A screening test for inherited disorder of purine metabolism. J. of Pediatrics 1968 73, 4, 583-592.

IMMUNOLOGICAL ASPECTS OF PURINE METABOLISM

J. Edwin Seegmiller, T. Watanabe, Max H. Shreier and
Thomas A. Waldmann
University of California San Diego, La Jolla, CA, U.S.A.
Basel Institute for Immunology, Basel, Switzerland
Metabolism Branch, National Cancer Institute, National
Institutes of Health, Bethesda, Maryland, U.S.A.

One of the newest and most rapidly developing areas of investigation in inherited metabolic disorders and certainly one of the most interesting and potentially useful is the role of purine metabolism in the immune response. The need of proliferating cells for a balanced supply of both purine and pyrimidine compounds for maintaining cell division and growth has long been known. Likewise the impairment of certain aspects of the immune response by purine derivatives such as azathioprine results in a well established therapeutic use in clinical medicine. But only recently has the importance of pathways of purine interconversion in maintaining the integrety of the immune system been revealed by the striking impairment of its function found in association with severe deficiencies of specific enzymes of purine metabolism.

The first enzyme defect to be found in association with severe immunodeficiency disease was reported by Dr. Eloise Giblett *et al.* in 1972. As so often happens in science, this discovery resulted from serendipity rather than from an organized attempt to study the immunodeficiency diseases. Dr. Giblett was exploring the use of isoenzymes of adenosine deaminase (ADA) to identify the range of genetic polymorphism to be found in human populations. In the course of these studies she found two blood samples with no detectable ADA activity, both of them from children with severe combined impairment of both T- and B-cell function. The rapid progress that has been made will be more understandable and meaningful if we first review briefly our knowledge of the immune system and the ways its function can be impaired to produce the state of immunodeficiency (Eisen 1974).

Our body of knowledge of the composition and functioning of

the immune system is a relatively recent acquisition. The plasma
cell of the blood was first identified as the source of antibody
formation and secretion in the 1940's. In the following decade
came the proposal that the immune system's response to an antigen
involves primarily an amplification of the production of a pre-
existing antibody molecule and that any given cell produces only
one kind of antibody. These concepts were verified in the 1960's
with the further demonstration that all plasma cells originate by
proliferation of small lymphocytes each of which exhibits samples
on its surface of the particular antibody molecule it is programmed
to produce.

The stimulus to proliferation involves a fitting of an antigen
to an antibody on display on the cell surface. This fitting occurs
entirely by the chance encounter of the antigen with one or more of
the myriad of antibody shapes on display. The closer the fit of
the antigen to the surface antibody the stronger the stimulus will
be to trigger the proliferation of that particular cell. The proliferation is rapid and requires a complex interaction of at least
three different types of cells.

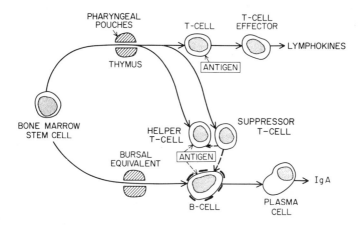

Fig. 1   Scheme of the cellular origin, differentiation and proliferation required for development of a specific immune response.
Reproduced from Waldmann *et al.* (1974) with permission of the publisher.

The identification of these cell types was another of the accomplishments of the 1960's and they evolved from studies of the embryological origin of the immune system in birds and in mammals. Cells of the blood-forming system as well as those of the immune system have a common origin in the undifferentiated stem cell of the bone marrow and fetal liver. A portion of these cells migrate into the thymus where, under the influence of thymic hormones and possibly other local factors, they develop into T-lymphocytes with a variety of very specialized functions in the immune system. They are responsible for cell mediated immunity and in this role can be likened to the infantry of an army ready for hand to hand combat with invading organisms particularly intracellular parasites, viruses and fungi. They also maintain surveillance of other cells and recognize and destroy most tumor cells or grafted cells foreign to the host. Their specificity is primarily for recognition of large molecules, particularly those on the surface of cells and they are responsible for the delayed type of hypersensitivity, for some drug allergies and some types of autoimmune diseases. In addition sub-classes of T-cells have a role in the differentiation, stimulation to proliferation or even the suppression of the other major class of lymphoid cells designated as B-lymphocytes, shown in Fig. 1.

B-lymphocytes originate in the bone marrow and with antigenic stimulation and with the help of T-cells and fixed macrophages proliferate to form the plasma cells responsible for secretion of antibodies which constitute the humoral immunity. In the battle against invading organisms B-cells are the counterpart of the army's long range artillary since the antibodies attack organisms at the point of invasion quite remote from their plasma cells of origin. B-cells are thus responsible for our developing immunity to viruses and bacteria after an initial encounter and for maintaining our resistance to a wide range of encapsulated bacteria that produce acute localized or "pyogenic" infections. In general the type of antigens to which B-cells respond are relatively small molecules or portions of a molecule as contrasted to the large molecules to which T-cells are responsive.

Complement is yet another component of the immune system consisting of 11 different proteins with a cascading interaction. Complement greatly amplifies the action of antibody-antigen complexes by combining with them. They exert their effect primarily on cell membranes causing lysis of some types of cells and aberrations of function in others such as the release of histamine from mast cells, increased permeability of small blood vessels, directed migration of polymorphonuclear leukocytes, increased phagocytic activity by leukocytes and macrophages and destruction by lysis of bacterial cells. Our concept of the full role of complement action has been significantly expanded in recent years by the identification of the clinical deficiency states of the various components. A decreased resistance to infection is the clinical consequence of genetically determined deficiency of some of these components such as C3 or C5 or of an impairment of phagocytosis. However, we must limit our discussion today to consideration of the role of aberrations of purine metabolism of the lymphoid cellular components in the generation of immunodeficiency diseases.

DEFECTS IN PURINE ENZYMES ASSOCIATED WITH SEVERE IMMUNODEFICIENCY

A severe impairment of the immune system has been found in association with a gross deficiency of the following two sequential enzymes of purine metabolism:

$$\text{Ad-R} \xrightarrow[\text{(ADA)}]{\text{Adenosine Deaminase}} \text{Hx-R} \quad + \quad \text{NH}_3$$

Ad-R   Hx-R   NH$_3$
Adenosine   Inosine   Ammonia

$$P_i + \text{Hx-R} \xrightarrow[\text{(PNP)}]{\text{Phosphorylase}}^{\text{Purine Nucleoside}} \text{Hx} \quad + \quad \text{R-1-P}$$

P$_i$ + Hx-R   Hx   + R-1-P
    Inosine   Hypoxanthine   Ribose-1-
Phosphate

Since the original report by Giblett *et al.* in 1972 ADA deficiency has been found in over two dozen children with severe combined immunodeficiency disease (Hirschhorn *et al.* 1975, Hitzig 1976, Parkman *et al.* 1975, Meuwissen and Polara, 1974). Several reviews have been written (Meuwissen *et al.* 1975 a,b, Bergsma 1976, Seegmiller *et al.* 1977). Affected children show impairment of both T- and B-cell function although in some children B-cell function may be less severely curtailed. The majority of affected children die of overwhelming infection within the first year or so of life unless elaborate precautions are taken to isolate them from infectious agents or they are converted to chimeras by giving them a graft of bone marrow cells or even embryonic liver cells from a donor suitably matched for major histocompatability types or minimal reaction in the mixed lymphocyte response as an index of compatability.

A gross deficiency in the next enzyme in the reaction sequence, purine nucleoside phosphorylase (PNP), was reported in 1975 by Giblett *et al.* in a child showing an isolated T-cell dysfunction. Two additional patients have since been found each with T-cell dysfunction and are being reported at this meeting.

POSSIBLE ROLES OF ADA AND PNP IN THE METABOLISM OF NORMAL CELLS

The marked toxicity of adenosine particularly for cells of lymphoid origin was presented by Dr. Snyder *et al.* earlier in our conference. In view of this marked toxicity the most reasonable

role to assign to ADA and PNP is to protect against this effect of
adenosine by preventing its accumulation. At low concentrations of
adenosine preference is given to the salvage pathway for rescuing
adenosine by phosphorylation to adenylic acid catalyzed by the
enzyme adenosine kinase. The high priority results from a very
high affinity of this enzyme for adenosine with a K_i of 2.0 μM
(Schnebli et al. 1967). By contrast, the lower affinity of ADA
with a K_m of 25 to 40 μM (Agarwal et al. 1975, Rossi et al. 1975)
and its higher capacity (V_{max}) seems well suited for rapid removal
of any excessive quantities of adenosine that might be presented
to the cell either from extracellular sources or from increased
breakdown of adenylic acid by a 5'-nucleotidase. Dr. Snyder will be
presenting further data in support of this role.

Adenosine Toxicity as a Possible Mechanism of Immunosuppression in
                    ADA and PNP Deficiency

     The rapid proliferation of lymphoid cells within the body with
a generation time of around 8 hours conceivably could make them
especially vulnerable to adenosine toxicity. In keeping with this
view is the presence of ADA in higher activity in spleen than in
any other organ in a variety of mammals (Brady and O'Donovan 1965).
A possible physiological role for ADA in the immune response is
suggested by increases in ADA activity in lymph and cells draining
a stimulated sheep lymph node in $situ$ at the peak of antibody
production (Hall 1965). ADA is abundant in human serum and even
more so in the fetal-calf serum commonly used in cell culture, but
only a very low activity is present in horse serum.

     A possible unifying theory to account for the immunodeficiency
observed in both ADA and PNP deficiency is suggested by some of the
properties of the enzyme ADA. This enzyme is very susceptible to
product inhibition. Therefore, accumulation of the reaction product,
inosine, exerts an inhibitory action with a K_i of 116 μM (Agarwal et a
1975, 1976).   If inosine and guanosine kinase are indeed absent in
all human cells (Friedmann et al. 1969), this accumulation should
be greatly enhanced in immunodeficient patients showing a deficiency
of the enzyme PNP required for further degradative metabolism of
inosine. The resulting secondary inhibition of ADA could allow an
accumulation of adenosine although it remains to be demonstrated.
The isolated T-cell dysfunction in this disease could then be (Fig. 16
explained as yet another example of adenosine toxicity, perhaps of
a less severe degree than observed in the primary genetic deficiency
of ADA. The theory would therefore require the T-cell to be most
susceptible to adenosine toxicity. The dysfunction of both T- and
B-cells in ADA deficiency would therefore reflect a more severe
loss of ADA activity in the primary genetically determined ADA
deficiency with a greater adenosine accumulation and toxicity in
this type of disease. An accumulation of inosine in affected

patients has been noted by Dr. Stoop and associates (personal communication 1976) and was presented by Dr. David Martin at a recent CIBA Symposium (1977) and will be discussed in greater detail by Dr. Hamet *et al.*, Dr. Wadman *et al.*, and Dr. Cohen *et al.* later in the program (pp. 471-480).

Hepatic Adenosine as a Source of Purine Nutrients for Other Tissues

Adenosine generated by the liver is a probable precursor of nucleotides of erythrocytes (Lerner and Lowy 1974). Considerable evidence has accumulated of a role for erythrocytes in providing supplementary nutrition of purine compounds to other tissues of the body (Pritchard *et al.* 1970).

Adenosine as a Modulator of Hormone Action and Mediator of Nerve
Transmission

Another possible role of ADA and adenosine as components of a system for the modulation of hormone action has been suggested by Fain and Weiser (1975) from studies of fat cells. Evidently fat cells secrete adenosine in culture which decreases both the intracellular cyclic adenosine-3',5'-phosphate and the rate of lipolysis. Addition of ADA to the culture medium counters both of these effects. A physiological role for adenosine in producing a vasodilation of coronary vessels in response to anoxia has also been presented (Gerlach, 1963). A role for ATP and possibly adenosine as a neurotransmitter for "purinergic nerves" has also been proposed by Burnstock (1977).

MODEL SYSTEMS FOR THE STUDY OF THE EFFECT OF ADA DEFICIENCY ON
LYMPHOID CELLS

Since cell lines of lymphoblasts deficient in ADA are not yet available, the potent inhibitors of ADA coformycin (Ohno *et al.* 1974) and erythro-9-(2-hydroxy-3-nonyl) adenine (EHNA) (Schaeffer and Schwender 1974) have been used to simulate the metabolic effects of ADA deficiency in a variety of types of cells. We have studied the effects of these inhibitors on three different types of lymphoid cells.

In a proliferating cell line of human lymphoblasts, Dr. Snyder and Dr. Hershfield found the inhibitors greatly potentiated the toxic effects of adenosine (Snyder *et al.* p. 30). The observed effects included a profound decrease of intracellular phosphoribosyl-1-pyrophosphate (PP-ribose-P), a marked decrease of intracellular pyrimidine nucleotides, a cessation of growth and excretion of orotic acid by the cells. Uridine was able to counteract to some extent the inhibition of lymphoblast growth produced by a combination of EHNA and adenosine.

Peripheral blood lymphocytes stimulated by a mitogen comprised the second model system. Adenosine produced a marked inhibition of the uptake by the cells of ^{3}H-leucine and ^{3}H-thymidine used as the index of the proliferative process. This also was partially relieved by uridine (Carson and Seegmiller 1976) (Figure 2). However, uridine also expands the pools of pyrimidine nucleotides and thereby decreases the incorporation of ^{3}H-thymidine. Dr. Snyder will be reporting later in our program on the metabolism of adenosine in this type of stimulated cell (Snyder *et al*.p. 441).

The relative sensitivity of T- and B-cells to inhibition by adenosine was evaluated by use of mitogens which are specific stimulators for each of these two types of cells. As shown in Figure 3, adenosine produced a greater inhibition of the T-cells stimulated by concanavalin A than of B-cells stimulated by lipopolysaccharide.

The third system studied was the formation of antibodies by mouse spleen cells during culture *in vitro* with sheep erythrocytes as antigens first described by Mishell and Dutton (1967)(Fig. 4).

As shown in Figure 5, addition of adenosine at concentrations above 1.5-2.0 mM produced a marked inhibition of plaque formation. Quite unexpected, however, was a 2- to 10-fold stimulation in plaque formation in response to adenosine at around 1.0 mM. The response was enhanced even further by addition of 20-50μM mercaptoethanol. (Click *et al*.1972, Seegmiller *et al*. 1977).

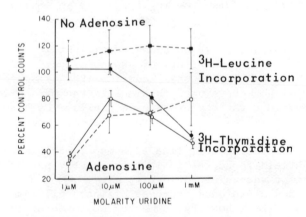

Figure 2. Reversal by uridine of adenosine toxicity in human lymphocytes stimulated by concanavalin A. Cells were cultured with the indicated concentrations of uridine, either with (o □) or without (● ■) 1.0 mM adenosine. (Carson and Seegmiller 1976). Reproduced with permission of the publisher.

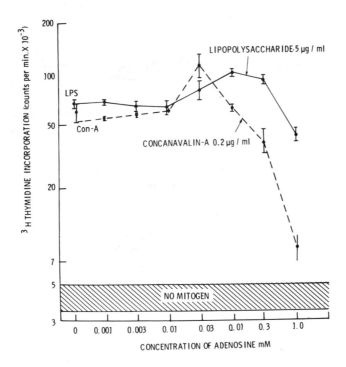

Figure 3. Inhibition by adenosine of ^{3}H-thymidine incorporation into mouse lymphocyte T-cells stimulated by concanavalin A as compared to B-cells stimulated by lipopolysaccharide. (Seegmiller *et al.* 1977.) Reproduced with permission of the publishers.

The effect of adenosine was most marked when it was added at the beginning of the incubation (Fig. 6). The addition of EHNA produced an even greater stimulation as well as an inhibition at lower concentrations of adenosine (Fig. 7). Inosine addition produced less stimulation than did adenosine but when added with adenosine significantly enhanced the inhibitory action of adenosine. This observation is quite in keeping with the possibility that inosine accumulation may lead to a secondary inhibition of ADA as outlined above. The study presented in Fig. 8 could therefore be a model of PNP deficiency with accumulated inosine producing a secondary inhibition of ADA.

CORRECTION OF ADA-ASSOCIATED IMMUNODEFICIENCY BY ENZYME REPLACEMENT

If the above systems using inhibitors of the enzyme ADA are indeed valid models of the immunodeficiency diseases produced by ADA or PNP deficiency we would expect a similar result in studies of lymphocytes of an affected child. Polmar *et al.* (1975) have now provided that data. At two weeks of age lymphocytes from an

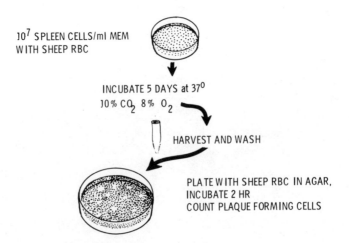

Figure 4. System for study of antibody synthesis *in vitro*. The mouse lymphocytes that developed the ability to form specific antibodies during 5-days incubation were detected by the lysis of sheep erythrocytes induced by the addition of complement. (Schreier and Nordin 1976).

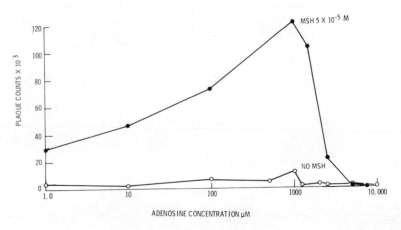

Figure 5. Effect of adenosine and of mercaptoethanol on the number of plaque-forming cells present on the fifth day of incubation of mouse spleen cells.

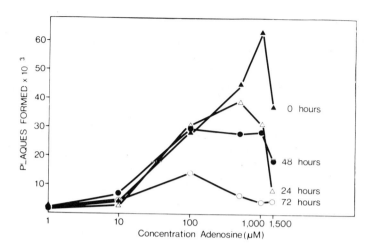

Figure 6.   The effect of time of addition of adenosine on the number of plaque-forming cells.

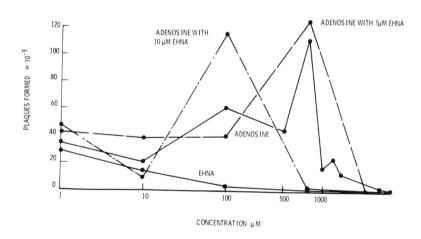

Figure 7.   Synergistic effect of EHNA and adenosine on plaque formation. Adenosine ●——●, Adenosine+ 1μM EHNA ●——●, Adenosine + 10 μM EHNA ●—·—●

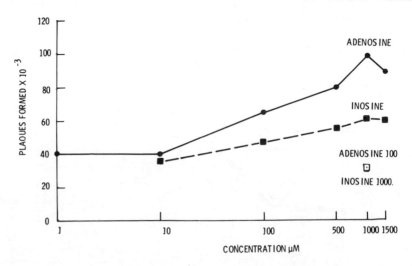

Figure 8. Effect of adenosine and inosine on the numbers of plaque-
forming cells. Inosine significantly potentiated the inhibition
produced by adenosine. (Seegmiller *et al*. 1977) Reproduced with
permission of the publishers.

ADA-deficient child showed only 25 per cent of the proliferative
response to stimulation with phytohemagglutinin (PHA) that was
shown by control cells from a normal individual. By six weeks of
age the lymphocyte count had decreased markedly and the response to
PHA was less than five per cent of normal. However, addition of
purified ADA to the culture medium produced a 2- to 6-fold increase
in incorporation of either ^{3}H-thymine or ^{3}H-leucine with mitogenic
stimulation. But addition of uridine had no effect on ^{3}H-leucine
incorporation. Likewise, inosine or guanosine failed to increase
the incorporation of ^{3}H-thymidine in response to mitogen stimulation
of either the patients' or control lymphocytes.

These results suggest that a lack of ADA *per se* rather than
a secondary depletion of pyrimidine nucleotides is responsible for
the deterioration in lymphocyte response. Even though an increase
in intracellular concentration of adenosine has still to be demonstrated in the mutant cells these results provide a strong argument
for the mechanism of the lymphopenia being a failure in the ability
of the lymphocyte to proliferate in response to normal physiological
stimuli or to mitogens in the absence of ADA. Presumably it may be
mediated by adenosine accumulation and toxicity.

The clinical improvement reported by Dr. Polmar *et al*. (1976)
in the child in response to enzyme replacement therapy *in vivo*
provides additional support for the proposed mechanism. After

receiving a transfusion of frozen irradiated erythrocytes, the ADA deficient child showed an increase in lymphocyte count to near normal values. The treatment also restored their ability to respond to mitogenic stimulation. In addition, the concentration of immunoglobulins increased and successful immunization with tetanus and diphtheria was achieved. Most significant, however, was the fact that the child remained free of infections with repeated transfusions for a full eight months up to the time of the report, thus providing clinical evidence of repair of the immunodeficiency state. This response has been seen so far in only one child. Conceivably, genetic heterogeneity could alter the effectiveness of the treatment in different patients. The results achieved with demonstration of a reversible immunosuppression hold forth the possibility of controlling the responsiveness of the immune system in a more rational manner and with more specific and possibly physiological agents in the future.

STIMULATION AND SUPPRESSION OF THE IMMUNE RESPONSE IN LESCH-NYHAN SYNDROME

Children with the Lesch-Nyhan syndrome show a gross deficiency in the enzyme hypoxanthine-guanine phosphoribosyltransferase (HPRT) (Seegmiller *et al.* 1967) which catalyzes the following reaction:

$$\text{Hx} \quad + \quad \text{PP-ribose-P} \xrightarrow{\text{Hypoxanthine-Guanine Phosphoribosyltransferase (HPRT)}} \text{Hx-R-P} + \text{PP}_i$$

| Hypoxanthine | Phosphoribosyl-1-pyrophosphate | Phosphoribosyltransferase (HPRT) | Inosinic Acid |

Frequent infections have not been an overt feature of the Lesch-Nyhan syndrome (Nyhan 1973). Nevertheless, some evidence has been found that the stem cells carrying a severe deficiency of HPRT may be at a selective disadvantage compared to normal stem cells. Since it is an X-linked disease, the random inactivation of one of the two X-chromosomes in the heterozygous female leads to the appearance of both normal and HPRT-deficient phenotypes in their cultured skin fibroblasts (Rosenbloom *et al.* 1968) but not in their erythrocytes. However, both phenotypes are present in erythrocytes of heterozygotes carrying less severe HPRT deficiency (Emmerson, p.359). In addition, colonies of HPRT-deficient stem cells cultured *in vitro* from bone marrow of affected children are smaller in size and less numerous than normal (McKeran 1977). In addition, Allison *et al.* (1975) reported a modest decrease in B-cells and in IgG concentration in plasma of three affected children.

Plasma Immunoglobulin Concentrations and Antibody Response *In Vivo*

We have examined various indices of the immune response in four children with the Lesch-Nyhan syndrome. Although two of the four showed concentrations of IgG below the range shown by 105 adults, the values were fully within the range (mean ± S.D.) found in 20 normal children less than 12 years of age (Fig. 9).

The antibody response to challenge with four different antigens was determined in 3 children with the Lesch-Nyhan syndrome as compared to the response of a group of 25 control subjects. The Lesch-Nyhan patients showed a full response to antigenic challenge (Fig. 10).

Lymphocyte Proliferation and Antibody Synthesis Induced by Mitogens
In Vitro

In Order to measure the immune responsiveness of children with the Lesch-Nyhan syndrome we examined the mitogenic response and antibody synthesis of their lymphocytes to a variety of agents during culture *in vitro*.

As shown in Fig. 11 the ability of the isolated lymphocytes of Lesch-Nyhan patients to synthesize immunoglobulins *in vitro* was not grossly impaired. In only one of the four patients was immunoglobulin synthesis appreciably below the 67 per cent confidence limits.

Response of Lymphocytes to Immunosuppressive Drugs
In Vitro

The HPRT enzyme is required for inhibition of the rate of *de novo* purine synthesis *in vivo* by azathioprine (Kelley *et al.* 1968) and *in vitro* by 6-mercaptopurine (6MP) a metabolite of azathioprine (Seegmiller *et al.* 1967). We were therefore interested in determining whether or not the HPRT enzyme was also required for the immunosuppressive action of this class of drugs. The drug 6-methyl-mercaptopurine riboside (6MMPR) is a potent inhibitor of purine synthesis *de novo* in both normal and HPRT-deficient fibroblasts (Seegmiller *et al.* 1967) since it is converted to its metabolically active nucleotide form by adenosine kinase. The effect of this agent was therefore compared with 6 MP which requires an intact HPRT enzyme to be converted to its metabolically active nucleotide form for inhibiting purine synthesis.

The inhibition by 6 MMPR of the blastogenic response of both the normal and Lesch-Nyhan lymphocytes to a variety of mitogens is shown in Fig. 12.

As was expected the Lesch-Nyhan lymphoblasts failed to show any inhibition of the blastogenic process by treatment with doses of 6MP that completely inhibited the process in normal cells (Fig. 13).

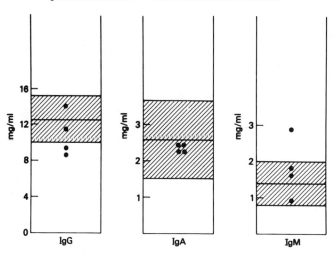

Figure 9. Concentrations of the various classes of immunoglobulins in plasma of four children with the Lesch-Nyhan syndrome. The mean values ± 1 S.D. found in 105 adults over 20 years of age are shown in the shaded regions ▨. The mean values found in 20 normal children <12 years of age were: IgM = 1.1 ± 0.4, IgA = 1.6 ± 1.1 and IgG = 8.9 ± 2.8 mg/ml. The procedure of Blaese *et al.* (1968) was used.

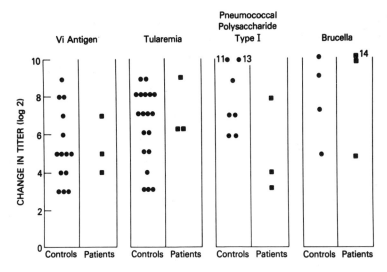

Figure 10. Antibody response to antigenic stimulation of patients with Lesch-Nyhan syndrome as compared to control subjects.

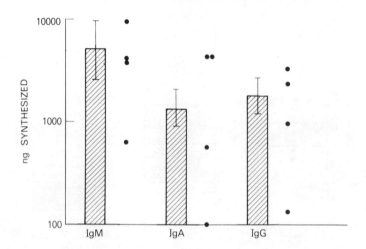

Figure 11. Synthesis of the various classes of immunoglobulins in ng/2 x 10^6 lymphocytes from children with the Lesch–Nyhan syndrome in response to stimulation by pokeweed mitogen as described by Waldmann *et al.* (1974).

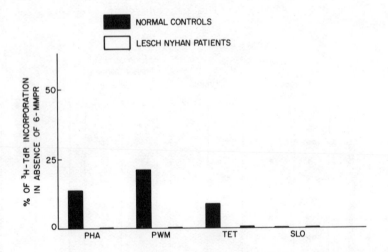

Figure 12. Effect of 6MMPR on the blastogenic response to a variety of mitogens in normal and Lesch–Nyhan lymphocytes using the procedure of Weiden *et al.* (1972). The mitogens used were: Phytohemagglutinin (PHA), pokeweed (PWM), tetanus (TET) and Streptolysin O (SLO).

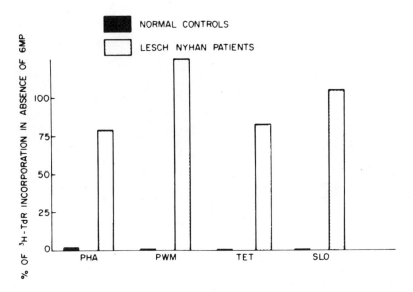

Figure 13.   Effect of 6 MP on the blastogenic response of normal and Lesch-Nyhan lymphocytes to a variety of mitogens using the procedure of Weiden *et al*. (1972).   See Figure 12 for legend.

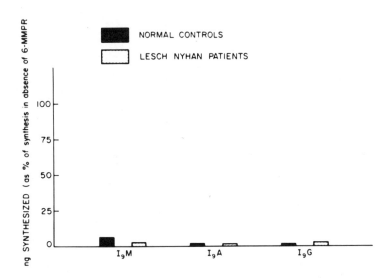

Figure 14.   Effect of 6MMPR on immunoglobulin synthesis by normal and Lesch-Nyhan lymphocytes.   The procedure of Waldmann *et al*. (1974) was used.

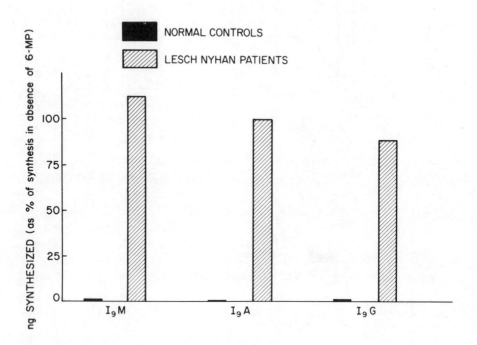

Figure 15.   Effect of 6 MP on immunoglobulin synthesis by normal and Lesch–Nyhan lymphocytes.   The procedure of Waldmann *et al.* (1974) was used.

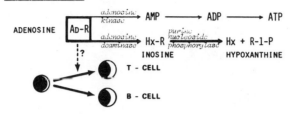

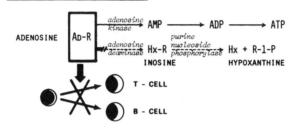

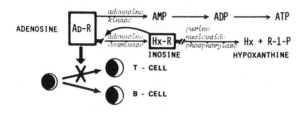

Figure 16. Hypothesis for a unitary mechanism for suppression of the immune system in hereditary deficiency of adenosine deaminase and of purine nucleoside phosphorylase by accumulation of adenosine and the resulting adenosine toxicity. The accumulation of adenosine in either type of mutant cell remains to be demonstrated. In the latter case inhibition of adenosine deaminase presumably is secondary to end-product inhibition by accumulation of inosine.

As might be expected, 6MMPR inhibited production of all classes of immunoglobulins in response to stimulation of both normal and Lesch-Nyhan lymphocytes by pokeweed mitogen *in vitro* (Fig. 14) while 6MP failed to inhibit production of immunoglobulins in lymphocytes from Lesch-Nyhan patients (Fig. 15).

SUMMARY

The development of our knowledge of the immune system has been reviewed and evidence presented of the need for a rapid rate of purine synthesis *de novo* for the proliferative events in this process.

The mechanism of the inhibition of the immune system in a model of ADA deficiency has been studied intensively and considerable indirect evidence obtained of adenosine toxicity as a possible mediator of a reversible inhibition of proliferation of T-cells and to a slightly lesser extent B-cells. A secondary inhibition of ADA by inosine accumulation in PNP deficiency is proposed as a unifying hypothesis in which a somewhat lesser adenosine toxicity would inhibit proliferation only of T-cells. The correction of the immune response by addition of ADA both *in vitro* and *in vivo* provides strong evidence in favor of this view.

In HPRT deficiency no evidence was found of a gross impairment o the immune system; however, the HPRT enzyme is required for inhibition of the immune response by 6MP in a variety of systems using different mitogenic stimuli.

ACKNOWLEDGEMENTS

This work was supported in part by United States Public Health Service grants AM-05646, AM-13622, GM-17702, and by grants from the National Foundation, the Kroc Foundation and the Josiah Macy Foundation.

REFERENCES

Agarwal, R. P., Sagar, S.M., and Parks, Jr.,R.P. (1975) Adenosine deaminase from human erythrocytes: Purification and effects of adenosine analogs. Biochem. Pharm., 24: 693-701.

Agarwal, R. P., Crabtree, G. W., Parks, Jr., R. P., Nelson, J. A., Keightley, R., Parkman, R., Rosen, F. S., Stern, R. C., and Polmar, S. H. (1976) Purine nucleoside metabolism in the erythrocytes of patients with adenosine deaminase deficiency and severe combined immunodeficiency. J. Clin. Invest. 57: 1025-1035.

Allison, A. C., Watts, R. W. E., Hovi, T., and Webster, A. D. B.
    (1975) Immunological observations on patients with Lesch-
    Nyhan syndrome, and on the role of de-novo purine synthesis
    in lymphocyte transformation. Lancet, 2: 1179-1182.
Bergsma, D. (1975) (Editor) *Immunodeficiency in Man and Animals.*
    Sinauer Associates, Inc., Sunderland, Mass (Birth Defects,
    Vol. II, No. 1).
Blaese, R. M., Strober, W., Brown, R. S., and Waldmann, T. A.:
    (1968) The Wiskott-Aldrich syndrome A Disorder with a possible
    defect in antigen processing or recognition. Lancet 1: 1056-1061.
Brady, T. G., and O'Donovan, C. I. (1965) A study of the tissue dis-
    tribution of adenosine deaminase in six mammal species. Comp.
    Biochem. Physiol., 14: 101-120.
Burnstock, G. (1976) Purinergic transmission. In *Handbook of
    Psychopharmacology* Vol. 5, Editors, L. L. Iverson, S. D.
    Iverson and S. H. Snyder. Plenum Publishing Corp., New York.
Burnstock, G. (1977) The purinergic nerve hypothesis. In *Purine
    and Pyrimidine Metabolism*, Elsevier, Amsterdam (In Press).
Carson, D. A., and Seegmiller, J. E. (1976) Effect of adenosine
    deaminase inhibition upon human lymphocyte blastogenesis.
    J. Clin. Invest., 57: 274-282.
Click, R. E., Benck, L, and Alter, B. J. (1972) Immune responses
    in vitro. I. Culture conditions for antibody synthesis. Cell.
    Immun., 3: 264-276.
Eisen, H. N. (1974) Immunology An introduction to molecular and
    Cellular Principles of the Immune Responses. Reprinted from
    Davis, Dulbecco, Eisen, Ginsberg, and Wood's *Microbiology*
    Second Edition, Harper and Row, Hagerstown. p. 475.
Fain, J. N., and Wieser, P. B. (1975) Effects of adenosine deaminase
    on cyclic adenosine monophosphate accumulation, lipolysis, and
    glucose metabolism of fat cells. J. Biol. Chem., 250: 1027-1034.
Friedmann, T., Seegmiller, J. E., and Subak-Sharpe, J. H. (1969)
    Evidence against the existance of guanosine and inosine kinase
    in human fibroblasts in tissue culture. J. Cell Res., 56:
    425-429.
Gerlach, E., Deuticke, B., and Dreisbach, R. H. (1963) Der
    Der Nucleotid-Abbau im Herzmuskel bei Sauerstoffmangel und
    seine mögliche Bedeutung für die Coronardurchblutung.
    Die Naturwissenschaften 6:228-229.
Giblett, E. R., Anderson, J. E., Cohen, F., Pollara, B., and
    Meuwissen, H. J., (1972) Adenosine-deaminase deficiency in
    two patients with severely impaired cellular immunity. Lancet
    2: 1067-1069.
Giblett, E. R., Ammann, A. J., Wara, D. W., Sandman, R., and Diamond,
    L. K., (1975) Nucleoside-phosphorylase deficiency in a child
    with severely defective T-cell immunity and normal B-cell
    immunity. Lancet, 1: 1010-1013.
Hall, J. G. (1963) Adenosine deaminase activity in lymphoid cells
    during antibody production. Austral. J. Exp. Biol. Med. Sci.
    41: 93-97.

Hirschhorn, R. (1975) Adenosine deaminase deficiency: Genetic and
    metabolic implications In *Combined Immunodeficiency Disease
    and Adenosine Deaminase Deficiency A Molecular Defect*
    (Meuwissen, H. J., Pickering, R. J., Pollara, B., and Porter,
    I. H., Eds.) Academic Press, New York, pp. 121-128.
Hitzig, W. (1976) Personal communication.
Kelley, W. N., Rosenbloom, F. M. and Seegmiller, J. E. (1967) The
    effects of azathioprine (Imuran) on purine synthesis in
    clinical disorders of purine metabolism. J. Clin. Invest.
    46: 1518-1529.
Lerner, M. H., and Lowy, B. A. (1974) The formation of adenosine
    in rabbit liver and its possible role as a direct precursor
    of erythrocyte adenine nucleotides. J. Biol. Chem., 249:
    959-966.
McKeran, R. O. (1977) The importance of purine biosynthesis *de novo*
    in the Lesch-Nyhan syndrome. In *Purine and Pyrimidine Metabol-
    ism*, Elsevier, Amsterdam (In Press).
Martin, D. W. (1977) Inosine accumulation in purine nucleoside
    phosphorylase deficiency. In *Purine and Pyrimidine Metabolism*,
    Elsevier, Amsterdam (In Press).
Meuwissen, H. J., and Pollara, B. (1974) Adenosine deaminase defi-
    ciency: the first inborn error of metabolism noted in
    immunodeficiency disease. J. Pediatr., 84: 315-316.
Meuwissen, H. J., Pickering, R. J., Pollara, B., and Porter, I. H.
    (Eds.) (1975) *Combined Immunodeficiency Disease and Adenosine
    Deaminase Deficiency A Molecular Defect.* Academic Press,
    New York (a)
Meuwissen, H. J., Pollara, B., and Pickering, R. J. (1975) Combined
    immunodeficiency disease associated with adenosine deaminase
    deficiency. J. Pediatr., 86: 169-181 (b)
Mishell, R. I., and Dutton, R. W. (1967) Immunization of dissociated
    spleen cell cultures from normal mice. J. Exp. Med., 126: 423-
    442.
Nyhan, W. L. (1973) The Lesch-Nyhan syndrome. Ann. Rev. Med., 24:
    41-60.
Ohno, M., Yagisawa, N., Shibahara, S., Kondo, S., Maeda, K., and
    Umezawa, H. (1974) Synthesis of coformycin. J. Amer. Chem.
    Soc., 96: 4326-4327.
Parkman, R., Gelfand, E. W., Rosen, F. S., Sanderson, A., and
    Hirschhorn, R. (1975) Severe combined immunodeficiency and
    adenosine deaminase deficiency. N. Engl. J. Med., 292: 714-719.
Polmar, S. H., Wetzler, E. M., Stern, R. C., and Hirschhorn, R.
    (1975) Restoration of in-vitro lymphocyte responses with
    exogenous adenosine deaminase in a patient with severe combined
    immunodeficiency. Lancet, 2: 743-746.
Polmar, S. H. Stern, R. C., Schwartz, A. L., and Hirschhorn, R. (1976)
    Enzyme replacement therapy for adenosine deaminase deficiency
    and severe combined immunodeficiency disease. Ped. Res.
    10: 392.

Pritchard, J. B., Chavez-Peon, F., and Berlin, R. D. (1970) Purines: Supply by liver to tissues. Am. J. Physiol., 219: 1263-1267.

Rosenbloom, F. M., Kelley, W. N., Henderson, J. F., and Seegmiller, J. E. (1967) Lyon hypothesis and X-linked disease. Lancet, 2: 305-306.

Rossi, C. A., Lucacchini, A., Montali, U., and Ronca, G. (1975) A general method of purification of adenosine deaminase by affinity chromatography. Int. J. Peptide Protein Res., 7: 81-89.

Schaeffer, H. J., and Schwender, C. F. (1974) Enzyme inhibitors. 26. Bridging hydrophobic and hydrophilic regions on adenosine deaminase with some 9-(2-hydroxy-3-alkyl) adenines. J. Med. Chem., 17: 6-8.

Schnebli, H. P., Hill, D. L., and Bennett, Jr., L. L. (1967) Purification and properties of adenosine kinase from human tumor cells of type H. Ep. No. 2. J. Biol. Chem., 242: 1997-2004.

Schreier, M. H., and Nordin, A. A. (1976) An evaluation of the immune response in vitro, In *B and T Cells and Immune Recognition* (Loor, F., and Roelants, Z. E., Eds.) (In Press) John Wiley and Sons, Ltd., Chichester, England.

Seegmiller, J. E., Rosenbloom, F. M., and Kelley, W. N. (1967) Enzyme defect associated with a sex-linked human neurological disorder and excessive purine synthesis. Science 155: 1682-1684.

Seegmiller, J. E., Watanabe, T., and Schreier, M. H., (1977) The effect of adenosine on the proliferation and antibody formation of lymphoid cells. In *Purine and Pyrimidine Metabolism*, Elsevier, Amsterdam (In Press).

Snyder, F. F., Mendelsohn, J., and Seegmiller, J. E. (1976) Adenosine metabolism in phytohemagglutinin-stimulated human lymphocytes. J. Clin. Invest. (In Press).

Stoop, J. W. (1976) Personal communication.

Waldmann, T. A., Durm, M., Broder, S., Blackman, M., Blaese, R. M., and Strober, W., (1974) Role of suppressor T cells in pathogenesis of common variable hypogammaglobulinaemia. Lancet, II: 609-621.

Weiden, P. L., Blaese, R. M., Strober, W., Block, J. B., and Waldmann, T. A., (1972) Impaired lymphocyte transformation in intestinal lymphagiectasia: Evidence for at least two functionally distinct lymphocyte populations in man. J. Clin. Invest. 51: 1319-1325.

PHOSPHORIBOSYLPYROPHOSPHATE (PRPP) AMIDOTRANSFERASE (EC 2.4.2.14)

ACTIVITY IN UNSTIMULATED AND STIMULATED HUMAN LYMPHOCYTES

J. ALLSOP & R. W. E. WATTS

MEDICAL RESEARCH COUNCIL CLINICAL RESEARCH CENTRE

WATFORD ROAD, HARROW, MIDDLESEX, HA1 3UJ

The mechanisms whereby new DNA synthesis is initiated after the exposure of lymphocytes to mitogens are poorly understood. It has been suggested that an influx of calcium ions, and increased levels of cyclic guanosine 3', 5'- monophosphate (cGMP) in the cells may be involved. Hovi, Allison & Allsop (1975) reported that the phosphoribosylpyrophosphate (PRPP) content of phytohaemagglutinin (PHA) stimulated lymphocytes increases transiently during the first hour of PHA-stimulation. They suggested either that this is a step in a chain of events which increases purine biosynthesis de novo and makes purine ribonucleotides available for DNA synthesis, or that a pulse of PRPP and purine biosynthesis is in some way needed to push the lymphocytes from the G_0 to the G_1 phase of the cell cycle.

The present work was undertaken to determine if either the early pulse of PRPP synthesis, or the later changes associated with PHA-stimulation involve detectable changes in phosphoribosylpyrophosphate (PRPP) amidotransferase (EC 2.4.2.14) activity. We have also compared: (i) the PRPP-amidotransferase activity of PHA-stimulated normal and HGPRT-deficient lymphocytes; (ii) the PRPP-amidotransferase activity of human peripheral blood granulocytes and unstimulated lymphocytes; (iii) the PRPP-synthetase (EC 2.7.6.1) activity and PRPP content of unstimulated and stimulated human peripheral blood lymphocytes.

MATERIALS AND METHODS

Lymphocytes were separated from buffy coats, each of which was derived from 5 healthy blood donors. These together with other

434

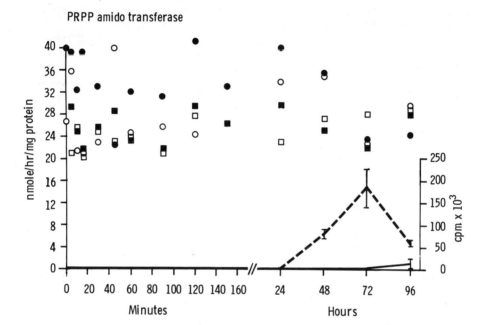

Fig. 1   The PRPP-amidotransferase activity of human lymphocytes
with and without PHA-stimulation.   The corresponding
[^{3}H]thymidine incorporation measurements are also shown.
         ○——○   Experiment 1, unstimulated lymphocytes
         □——□   Experiment 1, PHA-stimulated lymphocytes
         ●——●   Experiment 2, unstimulated lymphocytes
         ■——■   Experiment 2, PHA-stimulated lymphocytes
         I - - - I   [^{3}H]thymidine uptake with extreme range of
6 observations at each time point for stimulated cells.
         I——I   [^{3}H] thymidine uptake with extreme range of
6 observations for each time point for unstimulated cells.

special reagents were obtained from the sources listed by Hovi et al., (1975). The lymphocytes were separated from 10 ml blood samples from patients with the Lesch Nyhan syndrome and normal subjects. Granulocytes were obtained as a pellet from the lympho-cyte separations. Lymphocytes were maintained in short term tissue culture as described by Hovi et al., (1975).

PRPP-amidotransferase was determined by measuring the con-version of [U-^{14}C]glutamine to [U-^{14}C]glutamate in the absence of glutaminase activity (Wood & Seegmiller, 1973; Martin & Owen, 1972).

PRPP-and PRPP-synthetase were assayed as described by Fox & Kelley, 1971; and by Becker, Meyer & Seegmiller (1973) respectively. Protein was measured by the method of Lowry, Rosebrough, Farr & Randall (1951).

RESULTS

The PRPP-amidotransferase activity of the lymphocytes did not change during the short periods of PHA stimulation (up to 150 min) or in experiments lasting up to 96 hours during which cell transform-ation occurred as shown by the time course of the [^{3}H]thymidine incorporation into the cells (Fig. 1).

The PRPP-amidotransferase activities of the stimulated (72 hours exposure to PHA) and unstimulated control and Lesch-Nyhan patients lymphocytes were not appreciably different (Table 1).

Granulocytes and lymphocytes have similar PRPP-amidotransferase activities (Table 2).

The PRPP concentration and PRPP-synthetase activities of the stimulated lymphocytes were only studied at 24, 48, 72 and 96 hours after PHA-stimulation was begun. The PRPP synthetase activity of the PHA-stimulated cells remained approximately steady, but that of the unstimulated cells increased between 48 and 96 hours of incubation. The PRPP concentrations were barely increased after between 24 and 96 hours of incubation and the concentrations were generally higher in the unstimulated than in the stimulated cells.

DISCUSSION

Katanuma & Weber (1974) reported that the PRPP-amidotransferase activity of rat hepatoma cells was 2-3 fold greater than that of comparable non malignant cells. This contrasts with the present finding that PRPP-amidotransferase activity did not increase with PHA-induced lymphocyte transformation. Both of these groups of

TABLE 1

Phosphoribosylpyrophosphate (PRPP) amidotransferase (EC 2.4.2.14) activity of phytohaemagglutinin (PHA) stimulated and unstimulated lymphocytes from three patients with the Lesch–Nyhan syndrome and three simultaneously studied control subjects

(The peripheral blood lymphocytes were cultured for 72 hours in Dulbecco's medium containing foetal calf serum 10% (v/v) with $5\mu g/ml$ PHA/10^6 cells at $37°C$. The cells were disrupted by ultra-sonic vibrations (2 x 20 sec bursts at 8μ amplitude), and cell debris removed by centrifugation (30,000 x g_{Av} for 30 min). The results shown are for simultaneously studied Lesch Nyhan and control subjects' cells, each value being the mean of duplicate observations. (See text for further experimental details).

PRPP-amidotransferase activity (nmoles h^{-1} mg protein^{-1})			
Lesch Nyhan Patients		Control Subjects	
Unstimulated	PHA–stimulated	Unstimulated	PHA–stimulated
20.6	19.4	20.7	26.5
21.7	17.3	18.8	20.7
20.1	18.6	18.6	17.3

observations were made under saturating substrate conditions. They may point to fundamental differences between the two processes, and indicate caution in the extent to which the results of biochemical studies on lectin–stimulated lymphocytes should be directly extrapolated to other types of abnormal cell proliferation.

The present failure to observe increased PRPP–amidotransferase activity coincident with or following the pulse of PRPP accumul-ation during the first hour after PHA–stimulation which was observed by Hovi et al., (1975) may indicate that, PRPP synthesis is increased at this time for some purpose other than to provide a substrate for PRPP–amidotransferase. Alternatively, PHA–stimulation may be associated with a biochemical change which involves the utilization of less PRPP than previously, and that there is a temporary accumulation of this metabolite.

The present determinations were all performed at saturating substrate concentrations, so that increased PRPP–amidotransferase due to disaggregation of the enzyme by PRPP as described for the partially purified human placental enzyme by Holmes, Wyngaarden &

TABLE 2

Phosphoribosylpyrophosphate (PRPP) amidotransferase (EC 2.4.2.14)
activity of peripheral blood granulocytes and lymphocytes

[Both cell types were obtained from the same patient. The granulocytes
and lymphocytes were disrupted by ultrasonic vibrations (2 x 20 sec
bursts for lymphocytes and a single 20 sec burst at 8μ amplitude
for lymphocytes and granulocytes respectively. Cell debris removed
by centrifugation at 30,000 x g_{Av} for 30 min. The results shown
are the mean of duplicate determinations on the same batch of cells
(see text for further experimental details)].

PRPP-amidotransferase activity ($nmoles\ h^{-1}\ mg\ protein^{-1}$)	
Granulocytes	Lymphocytes
16.1	13.1
24.6	33.4
22.2	33.3
22.9	30.5
19.8	27.4
22.9	21.2

Kelley (1974) may have been masked by the assay procedure, and
further work is needed to determine if more direct evidence for
the occurrence of these conformational effects in intact cells can
be obtained.

The observations on PHA-stimulated lymphocytes presented in
Table 1 agree with the observation that long term lymphoblast lines
from two cases of the Lesch Nyhan syndrome had the same PRPP-amidotransferase activity as normal human lymphoblasts (Wood, Becker &
Seegmiller, 1973). Here again, allosteric changes induced in the
enzyme by PRPP may have been obscured by the use of saturating
concentrations of PRPP. Studies on cells from human lymphatic
leukaemia or lymphosarcoma would be of interest in view of the
results of Katanuma & Weber (1974).

Our finding of appreciable PRPP-amidotransferase activity in
human peripheral blood granulocytes as well as lymphocytes
indicates that these cells may also be of use in the study of
disordered purine metabolism in man. The occurrence of de novo
purine biosynthesis in these cells may be related to their need
for purine ribonucleotides (e.g. GTP) in connection with the bursts
of protein synthesis (new enzyme formation) which accompany phagocytosis. The activity of PRPP-amidotransferase in both granulocytes

and lymphocytes was inhibited by AMP (ID_{50} 3.0mM for lymphocytes). This ID_{50} value agrees with that obtained by Hill & Bennett (1969) for adenocarcinoma 755 cells. Azaserine also inhibited the enzyme with an ID_{50} value of 5.62 mM for both lymphocytes and granulocytes.

Our preliminary findings so far have failed to demonstrate a burst of PRPP-synthetase activity or of PRPP accumulation which is related in time to the occurrence of PHA-induced lymphocyte trans- formation as judged by the time of maximum [^{3}H]thymidine uptake. This may mean that, except for the early pulse of PRPP-accumulation described by Hovi et. al. (1975), the supply of and demand for PRPP are so finely balanced as to be undetectable, or that the critical changes in enzyme activity occur between 150 min and 24 hours after the beginning of PHA-stimulation that is, during a period of the transformation process which we have not so far studied.

The present observation that the specific activity of the enzyme (nmoles. h^{-1} mg protein^{-1}) is unaltered at any stage in the transformation process supports the view that modulation of PRPP- amidotransferase activity is due to allosteric effects. Studies with partially PRPP-amidotransferase have shown that PRPP is an allosteric activator of the enzyme (Holmes et al., 1974). Therefore the ultimate regulator of the rate of purine biosynthesis de novo may be the activity of PRPP-synthetase at some stage of the transformation process, but we have so far been unable to demonstrate this directly. Further studies in this area are in progress.

REFERENCES

BECKER, M. A., MEYER L. J. & SEEGMILLER, J. E. (1973). Gout with purine overproduction due to increased phosphoribosylpryophosphate synthetase activity. American Journal of Medicine, 55, 232-242.

FOX I. H. & KELLEY W. N. (1971). Human phosphoribosylpyrophosphate synthetase. Distribution, purification and properties. Journal of Biological Chemistry, 246, 5739-5748.

HILL D. L. & BENNETT L. L. (1969). Purification and properties of 5-phosphoribosylpyrophosphate amidotransferase from adenocarcinoma 755 cells. Biochemistry, 8, 122-130.

HOLMES, E. W.Jr., WYNGAARDEN, J. B. & KELLEY, W. N. (1974). Human glutamine phosphoribosylpyrophosphate (PP-ribose-P) amidotransfer- ase. Advances in Experimental Medicine and Biology, 41A, 43-53.

HOVI, T. ALLISON, A. C. & ALLSOP, J. (1975). Rapid increase in phosphoribosylpyrophosphate concentration after mitogenic stimulation of lymphocytes. FEBS Letters, 55, 291-293.

KATANUMA, N. & WEBER, G. (1974). Glutamine phosphoribosylpyrophosphate amidotransferase: increased activity in hepatomas. FEBS
letters, 49, 53-56.

LOWRY, O. H., ROSEBROUGH, W. J., FARR, A. L. & RANDALL, R. J. (1951)
Protein measurement with the folin phenol reagent. Journal of
Biological Chemistry, 193, 265-275.

MARTIN, D. W. & OWEN, N. T. (1972). Repression and derepression of
purine biosynthesis in mammalian hepatoma cells in culture.
Journal of Biological Chemistry, 247, 5477-5485.

WOOD, A. W., BECKER, M. A. & SEEGMILLER, J.E. (1973). Purine
nucleotide synthesis in lymphoblasts cultured from normal
subjects and in a patient with Lesch-Nyhan syndrome. Biochemical
Genetics, 9, 261-274.

WOOD, A. W. & SEEGMILLER, J. E. (1973). Properties of 5-phospho-
ribosyl-1-pyrophosphate amidotransferase from human lymphoblasts.
Journal of Biological Chemistry, 248, 138-143.

ADENOSINE AND GUANOSINE METABOLISM DURING PHYTOHEMAGGLUTININ INDUCED TRANSFORMATION OF HUMAN LYMPHOCYTES

FLOYD F. SNYDER[§], JOHN MENDELSOHN AND J. EDWIN SEEGMILLER

Department of Medicine, University of California,

San Diego, La Jolla, California 92093, U.S.A.

Reports of the human genetic deficiency of adenosine deaminase activity in association with severe combined immuno-deficiency disease (1,2) suggested adenosine deaminase activity may be necessary for lymphocyte response. We have examined the effect of adenosine and inhibitors of adenosine deaminase activity on lymphocyte transformation. Work in this laboratory has shown the inhibition of lymphocyte transformation by adenosine to be potentiated by inhibitors of adenosine deaminase activity in concanavalin A (3) or phytohemagglutinin (PHA) (4) stimulated human lymphocytes. The report of purine nucleoside phosphosphoryl-ase deficiency in association with defective T cell immunity (5) adds further impetus to the investigation of purine nucleoside metabolism in lymphocytes. The present report describes changes in adenosine and guanosine metabolism in both lysates and intact cells during PHA-induced transformation of human lymphocytes.

Purification, culture, and transformation of human lympho-cytes. Human lymphocytes were purified from blood of healthy donors as previously described (6,7). Freshly purified human lymphocytes were cultured in Eagle's minimum essential medium supplemented with 10% horse serum because of the negligible adeno-sine deaminase activity in serum from this source. Less than 5% of 50 μM adenosine was deaminated after 30 hours incubation in medium containing 10% horse serum (8). After 72 hours incubation, 90-95% of control and PHA-stimulated lymphocytes were viable. By morphologic examination unstimulated cultures had 2-3% blastic

[§]Present address: Biochemistry Group, Department of Chemistry, The University of Calgary, Calgary, Alberta, Canada.

cells. PHA-stimulated cultures had 40-70% blasts, and stimulated cultures had a 20-50 fold increase in {³H}thymidine incorporation into acid-precipitable material (Table 1).

Effect of adenosine and inhibition of adenosine deaminase activity upon lymphocyte transformation. An increasing amount of evidence indicates that inhibition of adenosine deaminase activity potentiates adenosine inhibition of mitogen induced lymphocyte transformation (3,4,9) or function (10). A concentration of 50 μM adenosine was found to arrest growth of WI-L2 cultured human lymphoblasts only when added together with an adenosine deaminase inhibitor (11).

There were no effects of adenosine, 50 μM, upon the response of human lymphocytes to PHA, measured 72 hours after the addition of adenosine and PHA. The effect of the tight binding adenosine deaminase inhibitor, coformycin (12,13), was also examined. Inhibition of greater than 95% adenosine deaminase activity by coformycin, 1 μg/ml, did not block lymphocyte transformation by criteria of morphology or thymidine incorporation into DNA (Table 1). The combination of coformycin and adenosine, however, produced a substantial reduction in the incorporation of thymidine into DNA, viable cell count and morphologic transformation. These results demonstrate that low concentrations of adenosine are toxic to lymphocytes stimulated to divide in the presence of the adenosine deaminase inhibitor, coformycin.

TABLE 1

EFFECT OF ADENOSINE AND COFORMYCIN ON PARAMETERS OF HUMAN

LYMPHOCYTE TRANSFORMATION

PHA	ADDITIONS	{³H}TdR→DNA CPM/1 ML CULTURE	MORPHOLOGIC TRANSFORMATION % BLASTS	VIABILITY %
-	NONE	1,780	<4	90-100
+	NONE	66,400	40-70	90-100
+	COFORMYCIN	67,700	40-75	75-80
+	ADENOSINE	67,100	35-70	75-80
+	ADENOSINE + COFORMYCIN	3,390	30-35	∿35

Lymphocytes were incubated in the presence and absence of PHA, 1 μg/ml, for 72 hours. Coformycin, 1 μg/ml; adenosine, 50 μM.

Effect of phytohemagglutinin-induced transformation on enzyme activities in lymphocyte extracts. Adenosine may be phosphorylated to AMP by adenosine kinase or converted to inosine by adenosine deaminase. Adenosine is also a potential precursor of adenine (14). Radiochemical assays of adenosine kinase, adenosine deaminase and purine nucleoside phosphorylase activities in extracts and intact cells are reported elsewhere (4,14). Of the three possible routes of adenosine metabolism in extracts of freshly purified lymphocytes, the ratios of the specific enzyme activity for adenosine deaminase: adenosine kinase: adenosine cleavage, were 1.0: 0.12: 0.0015.

The specific activities of three enzymes of purine nucleoside metabolism were measured in lymphocyte lysates after 72 hours incubation in the presence and absence of PHA. In lysates of PHA-stimulated cultures the specific activity of adenosine kinase and purine nucleoside phosphorylase remained essentially unchanged (Table 2). The specific activity of adenosine deaminase decreased 55% in PHA-stimulated cultures compared to control cells (Table 2). Other activities reported not to change during PHA-induced transformation of human lymphocytes are PP-ribose-P synthetase (4,15) and amidophosphoribosyltransferase (15). Adenosine deaminase is apparently subject to a different rate of synthesis or catabolism than several other enzymes during lymphocyte transformation.

Effect of phytohemagglutinin-induced transformation on adenosine and guanosine metabolism by intact lymphocytes. Additional studies were conducted with intact lymphocytes because activities measured in lysates may bear little correspondance to the rates of nucleoside metabolism in the intact cell where transport, competition for common substrates, and regulation of a given activity all influence the final rate. Incubation of lymphocytes for 72 hours with PHA produced a 12-fold increase in the rate of deamination of adenosine (Fig. 1), despite the decrease in specific activity in lysates. The increase was apparent as early as 3 hours and appeared to begin approaching maximal activity by 48 hours incubation with PHA. The phosphorylation of adenosine exhibited a similar increase in rate in response to incubation of lymphocytes with PHA, the overall increase being about 6-fold. The phosphorolysis of guanosine, representative of purine nucleoside phosphorylase activity, also increase approximately 6-fold after 72 hours culture with PHA. The actual activities for the metabolism of adenosine and guanosine in the intact cell (Fig. 1) were at most 25% of the activities measured in lymphocyte lysates.

We have examined the rates of adenosine metabolism as a function of adenosine concentration in intact lymphocytes. These studies showed the increased rates of adenosine metabolism by both phosphorylation and deamination reflected principally an increase in the maximal velocity of these processes. The nucleoside

TABLE 2

EFFECT OF PHYTOHEMAGGLUTININ-INDUCED LYMPHOCYTE TRANSFORMATION ON ENZYME ACTIVITIES IN CELL EXTRACTS

ENZYME	LYMPHOCYTE DONOR	ACTIVITIES -PHA nmoles/mg protein per min	ACTIVITIES +PHA	RATIO +PHA/-PHA	AVERAGE
ADENOSINE DEAMINASE	A	42.8	21.6	0.51	0.45
	B	57.4	22.1	0.39	
	C	48.1	20.5	0.43	
	D	56.4	26.5	0.47	
	E	55.8	21.4	0.38	
	F	64.5	32.5	0.50	
ADENOSINE KINASE	D	4.44	4.75	1.07	0.95
	E	5.09	4.88	0.96	
	G	7.80	6.33	0.81	
PURINE NUCLEOSIDE PHOSPHORYLASE	B	52.4	47.0	0.90	0.91
	C	49.3	49.0	0.99	
	E	65.2	54.4	0.83	
	F	62.9	62.0	0.99	
	G	61.5	51.4	0.84	

Extracts of freshly purified lymphocytes incubated for 72 hours in the presence and absence of 1 µg/ml phytohemagglutinin were pre-pared and enzyme activities were assayed in the 20,000 × g supernatant.

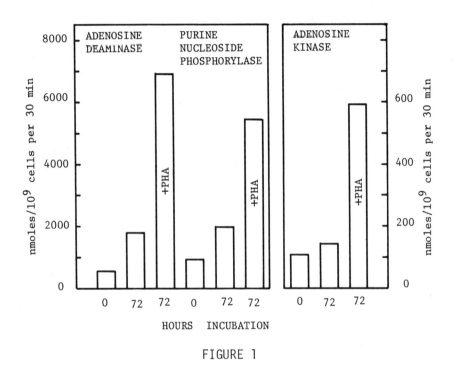

FIGURE 1

PHYTOHEMAGGLUTININ-INDUCED CHANGES IN PURINE NUCLEOSIDE METABOLISM

BY INTACT LYMPHOCYTES

The deamination or phosphorylation of {^{14}H}adenosine, 85 μM, and the phosphorolysis of {^{14}C}guanosine, 200 μM, by intact lymphocytes were measured before and after 72 hours culture in the absence or presence of PHA, 1 μg/ml.

transport inhibitor, nitrobenzylthioinosine (16), reduced the rate
of adenosine metabolism in PHA-stimulated lymphocytes to that of
unstimulated lymphocytes. These results suggest an increased rate
of entry for adenosine in PHA-stimulated lymphocytes.

Previous studies in other cells showed the relative amount of
adenosine metabolized via phosphorylation or deamination to be
dependent on the adenosine concentration (14,17). These findings
may be understood in terms of the lower Michaelis constant of 2 μM
for adenosine with adenosine kinase (17,18) than 25-40 μM with
adenosine deaminase (17,19,20,21) and the relative amounts of each
activity. In unstimulated lymphocytes after 72 hours culture the
rate of phosphorylation was approximately 0.05 that of deamination
over a range of adenosine concentrations, 1-100 μM. In contrast
for PHA-stimulated lymphocytes at 72 hours culture, the ratio of
adenosine phosphorylation to deamination was 1.5, 1.0, and 0.4 for
1, 5, and 10 μM adenosine respectively. A greater proportion of
adenosine was phosphorylated than deaminated below 5 μM adenosine
in the mitogen stimulated lymphocytes. These differences in
adenosine metabolism between lymphocytes cultured in the presence
and absence of PHA may represent a characteristic change between a
resting and mitogen stimulated lymphocyte.

We have also examined the possibility that the hereditary
deficiency of adenosine deaminase activity may reflect altered
kinetic properties of the enzyme with respect to adenosine. In
heterozygote lymphocytes, erythrocytes and fibroblasts, having
1/3 to 1/2 normal activity, no evidence of altered substrate
affinity for adenosine with adenosine deaminase activity was
detected (22).

These studies illustrate the dependence of the human lymphocyte on adenosine deaminase activity for the metabolism of adenosine. We have recently demonstrated the adenosine-mediated growth
inhibition of the human lymphoblast WI-L2 line to be associated
with a marked reduction in intracellular PP-ribose-P concentrations
and purine and pyrimidine nucleotides dependent upon PP-ribose-P
for their synthesis (8,11). The present report shows a concentration of adenosine, which alone had little effect, to be toxic
and to reduce PHA-induced transformation of human lymphocytes,
only in the presence of the adenosine deaminase inhibitor,
coformycin. The hereditary deficiency of adenosine deaminase may
also potentiate the growth inhibitory and toxic effects of
adenosine on human lymphocytes.

References:

(1) Giblett, E.R., Anderson, J.E., Cohen, F., Pollara, B. and
     Meuwissen, H.J. (1972) Lancet 2, 1067-1069.

(2)   Dissing, J. and Knudson, B. (1972) Lancet 2, 1316-1318.
(3)   Carson, D.C. and Seegmiller, J.E. (1976) J. Clin. Invest. 57, 274-282.
(4)   Snyder, F.F., Mendelsohn, J. and Seegmiller, J.E., J. Clin. Invest (In Press).
(5)   Giblett, E.R., Ammann, A.J., Wara, D.W., Sandman, R. and Diamond, L.K. (1975) Lancet 1, 1010-1013.
(6)   Mendelsohn, J., Skinner, A. and Kornfeld, S. (1971) J. Clin. Invest 50, 818-826.
(7)   Mendelsohn, J., Multer, M.M. and Boone, R.F. (1973) J. Clin. Invest. 52, 2129-2137.
(8)   Snyder, F.F. and Seegmiller, J.E., FEBS Letters (In Press).
(9)   Fox, I.H., Keystone, E.C., Gladman, D.D., Moore, M. and Cane, D. (1975) Immunological Comm. 4, 419-427.
(10) Wolberg, G., Zimmerman, T.P., Hiemstra, K., Winston, M. and Chu, L.-C. (1975) Science 187, 957-959.
(11) Snyder, F.F., Hershfield, M.S. and Seegmiller, J.E., 2nd International Symposium on Purine Metabolism in Man.
(12) Sawa, T., Fukagawa, Y., Homma, H., Takeuchi, T. and Umezawa, H. (1967) J. Antibiot. Ser. A. 20, 227-231.
(13) Cha, S., Agarwal, R.P. and Parks, Jr., R.E. (1975) Biochem. Pharmacol. 24, 2187-2197.
(14) Snyder, F.F. and Henderson, J.F. (1973) J. Biol. Chem. 248, 5899-5904.
(15) Wood, A.W., Astrin, K.H., McCrea, M.E. and Becker, M.A. (1973) Fed. Proc. 32, 652.
(16) Paterson, A.R.P. and Oliver, J.M. (1971) Can. J. Biochem. 49, 271-274.
(17) Meyskens, F.L. and Williams, H.E. (1971) Biochim. Biophys. Acta 240, 170-179.
(18) Schnebli, H.P., Hill, D.L. and Bennett, Jr., L.L. (1967) J. Biol. Chem. 242, 1997-2004.
(19) Osborne, W.R.A. and Spencer, N. (1973) Biochem. J. 133, 117-123.
(20) Agarwal, R.P., Sagar, S.M. and Parks, Jr., R.E. (1975) Biochem. Pharmacol. 24, 693-701.
(21) Rossi, C.A., Lucacchini, A., Montali, V. and Ronca, G. (1975) Int. J. Peptide Prot. Res. 7, 81-89.
(22) Snyder, F.F., Scott, D.L., and Seegmiller, J.E. (1974) Prog. and Abs. Amer. Soc. Human Genetics 82A.

Acknowledgements: This work was supported in part by USPHS grants CA 11971, RCDA CA 70891, AM-13622, AM-05646, GM-17702 and by grants from the National Foundation and the Kroc Foundation.

ADENINE AND ADENOSINE METABOLISM IN PHYTOHEMAGGLUTININ (PHA) - STIMULATED AND UNSTIMULATED NORMAL HUMAN LYMPHOCYTES

Kari O. Raivio and Tapani Hovi

Children's Hospital and Department of Virology

University of Helsinki, Helsinki, Finland

Several observations suggest that purine nucleotide synthesis and degradation play an important role in lymphocyte metabolism, particularly in the process of mitogenic activation. Much of the relevant evidence is based on human mutations affecting purine metabolism and lymphocyte function. Adenosine deaminase (ADA) deficiency is associated with a severe functional defect of both T and B lymphocytes (1), whereas nucleoside phosphorylase defect affects mainly T lymphocyte function (2). Adenosine itself strongly inhibits the proliferation of normal lymphocytes at concentrations of 10^{-4}M or higher (3), whereas lower concentrations are stimulatory (4). The inhibition caused by adenosine is potentiated by blocking ADA with coformycin (5) or erythro-9-(2-hydroxy-3-nonyl) adenine hydrochloride (4). It has not been established, whether the adenosine effects are mediated via a specific receptor linked to cyclic AMP production, or whether cellular metabolism of adenosine is involved.

We have evaluated the metabolism of adenosine and adenine in normal human peripheral blood lymphocytes in relation to mitogenic activation. The two compounds studied represent the two basic mechanisms of purine salvage, nucleoside kinase and purine phosphoribosyltransferase reactions.

METHODS

Lymphocytes were prepared from samples of peripheral blood by Ficoll-Isopaque centrifugation. The cell suspension was then passed through a Nylon wool column, and after this enrichment procedure over 90% of the cells were T lymphocytes. The reason for using a more homogeneous preparation in the experiments was that differences in purine metabolism evidently exist between T and B lymphocytes as well as monocytes.

The purified lymphocytes were incubated in medium RPMI 1640, usually containing 5% horse serum, with or without PHA for 48-72 hr. After this period, the cells were either sonicated and assayed for ADA, adenine phosphoribosyltransferase (APRT), and adenosine kinase (AK) activity, or incubated in Krebs-Ringer phosphate medium with ^{14}C-adenosine or ^{14}C-adenine. After the incubation, the purine compounds were extracted with perchloric acid and separated using thin-layer chromatography (6). The relative rates of the various metabolic pathways were evaluated on the basis of isotope incorporation into the relevant compounds. The results are usually expressed as cpm incorporated per 10^6 cells over a given time. Since intracellular adenine concentrations are generally considered to be negligible, metabolic rates in molar terms have been calculated in some cases.

RESULTS AND DISCUSSION

Adenine Metabolism

Freshly isolated lymphocytes utilized ^{14}C-adenine slowly (150-270 nmoles/hr/10^9 cells), and the rate for hypoxanthine was lower still. PHA-stimulated cells showed an increased rate of adenine metabolism, whereas lymphocytes incubated for an equivalent time without PHA had a depressed rate.

The metabolic fate of adenine was almost exclusively nucleotide synthesis. Over 80% of the adenine metabolized in 2 hr was recovered in the adenine nucleotides, with some degradation to inosine and hypoxanthine (Fig. 1). Adenosine or IMP did not become significantly labelled in the course of the 2 hr incubation. Maximal nucleotide synthesis was achieved at 40 uM adenine, with no significant further increase when higher concentrations were used (Fig. 2).

The effect of incubation of lymphocytes with or without PHA on the activity of APRT was examined. The pattern of change depends on the basis of reference used for expressing enzyme activity. Enriched T lymphocyte preparations show a considerable loss of cells over the first 24-48 hr of incubation with PHA, followed by an increase in the number and especially the size of the surviving cells. APRT activity, expressed as nmoles/min/mg protein, decreases by 27% in PHA-stimulated and by 76% in unstimulated lymphocytes over 4 days. However, when expressed on a per cell

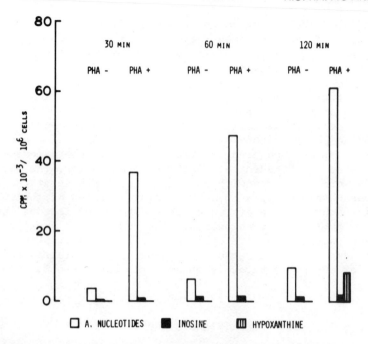

Fig. 1. Radioactivity in adenine derivatives in stimulated and unstimulated lymphocytes. Initial adenine concentration 36 uM.

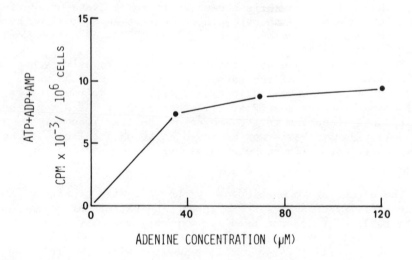

Fig. 2. Adenine nucleotide synthesis from ^{14}C-adenine in lymphocytes stimulated with PHA. Incubation time 2 hr.

basis, the stimulated cells show a marked increase with a peak on
day 2, whereas in unstimulated cells APRT activity steadily
decreases (Fig. 3).

Our results show that purine bases are not efficiently converted into nucleotides in normal lymphocytes. There are at least
three possible reasons for this: slow transport, low enzyme
activity, and lack of cosubstrate phosphoribosylpyrophosphate
(PRPP) for the reaction. Since PHA-stimulation is known to
increase intracellular PRPP concentrations (8) and nucleotide
synthesis from adenine (Fig. 1), the third possibility seems
plausible. Even though APRT activity is also increased by PHA,
the importance of this change for adenine metabolism is less clear.
The maximal rate of adenine metabolism measured, 11 nmoles/min/10^9
cells, represented less than 1% of the simultaneous APRT activity.
Therefore, low enzyme activity probably does not become limiting
for adenine metabolism.

Adenosine Metabolism

Adenosine was metabolized more rapidly than adenine under all
the conditions tested. Freshly isolated lymphocytes usually had
a somewhat faster rate than PHA-stimulated cells, whereas cells

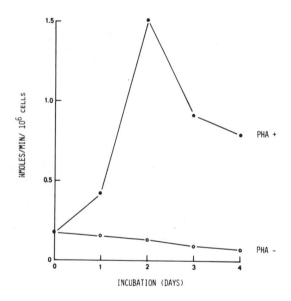

Fig. 3. Adenine phosphoribosyltransferase activity in lymphocytes
incubated with or without phytohemagglutinin (PHA).

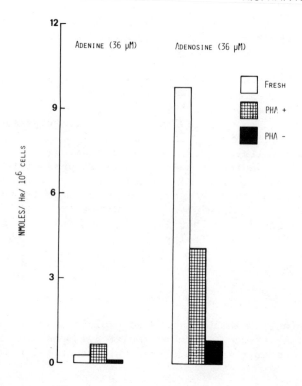

Fig. 4. Metabolism of ^{14}C-adenine and ^{14}C-adenosine in normal lymphocytes. Time of incubation 1 hr.

incubated without PHA utilized very little adenosine (Fig. 4).

The metabolic fate of adenosine depends on the initial concentration. Nucleotide synthesis accounted for 25-30% of the adenosine metabolized at concentrations around 10 uM, but when these were elevated above 40 uM, there was no further increase in adenine nucleoformation, and the excess adenosine was deaminated (Fig. 5). The labelling pattern in cells incubated without PHA was similar to that in stimulated cells, but both nucleotide synthesis and adenosine deamination, represented by inosine plus hypoxanthine, were considerably lower.

The distribution of the labelled deamination products between cells and medium was evaluated by extracting the total incubation mixture with perchloric acid as usual and comparing the radioactivity in the metabolic products to a parallel extract prepared after washing the cells with buffer. The washing resulted in

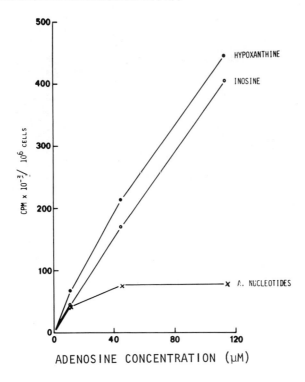

Fig. 5. Conversion of ^{14}C-adenosine into derivatives in lymphocytes. Time of incubation 1 hr.

total disappearance of inosine and hypoxanthine (Fig. 6), which suggests that they are extruded from the cell immediately after being formed, or that they exist in an easily mobilizable pool.

The effect of incubation with or without PHA on ADA activity of lymphocytes was studied. As in the case of APRT, enzyme activity per cell increased to a maximum at day 2 in the PHA-stimulated cells, whereas unstimulated cells lost activity (Fig. 7). A similar pattern was observed also for adenosine kinase.

[14]C-adenosine was used in most of these experiments as a tracer, ie. the concentrations were below 5×10^{-5} M and lower than those found to inhibit lymphocyte proliferation (4). Nucleotide synthesis was invariably more efficient than from the purine bases. Since the intracellular adenosine pool is probably larger than that of adenine, a difference in pool sizes cannot account for the difference in nucleotide synthesis.

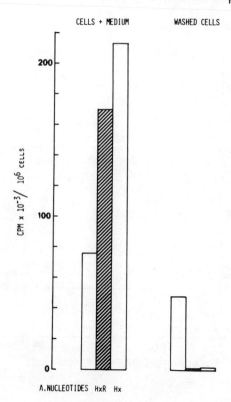

Fig. 6. Radioactivity in adenine nucleotides, inosine, and hypoxanthine in the total incubation mixture compared to the cells
alone.   Time of incubation 1 hr.

     The mechanism of adenosine toxicity to lymphocytes remains a
problem.   Excessive nucleotide synthesis does not appear to be
responsible, since maximal rates can be demonstrated at adenosine
concentrations not inhibitory to proliferation.   The deamination
products are rapidly extruded out of the cell.   This does not
exclude the possibility of harmful effects of inosine and hypoxanthine on lymphocyte function, but such effects have not been
demonstrated by adding the same compounds to lymphocyte cultures.

     The increased purine base utilization after PHA-stimulation
is evidence in favor of the concept that increased production of
PRPP is one of the early effects of PHA (8).

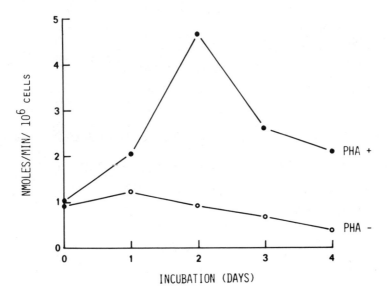

Fig. 7. Adenosine deaminase activity in lymphocytes incubated with or without PHA.

Acknowledgements. These studies were supported in part by the Sigrid Juselius Foundation and the Foundation for Pediatric Research in Finland.

REFERENCES

1. Meuwissen, H.J., Pollara, B., and Pickering, R.J.: J. Pediat. 86: 169, 1975.

2. Giblett, E.R., Amman, A.J., Sandman, R., Wara, D.W., and Diamond, L.K.: Lancet, 2: 1010, 1975.

3. Smith, J.W., Steiner, A.I., and Parker, C.W.: J. Clin. Invest. 50: 442, 1971.

4. Carson, D.A., and Seegmiller, J.E.: J. Clin. Invest. 57: 274, 1976.

5. Hovi, T., Smyth, J.F., Allison, A.C., and Williams, S.C.: Clin. Exptl. Immunol. 23: 395, 1976.

6. Raivio, K.O., and Seegmiller, J.E.: Biochim. Biophys. Acta, 299: 273, 1973.

7. Hovi, T., Allison, A.C., and Allsop, J.: FEBS Lett. 55: 291, 1975.

8. Hovi, T., Allison, A.C., Raivio, K.O., and Vaheri, A.: in Purine and Pyrimidine Metabolism, Ciba Foundation Symposium 1976 (in press).

ADENINE AND ADENOSINE METABOLISM IN LYMPHOCYTES

DEFICIENT IN ADENOSINE DEAMINASE (ADA) ACTIVITY

Kari O.Raivio, A.L.Schwartz, R.C.Stern and
S.H.Polmar
Children's Hospital, University of Helsinki,
and Departments of Pediatrics and Pharmacology,
Case Western Reserve University,Cleveland,Ohio

A number of patients with the clinical picture of
severe combined immunodeficiency have been shown to have
deficient activity of ADA in several tissues. Clinically
they cannot be reliably differentiated from patients
with normal activity, and the biochemical pathology of
the disorder remains unclear. Several explanations for
the abnormal lymphocyte function have been proposed,
most of them based on the postulated harmful effects of
adenosine. In vitro studies have indicated that adenosine at low concentrations stimulates but at higher concentrations inhibits lymphocyte transformation (1,2).
In the presence of coformycin, a potent inhibitor of
ADA, the toxic effects of adenosine are potentiated (1).
The mechanism of these phenomena has not been established.
Adenosine is capable of increasing intracellular cyclic
AMP levels in lymphocytes, and since cAMP under other
conditions is known to inhibit transformation, this is
a possible mechanism (3). Another possibility is pyrimidine starvation, known to be induced by adenosine (4).
Whatever the pathogenesis, replacement therapy with
infusions of frozen irradiated red cells and plasma
seems to bring about at least partial immunological re-
constitution and clinical improvement (5).

Adenosine metabolism in lymphocytes has not been
characterized in detail, and the role of purine reutilization in general is unknown. Enzyme studies have suggested that normal lymphocytes are incapable of purine

synthesis de novo (6) and hence apparently dependent on the salvage pathways for the maintenance of their nucleotide levels. Opposite views have also been presented (2). In erythrocytes and lymphocytes from a patient with ADA deficiency, intracellular adenine nucleotide concentrations were about twice the normal level, and elevated adenosine concentrations were shown in plasma (7).

The purpose of this work was to study the metabolism of adenine and adenosine in freshly isolated lymphocytes from a patient with SCID-ADA deficiency as well as from normal individuals. Unstimulated cells were used because the mutant cells are known to respond poorly to mitogenic stimulation. Adenine was chosen to represent purine base reutilization because preliminary work had shown that its uptake by lymphocytes, though relatively slow, was better than that of hypoxanthine or guanine. Furthermore, we hoped to be able to demonstrate accumulation of adenosine in the ADA-deficient cells by labelling predominantly the adenine nucleotide pool. Adenosine was studied to evaluate, if the actual rates of metabolism correlated with known enzyme activities in normal and ADA deficient cells.

METHODS

Lymphocytes were isolated by Ficoll-Isopaque centrifugation from peripheral blood samples of a normal adult female and from a patient with SCID-ADA deficiency. The latter sample was drawn at the start of a transfusion of frozen irradiated erythrocytes, by which method this patient has now been treated for several months (5).

The separated lymphocytes were washed twice with Krebs-Ringer phosphate buffer containing 5.5 mM glucose (KRPG) and suspended in the same buffer. The cells were then preincubated with shaking at $37^{\circ}C$ for 15min, after which ^{14}C-adenine or ^{14}C-adenosine was added and the incubation continued for the periods given. The total volume of 0.5ml contained 2×10^6 cells. The reactions were stopped, purine compounds extracted, separated and counted as previously described (8). Each experiment was performed in duplicate, and a blank value, obtained by adding the perchloric acid before the isotope, was subtracted from the radioactivity in each compound.

The rate of metabolism of substrates and formation
of products is usually expressed as cpm/10[6] cells over
a specified time. Since the pools of adenosine and especially adenine in mammalian cells have been considered
to be very low, certain values have been calculated in
terms of nmoles/10[6] cells, fully realizing that such
values represent minimal estimates of the true rates.

RESULTS AND DISCUSSION

Adenine Metabolism

The conversion of ^{14}C-adenine into other purine
compounds was studied using increasing concentrations
of the precursor and a constant incubation time of 2hr
(Fig. 1). At all concentrations, ^{14}C-adenine nucleotide
synthesis was approximately twice as high in the ADA-
deficient than in normal cells. It is evident that by
increasing the concentration of adenine above 35 uM very
little increase in nucleotide synthesis was obtained.

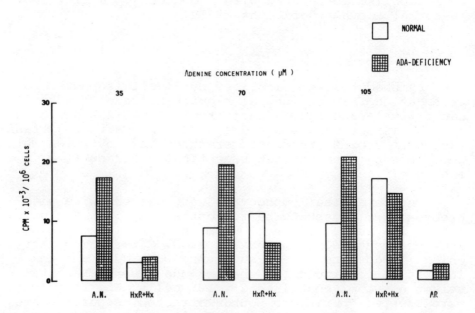

Fig.1. Conversion of ^{14}C-adenine to adenine nucleotides
(A.N.) and inosine plus hypoxanthine (HxR + Hx) by normal and ADA-deficient lymphocytes at 3 different adenine
concentrations. Time of incubation 2 hr.

There was some breakdown of adenine nucleotides, as evidenced by the radioactivity in inosine plus hypoxanthine, predominantly the latter. IMP became slightly labelled at the highest concentration of ^{14}C-adenine used, but the small difference in adenosine labelling in the mutant compared to normal cells (Fig.1) was not significant.

The cause of the increased conversion of ^{14}C-adenine into nucleotides in ADA-deficient lymphocytes is not clear. The activity of APRT in sonicated and dialyzed extract from these cells was 2.95 nmol/min/mg protein, which is the same as in normal lymphocytes (mean ± SD 3.0 ± 1.4 nmol/min/mg protein). Besides, the enzyme activity, measured under optimal conditions, is over one hundred times higher than the observed rate of adenine metabolism. Hence, altered enzyme activity does not seem to provide an explanation. Since other data suggest that PRPP availability may regulate both the rate of de novo synthesis and purine base reutilization in lymphocytes (2), a higher level of PRPP in the mutant cells would account for the more rapid adenine metabolism. Whether increased PRPP levels do exist and by what mechanism remains to be studied.

The rate of nucleotide synthesis from adenine was so low even in the ADA-deficient cells that the further metabolism of adenine nucleotides was difficult to evaluate. Specifically, no accumulation of adenosine was evident, suggesting that the small amounts of inosine and hypoxanthine formed were derived via AMP deaminase.

Adenosine Metabolism

[14]C-adenosine was rapidly metabolized by normal lymphocytes, whereas slow disappearance of the labelled precursor was demonstrated in the mutant cells (Fig. 2). The predominant fate of adenosine in the normal cells was deamination and, as expected, the radioactivity in inosine and hypoxanthine far exceeded the corresponding values in the mutant cells (Fig. 3). This correlated with the observed enzyme activities: ADA was unmeasurably low in the extract from the mutant cells, whereas the activity in normal lymphocytes was 11.2 ± 2.5 nmol/ min/mg protein (mean ± SD). The lower limit of detection of the assay was approximately 0.3 nmol/min/mg protein.

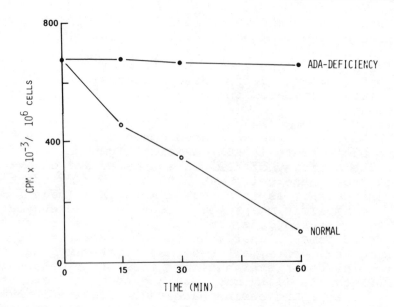

Fig. 2. Disappearance of ^{14}C-adenosine upon incubation with normal or ADA-deficient lymphocytes. Initial adenosine concentration 40 uM.

Adenine nucleotide synthesis was clearly higher in the ADA-deficient than in the normal lymphocytes (Fig. 4). Even though adenosine pools were not determined in the two cell types, a difference in this respect probably cannot explain the observation. In view of the nature of the enzyme defect and findings in red cells and plasma (7), one would expect that the adenosine pool is, if anything, larger in the mutant cells that are unable to catabolize the nucleoside normally. This would result in a greater underestimation of nucleotide synthesis in the ADA deficient then in normal lymphocytes.

The finding of increased nucleotide synthesis from adenosine is not surprising. It can be explained on the basis of persistently higher intracellular concentrations in the mutant cells, providing substrate for the adenosine kinase reaction.

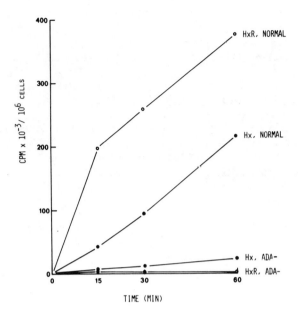

Fig.3. Catabolism of ^{14}C-adenosine (40 uM) to inosine
(HxR) and hypoxanthine (Hx) in normal and ADA-deficient
lymphocytes.

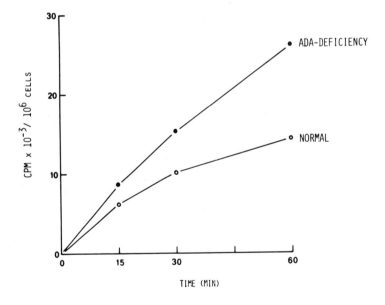

Fig.4. Adenine nucleotide synthesis from ^{14}C-adenosine
(40 uM).

REFERENCES

1. Carson, D.A. and Seegmiller, J.E.: J.Clin.Invest 57:274, 1976

2. Hovi, T., Allison, A.C., Raivio, K.O. and Vaheri A.: in Purine and Pyrimidine Metabolism, Ciba Foundation Symposium (in press).

3. Wolberg, G., Zimmerman, T.P., Hiemstra, K., Winston, M. and Chu, L.-C.: Science, 187:957, 1975.

4. Ishii, K. and Green, H.: J.Cell Sci. 13:429,1973

5. Polmar, S.H., Stern, R.C. Schwartz, A.L. and Hirschhorn, R.: Pediat.Res. 10:392, 1976 (Abstr.)

6. Scholar, E.M. and Calabresi, P.: Cancer Res. 33:94, 1973.

7. Schmalstieg, F.C., Goldman, A.S., Miles, G.C., Monahan, T.M., Nelson J.A. and Goldblum, R.M.: Pediat. Res. 10:393, 1976 (Abstr.)

8. Raivio, K.O. and Seegmiller, J.E.: Biochim. Biophys.Acta, 299:273, 1973.

IMMUNOREACTIVE ADENOSINE DEAMINASE (ADA) IN CULTURED FIBROBLASTS FROM PATIENTS WITH COMBINED IMMUNODEFICIENCY DISEASE

Dennis A. Carson, Randall Goldblum, Richard Keightley and J. E. Seegmiller

From the Department of Medicine, University of California at San Diego, La Jolla, California 92093, the Department of Clinical Research, Scripps Clinic and Research Foundation, 476 Prospect Street, La Jolla, California 92037, and the Departments of Pediatrics, University of Texas Medical Center at Galveston, Galveston, Texas 77550 and at San Antonio, San Antonio, Texas 78284

Since 1972, an association between adenosine deaminase (ADA) deficiency and severe combined immunodeficiency disease has been made in at least fourteen children (1, 2). Despite the absence of ADA in all tissues, only lymphoid growth is severely retarded (2). Recent in vitro experiments have shown that ADA inhibitors can depress the response of peripheral blood lymphocytes to mitogens, and sensitize them to the toxic effects of adenosine (3-5). Polmar and co-workers have reported that the addition of calf or human ADA to lymphocyte cultures from a patient with severe combined immunodeficiency disease and ADA deficiency partially restored their ability to proliferate when stimulated with mitogens (6). If confirmed, these results strongly suggest that ADA deficiency is causally related to immunodeficiency.

Hybridization studies have tentatively assigned the structural gene for ADA to a single locus on chromosome 20 (7). With the above knowledge, one can postulate several genetic mechanisms by which ADA deficiency could be associated with immunodeficiency, as shown in Table I.

Table I

Association of ADA Deficiency
with Combined Immunodeficiency Disease:
Possible Genetic Mechanisms

1.   Chromosomal deletion.
2.   Regulatory gene mutation, specific or non specific.
3.   Structural gene mutation
     a.   normal or increased number of enzyme
          molecules with decreased catalytic activity
          per molecule.
     b.   decreased number of enzyme molecules with
          normal or decreased catalytic activity per
          molecule.
4.   Post-translational enzyme modification.

Experiments to distinguish among these possibilities require
first a sensitive and specific enzyme assay to detect small amounts
of ADA activity, and secondly, a method for accurately measuring
catalytically inactive enzyme. In this regard, Van der Weyden
et al, using a sensitive radio-assay found 0.5% of the normal
activity of ADA in a splenic homogenate from a patient with
combined immunodeficiency disease (8). Chen and co-workers
reported 10% of the normal activity of ADA, with altered electrophoretic mobility, in cultured fibroblasts from an immunodeficient
patient (9). Hirschhorn et al found up to 25% of the normal levels
of ADA in two fibroblast lines from ADA deficient patients, again
with altered electrophoretic mobility (10).

To measure ADA levels, the latter two authors used a coupled,
spectrophotometric assay, in which nucleoside phosphorylase,
xanthine oxidase and adenosine are added to a cell extract (11).
Uric acid production subsequently is measured spectrophotometrically by an increase in absorbance at 293 nm. On theoretical
grounds, this assay is not entirely specific for ADA, but also
could measure adenosine phosphorylase, a ubiquitous mycoplas-
mal enzyme (12), since adenine, the product of the latter enzyme,
is oxidized by xanthine oxidase, accompanied by an increase in
absorbance at 293 nm (13). One therefore cannot be certain that
these authors were measuring human ADA rather than mycoplas-
mal adenosine phosphorylase. Even in the patient studied by Van
der Weyden, the possibility remains that the small amount of

activity measured was not ADA, but another enzyme not normally detected with a small but finite ability to deaminate adenosine, whose levels had increased as a result of the disease state. One cannot assume that very low levels of catalytic activity in crude cell extracts are in every case unique to a particular protein.

To extend the results of previous workers, we therefore chose to study not only the catalytic activity, but also the antigenic determinants associated with the ADA found in cultured fibroblasts from immunodeficient patients. To measure ADA activity in crude extracts prepared from these cell lines, we used a radio-assay which directly monitors the conversion of $8-^{14}C$ adenosine to inosine by thin layer chromatography of the reaction mixture on cellulose plates (4). The assay is performed in a phosphate free buffer to minimize any adenosine phosphorylase or nucleoside phosphorylase activity.

Table II shows the specific activity of ADA in erythrocyte, fibroblast and lymphoblast lysates obtained from normal subjects and patients heterozygous and homozygous for ADA deficiency. As can be seen, the fibroblast lines established from patients homozygous for ADA deficiency had catalytically detectable ADA, but less than 1% of the normal levels.

Table II

Specific Activity of ADA in Fibroblast,
Lymphoblast, and Red Cell Lysates

	Erythrocytes[*]	Fibroblasts	Lymphoblasts[‡]
Normals	0.714 ± 0.41	12.62 ± 3.50	23.8 ± 5.89
	(n = 8)	(n = 8)	(n = 3)
Heterozygotes	0.479 ± 0.17	4.79 ± 2.86	10.6
	(n = 3)	(n = 7)	(n = 1)
Homozygotes	0.019	0.048 ± 0.036	Not available
	(n = 1)	(n = 3)	

[*]The specific activities are expressed as the mean enzyme units per gram protein ± S.D. for the indicated number of patients. One enzyme unit equals one micromole inosine formed per minute at adenosine concentration of 80 μM.
[‡]Activities in long term human lymphoblast lines established from peripheral blood.

We then proceeded to develop a quantitative radio-immuno-
assay for the measurement of immunoreactive ADA in these fibroblast lines. Rabbits were immunized with a 45,000 fold purified
preparation of human erythrocyte ADA, prepared as described by
Osborne (14) and Agrawal (15), combined with chromatography on
an adenosine-agarose column. The resulting antisera could completely precipitate the ADA activity in normal, and heterozygous
fibroblast lysates, and the small amount of residual catalytic
activity in the three homozygous ADA deficient fibroblast extracts.
Pre-immune serum from the same animals had no effect on enzyme
activity. In addition, the anti-ADA antibodies, although not mono-
specific, did not reduce adenine phosphoribosyl transferase or
hypoxanthine-guanine phosphoribosyl transferase activities from
the same fibroblast lines.

Routinely, antibody bound and free ADA were separated by
precipitation with 33% saturated ammonium sulfate (SAS). By
measuring the residual enzyme activity in the supernatant after
the addition of normal or immune globulin, plus SAS to 33% saturation, we could determine the percent added enzyme specifically
precipitated by antibody. As shown in Figure 1, the percentage
of added enzyme precipitated by a constant dilution of rabbit antiserum was inversely proportional to the number of immunoreactive
units in the reaction mixture.

When an enzyme deficient fibroblast extract was added to a
constant amount of rabbit antibody and normal fibroblast lysate,
the percent enzyme precipitated decreased in proportion to the
number of inactive, immunoreactive enzyme units in the mutant.
Using the standard curve, we could then calculate the percent
normal cross-reactive material in the ADA deficient fibroblast
lines (Table III). Fibroblasts from immunodeficient patient L. L.
had 35% of the normal level of immunoreactive enzyme. The
fibroblasts of immunodeficient patients J. D. and M. R. had less
than 5% of the normal level of antigen, the lower limits of the
assay system under the conditions used. As mentioned above,
the less than 1% residual catalytic activity from all three patients
was easily detectable by radio-assay, and was antibody precipitable
in every case.

Figure 1

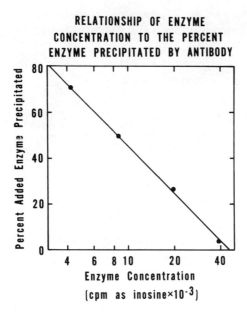

RELATIONSHIP OF ENZYME CONCENTRATION TO THE PERCENT ENZYME PRECIPITATED BY ANTIBODY

To duplicate tubes containing a 1:200 dilution of pre-immune or immune rabbit antibody globulin was added varying dilutions of a normal fibroblast extract. After an overnight incubation at 4°C, antibody bound enzyme was precipitated with 33% SAS, and the percent added enzymatic activity precipitated determined by radio-assay.

Table III

Quantitation of Cross-Reactive Material (CRM)
in Immunodeficient Fibroblast Extracts

Sample	Percent Control Enzyme Activity[*]	Percent Control CRM[‡]
L. L.	0. 32	35 ± 16[#]
M. R.	0. 28	$\lt 5$
J. D.	0. 55	$\lt 5$

[*]Determined by radio-assay in Tris-saline at an adenosine
concentration of 80 μM.
[‡]Calculated using a standard CRM curve in which the percent
added enzyme precipitated by a constant dilution of rabbit
antibody is proportional to the number of immunoreactive
enzyme units in the reaction mixture.
[#]Arithmetic mean of four separate experiments $\pm$ S. E. M.
Each experiment used a different fibroblast line as a standard.

On the basis of these results, it is highly improbable that ADA
deficient patients are homozygous for a chromosomal deletion,
since the fibroblasts contain catalytically active enzyme which
shares antigenic determinants with normal ADA. In one cell line
we could demonstrate with certainty an excess of antigenically
reactive over catalytically active enzyme. In this patient, a
mutation in the structural gene for ADA, or a post-translational
structural modification, is likely to be the cause of the enzyme
deficiency. In two other fibroblast lines, we did not find significant excess of immunoreactive over catalytically active enzyme.
The interpretation of these results is uncertain. There could be
a concomitant loss of antigenic and catalytic activity secondary to
a structural mutation. We do not favor this explanation, since
theoretically the number of antigenic determinants on a protein
in most cases should exceed the number of sites necessary for
the maintenance of catalytic activity. More likely, the reduced
levels of cross-reactive material in these immunodeficient
patients is the result of mutations which alter the rate of enzyme
synthesis or breakdown for a variety of reasons (16).

To summarize, we have demonstrated that cultured fibroblasts from immunodeficient patients contain ADA as measured both catalytically and antigenically. In one patient the amount of antigenically detectable enzyme exceeded the amount of catalytically active enzyme; in two others no significant excess of antigen could be demonstrated. These results suggest that differing genetic events, affecting both ADA structure, and the rate of enzyme synthesis and breakdown, can cause ADA deficiency and combined immunodeficiency disease.

REFERENCES

1. Giblett, E. R., Anderson, J. E., Cohen, F., Pollara, B. and Meuwissen, H. J. Lancet 2: 1067, 1972.

2. Pickering, R. J., Pollara, B. and Meuwissen, H. J. Clin. Immunol. Immunopathol. 3: 310, 1974.

3. Wolberg, G., Zimmer, T. P., Hilmstra, K., Winston, M. and Chu, L. Science 187: 957, 1975.

4. Carson, D. A., and Seegmiller, J. E. J. Clin. Invest. 57: 274, 1976.

5. Hovi, J., Smyth, F., Allison, A. C., and Williams, S. C. Clin. Exp. Immunol. 23: 355, 1975.

6. Polmar, S., Wetzler, E., Stern, R., and Hirschhorn, R. Lancet 2: 743, 1975.

7. Ruddle, F. H. Nature 242: 165, 1973.

8. Van der Weyden, M., Buckley, R. H., and Kelly, W. N. Biochem. Biophys. Res. Comm. 57: 590, 1974.

9. Chen, S. H., Scott, C. R., and Swedberg, K. R. Am. J. Hum. Genet. 27: 46, 1975.

10. Hirschhorn, R., Beratis, N., and Rosen, F. S. Proc. Nat. Acad. Sci. USA. 73: 213, 1976.

11. Hopkinson, D. A., Cook, P. J. C., and Harris, H. Ann. Hum. Genet. 32: 361, 1969.

12. Hatanaka, M. , Del Guidice, R. , and Long, C.   Proc. Nat.
     Acad. Sci. USA. 72: 1401, 1975.

13. Klenow, H. Biochem. J. 50: 404, 1952.

14. Osborne, W. R. A. , and Spencer, N. Biochem. J. 133:
     117, 1973.

15. Agrawal, R. P. , Sagar, S. M. , and Parks, R. E. , Jr.
     Biochem. Pharm. 24: 693, 1975.

16. Capecchi, M. R. , Capecchi, N. E. , Hughes, S. H. , and
     Wahl, M. Proc. Nat. Acad. Sci. USA. 71: 4732, 1974.

URINARY PURINES IN A PATIENT WITH A SEVERELY DEFECTIVE T CELL IMMUNITY AND A PURINE NUCLEOSIDE PHOSPHORYLASE DEFICIENCY

S. K. Wadman, P. K. de Bree, A. H. van Gennip, J. W. Stoop,
B. J. M. Zegers, G. E. J. Staal*, L. H. Siegenbeek van Heukelom*
University Children's Hospital Het Wilhelmina Kinderzieken-
huis, Nieuwe Gracht 137, Utrecht (The Netherlands)
* Medical Enzymology Dept., University Hospital, Utrecht

Recently a case of purine nucleoside phosphorylase (NP) deficiency
was diagnosed in our clinic, which motivated us to start new studies in the
immunological and enzymological field. We also decided to study quantitatively the patient's purine excretions, reflecting the effect of NP deficiency
on overall purine metabolism. The results of this work are given in the
present paper.

CASE REPORT

The patient, a girl, was born on 28th January 1975 after an uneventful
pregnancy and delivery as the fourth child of healthy unrelated parents.
The first and second child of these parents suffered from a selective
cellular immunodeficiency; they had, as far as could be investigated, a
normal humoral immune response. The eldest girl died from a lymphosarcoma at the age of 3 years, the second child died from a graft-versus-
host reaction (1) at $1\frac{1}{2}$. The third child is a healthy boy.

The cord blood lymphocytes of the child under study were normal in
number and function. In the first year of life the number of lymphocytes
decreased (to about $400/mm^3$), as well as the percentage of T lymphocytes
and the PHA responsiveness. At the age of 15 months the latter phenomenon
is only slightly positive: about 10% of normal. Immunoglobulin levels developed normally.

Although the patient did not suffer from severe infections until now,
it is obvious that she displays, as her two sisters, a selective cellular
immunodeficiency (2).

ENZYME DETERMINATIONS

NP activity was determined according to (3) and (4) at $37^{\circ}C$ in a final

Table 1   NUCLEOSIDE PHOSPHORYLASE ACTIVITIES IN
             HAEMOLYSATES OF THE FAMILY AND A CONTROL,
             expressed as μmoles/min/gram Hb (T = 37°C)

	inosine	xanthosine	guanosine
Father	14.7	1.4	1.8
Mother	13.0	1.5	2.0
Brother	13.9	1.4	2.2
Patient	0	0	0
Control	26.1	2.6	3.7

volume of 3 ml 0.1 M phosphate buffer pH 7 containing inosine (25 μ M) or
guanosine (125 μ M) or xanthosine (125 μ M). The decrease in absorbance
was followed at 295 nm with a Perkin Elmer spectrophotometer.

Results of NP determinations in haemolysates are reported in table 1.

The enzyme activity of the patient is zero for all three substrates,
while the enzyme activity in the father, mother and brother is in the heterozygote range compared with the activity in the control. When normal
haemolysate was mixed with haemolysate of the patient no inhibition at all
could be demonstrated. This excludes the presence of an inhibitor for NP
in the red cells of the patient. The physico-chemical properties of the NP
of the parents and of the brother will be published in detail elsewhere. In
the lymphocytes of the patient also no NP could be detected. In the frozen
stored cord blood lymphocytes no NP activity could be demonstrated. Control experiments proved that storage of lymphocytes had no influence on
NP activity.

With 2'-deoxyinosine and 2'-deoxyguanosine as substrates no activity
was detected in the patient's haemolysate, whereas a control responded
well.

Adenosine deaminase (ADA) activity in the patient's haemolysate
(determined according to Kalckar (4) at 37°C) was increased. The parents
and brother showed normal ADA values.

Incubation of erythrocytes with D-^{14}C adenosine in glucose containing
tyrode buffer of pH 7.3 at 37° suggested also a defect at the level of NP.
Adenosine (at 50 μ M and a 500 μ M level) was converted to inosine, but practically no radioactive hypoxanthine was formed.

Normal activities were found in the patient's haemolysates of:
hypoxanthine phosphoribosyl transferase, adenine phosphoribosyl transferase and adenosine kinase and PP-ribose-P synthetase (5).

URINARY PURINES

Urinary purines were analyzed by conventional automated cation
exchange column chromatography. We used a Technicon TSM1 autoanalyzer,

Figure 1   CATION EXCHANGE CHROMATOGRAPHY
OF URINARY PURINES IN PATIENT R.V.

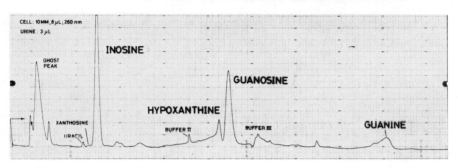

combined with a Schoeffel model SF 770 (uv) spectroflow monitor (10 mm
cell) an Infotronics CRS 309 integrator and a printer. The column
(1.69 cm, diam. 5 mm) was packed with Technicon spherical resin Chromo-
beads C3 (12 μ). Elution was performed with:

	Na$_3$citrate-HCl	pH	temp.	time
Buffer I	0.2 N	3.25	50°	60 min.
Buffer II	0.2 N	4.00	50°	45 min.
Buffer III	0.2 N	6.50	50°	104 min.

The urine (3-10 μl) was applied directly on the cartridge, equilibrated
with buffer I. Pump rate was 27 ml/hr. A typical chromatogram is shown
in figure 1.

Serum was deproteinized with 40 mg sulfosalicylic acid per ml;
0.1 ml of the supernatant was applied on the cartridge.

From table 2 we can see that the patient excreted excessive amounts
of inosine and guanosine, the guanosine level being about half of that of
inosine. After acid hydrolysis hypoxanthine and guanine were liberated.

Hypoxanthine and guanine were also found. These compounds are probably artefacts (for the greater part at least), formed by decomposition of
deoxynucleosides on the cation exchanger, as was also seen by others (6).
This decomposition caused also the asymmetry of their chromatographic
peaks. Moreover we detected deoxyribose after cation exchange treatment
of the urine (by 2-dimensional thin layer chromatography).

Xanthosine was low or absent (maximal 0.17 mmol/g creatinine)
and no adenosine was detected. Xanthine (coeluting with hippuric acid)
seemed not to be increased.

Table 2   URINARY PURINES IN R. V. AS MMOL/G CREATININE
creatinine as mg/24 hr

	ino-sine	guano-sine	hypoxan-thine	gua-nine	uric acid	creat-inine
Free diet						
5-2-'76	18	7.4	1.0	1.3	0.09	–
20/21-2	15	8.5	0.8	1.3	?	44
28/29-4	15	8.8	2.1	2.5	0.19	102
Purine restricted						
4/5-5	14	5.7	1.5	1.7	0.43	45
9/10-5	13	5.7	2.7	2.7	0.19	115
18/19-5	18	8.8	2.9	2.6	0.07	112
21/22-5	17	7.8	5.8	2.6	0.13	71
25/26-5	19	8.5	2.2	2.0	0.04	75
mean	16.1	7.7	2.4	2.1	0.16	81

Pyrimidines were present. For pseudo uridine a highest value of 0.82, for uracil of 0.49 and for uridine of 0.12 mmol/g creatinine was found (possibly exaggerated due to non-specific uv-detection).

Concentrations in a serum sample were: inosine 13.9, guanosine 5.6, hypoxanthine 5.2, guanine 2.7 μ mol/l. Hypoxanthine and guanine probably arise from decomposition of (deoxy)nucleosides in the sulfosalicylic acid containing solution.

Uric acid (determined enzymatically with uricase) was extremely low, except one value, which was considered to be doubtful and therefore neglected.

Also serum uric acid (uricase) was very low: mean 0.03 mmol/l, range 0.01 - 0.06   N = 10   normal: 0.12 - 0.33 mmol/l.

Dietary restriction of purines had little or no influence on excretory levels.

From table 3 it can be seen that the parents and brother, all heterozygotes according to the enzyme determinations, excrete inosine and a trace of guanosine. Also a peak of hypoxanthine was present in the chromatogram. It remains to be investigated whether these excretions reflect the heterozygous state as normal children show a small peak in the position of inosine.

COMMENTS

In the present stage of our investigations it is difficult to give definite interpretations of abnormalities of overall purine metabolism in the patient.

Table 3   URINARY PURINES IN FAMILY MEMBERS
           AS MMOL/G CREATININE

	ino-sine	guano-sine	hypoxan-thine	gua-nine	uric acid
Father creat. 1995 mg/l	0.13	trace	0.04	n.d.	1.74
Mother creat. 1435 mg/l	0.20	trace	0.03	n.d.	4.35
Brother (heterozygote) creat. 1010 mg/l	0.07	trace	0.09	n.d.	3.83

However, a few preliminary conclusions can be made.

There is an excessive overflow of the nucleosides inosine and gua-nosine, as can be expected in NP deficiency, but not of adenosine and xanthosine. Adenosine may largely be converted to inosine (compatible with increased ADA activity). For the present hypoxanthine and guanine are considered to be artefacts.

On the other hand uric acid excretion is very low, indicating that little free hypoxanthine or xanthine is available for oxidation.

The mean daily loss of nucleosides (hypoxanthine and guanine included) amounted 3.5 mmoles (∽81 mg creatinine). In a normal child of 10 kg the loss of purines as uric acid is some 1.5 mmoles. Therefore we conclude that the de novo synthesis of purines is increased in our patient. An increased PP-ribose-P (5) and an increased glutathione reductase activity, found in the patient's haemolysate, is in agreement with this hypothesis. Whether there are deficiencies of the purine nucleotides still has to be determined.

Also unexplained is the fact that dietary restriction of purines (from 2 to 1 mmol/day approximately), does not influence excretions. We wonder whether dietary purines are utilized for nucleotide synthesis, rather than oxidized to uric acid.

Finally we suppose that serum uric acid might be an easy parameter for a first screening of patients with symptoms of T cell dysfunction.

LITERATURE

1. Stoop, J. W., V. P. Eysvoogel, B. J. M. Zegers, B. Blok-Schut, D. W. van Bekkum and R. E. Ballieux, Clin. Immunol. Immunopath. 1976 (in press)
2. Stoop, J. W., B. J. M. Zegers, G. E. J. Staal, L. H. Siegenbeek van Heukelom, S. K. Wadman and R. E. Ballieux, 1976 (to be published)

3. Edwards, Y.H., D.A.Hopkinson and H.Harris, Ann.hum.Genit.,
   1971, 34, 395
4. Kalckar, H.M.,   J.Biol.Chem., 1947, 167, 429
5. De Bruyn, C.H.M.M.,   Nijmegen, The Netherlands, personal
   communication
6. Bonnelycke, B.E., K.Dus and S.L.Miller,   Analytical Biochem.
   1969, 27, 262

A SECOND CASE OF INOSINE PHOSPHORYLASE DEFICIENCY WITH SEVERE T-CELL ABNORMALITIES

Hamet M., Griscelli C., Cartier P., Ballay J.,

de Bruyn C. and Hösli, P   CHU   Necker

156. rue de Vaugirard   75015 - Paris

Recently, Giblett et al. described a patient with severe T-cell immunodeficiency, associated with inosine phosphorylase deficiency (1). We found a similar case which detailed clinical stories will be published later.

Case report : The patient, a boy, was born in April 1974, and was the third of three children. The two others are girls. No data of severe infections or early deaths were found for children of the families of the mother or the father. The boy was immunized with BCG vaccine at two months without abnormal reaction. No complications occured after administration of killed vaccines (tetanus, diphteria) and oral live attenuated (Sabin) poliomyelitis vaccine. During the first year of life, the only infections were pharyngitis. Vaccinia immunization, performed in october 1975, was followed, after the 15 days, by spreading and generalized necrotic lesions. The child had a lymphopenia (white blood cells count 12 300 per mm^3 with 4 % lymphocytes). Serum immunoglobulin levels were normal :

IgA 80 mg/100 ml, IgG 1100 mg/100 ml, IgM 150 mg/100 ml.
Investigations of blood group allohaemaglutinins revealed a severe T-cell deficiency. This status in a child apparently normal
in the first months of life was compatible with an enzymatic
defect. The patient was transfused twice with 250 ml of irradiated (5 000 rad) fresh blood. A fetal thymus was transplanted,
associated with hyperimmune antivaccinia serum and antibiotics.
Lymphocytes from a donor recently immunized with vaccinia
were transfused but finally, the condition was fatal in Decem-
ber 1975 at 20 months of age.

Adenosine deaminase and inosine phosphorylase were
tested by a radioenzymatic assay previously described (2). Erythrocyte adenosine deaminase activity is expressed in nmoles
inosine formed, by 1 ml of RBC per minute. Inosine phosphorylase activity is determined in both directions : from inosine and
hypoxanthine and gave the same results. In direction of inosine
synthesis, IP is expressed in umoles inosine formed by 1 ml of
RBC per minute.

Family studies are represented below.

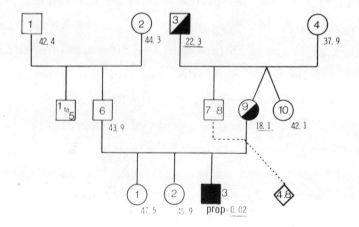

	IP (Hx $\longrightarrow$ Ino)	ADA
Normal values ($\pm 1\sigma$)	40.5 ± 3.1	494 ± 61
I_1	42.4	411
I_2	44.3	406
I_3	22.3	543
I_4	37.9	526
II_1	44.2	432
II_2	43.4	439
II_3	41.1	610
II_4	42.5	403
II_5	47.6	375
II_6 (father)	43.9	384
II_7	40.8	446
II_8	44.7	535
II_9 (mother)	18.1	366
II_{10}	42.1	436
III_1	47.5	377
III_2	45.9	629
III_3 (propositus)	0.026	539
III_4	38.4	606
III_5	43.0	529
III_6	39.9	475
III_7	40.2	467
III_8	49.9	626

Inosine phosphorylase and adenosine deaminase activities
in red-cell hemolysates

Table II

In the propositus, the RBC inosine phosphorylase activity is below 0.1 per cent of the normal. The same degree
of deficiency was detected in lymphocytes. Partial deficiency
was measured in the mother (45 % of normal) and maternal
grandfather (55 %). All other family members including the father were in the normal range (table II) : ADA activity of RBC
and lymphocytes were normal. The values of the Km for RBC
are for both the mother and the father in the normal range.

Inosine phosphorylase activity of fibroblasts were
studied by an ultramicrotechnique, described by Hösli and

de Bruyn (3), wich permits one to mesure enzyme activities at
the single cell level : a decrease of IP activity was detected
in the mother (48 % of normal), and an important residual activity in the propositus (32 %). However, this result is in con-
tradiction with IP determination by the classical method which
gives 0.33 nmoles/mg Proteine/min. (normal value =
18.1 $\overset{+}{-}$ 6.9), or 1.8 % of normal.

The patient described by Giblett, Amman and al.
was apparently homozygous with parents both heterozygous for
a probably autosomal recessive trait. The pedigree described here
suggest an autosomal recessive inheritance in the mother's family, the father being apparently normal. Blood red cell groups
of the patient and the father and their HLA antigens suggest
that the probability of exclusion of the father is very low.

1   Giblett, E., Amman, A.J., Wara, D., Sandman, R. and
     Diamond, L.K. 1975. Nucleoside phosphorylase deficiency
     in a child with severely defective T-cell immunity and nor-
     mal B-cell immunity. Lancet 1:1010.

2   Cartier, P. and Hamet, M. Clin. Chim. Acta 1976,
     à paraître.

3   De Bruyn, C., Hösli, P. and Oei, T. in "Aspects of pu-
     rine metabolism in man", de Bruyn, C., 1974 (Stichting
     Studentenpers Nijmegen, edit.).

REGULATION OF DE NOVO PURINE SYNTHESIS IN HUMAN AND RAT TISSUE:
ROLE OF OXIDATIVE PENTOSE PHOSPHATE PATHWAY ACTIVITY AND OF RIBOSE-
5-PHOSPHATE AND PHOSPHORIBOSYLPYROPHOSPHATE AVAILABILITY

O. Sperling, P. Boer, B. Lipstein, B. Kupfer, S. Brosh,
E. Zoref, P. Bashkin and A. de Vries

Tel-Aviv University Medical School, Rogoff-Wellcome
Medical Research Institute, Beilinson Medical Center,
Petah Tikva, and Department of Chemical Pathology, Tel-
Hashomer, Israel

5-Phosphoribosyl-1-pyrophosphate (PRPP) is a regulating substrate for the first committed and rate-limiting step in the pathway of purine nucleotide synthesis de novo. Studies in vivo and in
vitro demonstrated that in human and rat tissues lowering of PRPP
availability results in deceleration of purine production (1-4).
On the other hand, in subjects with deficiency of hypoxanthine-guanine phosphoribosyltransferase (HGPRT) in whose tissue PRPP accumulates due to decreased consumption, and in subjects with mutant
superactive PRPP synthetase who overproduce PRPP, there is an excessive production of purines (5-9).

PRPP is synthetized from ribose-5-phosphate (R-5-P), generated
by the pentose phosphate pathway (PPP), and from ATP, in a reaction
catalyzed by PRPP synthetase. This enzyme is allosterically activated by inorganic phosphorus (Pi) and inhibited by various nucleotides (10-12). At physiological tissue Pi concentration the enzyme
is not maximally activated and is susceptible to regulatory feedback
inhibition, therefore exhibiting intracellularly only a small fraction of its capacity. It was the aim of the present study to clarify whether, at the physiological conditions prevailing in various
tissues, the rate of activity of the oxidative PPP and the related
R-5-P availability are limiting for PRPP generation, and to re-
assess the regulatory role of PRPP availability in purine synthesis.

The tissues studied were cultured human fibroblasts and lymphoblasts, human peripheral leukocytes (13) and rat liver slices (14).
The rate of activity of the oxidative PPP was increased by the addition of methylene blue (MB) or of ascorbic acid (AA), and the

481

activity of PRPP synthetase was regulated by varying the Pi concentration in the incubation medium. PRPP availability was estimated
by the rate of [^{14}C]adenine incorporation into cellular nucleotides (8), and the rate of de novo purine synthesis by the rate of
[^{14}C]formate incorporation into purines (8).

Increasing the Pi concentration in the incubation medium resulted in increased availability of PRPP and in acceleration of
purine synthesis de novo in all tissues studied. The effect of increasing Pi concentration, from 1.4 mM (physiological extracellular) to 50 mM, on PRPP availability is demonstrated in Table 1.
The effect of increasing Pi concentration on the rate of purine
synthesis de novo is demonstrated in Table 2. The accelerating
effect of increasing Pi concentration on purine synthesis de novo
appears to operate through augmentation of PRPP availability, and
not by a direct action on glutamine PRPP amidotransferase, since
Pi has been reported to inhibit this enzyme (15). Indeed, it was
found in the present study with rat liver slices and cultured human lymphoblasts, that above a certain Pi concentration in the incubation medium (20 mM for the former, 30 mM for the latter), the
inhibitory effect of Pi on the purine synthesis pathway dominates
the acceleration brought about by the increase in PRPP availability.

The marked acceleration effect of increasing Pi concentration
on PRPP generation suggests that, at physiological intracellular Pi
concentration, the cellular availability of R-5-P is not limiting
for PRPP generation. Since, as found in our laboratory, increasing
Pi concentration does not affect the rate of activity of the oxidative PPP, the increased PRPP availability in the presence of high
Pi concentration reflects activation of PRPP synthetase only and
not increased R-5-P generation.

In the present investigation, the oxidative PPP accelerators
methylene blue (MB) and ascorbic acid (AA) were utilized for the
study of the effects of increasing R-5-P generation on PRPP availability and on the rate of purine synthesis de novo. MB, at 0.1 mM,
accelerated the rate of activity of the oxidative PPP in all tissues
studied, the acceleration ranging from 2.7 to 20 fold (Table 3). AA
affected the PPP less markedly, at 1.0 mM AA the acceleration of
the PPP being significant only in the cultured fibroblasts and lymphoblasts. In cultured fibroblasts, in which MB caused the greatest
acceleration of the oxidative PPP, but not in the other studied
tissues, the increased generation of R-5-P brought about by the MB-
induced acceleration of the oxidative PPP, was also manifest in a
significant increase in ribose-5-phopshate content (Table 4). That
the MB or AA-induced acceleration of the oxidative PPP increases
R-5-P availability also in the other studied tissues is evident
from the effects of MB and AA on PRPP availability at 50 mM Pi

TABLE 1

EFFECT OF Pi CONCENTRATION, METHYLENE BLUE (MB) AND ASCORBIC ACID (AA) ON PRPP AVAILABILITY

Tissue	Rate of [14C]adenine incorporation into cellular nucleotides[a]					
	at 1.4 mM [Pi]			at 50 mM [Pi]		
	Control	+MB (0.1 mM)	+AA (1.0 mM)	Control	+MB (0.1 mM)	+AA (1.0 mM)
Human						
peripheral leukocytes	1,000	1,000		5,250	6,750	
cultured fibroblasts	99.4	23	109.2	315	142.3	346
cultured lymphoblasts	46	29	24	92	151	180
Rat						
liver slices	400	400	420	1460	2200	1500

[a] for leukocytes in pmoles/ml packed cells/min (13); for cultured fibroblasts and lymphoblasts in pmoles/mg protein/min (8); for liver slices in pmoles/gm tissue/min (14).

TABLE 2

EFFECT OF Pi CONCENTRATION, METHYLENE BLUE (MB) AND ASCORBIC ACID (AA) ON THE RATE OF PURINE SYNTHESIS DE NOVO[a]

Tissue	Rate of $[^{14}C]$formate incorporation into purines[b]								
	at 1.4 mM [Pi]			at 30 mM [Pi]			at 50 mM [Pi]		
	Control	+MB	+AA	Control	+MB	+AA	Control	+MB	+AA
Human									
peripheral leukocytes	6.2	6.2	6.0	30	55		42	85	160
cultured fibroblasts	5.0	3.0	5.1	20	15	20	32	25	33
lymphoblasts	30.0	18.7	27.9	63	82	95	46	186	181
Rat									
liver slices	13.8	7.0	13.0	18	6	18	13	6	14

a MB– at 0.1 mM, AA– at 1.0 mM.

b In peripheral leukocytes into total purines (13), in dpm x $10^{-3}/2$x10^7 cells/h; in cultured fibroblasts and lymphoblasts into cellular purines (8), in dpm x 10^{-3}/mg protein/h; in rat liver slices into total purines (14), in dpm x 10^{-3}/50 mg tissue/h.

TABLE 3

EFFECT OF METHYLENE BLUE (MB) AND OF ASCORBIC ACID (AA) ON THE
RATE OF ACTIVITY OF THE OXIDATIVE PENTOSE PHOSPHATE PATHWAT (PPP)

Tissue	Increase in rate of oxidative PPP [a]	
	MB (0.1 mM)	AA (1 mM)
Human		
cultured fibroblasts	20 - fold	2.0 - fold
cultured lymphoblasts	10 - fold	1.7 - fold
peripheral leukocytes	5 - fold	1.1 - fold
Rat		
liver slices	2.7-fold	No effect

[a] Rate of activity of the oxidative PPP was estimated by the
release of $^{14}CO_2$ from $[1-^{14}C]$glucose (10,13,14)

TABLE 4

EFFECT OF METHYLENE BLUE (MB) ON CELLULAR R-5-P CONTENT[a]

Tissue	R-5-P content (nmoles/ml packed cells)	
	Control	+MB
		(0.1 mM)
cultured fibroblasts (5)	78.1	252.2
cultured lymphoblasts (4)	29.2	36.9
peripheral leukocytes (5)	112.0	135.0

[a] Cells were incubated at 1.4 mM Pi for 30 min with and without MB.
Values represent means. Numbers in parenthesis indicate number of
experiments.

concentration. At this high Pi concentration, at which PRPP synthetase is activated, MB increased PRPP availability in all tissues
studied, except for in cultured fibroblasts in which an inhibitory
effect of MB was observed. AA also caused increased PRPP availability at high Pi concentration, but only in cultured lymphoblasts to
a significant degree.

At the physiological 1.4 mM Pi concentration, MB or AA did not
accelerate PRPP generation nor the rate of purine synthesis de
novo in all tissues studied. The finding that increasing R-5-P a-
vailability at physiological conditions does not induce increased
generation of PRPP provides additional support for the suggestion
made above that R-5-P availability at physiological Pi concentration is saturating for PRPP generation.

The increased PRPP availability caused by MB and AA at high
Pi concentration was manifest in increased rate of synthesis de
novo in peripheral leukocytes and in cultured lymphoblasts. However, MB inhibited purine synthesis de novo in cultured fibroblasts
and in rat liver slices. The MB-induced inhibition of both salvage
and de novo purine nucleotide synthesis in cultured fibroblasts was
found to be associated, at low as well as high Pi concentration,
with increased cellular content of PRPP. The increase in PRPP content caused by MB at low Pi concentration is not compatible with
the above suggestion that at physiological conditions R-5-P is saturating for PRPP generation. On the other hand, the increase in
PRPP content in MB-treated fibroblasts could reflect, rather than
increased PRPP generation, an accumulation of this substrate due
to the inhibitory effect of MB on its utilization in both the salvage and de novo pathways of purine nucleotide synthesis.

In summary: the results obtained in the present study on various human and rat tissues confirm the previously known regulatory role of PRPP in purine synthesis de novo. On the other hand, the
results furnished evidence suggesting that tissue R-5-P availability at physiological conditions is saturating for PRPP generation,
and thus cannot be considered to regulate the rate of purine synthesis de novo.

ACKNOWLEDGEMENT

This study was supported in part by a research grant (No. 78)
from the United States - Israel Binational Science Foundation (BSF),
Jerusalem, Israel.

REFERENCES

1. Kelley, W.N., Fox, I.H. and Wyngaarden, J.B.   Biochim. Biophys.
   Acta, 215:512-516, 1970.
2. Fox, I.H., Wyngaarden, J.B. and Kelley, W.N.   New Engl. J. Med.,

283:1177–1182, 1970.

3. Boyle, J.A., Raivio, K.O., Becker, M.A. and Seegmiller, J.E. Biochim. Biophys. Acta, 269:179–183, 1972.

4. Rajalakshmi, S. and Handschumacher, R.E. Biochim. Biophys. Acta, 155L317–325, 1968.

5. Kelley, W.N., Greene, M.L., Rosenbloom, F.M., Henderson, J.F. and Seegmiller, J.E. Ann. Intern. Med., 70:155–206, 1969.

6. Kelley, W.N., Rosenbloom, F.M., Henderson, J.F. and Seegmiller J.E. Proc. Nat. Acad. Sci. U.S.A. 5–:1735–1739, 1967.

7. Sperling, O., Eilam, G., Persky-Brosh, S. and de Vries, A. Biochem. Med., 6:310–316, 1972.

8. Zoref, E., de Vries, A. and Sperling, O. J. Clin. Invest. 56:1093–1099, 1972.

9. Becker, M.A., Meyer, L.J., Wood, A.W. and Seegmiller, J.E. Arthr. and Rheum., 15:430, 1972.

10. Hershko, A., Razin, A. and Mager, J. Biochim. Biophys. Acta, 184:64–76, 1969.

11. Fox, I.H. and Kelley, W.N. J. Biol. Chem., 246:5739–5748, 1971.

12. Fox, I.H. and Kelley, W.N. J. Biol. Chem., 247:2126–2131, 1972.

13. Brosh, S., Boer, P., Kupfer, B., de Vries, A. and Sperling, O. J. Clin. Invest. In press, (August 1976).

14. Boer, P., Lipstein, B., de Vries, A. and Sperling, O. Biochim. Biophys. Acta, 432:10–17, 1976.

15. Holmes, E.W., McDonald, J.A., McCord, J.M., Wyngaarden, J.B. and Kelley, W.N. J. Biol. Chem., 248:144–150, 1973.

HEPATIC GLUCONEOGENESIS AND URATE FORMATION FROM VARIOUS NUCLEOSIDES

R. Haeckel

Institute for Clinical Chemistry, Medizinische Hochschule Hannover, D-3 Hannover 61

Nucleic acids and nucleotides are converted in the intestine to their corresponding nucleosides and partly to their purines. Both metabolites penetrate through the intestinal walls (1).

Purine bases and nucleosides from endogenous and exogenous sources which are transported to the liver are easily taken up by this organ and further metabolized to uric acid or utilized to form the corresponding nucleotides. Only uncomplete data on the rates of these metabolic pathways are published for the various purine bodies. When we started to measure these rates, we noticed that nucleosides caused a marked release of glucose from the perfused liver.

The perfusion procedure was that of Miller et al.(2) and Schimassek (3) as described elsewhere (4). The perfusate (100 ml) contained 30 g/l bovine albumin, 13 mg/l sodium ampicilline and bovine erythrocytes washed three times and taken up in Krebs-Ringer bicarbonate solution (haemoglobin concentration; 50 g/l). The perfusion experiment lasted 90 minutes. Substrates were added to the medium after 45 minutes. All

metabolites were measured enzymatically in a perchloric acid
extract from liver samples taken at the end of the perfusion
experiment with the freeze-stop technique as described recently (4). The concentration of uric acid, oxipurines and
inosine was determined by a NADP coupled reaction sequence
(5,6)., of allantoin according to Young et al. (7).

In the absence of any purine added to the perfusion medium
the <u>production of uric acid</u> is very low either in livers
from fed or fasted animals (table 1). With inosine the hepatic uric acid release rises to 98 in the fasted and to 112
nmol. min^{-1} . g^{-1} in the fed state. With IMP and adenosine
slightly lower rates are obtained than with inosine, GMP
yields the highest rate.

The maximal rate is always obtained during the first 15
minutes after the addition of the purines, whereas the <u>allan-
toin production</u> reaches its peak between the following 15
minutes (Table 2). During the period of observation allantoin
was formed from inosine 2.8 times as much as uric acid by
livers from fed animals.

When adenosine is converted to allantoin inosine, oxipurines,
uric acid and allantoin accumulate in the perfusion medium
(Fig. 1). Since the inosine concentration rises over 10 times
above that of hypoxanthine, the nucleoside phosphorylase reaction may be the rate limiting step.

In livers from rats fasted 48 hours the <u>glucose formation</u>
(Fig. 2) increased approximately fourfold over the endogenous
rate (control) in the presence of 10 mmol/1 GMP. With adenosine and inosine slightly lower rates of gluconeogenesis were
observed. The mono-phosphates of the corresponding nucleosides led to similar rates of glucose formation.

Table 1: Hepatic production of uric acid (nmoles . min⁻¹ . g wet weight⁻¹). Purines were added in a single dose (10 mmol/1) 45 minutes after the perfusion has been started.

purines added	without		inosine		IMP	adeno-sine	GMP
nutritional state	fed (n=5)	fasted[1] (n=5)	fed (n=5)	fasted (n=5)	fasted (n=2)	fasted	fasted
30th – 45th	+ 3(12)[3]	– 1(10)	– 2(3)	– 5(9)	– 4	– 2	+ 2
45th – 60th	+ 2(5)	– 1(6)	+ 112(42)	+ 98(64)	+ 72	+ 59	+ 246
60th – 75th	– 3(5)	– 1(1)	+ 97(29)	+ 88(20)	+ 49	+ 42	+ 185
75th – 90th	+ 3(5)	+ 6(7)	+ 76(26)	+ 63(12)	+ 40	+ 49	+ 135
Total amount produced between the 45th and 90th minute	+ 30	+ 45	+ 4275	+ 3735	+ 2415	+ 2254	+ 8485

1) rats fasted 48 hours   2) minute, perfusiontime   3) mean values with standard deviation and number of contributing values (n) in parenthesis.

Table 2: Hepatic production of allantoin (nmoles · min^{-1} · g^{-1} wet weight). Purines were added in a single dose (10 mmol/l) 45 minutes after the perfusion has been started.

perfusion time, minute	without purines added		with inosine	
	fed (n=5)	fasted[1] (n=4)	fed (n=4)	fasted (n=4)
30th - 45th	+ 1(13)[2]	+ 18(17)	+ 7(4)	9(10)
45th - 60th	+ 19(24)	+ 6(23)	+198(16)	+145(139)
60th - 75th	- 5(16)	+ 21(22)	+333(239)	+164(54)
75th - 90th	+ 10(4)	+ 14(11)	+302(177)	+138(66)
Total amount produced per g between the 45th and 90th minute	360	615	12495	6705

[1] rats fasted 48 hours   [2] mean values with standard deviation and number of contributing values (n) in parenthesis.

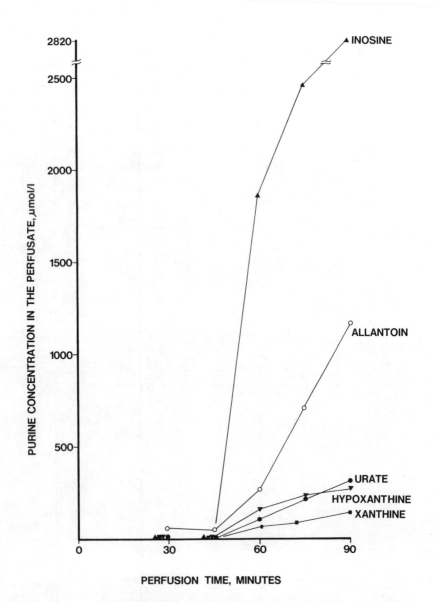

Fig. 1: The inosine, urate, oxipurine and allantoin production by the perfused liver from fed rats after adenosine
is added in a single dosage (5 mmol/l) 45 minutes after the
perfusion experiment is started.

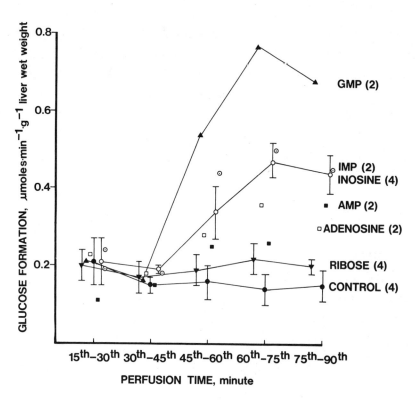

Fig. 2: The glucose formation by the perfused liver from rats
fasted 48 hours after nucleosides and ribose are added in a
single dosage (10 mmol/l) 45 minutes after the perfusion experiment is started. Vertical bars mean standard deviation, figures in parenthesis the number of experiments.

When the liver was taken from fed animals inosine led to
even higher rates of glucose production (Fig. 3). With adenosine similar rates were observed. These values exceed all
other reported so far for any gluconeogenic substrate. Purine
bases did not affect the hepatic glucose release.

The hepatic metabolite concentrations of the upper Embden
Meyerhof pathway were increased in the presence of nucleo-
sides(as shown for inosine in table 3). The concentration of
adenine - nucleotides (Table 4) and IMP were also elevated
when inosine was added.

Whereas the metabolic pathway from nucleosides to uric acid
in the rat liver is well established, the glucose formation
has to be discussed. Hunter and Jefferson (8) first reported
a glucose release of the perfused liver after addition of AMP,
but not with adenosine. These authors attributed this effect
to an intracellular rise of the AMP concentration which could
stimulate glycogenolysis. Although under the present experimental conditions, the AMP concentration is only slightly and
that of IMP 3 fold increased when inosine is added, the concentration of glycogen was not decreased (Table 4). Therefore,
glycogenolysis can not explain gluconeogenesis in the presence of nucleosides.

Livers from fed rats release about 100 μmol . g^{-1} glucose
during 45 minutes after inosine or adenosine is added to the
perfusate (endogenous amount in the absence of nucleosides :
11 μmol . g^{-1}). This amount exceeds the average glycogen
storage (Table 4).

Another explanation could be a glucose formation from the
ribose moiety of the nucleosides which are split either to
ribose or ribose-1-phosphate and the corresponding purine
base. Since the addition of ribose to the perfusion medium

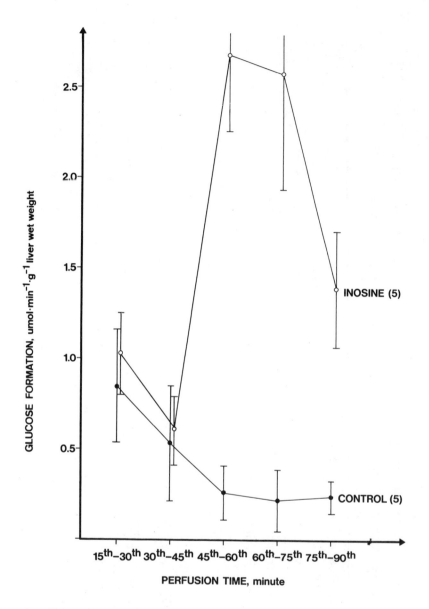

Fig. 3: The glucose formation by the perfused liver from fed rats after inosine is added in a single dosage(10mmol/1) 45 minutes after the perfusion experiment is started. Vertical bars mean standard deviation, figures in parenthesis the number of experiments.

Table 3: The concentration of hexose-6-phosphates and triose phosphates in liver samples taken at the end of the perfusion experiments.

Nutritional state of the rats	f e d		f a s t e d 48 hours	
Nucleosides added[1]	control	inosine	control	inosine
Glucose-6-phosphate	33 (8) [2]	277 (44)	<10	41 (2)
Fructose-6-phosphate	19 (8)	56 [3]	<10	<10
FDP	43 (39)	54 (6)	11 (5)	52 (9)
GAP[4]	41 (21)	35 (15)	25 (2)	15 (13)
DAP[4]	88 (37)	124 (74)	20 (9)	96 (12)
DAP + GAP + FDP	172	213	56	163

1) added in a single dosage (10 mmol/l) 45 minutes after the perfusion has been started
2) mean value (nmol · g^{-1} liver wet weight) from 4 experiments with standard deviation in parenthesis
3) mean value of 2 experiments
4) fructose-1,6-diphosphate, glyceraldehyde-3-phosphate, dihydroxyacetone-3-phosphate.

Table 4: The concentration of glycogen and some nucleotides in liver samples taken at the end of the perfusion experiments.

Nutritional state of the rats	f e d		f a s t e d 48 hours	
Nucleoside added [1]	control	inosine	control	inosine
ATP	2327 (348) [2]	3649 (480)	2203 (212)	2512 (330)
ADP	796 (47)	1100 (20)	939 (116)	1074 (102)
AMP	346 (94)	379 (58)	274 (42)	406 (167)
ATP+ADP+AMP	3469 (331)	5128 (445)	3416 (375)	3992 (358)
ATP/ADP	2.95 (0.58)	3.32 (0.38)	2.35 (0.22)	2.34 (0.42)
IMP	322 (85)	1039 (567)	462 (269)	1629 (356)
Glycogen [3]	63 800 (38 800)	90 000 (51 200)	367 (17)	424 (69)

1) added in a single dosage (10 mmol/l) 45 minutes after the perfusion has been started
2) mean values (nmoles · g^{-1} wet weight) from 4 experiments with standard deviation in parenthesis
3) Glucosyl units

did not increase the endogenous rate of gluconeogenesis it
could be assumed that ribose-1-phosphate is primarily formed by the nucleoside phosphorylase (E.C. 2.4.2.1) reaction.
The pentose phosphate may be converted to ribose-5-phosphate,
then through the non - oxidative branch of the pentose phosphate cycle to triose phosphate and finally to glucose
(Fig. 4). This pathway was also suggested in erythrocytes,
where hexose-6-phosphates accumulate in the presence of adenosine or inosine (9,10).

If this hypothesis is true, the sum of oxipurines, uric acid
and allantoin after adding adenosine or inosine should be
twice as much as the amount of glucose released. This, however, was only be observed with livers from fasted rats.
In the fed state, more glucose is formed than could be derived from the nucleosides added. In this case, a further
source for endogenous substrates which are converted to glucose has to be postulated.

In summary, this is the first report on a strong gluconeogenic effect of nucleosides.

These metabolites may reach the liver either from endogenous
or exogenous sources. In contrast to other gluconeogenic substances inosine increased the endogenous glucose formation 10
fold in the fed and only 3 fold in the fasted state. Since
nucleosides also lead to hepatic urate formation, they may be
a link between carbohydrate and purine metabolism which both
are disturbed in primary gout.

REFERENCES

1. Wilson, D.W. and Wilson, H.C. (1962), J. Biol. Chem. 237, 1643.

2. Miller, L.L., Bly, C. G., Watson, M.L. and Bale, W.F. (1951), J. Exp. Med. 94, 431.

3. Schimassek, H. (1963), Biochem. Z. 336, 460.

4. Haeckel, R. and Haeckel, H. (1972), Diabetologia 8, 117.

5. Haeckel, R. (1976), J. Clin. Chem. Clin. Biochem. 14, 101.

6. Haeckel, R. (1976), in preparation.

7. Joung, E.G., Conway MacPherson, C., Wentworth, H.P. and Hawkins, W.W. (1944), J. Biol. Chem. 152, 245.

8. Hunter, A.R. and Jefferson, (1969) L.S., Biochem. J. 111, 537.

9. Lionetti , F.J. and Fortier, N.L. (1963) Arch-Biochem. Biophys. 103, 15.

10. Bishop, C. (1964), J. Biol. Chem. 239, 1053.

INFLUENCE OF CARBOHYDRATES ON URIC ACID SYNTHESIS IN THE ISOLATED

PERFUSED CHICKEN LIVER

S. Boecker and H. Förster

Zentrum d. Biologischen Chemie d. Universität Frankfurt

6ooo Frankfurt/M., Theodor-Stern-Kai 7, BRD

A rise of uric acid concentration in serum after rapid infusion of fructose in human beings was described at the first time in 1967 and confirmed a short time later (4,12). Sorbitol causes a weaker effect than fructose. However after the intravenous application of xylitol a powerful effect on uric acid concentration in serum is found (5). In contrast to the glucose substitutes glucose itself causes no hyperuricaemia. Long lasting infusions and oral applications of fructose, sorbitol, and xylitol increase uric acid concentration in serum too (6).
The metabolic mechanism of glucose substitutes on purine metabolism can hardly be investigated in the usual laboratory animals because all mammals exept the primates disintegrate uric acid to allantoine by uricase, which can be detected only with great expense. To avoid these difficulties the isolated perfused chicken liver was choosen as a model for the uricotelic metabolism. Uric acid is the final product in uricotelic metabolism for an essential part of nitrogen metabolism. Therefore it can be assumed that in uricotelic metabolism the influence on uric acid metabolism is easier to prove.
In pretests chicken liver proved to be suitable for the perfusion of the isolated organ. The isolated perfused rat liver was studied in comparative tests with uricotelic metabolism. The perfusion of the isolated chicken liver was performed by a modification according to Wissdorfer et al. (13), while the perfusion of the rat liver was performed by a modification to Hems et al. (1o). Aged cattle erythrocytes dispenced in a polypeptide solution with substitution of electrolytes were used as perfusion medium.

Glucose (2o mM) is scarcely taken up from the perfusion medium, neither in perfusion of chicken liver (fig.1) nor in perfusion

Metabolism of glucose, fructose or xylitol by the isolated perfused chicken liver.

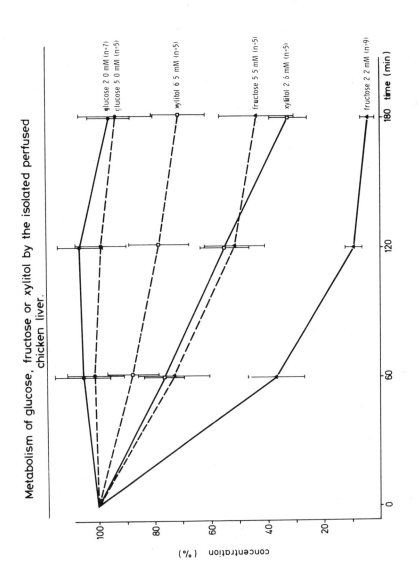

Fig. 1

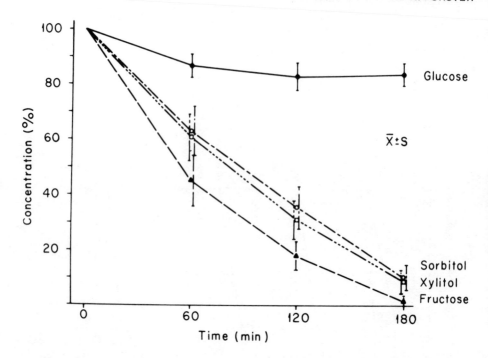

Fig. 2
Glucose (2o mM, n=8), fructose (22 mM, n=8), sorbitol
(22 mM, n=6), and xylitol (26 mM, n=8) taken up by isolated perfused rat liver.

of rat liver (fig. 2). On the contrary fructose (22 mM) is
metabolized during 3 hours lasting perfusion to nearly 1oo %. Xylitol (26 mM) is metabolized in the chicken liver by 7o %, and
in the rat liver by 9o %. When fructose (55 mM) or xylitol
(65 mM) are used in higher concentrations the degree of reduction in chicken liver is 6o % respectively 2o %. Obviously there
are quantitative differences in metabolism of xylitol and fructose
between rat and chicken. In both spezies fructose is metabolized
to the the same extent, while the metabolism of xylitol in chicken
is significantly limited.

There is an inverse ratio of catabolism of glucose substitutes and
glucose release of the perfused organ (fig. 3). Corresponding to
its small metabolic rate in chicken liver xylitol causes a retarded
gluconeogenesis in comparison to fructose.
The total production of glucose in rat liver (fig. 4) is relatively smaller than in chicken liver: 3 hours lasting perfusion with
fructose causes in rat liver about o.7 mM glucose release and in

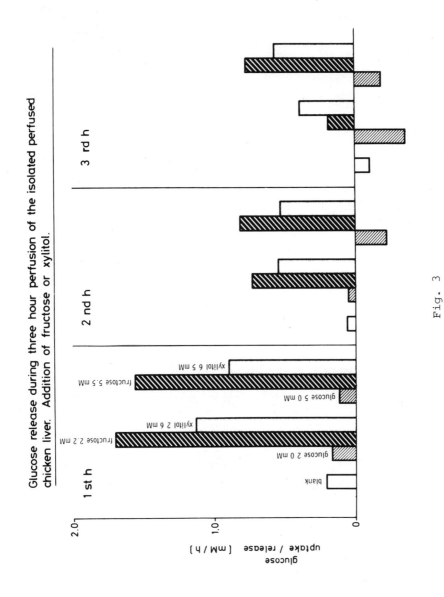

Glucose release during three hour perfusion of the isolated perfused chicken liver. Addition of fructose or xylitol.

Fig. 3

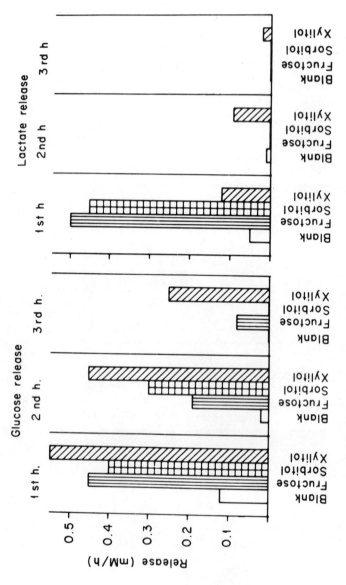

The effect of fructose, sorbitol, and xylitol on release of glucose and lactate by the isolated perfused rat liver.

Fig. 4

chicken liver 2.5 mM.

According to these results chicken liver is more qualified to transform glucose substitutes to glucose than rat liver. Together with
lactic acid a greater part of fructose and xylitol is probably used
for fatty acid synthesis in the rat liver. There is no other explanation for the differences between the uptake of glucose precoursers
and glucose release. Since in chicken liver a complete transformation of glucose substitutes to glucose occurs other metabolic pathways as fatty acid synthesis cannot play any important role. There
is a further difference between chicken and rat liver in respect
to the formation of lactic acid. Fructose and to a lower extent
xylitol cause a significant lactic acid release in rat liver. However the isolated chicken liver metabolises lactic acid during
uptake of fructose and xylitol. There is no release of lactic acid
into the perfusion medium (fig. 5).

Rapid infusion of fructose causes as well in human beings as in
rats a significant increase of uric acid concentration in serum.
Surprisingly perfusion of isolated chicken liver does not cause
this effect. In comparative tests without substrate and with glucose there is a release of uric acid o.1 mM (fig. 6) in the 1st
hour of perfusion. However perfusion with fructose causes only
uric acid release about o.o8 mM and perfusion with xylitol of about
o.o7 mM. If glucose, fructose or xylitol are used in higher concentrations the differences between glucose and glucose substitutes
are more evident during the whole perfusion. Hence fructose and
xylitol effect a significant inhibition of uric acid release in
isolated perfused chicken liver.

As toxic effects of glucose substitutes cannot be assumed, there
will be changes in the metabolic patterns. The question is therefore in which way glucose substitutes can inhibit uric acid synthesis in chicken liver.

It is known that the isolated rat liver cannot metabolise glucose
without insulin (8).Same is true for isolated chicken liver (2).
Fructose and xylitol are metabolized insulinindependent by the liver in a similar way as in human beings. In perfusion with glucose
or without substrate liver is in a marked catabolic status. The
uric acid release occuring in this case correlates to the urea
release of ureotelic metabolism in mammals (3). Consequently the
diminished uric acid release caused by fructose or xylitol is related to an improved energy situation in liver (9). Chicken liver
is able by utilisation of fructose or xylitol to save amino acid
from own proteins. The reduction of uric acid synthesis is equivalent to nitrogen-sparing effect of fructose and xylitol in
 mammals (1,11).

The great difference between metabolism of chicken and rats is
most evident in the relation between fructose and lactic acid. In
all test conditions including isolated liver fructose causes in
mammals lactic acidaemia. However isolated chicken liver releases

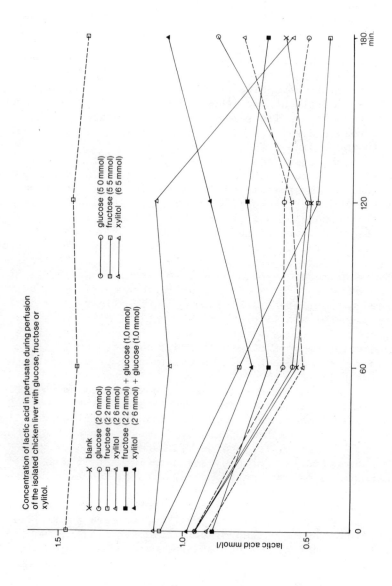

Concentration of lactic acid in perfusate during perfusion
of the isolated chicken liver with glucose, fructose or
xylitol.

×   blank
Θ   glucose   (2 0 mmol)
☐   fructose (2 2 mmol)
△   xylitol   (2 6 mmol)
■   fructose (2 2 mmol) + glucose (1.0 mmol)
▲   xylitol   (2 6 mmol) + glucose (1.0 mmol)

Θ   glucose   (5 0 mmol)
☐   fructose (5 5 mmol)
△   xylitol   (6 5 mmol)

Fig. 5

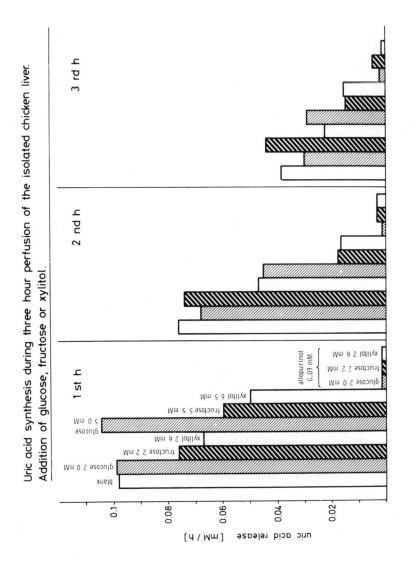

Uric acid synthesis during three hour perfusion of the isolated chicken liver. Addition of glucose, fructose or xylitol.

Fig. 6

no lactic acid during perfusion even with high dosed fructose . As shown in fig. 5 concentration of lactic acid is reduced. This mechnism is probably of particular physiological importance for aves. Lactic acid released from muscles can be metabolized to a greater extent by the liver, whereas the lactic acid release by liver seems to be minimal.

This proves differences of metabolism in chicken and in rat liver. It can be assumed that comparison of relationships between carbohydrates and other metabolic pathways is therefore not possible. The absence of comparability is characteristic of nitrogen metabolism. The disintegretion of glucoplastic amino acids serves in general for gluconeogenesis. In ureotelic organisms urea is synthesized, in uricotelic organisms uric acid. The availability of other glucose precursors leeds to a saving of proteins: in ureotelic organism the urea synthesis is diminished and in uricotelic organism the production of uric acid is diminished. The nitrogensparring effect of glucose substitutes fructose and xylitol leeds according to the corresponding conditions to different consequences. Therefore it is not possible to draw conclusions from results from influences of uric acid synthesis by carbohydrates in uricotelic organism to the situation in mammals espescially in human beings.

1. Aebi, M.: Mod. Probl. Paediat. 4, 5o3 (1959)
2. Barratt, E., P.S. Buttery, K.N. Boorman: Biochem. J. 144, 189
                                                                (1974)
3. Brown, G.W. in: Comparative biochemistry of nitrogen metabolism
   (Cambell, J.W.) London 197o, pp. 711
4. Förster, H., H. Mehnert: Lancet II, 12o5 (1967)
5. Förster, H., E. Meyer, M. Ziege: Klin. Wschr. 48, 878 (197o)
6. Förster, H., M. Ziege: Z. Ernährungswiss. 1o, 394 (1971)
7. Förster, H., D. Zagel: Dtsch. Med. Wschr. 99, 13oo (1974)
8. Förster, H. in: Sugars in nutrition (Sipple, H.L., K.W. McNutt)
   New York 1974, pp. 259
9. Heald, P.J.: Biochem. J. 86, 1o3 (1963)
1o.Hems, R., B.O. Ross, M.N. Barry, K.H. Krebs: Biochem. J. 1o1,
                                                                284 (1966)
11.Lang, K.: Klin. Wschr. 49, 233 (1971)
12.Perheentupa, J., K. Raivio: Lancet II, 528 (1967)
13.Wissdorf, H., H. Greyer, H. Lutz: Dtsch. tierärztl. Wschr. 78,
                                                                365 (1971)

This study was supported by the Deutsche Forschungsgemeinschaft.

STUDIES ON THE INFLUENCE OF ETHANOL AND OF LACTIC ACID ON URIC ACID METABOLISM

H. Hartmann and H. Förster

Zentrum der Biol. Chemie der Universität Frankfurt/M.

6000 Frankfurt, Theodor Stern Kai 7, BRD

It is known since long time that acute attacks of gout are often provoked by the consumption of large quantities of food and of alcoholic beverages (1). In 1962 Lieber et al. showed that serum uric acid levels were substantially higher during ethanol intoxication in alcoholic patients than several days later in the sober state (2). Administration of ethanol to volunteers resulted in a distinct increase in serum uric acid concentration of 1-3 mg/100ml (2). Additionally a rather small rise in blood lactic acid concentration was noted in man at periods of high blood levels of ethanol. Although lactic acid concentration increased merely by 1-1.5 mmol/l the effect on uric acid concentration was related by these authors to a decrease in renal uric acid excretion caused by the lacticacidemia (2). Especially Burch and Kurke (3) showed that renal excretion of uric acid was inhibited by the intravenous infusion of sodium lactate. During infusion of lactate serum uric acid concentration increased only by 0.4-0.5 mg/100 ml. However, in contrast to this renal mechanism discussed above Grunst et al. were able to show an increase in hepatic uric acid output during the application of ethanol in human volunteers (4).

The aim of our experiments was to study the influence of both: ethanol metabolism and lactic acid metabolism on uric acid synthesis.

In the former studies with human volunteers we only found a small increase in uric acid concentration during high blood ethanol concentrations (5). However, in our investigations blood ethanol levels remained lower than 200 mg/100 ml, whereas blood ethanol rose up to 300 mg/100 ml in the experiments of Lieber and al. (2). Uric acid concentration in our experiments rose by 0.4-1.0 mg per

100 ml during oral ethanol administration (fig.1). The renal excretion of uric acid was not much altered during the experimental
period. As was discussed earlier by us, blood lactic acid concentration seems to have only little influence on serum uric
acid concentration in man (5). During intravenous infusion of
respectively 25 g of the following substances in man ethanol as
well as lactate were without effect on serum uric acid concentration, whereas xylitol, fructose and sorbitol in this dose caused
an increase in serum uric acid concentration by 1-2 mg/100 ml (5).

We considered it very difficult to get a clearcut picture of the
metabolic events in studies with human volunteers. Therefore, we
looked for an animal model, that would be appropriate to study
the effects of ethanol and lactate on purine metabolism.
In most mammals uric acid is metabolized to allantoine. The
known analytical methods for determination of this substance
proved to be insufficient. However, uricase can be inhibited by
oxonate, a dihydroxy-monocarboxy-triazine. This compound was
introduced by Johnson et al. (6) in nutritional studies. Oxonate
is suitable for intravenous infusion also, because of its sufficient solubility. Uric acid assay by means of uricase was scarcely influenced by oxonate.
In the rat the intravenous administration of great amounts of
ethanol was without influence on blood lactic acid concentration
(7,8). Furthermore, disappearance of lactic acid was obviously
not very much retarded during simultaneous administration of
ethanol (fig.2). On the other hand intravenous ethanol did not
cause an increase in uric acid concentration in the rat, although
ethanol concentration rose up to 600 mg/100 ml. Because in studies with human volunteers red wine was possibly more effective
on uric acid synthesis than pure ethanol, the effect of combined
infusion of fructose and ethanol was investigated in the rat.
Fructose as well as ethanol in the dose used did scarcely influence uric acid concentration. The combined infusion of fructose and ethanol merely caused a neglectible rise in uric acid
concentration. On the other hand the combined infusion of ethanol
and of sodium lactate with insufficient amounts of oxonate were
accompanied by a greater increase in uric acid concentration.
The   high dosed intravenous   administration of sodium
lactate resulted in a large increase in uric acid concentration
(fig.3). This effect could not be attributed to an inhibition
of renal uric acid reabsorption. Nephrectomy led to a slight
additional increase in uric acid concentration. The total increase in uric acid concentration in relation to the blank experiment with infusion of Ringer-Solution remained unaltered
in nephrectomized animals receiving sodium lactate. The effect
of sodium lactate on uric acid synthesis showed a strong relation
to lactic acid concentration and not to the amount of lactate
infused. As is seen from fig. 4 the smaller amount of lactate

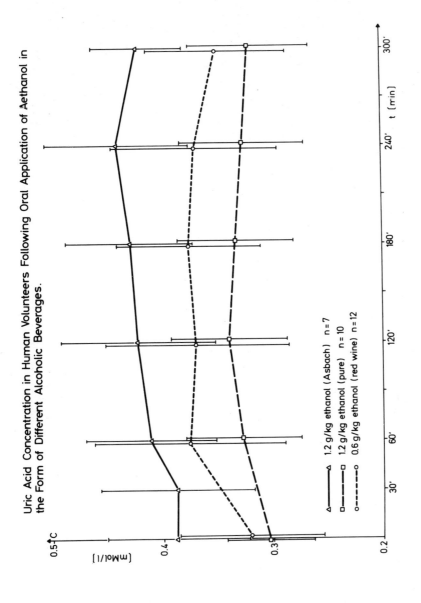

Fig. 1

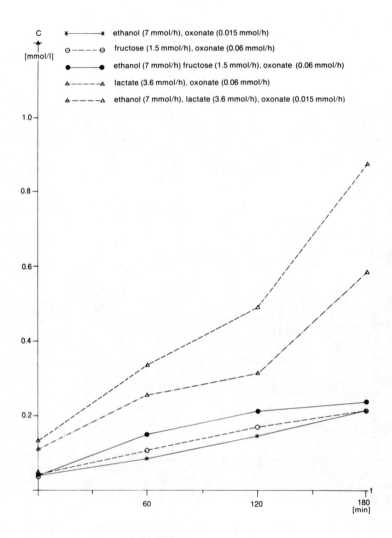

Fig. 2 Effect of continuous intravenous infusion
        of fructose, ethanol and lactate with oxo-
        nate on uric acid concentration.

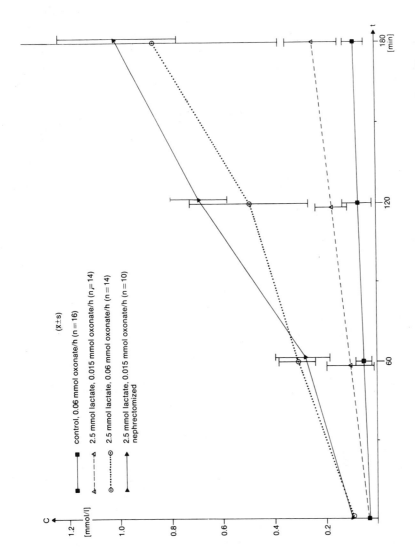

Fig. 3 Uric acid concentration during intravenous infusion of sodium lactate (2.5 mmol/h) and addition of oxonate.

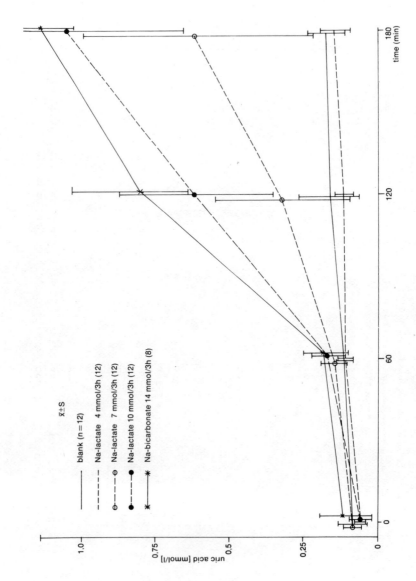

Fig. 4 Effect of continuous intravenous infusion of sodium lactate or sodium bicarbonate and of oxonate (o.o6 mmol/h) on uric acid concentration.

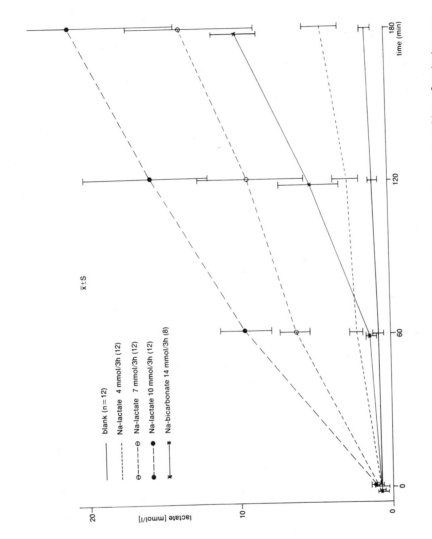

Fig. 5 Effect of continuous intravenous infusion of sodium lactate or sodium bicarbonate on blood lactate concentration.

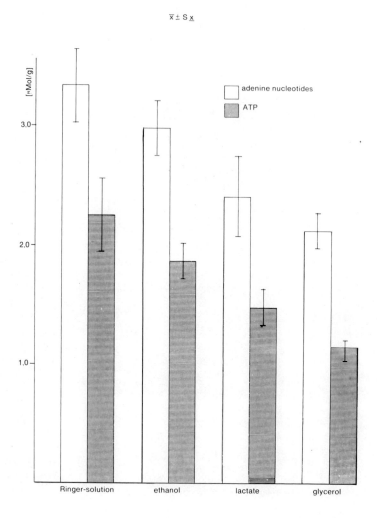

Fig. 6 Influence of 3 h intravenous infusion of
        different substances on the hepatic concen-
        trations of ATP and of adenine nucleotides
        in rats.

led only to a small increase in blood uric acid, whereas using
the higher doses of lactate, the increase in uric acid concentration was above proportion. With the highest dose of sodium lactate
uric acid concentration in serum was elevated nearly by 1 mmol/l
(fig. 4).
When sodium lactate is infused, lactate is taken up by several
tissues, especially by the liver. However, the sodium ion remains
in the circulation. Therefore, the consequence of a high dosed
sodium lactate infusion must be the development of an alkalosis,
despite elevated lactic acid concentration. Moreover, an increase
in lactic acid concentration is observed during high dosed sodium
bicarbonate infusion (fig.5). We propose that the metabolic
situation which develops following high dosed sodium lactate infusion or bicarbonate infusion, that means the increase in lactic
acid concentration in blood, should be called lactate alkalosis.
The question to be answered was whether lactic acid infusion or
sodium combined with alkalosis was the cause of hyperuricaemia.

For this purpose infusions of sodium bicarbonate were performed,
and as shown in fig. 4, an immense increase of uric acid was caused
by sodium bicarbonate as well as by sodium lactate. Therefore, it
can be concluded that extreme alkalosis is the cause of hyperuricaemia during infusion of sodium salts with metabolizable an-
ions. Under these extreme conditions sodium concentrations up to
2oo mmol/l were measured in serum, whereas chloride concentrations
were depressed to 7o mmol/l.
As carbone dioxide is volatile and excreted by the lungs, for compensation of alkalosis a nonvolatile anion is needed.Lactic acidaemia following sodium bicarbonate infusion is considered to be
a compensation mechanism. Lactic acid is formed in tissues and
accumulates in the blood for neutralisation of the large amounts
of sodium ions.
The original hypothesis was that an increase in uric acid concentration, following an infusion of fructose, is caused by a breakdown of preformed purine derivatives (9,10). To test this hypothesis freeze clamp determinations were performed in the liver. Fig.6
shows that the decrease in the concentration of adenine nucleotides
or ATP is only small during the infusion of sodium lactate. In
comparison the effect of glycerol on concentration of adenine nucleotides and of ATP is more pronounced whereas the effect on uric
acid synthesis is less than one half that of lactate (7,8) despite the large increase in uric acid concentration. Beyond this
the figure shows that ethanol application does not influence the
hepatic adenine nucleotide concentration.

It is concluded that increased uric acid synthesis during sodium
lactate infusion in the rat is not caused to a greater extent by
the breakdown of preformed adenine nucleotides in the liver. The
pronounced effect of sodium lactate on uric acid synthesis in the
rat seems to be an effect of sodium application or of alkalosis

since it is shown in a similar way by sodium bicarbonate. Because
the concentration of preformed purines is scarcely altered other
precursors of uric acid must be found. A full synthesis of purines
de novo beginning with ribose-phosphate seems unlikely because of
the large effects. Another possibility would be the breakdown of
ribonucleic acids in certain tissues. This speculation is not
proven until now, however, and further experiments have to be performed.

In normal human beings the intake of ethanol seems to cause a very
small increase in uric acid synthesis not connected with lactic
acid metabolism in any way. Unfortunately ethanol is without effect
on purine metabolism in the rat. Therefore this laboratory animal
is unsuitable for investigations of these relations between ethanol
and uric acid synthesis. However, it is questionable if the very
small effects on serum uric acid concentration caused by ethanol
in normal volunteers have a greater pathophysiological significance.
On the other hand it seems possible that certain individuals over-
react when treated with ethanol.

1. MacLachlan, M.J., G.P. Rodnan: Amer. J. Med. 42, 38 (1967)
2. Lieber, C.S., D.P. Jones, M.S. Losowsky, C.S. Davidson: J. Clin.
   Invest. 41, 1863 (1962
3. Burch, R.E., N. Kurke: Proc. Soc. Exp. Biol. 127, 17 (1968)
4. Grunst, J., G. Dietze, M. Wicklmayr, H. Mehnert: Verh. Dtsch.
   Ges. Inn. Med. 8o, 487 (1974)
5. Förster, H. in: Gastroenterologie und Stoffwechsel, ed: Becker,
   Witzstrock, Baden - Baden 1975, p. 28o
6. Johnson, W.H., B. Stavric, A. Chartrand: Proc. Soc. Exp. Biol.
   131, 8 (1969)
7. Förster, H., H. Hartmann: in preparation
8. Hartmann, H., I. Hoos, H. Förster: Arch. Pharmakol. 293,
   Suppl. 1, 235 (1976)
9. Perheentupa, K.O., J.R. Raivio: Lancet II, 528 (1967)
1o.Woods, H.F., G.M. Alberti: Lancet 11, 354 (1972)

Supported by the Deutsche Forschungsgemeinschaft (Fo 54/13)

CARBOHYDRATE INDUCED INCREASE IN URIC ACID SYNTHESIS. STUDIES IN HUMAN VOLUNTEERS AND IN LABORATORY RATS

H. Förster and I. Hoos

Zentrum d. Biologischen Chemie d. Universität Frankfurt

6000 Frankfurt/M., Theodor-Stern-Kai 7,BRD

Rapid intravenous infusions of fructose cause an increase in serum uric acid concentration (1). This surprising effect was confirmed in the same year (2) and supplemented by an hypothesis on this unexpected correlation between carbohydrate metabolism and purine metabolism. It was shown that rapid fructose infusions caused a decrease in adenine nucleotide concentration in the livers of both rats (2,3,4,5) and human beings (4). On the other hand in the course of long lasting fructose infusions decrease of liver adenine nucleotide concentration was minimum (5). Additionally the serum uric acid concentration was scarcely influenced by long lasting fructose infusions (6). In the following period it could be demonstrated that beyond fructose xylitol and sorbitol (fig.1) likewise cause an increase in uric acid concentration of man (7). Xylitol proved to be the most effective of them (8). Moreover, it was established that the carbohydrate induced increase in uric acid values of man was due to an increased de novo synthesis and was not caused by renal effects. Inhibition of the fructose induced (9) or xylitol induced (fig 2) rise in serum uric acid concentration of man can be achieved by allopurinol premedication. Allopurinol was given in doses of 3 times 2oo mg per person before starting the test. Additionally, renal uric acid excretion is significantly intensified during the high dosed rapid infusion of fructose, sorbitol and xylitol (8). It was soon found out that long lasting permanent infusion of fructose causes a moderate increase of uric acid synthesis only (6,1o).As shown in fig.3 the 48 h-infusion of o.25 g/kg.bw. of fructose per hour led to an initial increase in uric acid concentration of 2 mg/1oo ml, whereas in the further course of the fructose infusion uric acid level remained constant or even decreased. When xylitol was applied at the same dose the rise in uric acid concentration lasted all over the infusion period. However, despite an overall amount of

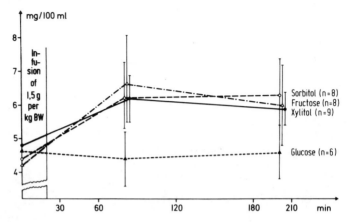

Fig. 1. Serum Uric Acid in Healthy Volunteers

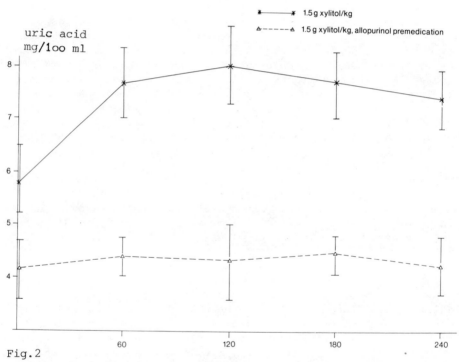

Fig.2
Increase in uric acid concentration in healthy volunteers (n=8)
caused by rapid intravenous infusion of xylitol and influence of
allopurinol premedication (x̄ s)

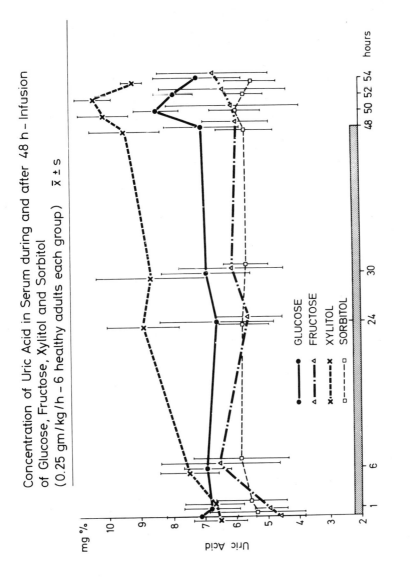

Concentration of Uric Acid in Serum during and after 48 h – Infusion of Glucose, Fructose, Xylitol and Sorbitol (0.25 gm/kg/h – 6 healthy adults each group) $\bar{x} \pm s$

Fig. 3

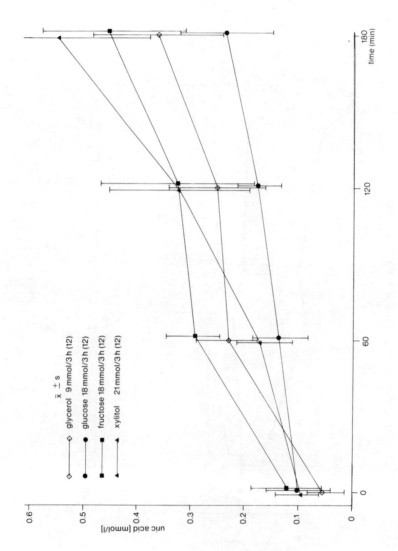

Fig. 4 Effect of continous intravenous infusion of glucose, fructose (18 mmol/ 3 h), xylitol (21 mmol/ 3 h) or glycerol (9 mmol/ 3 h) and of oxonate 0.06 mmol per h) on uric acid concentration.

12 g/kg.b.w. in course of 48 hours total increase in the case of
xylitol was 3 mg/100 ml only, whereas the rapid infusion of
1.5 g/kg.bw. was followed by an increase in uric acid of 2mg/100ml.
It should be mentioned that the situation is a different one for
intensive care patients.In this group of patients the high dosed
infusion of glucose substitutes results in very small increases in
uric acid concentration (in preparation).Since Burch and Kurke reported that renal excretion of uric acid had been slightly de-
minished after intravenous infusion of sodium lactate (11),it
should be emphasized that the increase in uric acid synthesis is
not related to blood lactic acid concentration in man.This results
from the following examples.Glucose causes a rise in lactic acid
concentration in man, but it is without any effect on uric acid
level. Fructose leads to an increase in uric acid concentration
and to an increase in lactic acid concentration. Xylitol is without effect on lactic acid values,but it has the most pronounced
effect on uric acid level. It proved that the course of fructose
induced hyperuricaemia renal uric acid excretion also increased,
despite the presence of fructose induced lacticacidaemia (8).
In laboratory rats infusion of sodium lactate was followed by an
increase in uric acid synthesis (12).
Most research on the mechanism of the glucose substituts induced
intensification of uric acid synthesis was performed in the experimental animals. Since in rat uricase cleaves urate to allantoine which is not easy to analyze, it seemed difficult to compare
quantitative data of metabolite concentrations in the rat liver
and data for uric acid in man. It became possible to measure
uric acid metabolism in the laboratory animal by use of intravenous
oxonate as uricase inhibitor (12).
Under simultaneous infusion of the glucose substitutes and of oxonate uric acid level raised by 4-6 mg/100 ml or 0.3-0.4 mmol/l
within three hours (fig.4).The same effect was attained by infusion
of glycerol. The increase of uric acid seen in the initial period
of infusion was higher both with fructose and glycerol than with
xylitol. On the other hand xylitol infusion was accompanied by
continuous rise in serum uric acid concentration. During the last
hour of fructose or of glycerol infusions uric acid concentration
increases again (fig 4).These different reactions to high dosed
infusions of the various substances may be due to the molecular
mechanism of their action. It seems possible that different mechanisms are involved in the effects of xylitol and fructose or
glycerol. The continuous rise in serum uric acid during xylitol
could be the consequence of an uniform mechanism, whereas two
different mechanisms are concerned with the increase in uric acid
synthesis caused by fructose or glycerol.
In the next series ofexperiments the effects of the glucose substitutes fructose, sorbitol, and xylitol on the concentration of
liver metabolites in comparison to glucose had been investigated
(fig.5). First of all high dosed rapid infusions of the different

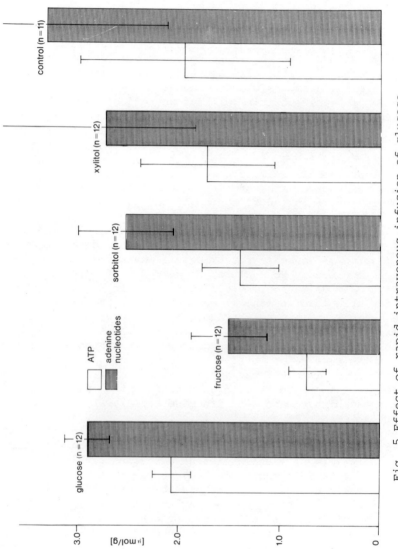

Fig. 5 Effect of rapid intravenous infusion of glucose, fructose, sorbitol (12 mmol/ 30 min) or xylitol (16 mmol/ 30 min) on hepatic concentrations of ATP or adenine nucleotides.

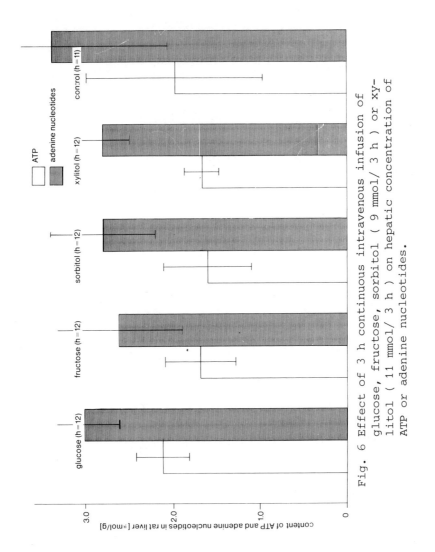

Fig. 6 Effect of 3 h continuous intravenous infusion of glucose, fructose, sorbitol (9 mmol/ 3 h) or xylitol (11 mmol/ 3 h) on hepatic concentration of ATP or adenine nucleotides.

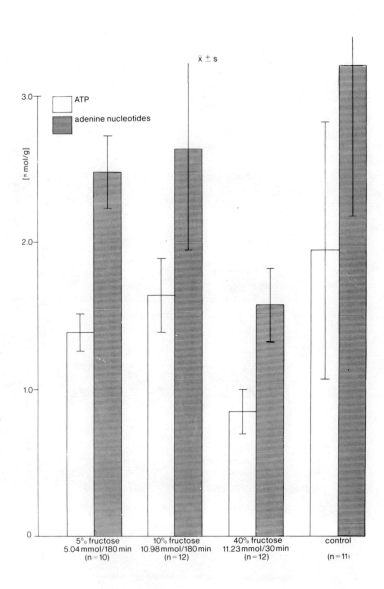

Fig. 7 Influence of fructose in different doses on hepatic
concentrations of ATP and of adenine nucleotides.

substances were used. Only fructose infusion led to substantially
lowered concentrations of ATP and of adenine nucleotides. Sorbitol
exerts a very small influence only. Xylitol was almost without
any effect.
In lower dosed continous infusion i.e. similar doses were given
over a period of three hours than of 3o minutes all the glucose
substitutes fructose, sorbitol, and xylitol showed neglectible
effects on concentration of ATP or of adenine nucleotides (fig.6).
A direct comparison of the fructose effects in relation to dose
and to speed of infusion is given with the next figure (fig.7).
Obviously the high dosed rapid infusion only was able to decrease
the levels of ATP and of adenine nucleotides to an higher extend.
The lower doses did not alter the concentration of ATP and of
adenine nucleotides.
In further studies we found that intravenous infusion of glycerol
led to a substantial suppression of hepatic concentration of ATP
and of adenine nucleotides (12). The effect of glycerol on this
hepatic compounds was comparable to the effect of fructose. As
shown in fig. 4 the mode of increase in serum uric acid during
fructose and glycerol infusion differs from that we found with
xylitol. It is concluded, therefore, that the increase in uric
acid synthesis in rats is partially due to an initial break down
of preformed purines. This effect is documented in uric acid
values in the first hour of the experiment. A further increase
in the third hour of the experiment should be due to another mechanism. The most likely one is synthesis of uric acid de novo.
On the other hand, using xylitol the continous increase in uric
acid synthesis during the whole experimental period of three hours
could be the consequence of a purine synthesis de novo.

We have found distinct differences between the effects of glucose
substitutes in man and in the experimental animal. One of the most
essential difference is the quantitative aspect. The increase of
uric acid in the rat of 4 - 6 mg/1oo ml has no equivalent in man
where the maximum increase within three hours is 1 -3 mg/1oo ml.
However, the urinary excretion of the relatively insoluble uric
acid is only a problem for man. Therefore, feed back mechanisms
in purine metabolism might be more effective in man as compared
to the experimental animal.

1. Förster, H., H. Mehnert, I. Alhough: Klin. Wschr.45, 436 (1967)
2. Perheentupa, K.O., J.R. Raivio: Lancet II, 528 (1967)
3. Woods, H.F., G.M. Alberti: Lancet II, 354 (1972)
4. Bode, J.C., O. Zelder, H.J. Rumpelt, U. Wittkamp: Europ. J.
   Clin. Invest. 3, 436 (1973)
5. Brinkrolf, H., K.H. Bässler: Z. Ernährungswiss. 11, 167 (1972)
6. Sahebjami, H., R. Scarlettar: Lancet I, 366 (1971)
7. Förster, H., E. Meyer, M. Ziege: Klin. Wschr. 48, 876 (197o)
8. Förster, H., S. Boecker, M. Ziege: Med. Ernährung 13,193 (1972)

9. Förster, H., H. Mehnert: Lancet II, 12o5 (1967)
1o. Förster, H., L. Heller, U. Hellmund: Dtsch. Med. Wschr. 99,
    1723 (1974)
11. Burch, R.E., N. Kurke: Proc. Soc. Exptl. Biol. 127, 17 (1968)
12. Hartmann, H., H. Förster: Arch. Pharmakol. 293, 235 (1976)

Supported by the Deutsche Forschungsgemeinschaft (Fo 54/13)

PHARMACOLOGICALLY INDUCED ERYTHROCYTE PP-RIBOSE-P

VARIATIONS

Giuseppe Pompucci, Vanna Micheli, Roberto Marcolongo

Institute of Biochemistry and Service of Rheumatology

University of Siena, Siena, Italy

Purine "de novo" biosynthesis appears to be regulated by PP-ribose-P-glutamine amido-transferase activity. This activity is, in turn, influenced by the concentration of PP-ribose-P which is not only a substrate but is also able to reverse inhibition by purine nucleotides on this enzyme. The apparent Km of human amidotransferase for PP-ribose-P is around 0.5mM (1,2). As this value is 10 to 100 times the intracellular concentration of PP-ribose-P (3), changes in PP-ribose-P concentration could well be responsible for a corresponding change in the rate of purine biosynthesis.

PP-ribose-P is synthetized from ribose-5-P and ATP. This derivation places purine biosynthesis in close relation to glucide metabolism and specially to the pentose phosphate pathway. It is very possible that variations in concentration of these pathway substrates increase not only the rate of PP-ribose-P but eventually the uric acid production as well.

With the intention of examining the effects of the administration of sugars on intracellular PP-ribose-P concentration and serum uric acid levels, using groups of three subjects each, we administered intravenously 500 mg/Kg of fructose, glucose, ribose, sorbitol, and xylitol to healthy volunteers, subsequently testing these subjects for erythrocyte concentration of PP-ribose-P and serum uric acid levels. A sixth group loaded with fructose and phosphate was also tested. Blood was drawn

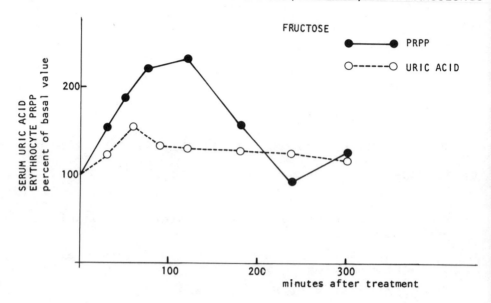

Fig. 1

several times up to 300 minutes after treatment.

The administration of fructose led to a significant increase in intracellular PP-ribose-P levels with a maximum increase to a mean of 230% of basal values in 120 minutes, with an increase in serum uric acid equal to 150% of basal values after only 60 minutes (Fig. 1).

A contemporaneous administration of fructose and phosphates produced no substancial changes in the situation already observed although the values of uric acid appear to be slightly lower (Fig. 2). Sorbitol, which is converted to fructose by sorbitol dehydrogenase, will also cause an increase but with an effect lasting over five hours after treatment (Fig. 3). However, while an increase in uric acid is noted immediately, the levels of PP-ribose-P initially show a sharp decrease of up to 40% of basal values. Xylitol (Fig. 4) produces more or less the same effect.

PHOSPHATE BUFFER + FRUCTOSE

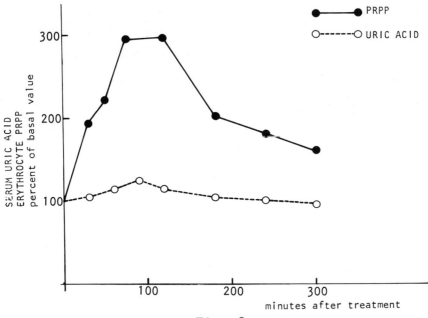

Fig. 2

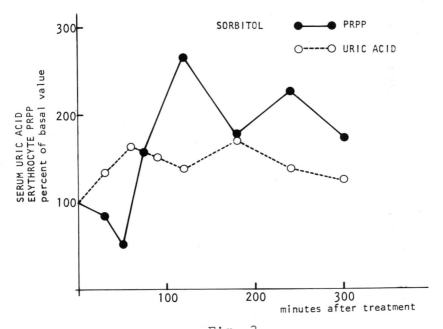

Fig. 3

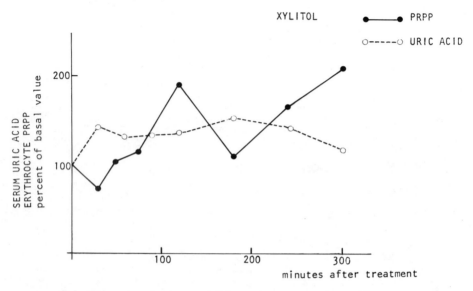

Fig. 4

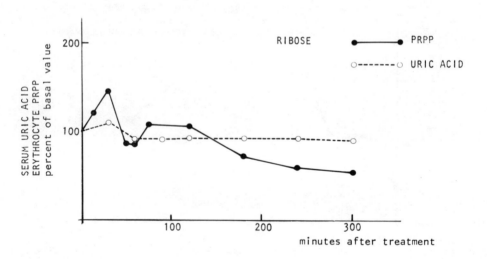

Fig. 5

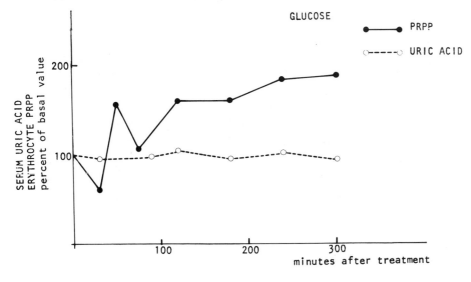

Fig. 6

Ribose seems to have no effect on uricemia while causing a rapid
initial increase in the levels of PP-ribose-P which subsequently
fall to a level of 50% of basal values (Fig. 5). The
effect of glucose is an increase in the levels of PP-
ribose-P after a slight initial decrease, but there is
no change in uricemia values (Fig. 6).

Excessive administration of glucides or closely
related compounds causes an increase in the levels of
PP-ribose-P which, in turn, cause an increase in the
rate of purine biosynthesis. An explanation for what
appears to be an exception in the case of ribose may
be that this pentose is rapidly phosphorilated, producing an increase in PP-ribose-P in less than 15 minutes.
In fact, experiments 2in vitro2 have shown that ribose
increases the concentration of PP-ribose-P (4). The subsequent decrease has not yet been explained.

The rapid increase which we observed in the serum
uric acid does not seem to be related to the increased
concentration of PP-ribose-P as the increase in serum
uric acid occurs either before or at the same time as
the increase in PP-ribose-P. It is probable that this
phenomenon is due either to an increased purine catabolism resulting from disinhibition of AMP-deaminase (5) or
to the fact that high levels of PP-ribose-P can be
reached in a far shorter period of   time
in the liver than in erythrocytes.

REFERENCES

1. Wood, A. W. and Seegmiller, J. E., J. Biol. Chem. 248, 138 (1973).
2. Holmes, E.W., McDonald, J. A., McCord, J. M. et al., J. Biol. Chem. 248, 144 (1973).
3. Fox, I. H. and Kelley, W. N., Ann. Inter. Med. 74, 424 (1971).
4. Fox, I. H. and Kelley, W. N., in Purine Metabolism in Man, Edited by D. Sperling, A. De Vries and J.B. Wyngaarden, vol. 1º p. 93, Plenum Press, New York (1974).
5. Bowering, J., Calloway, D. H., Margen, S. and Kaufmann, N. A., J. Nutr. 100, 249 (1970).

ISOLATED RAT LIVER CELLS - PURINE METABOLISM AND EFFECTS OF FRUCTOSE

Camilla Smith, Liisa Rovamo and Kari O. Raivio

Children's Hospital, University of Helsinki

Helsinki, Finland

Isolated rat hepatocytes have been found useful in the study of a number of problems of liver biochemistry (1,2,3). However, their properties with respect to purine metabolism have not been extensively studied. In our laboratory we have prepared suspensions of rat liver cells with good viability and well preserved energy metabolism which are suitable for purine studies.

Rat liver cells were prepared using the method of Seglen (4), with some modifications. The rat liver is first perfused in situ through the portal vein with warm, oxygen-saturated buffer to wash out the blood. The hepatic vein and inferior vena cava are cut to permit free outflow of the perfusion buffer. During this perfusion the liver is cut free from the rat. Next the buffer is changed to one containing collagenase which is also warm and oxygen-saturated. This enzymatic treatment with collagenase dissolves the intracellular binding substance of the liver and allows the parenchymal cells to be easily dispersed into a suspending medium. Purification of the parenchymal cells is achieved by filtration and differential centrifugation.

The final cell suspension used for experiments contains mostly single parenchymal cells at a density of 1 to 2×10^7 cells per ml. The cells were counted using a haemocytometer after mixing an aliquot of the cells with Trypan blue. Cell viability, as defined by the percentage of cells excluding Trypan blue, was about 85%. Electron micrographs of the cells from our preparation show intact mitochondria and other subcellular organelles. This also suggests that the cells are not damaged during the isolation procedure.

During incubation of rat hepatocytes, a high oxygen atmosphere is required for the maintenance of normal ATP/ADP ratios. Table 1 shows the effect of oxygen on incorporation of ^{14}C-adenine into adenine nucleotides and the ratio of radioactivity in ATP to ADP. Cells were incubated for 60 min in the presence of 100 μM ^{14}C-adenine. The incorporation of radioactivity into adenine nucleotides was not affected by oxygen but the ratio of ATP/ADP was high only in the presence of a high oxygen atmosphere. In a buffer containing 10mM phosphate, 5.5 mM glucose, with a high oxygen atmosphere, cell suspensions could be incubated for at least 4 hours with no loss of viability as determined by the exclusion of Trypan blue or reduction in the ratio of radioactivity in ATP to ADP.

The metabolism of the radioactive purine precursors hypoxanthine, guanine, adenine and adenosine was compared (Table 2). The initial concentrations were 100 μM for hypoxanthine, adenine and adenosine and 200 μM for guanine. After incubating the cells with the precursor, the purine components were extracted with perchloric acid and separated using thin layer chromatography (5). Hypoxanthine was rapidly metabolized primarily to allantoin. After 30 min of incubation, the radioactivity in purine nucleotides was only 9% of the total hypoxanthine metabolized to acid-soluble products. The ratio of radioactivity in adenine nucleotides compared to that in guanine nucleotides was 5.8.

Guanine was also rapidly catabolized to allantoin. Only 4% of the guanine was metabolized to guanine nucleotides after 30 min of incubation.

Table 1. Effect of oxygen on ^{14}C-adenine incorporation into nucleotides

	Nucleotide Synthesis nmoles /10^6 cells/hr	ATP/ADP
Low Oxygen Atmosphere	8.6	0.97
High Oxygen Atmosphere	8.8	4.8

Table 2. Metabolism of radioactive purine bases and
          adenosine in isolated rat liver cells

Precursor	Purine Component	Incubation Time (min)		
		15	30	60
		cpm /10^7 cells		
Hypoxanthine	A. Nucleotides	5500	4137	3662
	G. Nucleotides	901	775	640
	Hypoxanthine	12746	7784	348
	Allantoin	46276	49372	60129
Guanine	G. Nucleotides	9272	7073	6062
	Guanine	23377	7792	2869
	Allantoin	86774	158082	179273
Adenosine	A. Nucleotides	36349	33831	24148
	Adenosine	19484	5882	2045
	Inosine	1769	554	356
	Hypoxanthine	3154	160	1
	Xanthine	3401	2722	4088
	Uric Acid	1863	715	429
	Allantoin	24387	48081	55431
Adenine	A. Nucleotides	14888	20142	29882
	Adenine	82993	74148	56804
	Uric Acid	1044	1106	2066
	Allantoin	1533	3923	11086

Adenosine was less rapidly metabolized than hypoxanthine and guanine. After 15 min, the radioactivity
in adenine nucleotides was 49% of the total adenine metabolized and the radioactivity in catabolic products
was 48%. At 15 min there was significant radioactivity
in inosine, hypoxanthine, xanthine, and uric acid.

Adenine is metabolized to adenine nucleotides via
adenine phosphoribosyltransferase. Because adenine itself
is not catabolized, allantoin is derived only from catabolism of nucleotides; thus the production of allantoin
is slower than with the other purine precursors. Adenine
is a suitable precursor for prelabelling the adenine
nucleotide pool.

We have examined the effects of fructose on adenine
nucleotide catabolism in isolated rat liver cells. Fructose is known to increase purine nucleotide catabolism
in man and in rats (6). In the liver, fructose is rapidly

phosphorylated by fructokinase to fructose-1-phosphate
which reduces ATP levels. Fructose-1-phosphate is only
slowly metabolized so there is an accumulation of fruc-
tose-1-phosphate and a decrease in inorganic phosphate
levels. Enzyme studies suggest that a decreased concentration of ATP and inorganic phosphate increases the
activity of adenylate deaminase resulting in increased
purine nucleotide catabolism.

        For our experiments, liver cells were labelled with
[14]C-adenine, washed and resuspended in a medium containing fructose or glucose at a concentration of 5 mg/ml.
They were then incubated at 37°C. As shown in Fig.1,
in cells suspended in a fructose-containing medium there
was a decrease in radioactivity in adenine nucleotides.
The reduction was most rapid at the beginning of the
incubation and after 10 min there was no further decrease. At 10 min the radioactivity in adenine nucleotides
was 1/2 that at the beginning of the experiment. There
was an increase in the radioactivity in allantoin which
closely equals the loss of radioactivity from adenine
nucleotides. At 5 min there was significant radioactivity in hypoxanthine. In cells suspended in glucose, there
was only a small decrease in radioactivity in adenine
nucleotides and a small increase in allantoin.

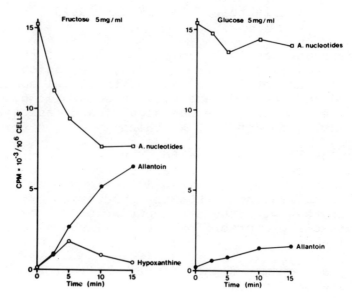

Fig.1. Effect of fructose on adenine nucleotide catabolism.

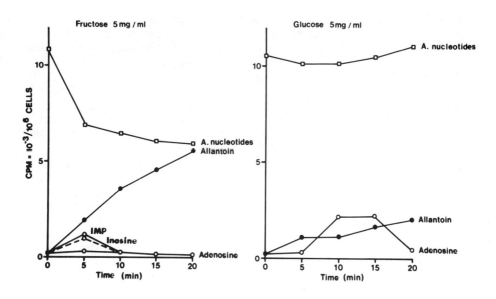

Fig. 2. Adenine nucleotide catabolism in the presence
of an adenosine deaminase inhibitor and 10mM phosphate.

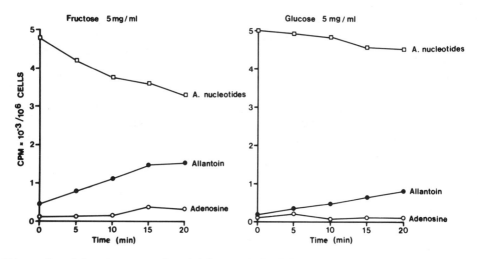

Fig. 3. Adenine nucleotide catabolism in the presence of
an adenosine deaminase inhibitor and 1mM phosphate.

There are two alternative routes of AMP catabolism
to inosinate - one via inosinate and one via adenosine.
By using an inhibitor of adenosine deaminase it should
be possible to assess the relative importance of these
two pathways. We have done fructose-induced purine nucleotide catabolism experiments in the presence of the
adenosine deaminase inhibitor 9-erythro- (2-hydroxy-3-
nonyl) adenine (7). As shown in Fig. 2, in cells suspended in fructose and 10mM phosphate, there was little
or no accumulation of labelled adenosine but there was
significant radioactivity in inosinate and inosine.
This confirms the previous hypothesis that fructose induced purine nucleotide degradation occurs mainly via
inosinate. For cells suspended in glucose and 10 mM
phosphate, there was significant accumulation of labelled adenosine at 10 and 15 min in the presence of
the adenosine deaminase inhibitor. At high ATP and phosphate concentrations, AMP seems to be dephosphorylated
to adenosine rather than deaminated to inosinate. The
reduction in adenosine at 20 min was likely due to re-
phosphorylation of adenosine to adenine nucleotides.

In cells suspended in fructose and 1mM phosphate,
there was little accumulation of adenosine in the presence of the adenosine deaminase inhibitor (Fig.3). For
cells suspended in glucose and 1 mM phosphate, adenosine
did not accumulate. This suggests that phosphate concentration is an important factor for the regulation of
adenylate deaminase in this system.

The main purpose of this paper has been to describe
our method of liver cell preparation and to demonstrate
its usefulness in purine research. In comparison to the
numerous isolated cell systems that have been studied,
the rat hepatocyte is thus far the only cell type expressing the entire purine degradation pathway including
uricase. The implications for biochemical and pharmacological studies should be obvious.

REFERENCES

1. M.E.M. Tolbert and J.N. Fain: J.Biol.Chem, 249: 1162-1166, 1973.

2. R. Berger and F.A. Hommes: Biochim.Biophys.Acta 333:535-545, 1974.

3. S.J. Pilkis, T.H. Claus, R.A. Johnson and C.R. Park: J.Biol.Chem. 250:6328-6336, 1975.

4. P.O. Seglen: Exptl.Cell Res. 82:391-398, 1973.

5. G.W. Crabtree and J.F. Henderson: Cancer Res. 31:985-991, 1971.

6. K.O. Raivio, M. Kekomäki and P. Mäenpää: Biochem. Pharmacol. 18:2615-2624, 1969.

7. H.J. Schaeffer and C.F. Schwender: J.Med.Chem. 17:6-8, 1974.

STUDIES ON THE EFFECT OF FRUCTOSE AND XYLITOL IN THE RAT LIVER: 5'-NUCLEOTIDASE, ADENOSINE DEAMINASE, DE NOVO PURINE SYNTHESIS

J. D. Schwarzmeier, M. M. Müller and W. Marktl

1st Dept. of Medicine
1st Dept. of Medical Chemistry
and Dept. of Physiology
University of Vienna, Austria

INTRODUCTION

Fructose as well as xylitol are both rapidly metabolized in liver the main product being glucose (1, 11). In high doses these sugar surrogates induce an increase of uric acid levels both on parenteral and oral application (4, 5, 8, 14, 16, 18, 20). The causes underlying the hyperuricemia thus produced are still unclear. Since the metabolism of both substrates involves a phosphorylation step, the liver cell ATP concentration is (at least temporarily) reduced. This reduction is accompanied by a depletion of ADP, AMP and inorganic phosphates. The loss of inorganic phosphates (P_i) is thought to activate catabolic enzyme reactions of purine metabolism which, in turn, would enhance purine nucleotide degradation (2, 5, 12, 17 - 20). If this hypothesis is correct, both fructose and xylitol should be expected to activate catabolic enzymes, such as 5'-nucleotidase and adenosine deaminase. This, in turn, would result in a secondary increase of de novo purine synthesis. To validate this assumption we studied the effects of the two sugar alcohols on 5'-nucleotidase and adenosine

deaminase activities in the rat liver in vivo and in
vitro. At the same time, the incorporation of the
purine precursor glycine in rat liver adenine
nucleotides was measured.

MATERIALS AND METHODS

Studies were carried out on male Wistar rats
with a body weight of 200 to 300 g fed on a
conventional laboratory diet. The animals fasted
for 20 to 24 hours before the experiments.
They were anesthetized with urethane (1.2 g/kg). With
the test subjects fully anesthetized, the femoral
vein was exposed and 50 μCi glycine (1-14C-glycine,
56 mCi/mMol, Radiochemical Centre Amersham, UK) were
injected via a venous catheter. Subsequently, 0.9 %
NaCl, fructose or xylitol was infused. The doses and
infusion rates of the two sugar alcohols used in the
experiments are inducated below. At the end of the
experiments, the rats were sacrificed and liver tissue
was rapidly removed and cooled in liquid nitrogen. The
tissue samples were homogenized and extracted in cold
0.3 N HClO$_4$. Adenine nucleotides were isolated from
the perchloric acid supernatant using a modification
of the technique reported by MARKO et al. (13) and
the activity due to the incorporation of labeled
glycine was measured. The specific activity was
expressed as dpm/μ mole adenine.

Enzyme activities were assayed under identical
conditions. Labeled glycine was, however, omitted
prior to NaCl, fructose and xylitol infusions. At the
end of the experiments, liver tissue was komogenized
in saccharose triethanol-amine buffer and centrifuged
for 20 min. at 16,000 x g and 4°C. Supernatants were
deep frozen and the following enzymes assayed on the
next day: adenosine deaminase (EC 3.5.4.4) (7, 15),
5'-nucleotidase (EC 3.1.3.5) (6), transketolase
(EC 2.2.1.1.)(39), transaldolase (EC 2.2.1.2) (10),
hypoxanthine-guanine phosphoribosyltransferase (HG-PRT,
EC 2.4.2.8)(10) and adenine phosphoribosyltransferase
(A-PRT, EC 2.4.2.7)(10). Some of the enzyme assays
were carried out in vitro after incubation of anuclear
rat liver homogenates with 0.9 % NaCl, fructose,
xylitol, fructose-1-P, fructose-6-P, glyceraldehyde-3-P.
Incubation data are listed in the legends; samples
were incubated at 37°C in normal air.

RESULTS

As can be seen in Fig. 1. fructose infusions at
a rate of 0.7 g/kg/30 min. were not found to result in
a significantly change of glycine incorporation into
free (acid soluble) adenine nucleotides of rat liver
both after 15 and 30 min. infusions. Rats given 0.9 %
NaCl served as controls. Livers removed 60 minutes
after the end of 30-minute infusions (total test time:
90 minutes) show a substantial and significant increase
of specific adenine nucleotide (SA-adenine) activities.
A similar SA-adenine pattern was found to be present
after fructose infusions at a rate of 2.4 g/kg/30 min.
Again, there was no significant rise after 15 or
30 minutes, while 90 minutes specimens revealed a
considerable increase of glycine incorporation.

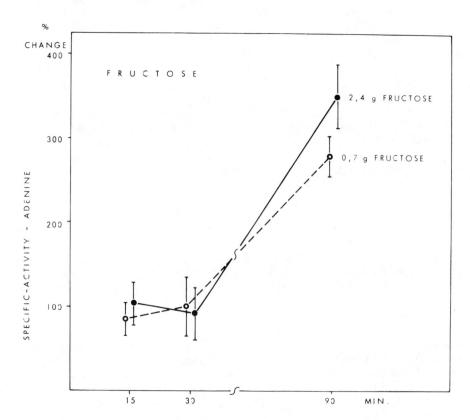

Fig. 1. ^{14}C-glycine incorporation into adenine
nucleotides of rat liver in vivo.

		ADA	5'-N	TK	TA	HG-PRT	A-PRT
0.9 % NaCl	x̄	16.3	0.32	12.3	14.3	120	44
	S.D.	1.6	0.09	3.2	1.7	25	8
0.7 g/kg fructose	x̄	23.5	0.48	12.1	13.0	116	49
	S.D.	3.0	0.12	2.6	3.4	24	11
	p	<0.005	<0.05				
2.4 g/kg fructose	x̄	32.3	1.44	12.1	14.5	119	63
	S.D.	3.4	0.36	3.3	1.1	23	20
	p	<0.001	<0.001				

Table 1. Enzyme activities in rat liver after infusion of 0.9 % NaCl, 0.7 and 2.4 g fructose/kg/30 min. Infusion time: 15 minutes. Mean values of 8 animals. Activities of adenosine deaminase (ADA), 5'-nucleotidase (5'-N), transketolase (TK) and transaldolase (TA) are expressed as mU/mg protein, HG-PRT and A-PRT as nmole/mg/h.

To identify any potential activation of enzyme
reactions in the pentose phosphate cycle, the salvage
pathway or the catabolic steps of purine metabolism,
which might precede the rise of SA-adenine activities,
transketolase (TK), transaldolase (TA), HG-PRT, A-PRT,
5'-nucleotidase (5'-N) and adenosine deaminase (ADA)
activities were assayed after 15-minute infusions.
Table 1 shows these enzyme activities in rat livers
after fructose application at a rate of 0.7 and 2.4
g/kg/30 min. As can be seen, TK, TA, HG-PRT and A-PRT
activities were found to be unchanged vs. NaCl treated
controls under the conditions described. 5'-N and ADA
activities, by contrast, were significantly increased,
the rise being more pronounced after 2.4 g fructose/kg.

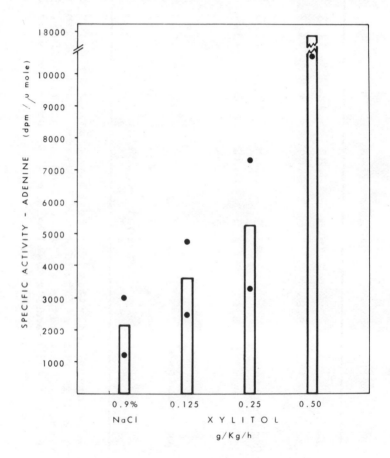

Fig. 2. ^{14}C-glycine incorporation into adenine
nucleotides of rat liver during xylitol infusions.

Fig. 2 shows the effect of variable doses of xylitol on 14C-glycine incorporation into rat liver adenine nucleotides. Rats receiving NaCl infusions again served as controls. At an application rate of 0.125 g/kg/h xylitol was found to result in a slight, but significant rise of SA-adenine after 90 minutes (30-minute infusion plus 60 minutes prior to removing the liver). On increasing the xylitol dose glycine incorporation continued to rise, the effect being most pronounced at 0.5 g/kg/h.

If infusions of 0.5 g xylitol/kg are given within a shorter time (Fig. 3), SA-adenine is found to be unchanged after 15 minutes, with a slight rise after 30 minutes and an substantial and significant increase of glycine incorporation in adenine nucleotides after 60 minutes.

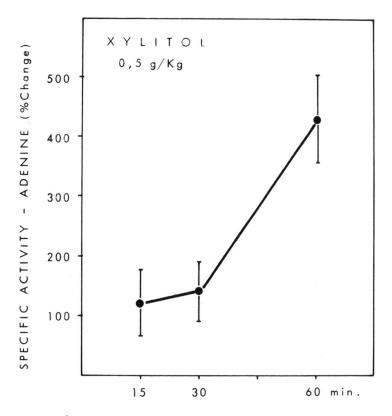

Fig. 3. ^{14}C-glycine incorporation into adenine nucleotides of rat liver in vivo.

		5'-N	ADA
Control	$\bar{x}$	0.32	16.3
(0.9 % NaCl)	S.D.	0.09	1.6
Xylitol	$\bar{x}$	0.65	20.7
(0.50 g/kg/h)	S.D.	0.15	2.8
	p	0.001	0.01

Table 2.   Activities of 5'-nucleotidase (5'-N) and
adenosine deaminase (ADA) in rat liver after infusion
of 0.9 % NaCl and 0.5 g/kg/h xylitol. Infusion time:
15 minutes. Mean values of 6 animals. Enzyme activities
are expressed as mU/mg protein.

A further series of tests served to identify any
potential increase of 5'-N and ADA activities after
15-minute infusions of 0.5 g xylitol/kg at a time,
where SA-adenine was still within normal. As can be
seen from table 2, there is an appreciable and
significant increase of 5'-N activity, while ADA
activities, though rising, are not significantly
changed.

In another series of experiments an attempt was
made to reproduce the enzyme activity changes observed
in vivo under in vitro conditions. For this purpose,
anuclear rat liver homogenates were incubated with
fructose, fructose-1-P, fructose-6-P, glyceraldehyde-3-P
and with xylitol. Table 3 lists the data obtained.
While fructose and xylitol in the concentrations
reported enhanced 5'-N and ADA activities, the other
substances were not found to have any effect on enzyme
activities.

DISCUSSION

The results of the present studies suggest that
both fructose and xylitol-induced hyperuricemias are
primarily caused by an activation of purine catabolism,
while the enhancement of de novo purine synthesis
appears to be a secondary factor. Under the conditions

		5'-N	ADA
Control	$\bar{x}$	0.75	18.6
(0.9 % NaCl)	S.D.	0.25	2.4
Fructose	$\bar{x}$	1.24	21.8
(0.5 mg/g)	S.D.	0.28	3.0
	p	0.01	
Fructose	$\bar{x}$	1.50	35.4
(1.0 mg/g)	S.D.	0.36	3.2
	p	0.01	0.001
Fructose-1-P	$\bar{x}$	0.68	17.8
(1.0 mg/g)	S.D.	0.31	2.6
Fructose-6-P	$\bar{x}$	0.95	14.7
(1.0 mg/g)	S.D.	0.30	2.8
Glyceraldehyde-3-P	$\bar{x}$	0.81	18.7
(0.1 mg/g)	S.D.	0.24	3.3
Xylitol	$\bar{x}$	1.12	28.9
(0.5 mg/g)	S.D.	0.26	3.6
	p	0.05	0.001

Table 3.   Activities of 5'-nucleotidase (5'-N) and
adenosine deaminase (ADA) in rat liver homogenates
after 15 min. incubation with fructose, fructose-1-P,
fructose-6-P, glyceraldehyde-3-P and xylitol respectively at 37°C. Mean values of 6 animals. Enzyme
activities are expressed as mU/mg protein. The amounts
of added substances are expressed as mg/g wet-weight
of rat liver.

chosen for our experiments, the 2 sugar surrogates
were found to enhance 5'-N and ADA activities within
a relatively short time (after 15-minute infusions).
By contrast, 1-14C-glycine incorporation into free rat
liver adenine nucleotides was still unchanged at that
time under identical in vivo conditions. For de novo
formation of adenine nucleotides a latent period was
apparently necessary. The effects both in terms of
enzyme activation and (delayed) de novo purine synthesi
were found to be most pronounced at the highest
fructose and xylitol concentrations.

The enhanced purine nucleotide degradation
associated with 5'-N and ADA activation might suggest
a compensatory increased recycling of free purine
bases through the salvage pathway. This mechanism is,
however, unlikely, since HG-PRT and A-PRT activities
failed to rise during the short time experiment with
fructose. The absence of transketolase and transaldolase changes under identical conditions apparently
also rules out an early activation of the pentose
phosphate cycle by fructose and xylitol.

The effects of fructose and xylitol on 5'-N and
ADA activities were also studied in vitro. Results
obtained with anuclear rat liver homogenates basically
substantiated the in vivo data. When fructose was
replaced by phosphorylated intermediated of fructose
metabolism for incubation, the activities of the
enzymes assayed were not found to increase. This
suggests that fructose metabolites have no direct
effect on 5'-N and ADA activities. As postulated by
other workers in this field, enzyme activation is more
likely to be due to a depletion of ATP and P_i (5, 12,
18 - 20). Much the same mechanism is apparently at
work in the development of xylitol-induced hyperuricemia.

Our results suggest an early activation of
catabolic enzyme reactions involved in purine
metabolism. This agrees well with clinical observations
and animal experiments which have shown serum uric acid
levels to rise rapidly after administration of
unphysiologically high fructose and xylitol doses.

Supported by "Fonds zur Förderung der wissenschaftlichen Forschung" (Austria), projects 1804 and
2746.

REFERENCES

1. Bässler, K. H., Stein, G., Belzer, W.:
   Biochem. Z. 346, 171 (1966)
2. Bode, L., Schumacher, H., Goebell, H., Zelder, O.,
   Pelzel, H.:
   Horm. Metab. Res. 3, 71 (1971)
3. Brand, K.:
   In: Bergmeyer, H. V.: Methoden der enzymatischen
   Analyse, 3rd ed., p. 672, Verlag Chemie, Weinheim,
   1974
4. Förster, H., Boecker, St., Ziege, M.:
   Med. u. Ernähr. 13, 193 (1972)
5. Fox, I. H., Kelley, W. N.:
   Metabolism 21, 713 (1972)
6. Gerlach, U., Hiby, W.:
   In: Bergmeyer, H. V.: Methoden der enzymatischen
   Analyse, 3rd ed., p.903, Verlag Chemie, Weinheim,
   1974
7. Giusti, G.:
   In: Bergmeyer, H. V.: Methoden der enzymatischen
   Analyse, 3rd ed., p. 1134, Verlag Chemie, Weinheim,
   1974
8. Heuckenkamp, P. V., Zöllner, N.:
   Lancet II, 808 (1971)
9. Horecker, B. L., Smyrniotis, P., Hurwitz, J.:
   J. biol. Chem. 223, 1009 (1956)
10. Kelley, W. N., Rosenbloom, F. M., Henderson, J. F.,
    Seegmiller, J. E.:
    Proc. Nat. Acad. Sci (USA) 57, 1735 (1967)
11. Leuthardt, F., Stuhlfauth, K.:
    Biochemische, physiologische und klinische Probleme
    des Fructosestoffwechsels. Thieme, Stuttgart, 1960
12. Mäenpää, P. H., Raivio, K. O., Kekomäki, M. P.:
    Science 161, 1253 (1968)
13. Marko, P., Gerlach, E., Zimmer, H. G., Pechan, I.,
    Cremer, Th., Trendelenburg, C.:
    Hoppe-Seylers Z. physiol. Chem. 350, 1669 (1969)
14. Perheentupa, J., Raivio, K. O.:
    Lancet II, 528 (1967)
15. Rothman, I. K., Zansani, E. D., Gordon, A. S.,
    Silber, R.:
    J. Clin. Invest. 49, 2051 (1970)
16. Sahebjami, H., Scalettar, R.:
    Lancet I, 366 (1971)
17. Schwarzmeier, J. D., Marktl, W., Moser, K.,
    Lujif, A.:
    Res. exp. Med. 162, 341 (1974)

18. Woods, H. F., Alberti, K. G. M. M.:
    Lancet II, 1354 (1972)
19. Woods, H. F., Krebs, H. A.:
    Biochem. J. <u>134</u>, 437 (1973)
20. Woods, H. F.:
    Nutr. Metabol. <u>18</u> (Suppl. 1), 65 (1975)

THE NADP/NADPH$_2$ RATIO IN ERYTHROCYTES OF HYPERURICEMIC PATIENTS, A POSSIBLE REGULATOR BETWEEN PURINE, CARBOHYDRATE AND LIPID METABOLISM

H. Greiling, J. van Wersch, C. Karatay

Department of Clinical Chemistry

Technical University Aachen, FRG

There is much evidence, that a disturbed carbohydrate and lipid metabolism is connected with a disturbed purine metabolism. Many studies are done about the interrelationship between the purine, lipid and carbohydrate metabolism, but the results are controversial.

The observed hyperuricemia in patients with the enzyme defect of glucose-6-phosphatase (glykogenosis, type I) results in an increased synthesis of glucose-6-phosphate which stimulates the glucose-6-phosphate-dehydrogenase reaction and therefore the 5-phosphoribosyl-1-pyrophosphate-synthesis too. On the other hand the increased NADPH$_2$ concentration activates the fatty acid synthesis.

In our studies we proved to find a correlation between the serum urate concentration and the NAD, NADPH$_2$, NADP, and NADPH$_2$ concentration in erythrocytes of hyperuricemic patients.

The NAD, NADH$_2$, NADP, and NADPH$_2$ in erythrocytes are estimated according to the method of Klingenberg (1).

In comparison to normal individuals we found in hyperuricemic patients an increased concentration of NADPH$_2$ and a decreased NADP/NADPH$_2$ ratio, as it is shown in table 1.

	normal (n = 19) $\bar{x} \pm 2s$	hyperuricemia (n = 16) $\bar{x} \pm 2s$
NADP [µMol/1]	$16,2 \pm 3,1$	$16,8 \pm 4,5$
NADPH$_2$ [µMol/1]	$12,7 \pm 2,9$	$20,7 \pm 8,7$
NADP/NADPH$_2$	$1,35 \pm 0,45$	$0,8 \pm 0,5$

Table 1: NADP, NADPH$_2$ concentrations and the NADP/NADPH$_2$ ratio of erythrocytes of normals and of patients suffering from hyperuricemia

Contrarily to these results the NAD/NADH$_2$ ratio however does not show an essential difference between normal individuals and patients with hyperuricemia (table 2)

	normals $\bar{x} \pm 2s$	hyperuricemic patients $\bar{x} \pm 2s$
NAD [µMol/1]	$29,1 \pm 5,6$	$27,7 \pm 5,8$
NADH$_2$ [µMol/1]	$4,2 \pm 2,5$	$4,5 \pm 3,6$
NAD/NADH$_2$	$7,5 \pm 3,8$	$6,8 \pm 4,2$

Table 2: The NAD, NADH$_2$ and NAD/NADH$_2$ ratio of normals and hyperuricemic patients

In table 3 you see a survey of our trial to correlate the NADP and NADPH$_2$ concentrations in erythrocytes with the urate and triglycerides concentration in blood serum of patients with elevated urate and triglycerides concentrations.
We could not find any correlation between the NADPH$_2$ and triglycerides concentration (coefficient of correlation 0,2), but there is a positive tendency to high urate concentrations in blood serum in connection with low NADP/NADPH$_2$ ratios in erythrocytes.

NADP $[\mu Mol/l]$	NADPH$_2$ $[\mu Mol/l]$	NADP/NADPH$_2$ $[\mu Mol/l]$	Urate $[\mu Mol/l]$	Triglycerides $[mMol/l]$
15, 8	33, 5	0, 47	470	2, 00
16, 4	37, 4	0, 44	571	4, 40
11, 2	24, 5	0, 46	464	5, 70
14, 6	17, 0	0, 86	482	8, 48
14, 7	17, 8	0, 83	791	2, 23
17, 5	15, 8	1, 11	494	5, 97
17, 3	17, 0	1, 02	637	1, 06
14, 5	29, 7	0, 49	613	0, 85
Normal values				
13, 1 - 19, 3	9, 8 - 15, 6	0, 90 - 1, 90	200 - 420	0, 84 - 1, 94

Table 3: Correlation between NAD, NADPH$_2$, NAD/NADPH$_2$
ratio and the urate and triglycerides concentration
in hyperuricemia

In figure 1 we make an attempt to explain the biochemical corre-
lationship between purine, carbohydrate and lipid metabolism,
regulated by the NADP/NADPH$_2$ ratio.
An activation of glucose-6-phosphate-dehydrogenase which results in a higher NADPH$_2$-concentration increases also the
synthesis of phosphoribosyl-1-pyrophosphate, which stimulates
urate synthesis. An increased synthesis of NADPH$_2$ increases
the synthesis of free fatty acids and therefore also the triglycerides synthesis. Gouty patients have decreased concentration
of dehydroepiandrosteron in plasma, erythrocytes and urine (2).
Dehydroepiandrosteron is an inhibitor of glucose-6-phosphate-
dehydrogenase and therefore also a regulator for the pentose-
phosphate-cycle. The decrease of dehydroepiandrosteron activates the glucose-6-phosphate-dehydrogenase and therefore also
the PRPP-concentration and finally the biosynthesis of urate.

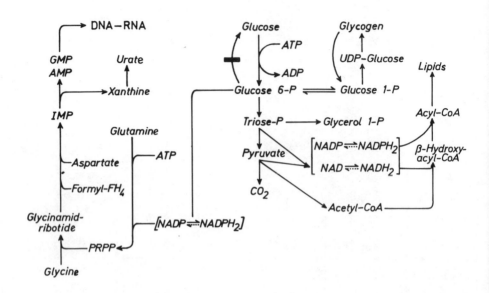

Figure 1: Biochemical pathways between purine; carbohydrate-
and lipid metabolism regulated by the NADP/NADPH$_2$
ratio

Perhaps results the increase of the glucose-6-phosphate-con-
centration in patients with glycogenosis I with the enzyme defect
of glucose-6-phosphatase in an increased rate of the glucose-6-
phosphate-dehydrogenase-reaction. The increased intracellular
glucose-6-phosphate-concentration therefore may be the cause
for the increased biosynthesis of 5-phosphoribosyl-1-pyrophos-
phate.
Further studies are however necessary to study the regulative
system between carbohydrate, lipid and purine metabolism on
the molecular level.

References:
1. M. Klingenberg in Methoden der enzymatischen Analyse II,
   Ed.: H. U. Bergmeyer, Verlag Chemie, Band II, 3. Aufl.,
   2094 (1974)

2. I. H. Casey, M. M. Hoffman, S. Solomon, Arth. and Rheum.
   11, 44 (1968)

HYPERTRIGLYCERIDAEMIA AND CHENODEOXYCHOLIC ACID

Roberto Marcolongo, Alessandro Debolini

Service of Rheumatology

University of Siena, Siena, Italy

Abnormalities of lipoprotein metabolism are frequently found in patient with primary gout. The most common lipid abnormality is hypertriglyceridaemia, independent of, and not secondary to, either obesity, alcohol consumption, or carbohydrate intolerance (1,2). On the other hand, it has been emphasized that endogenous hypertriglyceridaemia is a common metabolic disorder associated with an increased risk of coronary heart-disease (3,4). Thus, therapeutic efforts in this field seem to be indicated. However, the dietary and drug treatment is frequently unsatisfactory, as many patients do not respond adequately or manifest untoward side-effects. Recent — ly, clinical trials of chenodeoxycholic acid (C.D.C.A.), a gall-stone dissolving agent, have demonstrated a decrease in serum triglycerides in patients with varying degrees of hypertriglyceridaemia (5,6,7).

We have evaluated the triglyceride-lowering effect of C.D.C.A. in 27 subjects with hyperlipoproteinaemia type IV (aged 25-58) and in 12 patients with primary gout and associated hypertriglyceridaemia (aged 30-71). All patients had serum triglyceride values greater than 170 mg/100 ml. In the first group, the values ranged from 197 to 1270 mg/100 ml, in the second from 233 to 490 mg/100 ml. After a control period of 2 weeks, a dose of 500 mg of C.D.C.A. was given daily for two months. Triglyceride levels of fasting subjects were measured before treatment with C.D.C.A., then at monthly intervals during treatment, and finally after suspension of treatment. Treatment with C.D.C.A. significantly $(p < 0.001)$

TABLE

Serum triglycerides (mg/100 ml) before, during, and after treatment with C.D.C.A.

	baseline values	after 1 month C.D.C.A. treatment	after 2 months C.D.C.A. treatment	1 month after suspension of treatment
subjects with hyperlipoproteinaemia type IV	448.5 ± 225.0 (°)	265.4 ± 149.6	220.5 ± 125.0	191.8 ± 92.9
gouty patients with hypertriglyceridaemia	337.6 ± 90.8	199.5 ± 50.9	143.2 ± 24.4	132.2 ± 60.8
subjects with "very high" serum triglyceride levels	711.3 ± 233.5	358.5 ± 241.2	298.1 ± 192.9	255.2 ± 136.3
subjects with "increased" serum triglyceride levels	337.7 ± 91.5	226.2 ± 65.0	187.8 ± 65.6	165.0 ± 50.7
subjects aged under 45 yrs	376.8 ± 112.8	224.4 ± 115.9	180.4 ± 87.7	167.8 ± 79.4
subjects aged 45-65 yrs	484.2 ± 162.4	285.9 ± 170.3	240.6 ± 121.4	203.6 ± 65.7

(°) mean ± S.D.

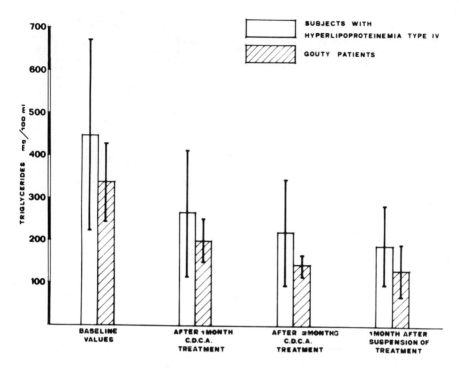

Fig. 1

reduced serum triglyceride levels from a baseline value of 443.5 ±
± 225.2 to 220.5 ± 125 mg/100 ml in subjects with hyperlipoproteinaemia type IV, and from 337.6 ± 90.8 to 143.2 ± 24.4 mg/100
ml in patients with primary gout, after 2 months of therapy (table
and Fig. 1). The differences between these two groups are not
statistically significant. The per cent decrease of serum triglycerides progressively increased during the period of drug administration and the effect persisted even one month after suspension
of treatment, in both groups examined. We confirmed the observation
of Miller and Nestel (7) that the per cent triglyceride reduction
is correlated with the pre-existing serum levels. The subjects with
very high serum triglyceride levels (exceeding 500 mg/100 ml) showed
a per cent decrease significantly (p < 0.02) greater than that
observed in the other subjects (Fig. 2). No significant differences
in per cent decrease of serum triglycerides at different time
intervals have been observed between subjects under the age of 45

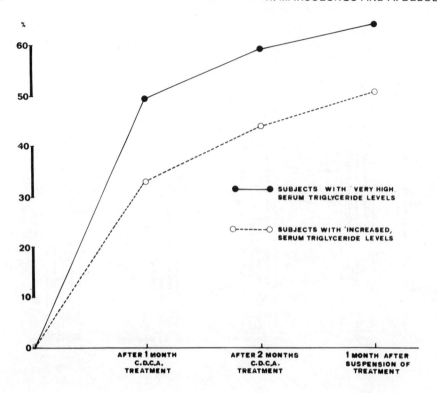

Fig. 2

and those aged 45–65 years. In all subjects studied, there was no significant change in mean serum cholesterol levels and in body weight during C.D.C.A. treatment.

At the present time, the mechanism of the triglyceride-lowering action of C.D.C.A. is unknown, but its effect could be due to either decreased synthesis or increased clearance (5,7). There is general agreement that serum triglycerides derive almost solely from only two sources, intestine and liver, and the likely cause of primary hypertriglyceridaemia is an increased endogenous synthesis of triglycerides (8). The observation that bile acids may inhibit triglyceride synthesis in animal liver and small intestine (7,9), and the absence of an increased plasma clearance of exogenous triglycerides by C.D.C.A. (7), suggest that a decreased triglyceride sinthesis could be the more likely explanation. There is additional evidence for a relationship between bile acids and triglyceride

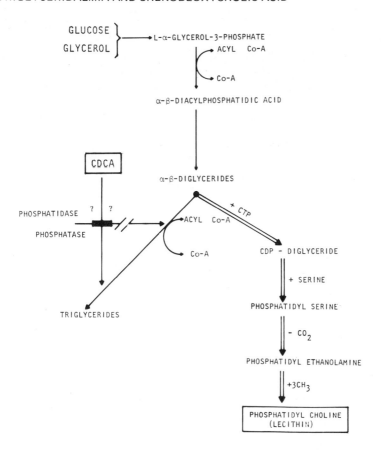

Fig. 3

metabolism. The treatment with cholestyramine and the ileal resec-
tion deplete the bile acid pool, giving a rise in serum triglyceride
levels (10,11,12). Furthermore, since bile acids increase liver
phospholipid synthesis (13), the effect of C.D.C.A. on serum triglyceride levels could occur through an increased utilization of
diglycerides, that are triglyceride and phospholipid precursors,
for this pathway (5), thus reducing endogenous triglyceride
synthesis (Fig. 3).

REFERENCES

1. Emmerson, B.T., Knowles, B.R., Metabolism, 1971, 20, 721.
2. Darlington, L.G., Scott, J.T., Ann.Rheum.Dis., 1972, 31, 487.

3. Carlson, L.A., Bottiger, L.E., Lancet, 1972, i, 865.

4. Goldstein, J.L., Hazzard, W.R., Schrott, H.G., Bierman, E.L.,
   Motulsky, A.G., J.Clin.Invest., 1973, 52, 1533.

5. Bell, G.D., Lewis, B., Petrie, A., Dowling, R.H., Brit.Med.J.,
   1973, 3, 520.

6. Hoffman, N.E., Hoffmann, A.F., Thistle, J.K., Mayo Clin.Proc.,
   1974, 49, 236.

7. Miller, N.E., Nestel, P.J., Lancet, 1974, ii, 929.

8. Fredrickson, D.S., Levy, R.I., Lees, R.S., New Engl.J.Med.,
   1967, 276, 34.

9. Bray, C.A., Gallagher, T.F., Proc.Soc.exp.Biol.Med., 1969,
   130, 175.

10. Jones, R.J., Dobrilovic, L., Circulation, 1969, 40, suppl. 3,
    12.

11. Dowling, R.H., Mack, E., Small, D.M., J.Clin.Invest., 1970,
    49, 232.

12. Press, M., Miller, P., Shimoyama, T., Kikuchi, H., Thompson,
    G.R., Europ.J.Clin.Invest., 1972, 2, 301.

13. Nillson, S., Schersten, T., Europ.J.Clin.Invest., 1970, 1, 109.

FATTY ACID COMPOSITION OF PLASMA LIPID FRACTIONS IN GOUT

A. Novak, E. Knesl, M. M. Müller, E. Kaiser

1st Department of Medical Chemistry, University

of Vienna, 9. Wahringerstrasse 10, Vienna, Austria

Studies of lipid metabolism in gout have dealt almost ex-clusively with demonstrating elevated levels of the total lipids and some of the major lipid classes. Gout is frequently associated with diabetes and atherosclerosis. These disorders too are characterized by hyperlipidemia. Although data are available for total fatty acid composition of gouty synovial fluid (1) and its matching serum, these findings do not readily lend themselves to comparison with results gathered from works on normal serum. The hyperlipoproteinemia characteristic of gout reflects an under-lying disorder in fat metabolism. This alone warranted the separate investigations of the component fatty acids in the major lipid fraction, fatty acids being important parts of lipid molecules. Since the lipid classes have different functions they are expected to have different component fatty acids.

SUBJECTS AND METHODS

The study comprises 53 persons, 28 suffering from primary gout (with high levels of uric acid and triglycerides), the rest apparently healthy, demonstrated by their clinical status and chemical values. Both groups are males, of the same age group and social background. No other matchings were attempted.

Blood samples were drawn after overnight fasting. Plasma were tested for uric acid levels by the method of Kageyama (2). Total lipids, total cholesterol, triglycerides, free fatty acids and phospholipids were determined using Boehringer kits.

For gas chromatographic studies the lipids were promptly extracted and purified according to Folch (3). Separation of the major classes* were carried out by thin layer chromatography. 20 x 20 cm plates were coated with 0.5 mm silica gel G slurry in water. Following activation the plates were pre-run in benzene. The extracts were applied as band and developed in two steps. The first run (10 cm) was done in hexane : diethyl ether : acetic acid (70 : 30 : 1.5). After brief drying the sheets were allowed to develope in benzene (15 cm). All tree eluents – pre-run too – contained 0.02% BHT (butylated hydroxytoluene) in order to lend stability against autoxidation while being exposed to air. The main section of the sheets were covered while spraying the rest with methanolic iodine. The protected bands were scraped off and eluted from the adsorbent using solvents of suitable polarity. The ester bonds were hydrolyzed in methanolic potassium hydroxide and re-esterified in methanolic bortrifluoride (4). The methyl esters were removed by heptane and after suitably concentrating the mixture was subjected to gas chromatography.

Chromatograms were obtained with a Perkin – Elmer F 11 instrument having dual flame ionization detectors and temperature programming. The stationary phases were ethylene glycol succinate and dietylene glycol succinate both coated on Chromosorb W/AW 80-100 mesh. These two types were necessary to resolve certain pairs of esters which were eluted at the same time on one or the other stationary phase. Peak identification was established by comparison of equivalent chain length with published values (5) or by applicati of log-plot correlation of Ackman (6). Beside these, standards from Applied Sci. Labs. were run occasionally.

RESULTS

The results of the two groups of individuals – gouty and contro are tabulated in detail in Tables 1-3 along with statistical evaluations. Table 1 demonstrates the basic differences between the two groups in lipid and uric acid contrations. Since no unusual and hitherto unknown acids were detected in the patients' serum, we have to consider all fatty acids encountered above 0.1% level, the more so because they are interconvertible.

As seen in Tables 2-3 the two groups posses very similar fatty acids composition in all four lipid classes. However, differences were found in CHE for arachidonic acid, with larger amounts in gouty persons, and in palmitoleic and stearic acids in which the difference were reversed (Table 2).

*The following abbreviations will be used in this text: CHE (=cholesterol ester), PL (=phospholipids), TG (=triglycerides), FFA (=free fatty acids)

Table 1. Plasma lipid and uric acid concentration in gouty and
normal persons.

		GOUTY (N=28)		NORMAL (N=25)	
		Mean	± SD	Mean	± SD
Uric acid	(mg/100 ml)	9.1	1.5	5.2	1.8
Total lipids	(mg/100 ml)	1225.	830	722	205
Triglycerides	(mg/100 ml)	349	35	149	57
Phospholipids	(mg/100 ml)	251	77	199	45
Cholesterol	(mg/100 ml)	250	74	224	42
Free fatty acids	(mVal/1)	0.36	0.19	0.25	0.13

In the fatty acids of TG the same result was found for linoleic acid (18:2 w 6*) with the reverse in palmitic acid. Myristic
and palmitoleic acids were also reduced in this fraction of gouty
serum. (Table 2)

In PL the differences between the fatty acid composition of
normal and sick persons were less pronounced. (Table 3)

In FFA fraction higher values were found for oleic acid in
gouty people. This is reflected in a rise of total unsaturated acids.
Concentration of linolenic acid was significantly higher (P<.01) in
the patients FFA fraction also. (Table 3)

When we combine the fatty acids into groups such as unsaturated
and monoenes and those belonging to linoleic and linolenic types we
find significant rise in the total unsaturated acids of the FFA
fraction and in the total acids of the linoleic series of CHE and
of TG.

---

*Convenient notation indicating chain length : number of
olefinic centers, and position of the ultimate double bond counted
from the terminal methyl group. All unsaturated acids mentioned
here are assumed to be cis isomers.

Table 2. Composition of cholesterol esters and triglycerides

(in % of total fatty acids)

Component fatty acids	CHOLESTEROL ESTERS				TRIGLYCERIDES			
	GOUTY		NORMAL		GOUTY		NORMAL	
	Mean	+SD	Mean	+SD	Mean	+SD	Mean	+SD
14:0	1.1	.7 *	1.6	.8	1.3	.7 *	1.8	.9
14:1					.2	.1	.2	.2
15:0	.4	.1	.2	.1	.4	.2	.2	.2
16:0	12.6	2.4	12.1	2.1	24.0	4.5 **	28.3	4.0
16:1 w 7§	3.7	2.0 **	5.3	1.9	4.5	1.2 **	5.4	1.0
16:2 w 4§	.3	.4	.3	.3				
17:0	.4	.5	.5	.6	.5	.3	.5	.4
17:1	.5	.3	.5	.4	.5	.4	.6	.6
18:0	1.6	1.0 *	2.5	.8	4.5	.7	4.7	1.1
18:1 w 9	19.3	2.1	19.1	2.3	39.0	3.3	38.2	2.8
18:2 w 6	45.0	5.8 *	43.3	3.8	17.2	4.1 ***	12.5	2.3
18:3 w 6	1.0	.5	.9	.5	.7	.4	.8	.4
18:4 w 3	.5	.4	.5	.7	.5	.4	.2	.4
18:3 w 3	.8	.3	.7	.3	1.1	.3	1.4	.4
20:0							.2	.2
20:1 w 9	.2	.2	.3	.1	.6	.2	.6	.3
20:2 w 6	.1	.1	.2	.1	.5	.2	.4	.5
20:3 w 3	.8	.2	.7	.4	.4	.3	.4	.3
20:4 w 6	6.8	3.2 ***	4.4	1.3	1.8	1.0	2.0	.8
20:4 w 3	.3	.3	.2	.2	.5	.6	.6	.5
20:5 w 3	1.1	.8	1.1	.7	.3	.3	.8	.9
21:1	.4	.3						
22:1 w 9	.4	.6	.5	.4				
22:2 w 6	.2	.2	.4	.3			.2	.2
22:4 w 3	.2	.2	.3	.3				
22:5 w 3	.3	.2	.2	.2	.2	.2	.3	.3
22:6 w 3	1.1	.8	1.4	.8	.7	.4	.6	.3
Total un-saturated	79.1	7.8	81.0	7.2	67.2	14.2 *	65.5	9.2
Total w 6	54.5	5.9 **	49.5	4.8	20.8	3.6 **	16.4	3.2

§ Other isomers may be present

* P < .05 ** P < .01 *** P < .001

Table 3. Composition of free fatty acids and phospholipids.

(in % of total fatty acids)

Component fatty acids	FREE FATTY ACIDS				PHOSPHOLIPIDS			
	GOUTY		NORMAL		GOUTY		NORMAL	
	Mean	+SD	Mean	+SD	Mean	+SD	Mean	+SD
12:0	.4	.4	.4	.4				
13:0	.2	.3	.4	.4				
14:0	1.3	1.1	1.9	1.5	.8	.7	.9	.8
14:1	.5	.5	.5	.7				
14:2					.3	.5	.2	.2
15:0	.8	.7	.5	.5	.5	.6	.7	.4
15:1	.4	.4	.4	.4				
16:0	21.3	2.8 *	22.0	3.3	24.9	4.9	22.5	4.4
16:1 w 7§	4.4	1.5	5.2	1.4	2.9	1.3 *	2.1	1.0
17:0	.6	.5	.7	.8	.5	.6	.3	.4
17:1	.8	.5	.8	.5	1.1	1.3	.9	.6
18:0	10.9	3.1	11.8	2.9	17.1	2.2	17.3	2.0
18:1 w 9	33.3	7.3 *	28.8	5.2	14.7	3.7 *	17.1	3.9
18:2 w 6	13.1	3.4	13.4	3.5	16.5	3.2 *	19.3	5.3
18:3 w 6	.5	.4	.7	.4	.6	1.4	.5	.6
18:3 w 3	1.2	.5 **	.9	.4	1.3	1.2	1.8	1.3
18:4 w 3	.3	.4	.5	.4	.4	.3	.4	.3
20:1 w 9§	.5	.3	.5	.3	.5	.6	.5	.6
20:2 w 6	.7	.5	.8	.4	.5	.4	.3	.3
20:3 w 3	.5	.4	.5	.4	1.4	.8	1.0	.8
20:4 w 6	2.5	1.4	2.1	1.4	8.1	2.7	7.1	2.7
20:4 w 3	.6	.8	.5	.4	1.0	1.3	.8	.9
20:5 w 3	.5	.3	.7	.7	.7	.6	.4	.4
22:1 w 9	.7	.7	.7	.7				
22:4 w 6	.3	.3	.4	.3	.4	.5	.3	.4
22:4 w 3					.2	.3	.3	.2
22:5 w 3	.3	.3	.3	.3	.5	.7	.4	.6
22:6 w 3	1.2	1.6	1.4	1.4	1.5	.9	1.7	1.4
24:1 w 9					.6	.6	.4	.4
Total un-saturated	67.0	8.3**	60.6	8.0	54.2	4.7	56.2	4.1
Total w 6	17.6	9.6	17.9	8.5	26.6	3.4	27.7	3.0

§ Other isomers may be present

* $P < .05$ ** $P < .01$

DISCUSSION

The results of lipid status essays clearly bear out the differences between those afflicated with gout and those not. As mentioned no aberrant fatty acids were detected, thus excluding their connection in disorders of this kind. The fatty acid pattern found by us in different serum lipid fractions of normal persons compare well with the values in literature (7, 8, 9, 10). However it must be stated that such detailed studies of isomers of unsaturate fatty acids in human serum have not yet been published. Possibly because of imperfect gas chromatographic partitioning many of these compounds were either not mentioned or were else presented with othe acids.

Compared to normal persons, gouty patients show significantly higher amounts of linoleic acid in TG. It is interesting that according to Dyerberg (9) dietary influence is indicated in Eskimos living in Denmark, by the similar high content of this acid in serum TG.

The slight differences among the fatty acids of PL fraction fro gouty and normal groups raise doubts as to the involvement of fatty acid variability in hyperuricemia. While deviations were emphasized in all fraction the similarities are great between the fatty acid spectra of normal and afflicted persons. In Tables 2-3 we see side by side total unsaturated and total w 6 (linoleic type) acids in the lipid fractions of the two groups. The other types not shown, are those belonging to linolenate and oleate families. They vary even to a lesser degree.

Schrade, and later others, found reduced levels of poly-unsaturated acids in sera of diabetics (11, 12), whereas we demonstrated normal concentrations of these acids in gouty persons. However, like the other authors, we too found a decrease in palmitoleic acid (16:1 w 7) in FFA fractions of gouty patients.

Coronary patients tend to have lower levels of linoleic acid and its metabolic product arachidonic acid in CHE fraction, than do healthy persons (13, 14, 15, 16). We failed to demonstrate a likewise diminished level of these acids in the gouty group. The postulated direct correlation between gout and atherosclerosis on the basis of a common disturbance of lipid metabolism, could lead one to expect similar fatty acid pattern of the lipid classes in both diseases. In accordance with the epidemiological investigations of Hall (17) and later Myers (18) the data presented here do not favor the assumption of a common basic disorder of lipid metabolism in gout and atherosclerosis.

This study was supported by a grant of "Fonds zur Förderung der wissenschaftlichen Forschung Österreichs" (project 1804)

REFERENCES

1. Kim,C.H, Cohen,A. (1966) Proc. Soc. Exp. Biol. Med. 123,77.
2. Kageyama, N. (1971) Clin. Chim. Acta 31, 421.
3. Folch,J., Lees, M., Sloane-Stanley,G.H. (1957) J.Biol. Chem. 226, 497.
4. Metcalfe, L.D., Schmitz,A.A. (1961) Anal.Chem. 33,363.
5. Hofstetter, H.M., Sen,N., Holman,R.T. (1965) J.Am.Oil Chem. Soc. 42, 537.
6. Ackman,R.J. (1963) J. Am. Oil Chem. Soc. 40, 558.
7. Wene,J.D., Connor, W.E., DenBesten,L. (1975) J. Clin. Invest. 56, 127.
8. Gnauck, G., Thoelke,H., Zegenhagen,R., Singer,P., Honigmann,G. (1974) Dtsch. Ges. wesen 29, 1406.
9. Dyerberg, J., Bang, H.I., Hjørne, N. (1975) Am. J. Clin. Nutr. 28, 958.
10. Hallgren, L. (1960) J.Clin. Invest. 39, 1424.
11. Schrade, W., Bohle, E., Biegler, R., Harmuth, E. (1963) Lancet 1, 285.
12. Skorepa, J., Fucik, M., Hrabak, P., Mares,P., Todorovicova, H. (1970) Sb. Lek. 72, 166.
13. Schrade, W., Bohle, E., Biegler, R., Ullrich, T+B. (1960) Klin. Wochschr. 38, 729.
14. Boettcher, C.J.F., Woodford, F.P., (1961) J. Atherosclerosis Res. 1, 434.
15. Kingsbury, K.J., Morgan, D.M., Aylott, G., Burton, P., Emmerson, R., Robinson, P.J. (1962) Clin. Sci. 22,161.
16. Alimova, E.K., Endakova, E.K. (1970) Vop. Med. Chim. 16, 310.
17. Hall, A.P. (1965) Arthritis, Rheum. 8, 846.
18. Myers, A., Epstein, F.H., Dodge, H.J., Mikkelsen, W.M. (1968) Am. J. Med. 45, 520.

AN IMPROVED METHOD FOR THE ELECTROPHORETIC SEPARATION AND HISTO-
CHEMICAL IDENTIFICATION OF 5-PHOSPHORIBOSYL-1-PYROPHOSPHATE
SYNTHETASE USING MICROGRAM QUANTITIES OF CRUDE CELL EXTRACTS

Roger V. Lebo and David W. Martin, Jr.

University of California, San Francisco
Departments of Medicine and Biochemistry and Biophysics
San Francisco, California 94143

The enzyme, 5-phosphoribosyl-1-pyrophosphate synthetase, is
required for de novo purine and pyrimidine biosynthesis and purine
salvage and is rate limiting for purine synthesis, a property responsible for its being of particular interest in purine overproduction states. Consequently this enzyme is of central importance
for cellular growth and development. The histochemical stain which
we and others have previously used to detect the enzyme activity
has been modified to be catalytically rather than stochiometrically
dependent upon the formation of ATP. In addition, an electrophoresis system has been developed which separates human from mouse and
hamster PribosePP synthetase while maintaining nearly complete enzyme activity.

ELECTROPHORESIS

5-Phosphoribosyl-1-pyrophosphate synthetase is an unstable enzyme which requires appropriate buffer conditions to maintain its
activity during storage and manipulation. Since the human enzyme
activity is far less stable than that of rodent enzymes, stability
studies were conducted to maximize the recovery of the human
PribosePP synthetase. The electrophoresis buffer which is used
consists of 250 mM Tricine, 1.0 mM $MgCl_2$, 1.0 mM H_3PO_4, 0.05 mM
ribose-5-phosphate, and 3.0% Nonidet, titrated to pH 6.8 with
potassium hydroxide. The organic buffer Tricine; the divalent metal
cation, Mg^{++}; and the anion, PO_4^{3-}, stabilize the enzyme. The high
concentration of Tricine and the nonionic detergent, Nonidet, improve the solubility of the enzyme during electrophoresis. The substrate ribose-5-phosphate improves stability while it does not

interfere with the subsequent histochemical stain. Mg^{++}, PO_4^{3-} and ribose-5-phosphate are maintained at low concentrations to enhance any differences in binding which could then be detected by altered mobility between samples. The pH 6.8 is sufficient to maintain enzyme activity and solubility during migration yet close enough to the pI (~5) to discriminate between enzymes with minimally altered pI's.

Electrophoresis is accomplished in a commercially available apparatus (H. Holzel, Bernoderweg, West Germany) which maintains the temperature at 2°C by circulating 50% cooled ethylene glycol within the metal block upon which the cellogel strips are supported. Evaporation is prevented by applying a lid with an attached rubber gasket sealing the cellogel in a small chamber of air.

Cellogel strips (5 X 28 cm) are equilibrated with buffer overnight and preelectrophoresed for 15 min at 300 volts before adding the samples. Extracts of lymphocytes and rodent cells are prepared by adding to the cell pellet an equal volume of 3% Nonidet in 25 mM Tricine-KOH (pH 7.0), vortexing, freeze-thawing, and centrifuging at 400 X g for 10 min to remove particulate matter. Red blood cell samples are prepared by passing the hemolysate through a small column of glass wool to remove ghosts. Up to nine 0.25 microliter samples are applied to 5 cm wide cellogel strips and electrophoresis follows at 300 volts for 5 hours. Of the applied enzyme activity, 60% to 75% remains following electrophoresis.

HISTOCHEMICAL STAIN

Following electrophoresis, the cellogel is applied to a 3 MM Whatman paper sheet saturated with the histochemical stain. The enzyme may be visualized in 1 to 2 hours. The stain (Figure 1) depends on the formation of ATP from PribosePP and AMP by driving the reaction in the reverse direction. ATP may then be detected by a series of additional reactions. Hexokinase phosphorylates glucose to form glucose-6-phosphate removing the phosphate from ATP to form ADP. Glucose-6-phosphate is then dehydrogenated to 6-phosphogluconate while simultaneously reducing NADP. An hydride ion is then passed to phenazine methosulfate and finally to nitro blue tetrazoleum. The last reaction results in the formation of a blue formazan precipitate which can be detected visually or spectrophotometrically. We have increased the sensitivity of the stain 100-fold by recycling the ADP formed by the hexokinase reaction to ATP using pyruvate kinase and the high energy substrate, phosphoenol pyruvate. Thus, once a single ATP molecule is formed by PribosePP synthetase, the reaction proceeds without net consumption of the ATP, but with continual production of the blue formazan. In 1.5 hours the stain is capable of detecting 50-100 picomoles of ADP.

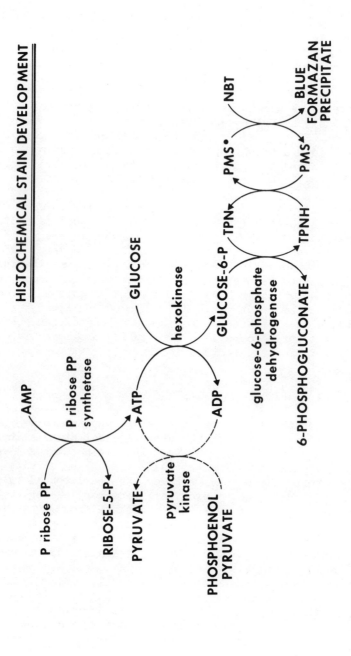

Figure 1. The histochemical stain depends upon the formation of ATP by PribosePP synthetase. ATP is then detected by a series of subsequent reactions which produce a blue formazan precipitate. The pyruvate kinase reaction, which increases the stain sensitivity by regenerating ATP from ADP, is indicated by broken arrows.

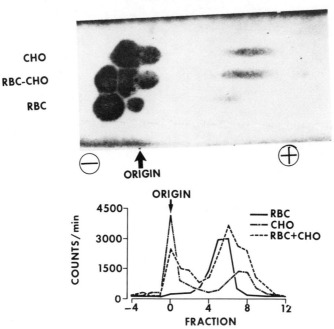

Figure 2. This illustrates the histochemical staining pattern obtained following the electrophoresis of a cultured Chinese hamster ovary (CHO) cell extract, a human white blood cell extract, and a mixture of each (top). The specificity of the histochemical stain was tested using the orotidine-5-phosphate pyrophosphorylase assay for PribosePP synthetase (bottom). While the stain is not entirely specific for PribosePP synthetase, the enzyme migrates anodally to a region where the histochemical stain does correspond to enzyme activity.

When creatine kinase and creatine phosphate are substituted for pyruvate kinase and phosphoenol pyruvate, the stain is 10-fold more sensitive. However, the creatine kinase enzyme is less stable when obtained commercially.

The stain is not entirely specific for 5-phosphoribosyl-1-

P ribose PP SYNTHETASE SCREENING IN RBC

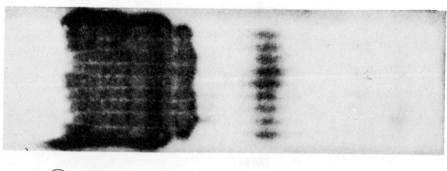

 ORIGIN

Figure 3. This is the histochemical pattern which developed following the electrophoresis of red blood cell extracts from a series
of seven patients. In this case no difference in mobility has been
detected. One sample is routinely spotted in triplicate at the top,
center, and bottom positions.

pyrophosphate synthetase as can be seen in Figure 2. Much of the
stain developing at the origin is not specific according to the
radiochemical assay for PribosePP synthetase used in our laboratory.
However, the enzyme migrates 10-12 cm toward the anode to a region
where nonspecific staining is not observed. As can be seen in the
figure, the histochemical stain correlates with the enzyme activity
detected by the radiochemical assay.

We are currently screening red blood cell samples obtained from
a random population of hospital patients for altered electrophoretic
mobilities of this enzyme activity (Figure 3). Of the first
seventy-five samples tested, one had an altered enzyme mobility.
This patient is hyperuricemic but severe renal failure complicates
the picture and prevents any conclusions from being drawn.

In conclusion, a procedure has been developed for the routine
characterization of 5-phosphoribosyl-1-pyrophosphate synthetases
from different humans and different species. An electrophoretic
system which maintains and separates the enzyme activity from crude
cell extracts and an improved histochemical stain which detects
this activity have contributed to this development.

A NEW ASSAY FOR THE DETERMINATION OF PHOSPHORIBOSYLAMINE

George L. King and Edward W. Holmes

Duke University Medical Center, Durham, North Carolina

27710

INTRODUCTION

Glutamine phosphoribosylpyrophosphate amidotransferase (amido-phosphoribosyltransferase, E.C.2.4.2.14) catalyzes the initial reaction in purine biosynthesis de novo (Goldthwait, 1956; Hartman and Buchanan, 1957). The substrates of this reaction are phosphoribosylpyrophosphate (PP-ribose-P), glutamine and H_2O; the products are phosphoribosylamine (PRA), glutamate and pyrophosphate (PPi). The regulatory properties of amidophosphoribosyltransferase suggest that it may be important in the control of purine biosynthesis (Wyngaarden, 1972; Holmes, et al., 1973; Holmes, Wyngaarden and Kelley, 1973), and as a consequence there has been considerable interest in the study of this enzyme.

Amidophosphoribosyltransferase has been assayed by a variety of techniques. Determination of PRA is the most desirable assay since this is the only product unique to purine biosynthesis. Due to the lability of PRA (Goldthwait, 1956), it has been necessary to use an assay for the determination of this product (Hartman, Levenberg and Buchanan, 1956; Nagy, 1970; Nierlich and Magasanik, 1965) which is coupled with the second enzyme of the purine biosynthetic pathway, phosphoribosylglycinamide (PRG) synthetase (E.C.6.3.1.3). Since the substrates for PRG synthetase (ATP and glycine) as well as additional enzymes and substrates for the determination of PRG must be included in the reaction mixture, this assay has not been entirely satisfactory for routine studies.

Consequently, other assays for amidophosphoribosyltransferase have been employed, such as the determination of glutamate, either as a product of radiolabeled glutamine (Hartman, 1963a) or as a substrate for glutamate dehydrogenase (Hartman and Buchanan, 1957). However, there are several limitations in this assay. For example, it is not possible to evaluate the role of NH_3 as a potential substrate for amidophosphoribosyltransferase, and the presence of glutaminase activity in crude enzyme preparations may obscure the amidophosphoribosyltransferase activity. Other assays have relied upon the detection of PPi (Hartman, 1963a), but the sensitivity of this method has precluded its general use. An assay based on the disappearance of PP-ribose-P has also been described (Goldthwait, 1956; Henderson and Khoo, 1965). However, objections to measuring the disappearance of a substrate and the complexity of coupled enzyme reactions have discouraged the routine use of this assay.

For these reasons we have been interested in developing a sensitive and direct assay for PRA. In this report an assay based on the formation of a stable compound between the ribose-5-phosphate moiety of PRA and [^{35}S]cysteine is described.

METHODS

Enzyme Purification

Amidophosphoribosyltransferase was purified 15-fold from human placenta, as described by Holmes et al. (Holmes, Wyngaarden and Kelley, 1973). This purification involved adsorption of the enzyme to DEAE cellulose and elution with a KCl gradient. The enzyme preparation contained no detectable nucleotidase or glutaminase, and it was stable to storage at $-70°$ for greater than two months (Holmes, Wyngaarden and Kelley, 1973).

Ribokinase (E.C.2.7.1.15) was partially purified from bovine liver by the method of Agranoff et al. (Agranoff and Bruer, 1956).

Enzyme Assays

(1) Amidophosphoribosyltransferase

(A) [^{35}S]cysteine. The [^{35}S]cysteine used for this assay was prepared in the following manner: 10 μl of [^{35}S]cystine stock solution, 25.8mM, were mixed with 250 μl of 0.01N HCl and to this were added 750 μl of 0.5M Tris-HCl buffer, pH 9.0, containing 3mM dithiothreitol (DTT) and 250mM L-cysteine. For assay of amidophosphoribosyltransferase 40 ul of [^{35}S]cysteine solution and 50 μl of enzyme preparation, specific activity of 58 nmoles/hr/mg of protein, were added to 50 μl of substrate solution to yield a final reaction

volume of 140 µl. The standard reaction mixture contained a final
concentration of 5mM PP-ribose-P, 5mM MgCl$_2$, 5mM glutamine or 80mM
NH$_4$Cl, 10mM beta-mercaptoethanol, 0.86mM DTT and 53mM [^{35}S]cysteine
in 100mM Tris-HCl buffer, pH 8.6. A glutamine or NH$_4$Cl blank was
employed for each assay. The reaction was stopped by the addition
of 10 µl of 200mM disodium EDTA and the incubation was continued for
another 15 minutes at 37°. Twenty microliters of the reaction
mixture and 10 µl of carrier ribose-5-phosphate/cysteine were
spotted on Whatman 3MM chromatography paper, and the paper was
electrophoresed for 35 minutes (88 volts/cm) in 25mM sodium citrate
buffer, pH 3.5. The cysteine, ribose-5-phosphate, and ribose-5-
phosphate/cysteine spots were identified with ninhydrin or aniline
spray following development at 70° for 15 minutes. The relative
electrophoretic mobilities of these compounds were 0.14, 1.0, and
0.76, respectively. The ribose-5-phosphate cysteine spots were cut
out and counted at 38% efficiency in a Packard Tri-Carb liquid
scintillation spectrometer.

    (B) [^{14}C]glutamine. This assay has been described and
the stoichiometry characterized in a previous publication (Holmes,
et al., 1973).

Preparation of [^{3}H]Ribose-5-phosphate

    [^{3}H]Ribose-5-phosphate was synthesized from [^{3}H]ribose with the
partially purified preparation of ribokinase. At the conclusion of
the reaction an equal volume of 100% ethanol was added and the
reaction mixture placed in boiling H$_2$0 for 60 seconds. The protein
was removed by centrifugation at 10,000 X g for 15 minutes at 4°.
The ribose and ribose-5-phosphate were separated in the electrophoresis system described above (25mM sodium citrate buffer, pH 3.5)
and the respective areas of the chromatogram were cut out and the
paper digested with 60% perchloric acid - 30% H$_2$0$_2$ (Mahin and
Lofberg, 1966). The [^{3}H] samples were counted at 15.9% efficiency
in Aquasol. Utilizing this system it was confirmed that 94% of the
[^{3}H]ribose was converted to [^{3}H]ribose-5-phosphate.

Autoradiography

    This experiment was performed by spotting 20 ul aliquots of
the reaction mixture containing the indicated carbohydrate and [^{35}S]
cysteine in 100mM Tris-HCl buffer, pH 9.0 on Whatman 3MM chromatography paper. Electrophoresis in sodium citrate buffer was performed
as described above. After the paper was dried, it was placed in an
x-ray casset and the film was developed for 9 days. Only two areas
of radioactivity were identified. One had a relative electrophoretic mobility of 0.14 (corresponding to cysteine) and the other had

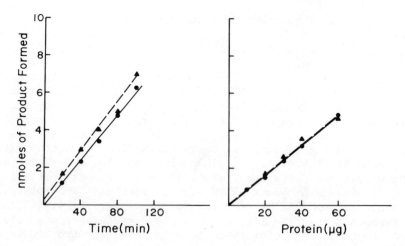

Fig. 1. Comparison of [35S]cysteine and [14C]glutamine assays.
Left panel depicts product formed relative to time of incubation at
37°; right panel depicts product formed relative to enzyme protein.
PRA (●--●); glutamate (Δ--Δ).

a relative electrophoretic mobility similar to that of the ribose-5-
phosphate/cysteine compound.

RESULTS

Enzyme Assays

Figure 1 compares the [14C]glutamine and the [35S]cysteine
assays for amidophosphoribosyltransferase. The reaction conditions
were identical except for the radioactive isotopes of glutamine or
cysteine used. As shown in Figure 1, both assays were linear with
respect to time of incubation and protein concentration used. The
stoichiometry of the reaction was the same with both assays.

Inhibition of amidophosphoribosyltransferase by purine ribonucleotides was compared with these two assays. Inclusion of 5mM
AMP in the reaction mixture resulted in 22% and 24% inhibition of

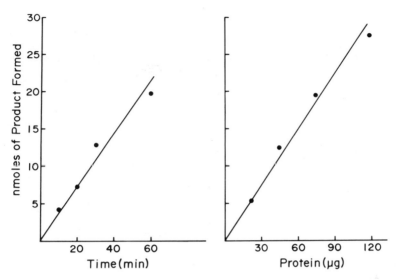

Fig. 2. NH$_3$ utilization by amidophosphoribosyltransferase. Left panel depicts PRA produced relative to time of incubation at 37°; right panel depicts PRA produced relative to enzyme protein.

enzyme activity in the [^{14}C]glutamine and [^{35}S]cysteine assays, respectively. Five millimolar GMP produced 25% and 19% inhibition, respectively.

Figure 2 demonstrates that glutamine may be replaced by NH$_3$ for the synthesis of PRA. Approximately 2 times more PRA was synthesized when NH$_3$, rather than glutamine, was used as substrate.

Characterization of the Reaction Between PRA and Cysteine

Under the conditions routinely used in the enzymatic assay for amidophosphoribosyltransferase, the reaction between PRA and cysteine was complete in 20 min. Cysteine could not be replaced by cystine in this reaction. While 100mM beta-mercaptopropionic acid did not

TABLE 1

REACTION OF CYSTEINE WITH CARBOHYDRATE COMPOUNDS

Compound*	Product[‡] (nmoles)
Aldopentose	
Phosphoribosylamine	211
Ribose-5-phosphate	200
2-Deoxy-Ribose-5-phosphate	209
Arabinose-5-phosphate	231
Ribose-1-phosphate	0.0
Xylose-1-phosphate	0.0
PP-ribose-P	4.5
Ribose	0.0
Adenosine-5'Monophosphate	0.0
Adenosine-5'Diphosphate	0.0
Adenosine-5'Triphosphate	0.0
Guanosine-5'Monophosphate	0.0
Guanosine-5'Diphosphate	0.0
Ketopentose	
Ribulose-5-phosphate	12.0

*250 nmoles of carbohydrate and 4600 nmoles of [^{35}S]cysteine were incubated at 37° for 30 min. in 100 μl of 100mM Tris-HCl buffer, pH 9.0.

inhibit the reaction between cysteine and ribose-5-phosphate, 100mM thioethanolamine produced 82% inhibition of the formation of the ribose-5-phosphate/cysteine compound.

To characterize the reaction between PRA and cysteine it was necessary to synthesize PRA (Malloy, Sitz and Schmidt, 1973). In preliminary experiments it was found that cysteine reacted equally well with PRA and ribose-5-phosphate. This is also demonstrated in Table 1. Because of the ready availability of ribose-5-phosphate, this compound was used for additional studies to characterize this reaction.

TABLE 1 (Cont'd)

REACTION OF CYSTEINE WITH CARBOHYDRATE COMPOUNDS

Compound*	Product[‡] (nmoles)
Aldohexose	
Glucose 6-phosphate	21.0
Glucose-6-sulfate	10.5
Glucose-1-phosphate	0.0
Glucose	0.0
2-Deoxy-Glucose	0.0
Galactose-6-phosphate	50.5
Galactose-1-phosphate	0.0
Mannose-6-phosphate	71.0
Mannose-1-phosphate	0.0
Ketohexose	
Fructose-6-phosphate	0.0
Fructose-1-phosphate	0.0
Fructose	0.0
Sedoheptulose	
Sedoheptulose-7-phosphate	10.5
Sedoheptulose-1,7-diphosphate	72.5

[‡]20 μl of reaction mixture were spotted on Whatman 3MM chromatography paper and electrophoresed in sodium citrate buffer. The area of radioactivity was located by radioautography, cut out, and counted in a scintillation counter as described in the methods section.

Other experiments revealed that the following constituents routinely present in the enzymatic assay for amidophosphoribosyltransferase did not interfere with the formation of the ribose-5-phosphate/cysteine compound: 100mM KCl, 10mM glutamate, 10mM glutamine, 60mM beta-mercaptoethanol, 10mM MgCl$_2$, 20mM EDTA, or 50 μg of enzyme protein.

A survey study of the reaction between cysteine and a number of carbohydrate compounds is presented in Table 1. The compound formed from the reaction between [35S]cysteine and the carbohydrate was identified by radioautography. It was assumed that the area of radioactivity, other than that attributed to the cysteine spot,

represented the carbohydrate/cysteine compound. The relative electrophoretic mobility of the carbohydrate derivatives was similar to
that observed with the ribose-5-phosphate/cysteine compound following development of the paper with ninhydrin spray.

Cysteine reacted equally well with PRA, ribose-5-phosphate,
2-deoxy-ribose-5-phosphate, and arabinose-5-phosphate. There was
essentially no reactivity with ribose. The substitution of a phosphate, pyrophosphate or purine ring at C-1 completely blocked the
reaction with cysteine. However, the hydroxyl group at C-2 did not
appear to be critical for this reaction. Cysteine demonstrated the
following order of reactivity towards carbohydrate compounds:
aldopentose-5-phosphate > aldohexose-6-phosphate > sedoheptulose-7-
phosphate. Ketopentoses and ketohexoses did not readily react with
cysteine.

TABLE 2

DOUBLE LABELING EXPERIMENT

	[^{3}H] (nmoles/ section of paper)	[^{35}S] (nmoles/ section of paper)	Relative Electrophoretic Mobility (paper section)
[^{3}H] ribose-5-phosphate + unlabeled cysteine	4.0	–	0.76
unlabeled ribose-5-phosphate + [^{35}S] cysteine	–	4.5	0.76

For these studies [^{3}H] labeled or unlabeled ribose-5-phosphate
was incubated at 37° for 30 min. in 100mM Tris-HCl buffer, pH 9.0,
with unlabeled or [^{35}S] labeled cysteine, respectively. Twenty
microliters of each reaction mixture and 20 μl of carrier were spotted on Whatman 3MM paper, electrophoresed in sodium citrate buffer,
and the ribose-5-phosphate/cysteine area was identified and cut out.
As indicated in the methods section, the relative electrophoretic
mobility of the ribose-5-phosphate/cysteine compound was 0.76. The
paper was digested with perchloric acid/H_2O_2 and counted with 15.9%
efficiency for [^{3}H] and 38% efficiency for [^{35}S].

Analysis of the Ribose-5-phosphate/cysteine Compound

To substantiate the stoichiometry of the reaction between ribose-5-phosphate and cysteine a double labeling experiment was performed (Table 2). Following incubation of [^{35}S]cysteine with unlabeled ribose-5-phosphate or [H^3]ribose-5-phosphate with unlabeled cysteine, the resultant compound formed from the reaction between these two molecules was isolated by high voltage electrophoresis (see Methods for relative electrophoretic mobilities). That section of the chromatogram corresponding to the ribose-5-phosphate/cysteine compound contained 4.0 nmoles of [^{3}H]ribose-5-phosphate and 4.5 nmoles of [^{35}S]cysteine.

DISCUSSION

The present report describes a new assay for amidophosphoribosyltransferase based on the formation of a stable compound between the ribose-5-phosphate moiety of PRA and cysteine. This assay is comparable in sensitivity to the [^{14}C]glutamine assay and in addition it permits NH$_3$ as well as glutamine utilization for substrate. Stoichiometric analysis reveals that one mole of PRA is generated for each mole of glutamate produced. These data are consistent with earlier studies which have demonstrated that one mole of glutamate is produced for each mole of PP-ribose-P and glutamine consumed in the amidophosphoribosyltransferase reaction (Holmes, et al., 1973). In the enzymatic reaction it is not known whether [^{35}S]cysteine reacts directly with PRA or ribose-5-phosphate which is rapidly formed from the hydrolysis of PRA (Goldthwait, 1956). Since other studies reveal that cysteine reacts equally well with these two compounds, this question has not been pursued in the present study.

Substitutions at the C-1 and C-5 positions of ribose are critical to the formation of the ribose-5-phosphate/cysteine compound. There is a requirement for a phosphate at the C-5 position, and the substitution of a phosphate, pyrophosphate, or purine ring at C-1 results in marked loss of reactivity of the pentose towards cysteine. In addition, the ketoisomer, ribulose-5-phosphate, reacts poorly with cysteine. The observation that cystine cannot replace cysteine in this reaction indicates the need for a reduced sulfhydryl group. The observation that beta-mercaptoethanol and 3-mercaptopropionic acid did not inhibit the formation of the ribose-5-phosphate/cysteine compound, while a concentration of thioethanolamine equal to that of cysteine produced inhibition, suggests that the amino group of cysteine may also be important in the reaction between ribose-5-phosphate and cysteine. These findings indicate some of the requirements for the reaction between ribose-5-phosphate and cysteine, and the studies with radiolabeled cysteine and ribose-5-phosphate confirm that the compound consists of one molecule of ribose-5-phosphate and one molecule of cysteine.

One of the major advantages of this new assay is the potential it provides for evaluating the role of NH_3 utilization in the synthesis of PRA. Previous studies of amidophosphoribosyltransferase from chicken liver (Hartman and Buchanan, 1957; Hartman, 1963a, Hartman, 1963b),pigeon liver (Wyngaarden and Ashton, 1959), and human cells (Reem, 1972) have indicated that NH_3 may replace glutamine as a substrate, and the present study confirms these data. The direct quantitation of PRA formation by the [^{35}S]cysteine assay should be useful in the kinetic evaluation of NH_3 utilization by human amido-phosphoribosyltransferase. Such studies may help to elucidate the role of NH_3 in purine biosynthesis, since a recent report by Sperling, Wyngaarden and Starmer has suggested that NH_3 may be directly incorporated into the purine ring in man (Sperling, Wyngaarden, and Starmer, 1973).

REFERENCES

Agranoff, B. W., and Bruer, R. O. 1956. J. Biol. Chem. 219: 221-229.

Goldthwait, D. A. 1956. J. Biol. Chem. 232: 1051-1068.

Hartman, S. C. 1963a. J. Biol. Chem. 238: 3024-3035.

Hartman, S. C. 1963b. J. Biol. Chem. 238: 3036-3047.

Hartman, S. C., and Buchanan. 1957. J. Biol. Chem. 233: 451-461.

Hartman, S. C., Levenberg, B., and Buchanan, J. M. 1956. J. Biol. Chem. 221: 1057-1070.

Henderson, J. B., and Khoo, M.K.Y. 1965. J. Biol. Chem. 240: 3104-3109.

Holmes, E. W., McDonald, J. A., McCord, J. M., Wyngaarden, J. B. and Kelley, W. N. 1973. J. Biol. Chem. 248: 144-150.

Holmes, E. W., Wyngaarden, J. B., and Kelley, W. N. 1973. J. Biol. Chem. 248: 6035-6040.

Lowry, O. H., Rosebrough, N. J., Farr, A.L., Randall, R. J. 1951. J. Biol. Chem. 193: 265-275.

Mahin, D. T., and Lofberg, R. T. 1966. Anal. Biochem. 16: 500-509.

Malloy, G. R., Sitz, T. O., and Schmidt, R. R. 1973. J. Biol. Chem. 248: 1970-1975.

Nagy, M. 1970. Biochim. Biophys. Acta 198: 471–481.

Nierlich, D. P., and Magasanik, B. 1965. J. Biol. Chem. 240: 358–365.

Reem, G. H. 1972. J. Clin. Invest. 81: 1058–1062.

Sperling, O., Wyngaarden, J. B., and Starmer, C. R. 1973. J. Clin. Invest. 52: 2468–2485.

Wyngaarden, J. B. 1972. Current Topics in Cellular Regulation 5: 135–176.

Wyngaarden, J. B., and Ashton, D. M. 1959. J. Biol. Chem. 234: 1492–1496.

FACILITATED PURIFICATION OF HYPOXANTHINE-PHOSPHO-RIBOSYLTRANSFERASE BY AFFINITY CHROMATOGRAPHY.[+]

Wolf Gutensohn

Institut für Anthropologie und Humangenetik der

Universität, D 8000 Munich, FRG

In recent years quite a number of different methods for the purification of hypoxanthine-phosphoribosyltransferase (HPRT, EC 2.4.2.8) have been developed. They may include ion-exchange chromatography, gel-filtration, preparative isoelectric focusing, preparative polyacrylamide-gel-electrophoresis, or an additional heat-treatment as described by Olson and Milman (1). All these procedures have in common: They need a lot of starting material, are tedious and time-consuming, bring about severe losses of enzyme activity due to handling of the labile enzyme in dilute solution and finally end up with very little pure enzyme. People faced with this problem for some time had in mind that affinity chromatography might be a better method and that GMP with its high affinity for the enzyme might be a suitable ligand.

GMP bound to a matrix via its 2-amino-group (as reported by others) did not give an active adsorbent which is not so astonishing, since we know that a free 2-amino-group on GMP is essential for tight binding to HPRT. We first tried to couple periodate-oxidized GMP to CNBr-activated Sepharose 4B via adipic acid dihydrazide by the method of Wilchek (2). There was some retention of HPRT on this type of adsorbent, but it was not quantitative and therefore this procedure was abandoned.

The method of choice was then published by Hughes and coworkers (3). Here periodate-oxidized GMP is coupled to activated Sepharose 4B via imino-bis-propylamine and the labile Schiff-base is reduced by $NaBH_4$. For the preparation of this

affinity adsorbent we closely followed the procedure given by
the authors just cited. For the purification of HPRT on this
affinity-column the following modification was introduced:
- A crude protein mixture (see below) is applied to the column
  and a large peak of unretained, enzymatically inactive protein
  is first eluted with a low-salt buffer (50 mM TRIS-HCl (pH 7.4)
  10 mM $MgCl_2$/ 1 mM mercaptoethanol/ 25 mM KCl).
- A second inactive protein peak is then eluted with a high-salt
  buffer (50 mM TRIS-HCl (pH 7.4)/ 10 mM $MgCl_2$/ 1 mM mercap-
  toethanol/ 1.2 M KCl).
- However, instead of continuing elution with this high-salt buffer
  (as in Hughes' procedure) whereupon HPRT appears as a rather
  broad peak, we go back to the low-salt buffer and add 5 mM
  GMP. The elution front of the GMP can easily be monitored
  by the OD 260 nm or 280 nm. Immediately with this front HPRT
  is now eluted as a very sharp peak, usually within 10 - 15 ml.
The enzymatically active fractions are then concentrated and
freed from GMP by pressure-ultrafiltration in a small 10 ml-
Amicon cell.

    What are the advantages of this procedure and the modifi-
cation?
1) Preparation of the protein samples for this affinity chromatography can be done by rapid batch-procedures exclusively
which need no checkup by enzyme assays. For example:
- Material from rat-brain is carried through the heat-step procedure, basicly as described (4). Brains are homogenized, a
  soluble supernatant obtained by centrifugation, an ammonium-
  sulfate cut (25 - 60% saturation) is carried out , the material
  is taken up in buffer, dialyzed and heat-treated (65° for 8 min).
  Precipitated protein is removed by centrifugation and the clear
  supernatant is applied to the affinity column.
- Human erythrocytes are washed in isotonic solution and the
  buffy coat is removed. 1 vol of packed cells is lysed in 4 vol
  of 10 mM TRIS-HCl (pH 7.4) and the ghosts are separated by
  centrifugation. HPRT in the lysate is separated from the bulk
  of hemoglobin by a published (5) batch-procedure with DEAE-
  cellulose. The final eluate is concentrated and freed from KCl
  by pressure-ultrafiltration and is then applied to the affinity-
  column.
2) As already mentioned by Hughes even from crude starting
material (e.g. erythrocyte-lysate without any further treatment)
HPRT can be purified on the affinity column with good success
and an example will be given below.

3) In contrast to Hughes procedure the enzyme in the active fractions is not so highly diluted and in addition is protected from inactivation by the GMP in the buffer. So for the subsequent concentration by ultrafiltration we need not add a foreign protein - like cytochrome c or serum albumin - to protect the HPRT and we obtain the enzyme in pure and homogeneous form.
4) In contrast to Hughes´ observation we could reuse the same column frequently after regeneration in high-salt buffer.
5) This is a very rapid method giving you pure enzyme within 2 - 3 days.

How do the enzyme preparations look like ?
- From rat brain we got a 1000 - 2000-fold purification over the crude homogenate and an overall yield of enzyme activity of about 10%.
- From human erythrocyte-lysate (prepared as described above) we got up to 8000-fold purification and yields of about 10%, and that compares favorably with other procedures (6).
- When the lysate is put directly on the column, purification is about 1000-fold and yields are up to 25%.

The homogeneity of the enzyme preparations obtained this way is demonstrated in the densitograms in Fig. 1. The only real difference between the rat and the human enzyme is revealed in the nondenaturing, discontinuous electrophoresis system. Whereas the rat enzyme gives a single sharp band, human HPRT splits up into a cluster of 4 - 5 closely adjacent bands, no longer resolvable by densitometry, of similar intensity and identical distances between bands.

From the SDS gels the subunit molecular weight can be calculated and was found in the range of 25 000 - 27 000 Daltons for rat and human enzyme. In a recent discussion by Strauss (7) the question of the molecular weight of the "genetically coded" or "primary" subunit of HPRT has been raised. The experimental conditions in our study, especially during electrophoresis, are not exactly comparable to those of Strauss. However, although preparation of HPRT by this affinity chromatography seems rather mild and rapid, we have never observed a subunit molecular weight higher than 27 000.

Although these preparations look nicely homogeneous the question is: Is HPRT really the only enzyme retained by this type of affinity-column ? There are of course many enzymes which bind GMP and we could not look for all of them. Nevertheless, the behavior of guanylate-kinase (EC 2.7.4.8) on this

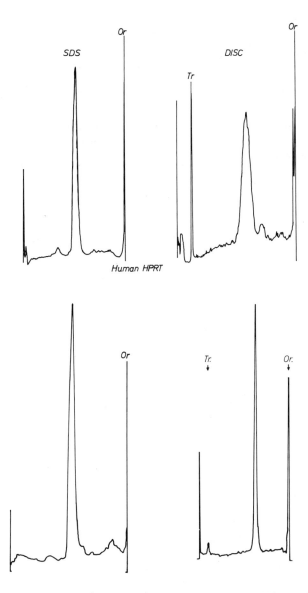

Fig. 1. Densitometry (550 nm) of polyacrylamide-gels stained with Coomassie blue. Upper panel: Human HPRT. Lower panel: Rat-HPRT. Left: 10% SDS-gels (Weber-Osborn system). Right: Discontinuous, nondenaturing system (Davis, Ornstein) in 7% gels. Or = origin of gel, Tr = tracking dye (after destaining visible only in nondenaturing gels). 5 - 10 ug of protein from purified HPRT were separated on each gel.

affinity-column was studied in detail. When a protein-sample is heat-treated - as we usually did with the brain preparatione - guanylate-kinase activity is completely destroyed and therefore can be neglected in the subsequent affinity chromatography. How ever, if there is no heat step included, as usually with the erythrocyte preparations, guanylate-kinase is applied to and retained by the GMP-agarose. Guanylate-kinase activity then appears as a sharp peak at the trailing edge of the protein peak eluted with the high-salt buffer, before and well separated from HPRT which only appears in the GMP-buffer front. This clean separation is observed with rat brain as well as with human erythrocyte preparations. It is clear that the high-salt buffer does elute contaminating proteins together with guanylate-kinase from the affinity-column and that this enzyme preparation cannot yet be homogeneous. This can also be shown by SDS-gel-electrophoresis. Nevertheless, we have obtained a purification of about 300-fold for rat brain and of 1200-fold for human erythrocyte-guanylate-kinase. Although this has not been studied in detail, one should think that appropriate variations in the elution-conditions might render this procedure very useful for the complete purification of guanylate-kinase too.

Literature

+) For further details of this method see: Gutensohn, W., Huber, M. & Jahn, H. (1976) Manuscript submitted to Hoppe-Seyler's Z. Physiol. Chem.

1) Olsen, A.S. & Milman, G. (1974) J. Biol. Chem. 249, 4030 - 4037.

2) Lamed, R., Levin, Y. & Wilchek, M. (1973) Biochim. Biophys. Acta 304, 231 - 235.

3) Hughes, S.H., Wahl, G.W. & Capecchi, M.R. (1975) J. Biol. Chem. 250, 120 - 126.

4) Gutensohn, W. & Guroff, G. (1972) J. Neurochem. 19, 2139 - 2150.

5) Rubin, C.S., Dancis, J., Yip, L.C., Nowinski, R.C. & Balis, M.E. (1971) Proc. Natl. Acad. Sci. 68, 1461 - 1464.

6) Arnold, W.J. & Kelley, W.N. (1971) J. Biol. Chem. 246, 7398 - 7404.

7) Strauss, M. (1975) BiochimBiophys. Acta 410, 426 - 430.

ULTRAMICROCHEMISTRY: A CONTRIBUTION TO THE ANALYSIS OF PURINE

METABOLISM IN MAN

P. Hösli[*] and C.H.M.M. de Bruyn[o]

[*]Department of Molecular Biology, Institut Pasteur,
25, Rue du Dr.Roux, Paris 15, France
[o]Department of Human Genetics, Faculty of Medicine,
University of Nijmegen, Nijmegen, The Netherlands

Since 1963 a simple general methodology has been developed
for ultra-micro enzyme assays of cultured fibroblasts (1,2,3).
These techniques permit to isolate from cell cultures grown in a
Plastic Film Dish (PFD;2) a small number of visually selected,
lyophilised cells.

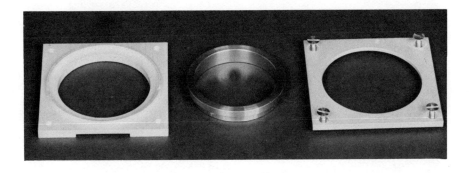

Figure 1.
The three parts of the Plastic Film Dish (PFD).
Left: top-plate; middle: ring; right: base plate.

Figure 2.
Assembled PFD seen from above.

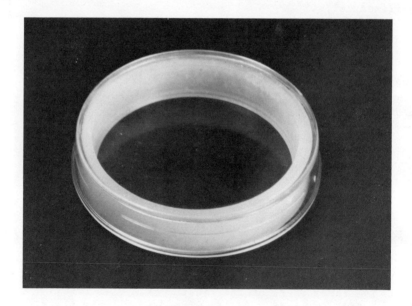

Figure 3.
Disposable version of the PFD.

A PFD is comparable to a Petri dish, with a completely flat, transparent plastic film bottom. It is commercially available from Tecnomara (Zürich, Switserland). It consists of a reusable tripartite holder and a disposable plastic film (fig.1). The ring (fig.1, middle) makes the wall and the completely stretched film the bottom (fig.2). A disposable version of this PFD (fig.3),developed in our laboratory in the past few years,will also become commercially available soon. Cells are cultivated on a 4 to 50 u thick biologically inert plastic film support.

After lyophilisation -to obtain maximal enzyme activity- the cells can visually be selected and cut out free-hand under a stereomicroscope. The single or small numbers of cells on the small plastic leaflet (fig.4) can be assayed for enzyme activity in Parafilm Micro Cuvettes (PMC's;2). These PMC's are very small disposable incubation vessels with volumes of approximately 0.3 ul. Immediately before use they are moulded in a strip of conventional parafilm. They are filled with 0.3 ul of incubation mixture with the aid of a constriction pipette. After the plastic leaflets, carrying the cells to be assayed, have been brought into the PMC's a second parafilm is layered on top of the first (in which the PMC's were moulded) for sealing (fig.5).

These techniques have originally been introduced for enzyme activity measurements with fluorogenic substrates, e.g. lysosomal

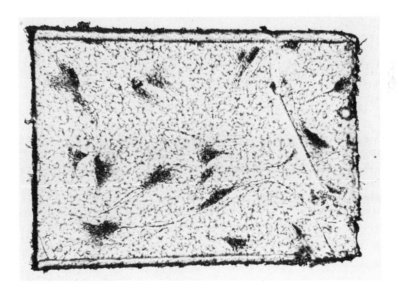

Figure 4.
Plastic film leaflet with a counted number of morphologically typed, lyophilised fibroblasts.

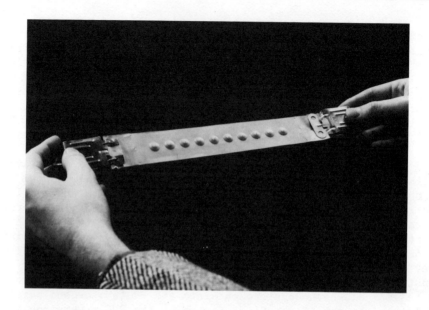

Figure 5.
A series of ten Parafilm Micro Cuvettes (PMC's), ready for incubation in a waterbath.

hydrolases. In this way rapid prenatal diagnosis (3,4,5) and genetic
complementation analysis (2,6,7) could be carried out with minimal
amounts of material.

    Recently these techniques have been adapted for the assay of
enzymes involved in the purine salvage pathway, using natural radioactive substrates (8,9). Depending on the predetermined stability
of the enzyme under study an incubation time is used as long as is
necessary to obtain a signal: noise ratio of at least 3:1. The content of a PMC is then pushed out on a chromatographic paper strip
and separation, identification and quantification of substrates
and products is carried out in conventional ways (8,9).With this set
up several enzymes of the purine interconversions have been scaled
down to the single cell level (HG-PRT; A-PRT; ADA; PNP; ref.9,10).

    A principal adavantage of ultra-micromethods is that a favour-
able signal: noise ratio can be obtained. The noise, "blank", in an
enzyme reaction is almost invariably due to the amount of substrate
employed. The signal: noise can be improved significantly by decreasing the incubation volume (lower blank value) and prolonging
the incubation time (stronger signal). An example is given in table I.

Table I

incubation volume	1 ml	0.1 µl	0.1 µl
incubation time	12 min.	12 min.	600 min.
signal: noise ratio	1:10,000	1:1	50:1

Example of the effect of incubation volume and time on the signal: noise ratio in assays with very limited amounts of material (saturating conditions). The signal:noise ratio of 1:10,000 (left hand column) is a hypothetical value.

Besides of this, further major advantages include:
1. very limited amounts of material can be handled which permits e.g. genetic complementation analysis of mutants involved in the purine salvage pathway, because this method works without the establishment of vital clones. No selective cell culture procedures are needed: fused cells can directly be assayed;
2. the enzyme activities are expressed per cell and not in terms of the total cell protein, the latter being a notoriously bad parameter, fluctuating depending on the cell culture dynamics. Therefore, with the microtechniques gene dosage effects can be assayed very accurately, which permits e.g. to carry out regulation studies with respect to purine metabolism;
3. single, morphologically defined cells can be visually selected and isolated from a cell mixture e.g. from an amniotic fluid culture. This is very essential since the different cell types in amniotic fluid cultures (fibroblasts, epithelial and intermediate cells) display different enzyme activities. With the present ultramicrotechniques normal, heterozygotes and homozygous affected individuals can unambiguously be discriminated (4,7).

An example of a rapid prenatal diagnosis of the Lesch-Nyhan syndrome (associated with severe HG-PRT deficiency) with in total only 50 amniotic fluid derived fibroblasts is shown in table II.

It should be stressed that the methods described in this paper provide a general approach to ultra-microchemistry. In general it can be stated that any assay of an enzyme of purine metabolism can readily be adapted to the single or few cell level if radioactive substrates are available.

Table II

	HG-PRT activity x 10^{-13} moles/cell/ hour .
amniotic fluid derived fibroblasts (pregnancy at risk)	0.07 (0.03-0.12)
control amniotic fluid derived fibroblasts	0.90 (0.80-1.25)
control skin derived fibro-blasts	1.43 (1.25-1.58)

Prenatal diagnosis of a case of Lesch-Nyhan syndrome (severe HG-PRT deficiency). Enzyme activity was determined according to de Bruyn et al. (9). Values given are mean values from experiments with 5 fibroblasts per incubation, carried out in tenfold. Values in brackets indicate the actual range.

REFERENCES

1. Hösli P. (1967) Anat.Anz.120, 583.
2. Hösli P. (1972) Tissue Cultivation on Plastic Films; Tecnomara, Zürich.
3. Hösli P. (1972) Prenat.Diagn.Newsletter (Canada) 1, 10.
4. Hösli P. (1974) In: Birth Defects (A.Motulsky and W.L.Lenz, Eds.) pp.226. Excerpta Medica, Amsterdam.
5. Hösli P. (1974) In: Proc.Third Intern.Congr. Intern.Assoc. for Scient.study of Mental Deficiency (D.A.A. Primrode,Ed.) pp. 394. Polish Medical Publishers, Warsaw.
6. Hösli P. (1972) Bull. Europ.Soc.Hum.Gen.pp.32.
7. Hösli P. (1976) In: Current Trends in Sphyngolipidosis Research and Allied Disorders (B.W.Volk and L.Schneck,Eds.) pp.1 Plenum Press, New York.
8. Hösli P., de Bruyn C.H.M.M. and Oei T.L. (1974) In: Purine Metabolism in Man (O.Sperling, A.de Vries and J.B.Wijngaarden,Eds.) pp. 811. Plenum Press, New York.
9. de Bruyn C.H.M.M., Oei T.L. and Hösli P. (1976) Biochem.Biophys. Res. Commun. 68, 483.
10. Uitendaal M.P., de Bruyn C.H.M.M., Oei T.L. and Hösli P. (1976) These Proceedings.

ULTRAMICROCHEMICAL STUDIES ON ENZYME KINETICS

M.P.Uitendaal[*], C.H.M.M. de Bruyn[*], T.L.Oei[*] and P.Hösli[o]

[*] Dept.Hum.Genetics, University of Nijmegen,The Netherlands

[o] Dept.Molecular Biology, Institut Pasteur, Paris,France

INTRODUCTION

Methods to measure enzyme activities in very small numbers of cells making use of radioactive substrate have been introduced for enzymes of the purine salvage pathway (1).

Two other enzymes of the salvage pathway, nucleoside phosphorylase (NP; EC 2.4.2.1) and adenosine deaminase (ADA; EC 3.5.4.4) have drawn our attention. Deficiency of ADA is shown to be associated with severe combined immunodeficiency (2) and ADA activity is also aberrant in various forms of leukemia (3,4). Recently, a deficiency of NP was reported to be associated with T-cell immunity defects (5).

In this communication a method to determine enzyme properties, such as K_M values and pH optimum, with only 125 cells is reported.

METHODS

The equipment and techniques for the ultramicrochemical measurements with the use of Plastic Film Dishes (PFD's) and Parafilm Micro Cuvettes (PMC's) have been described previously (6).

The reaction mixture for the NP reaction with hypoxanthine as a substrate (NP-Hx) contained 0.17 M Tris-HCl buffer (pH 7.0), 0.5 °/oo BSA, to prevent surface denaturation of the enzyme, 0.8 mg/ml penicillin, 0.8 mg/ml streptomycin, 3.3 mM EDTA to inhibit the hypoxanthine phosphoribosyl transferase reaction, 1.7 mM inorganic phosphate and 0.133 mM ^{14}C-labeled hypoxanthine. For the reverse NP reaction with inosine as substrate (NP-Ino) the reaction mixture was 0.17 M Tris-HCl buffer (pH 7.4), 0.5 °/oo BSA, 0.8 mg/ml penicillin, 0.8 mg/ml streptomycin, 3.3 mM EDTA, 1.7 mM ribose-1-phosphate and 0.267 mM ^{14}C-labeled inosine. Reactions were terminated by pushing out the total contents (0.3 µl) of the PMC's onto Whatman 3MM paper strips (1). Substrate and reaction products were separated by descending chromatography with a 0.5 N ammonia, 0.05 N EDTA solution as eluent.

The ADA reaction mixture contained 0.17 M Tris-HCl buffer (pH 7.4), 0.5 °/oo BSA, 0.8 mg/ml penicillin, 0.8 mg/ml streptomycin, 3.3 mM EDTA and 0.133 mM ^{14}C-labeled adenosine. Separation of substrate and product was performed on DEAE paper with 1mM ammonium formate as eluent.

Unlabeled reference compounds were cochromatographed to visualise after separation the spots under U.V. light. These spots were cut out and counted in a liquid scintillation counter (1).

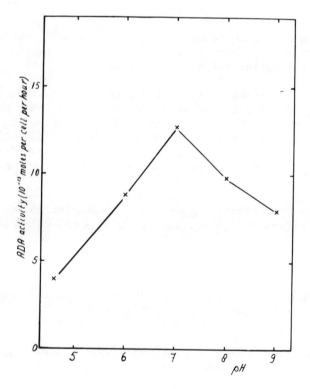

Figure 1.
pH dependence of ADA from normal human fibroblasts. Each point represents an average of five measurements.

RESULTS AND DISCUSSION

In the measurements presented here plastic leaflets cut out from the bottom of a PFD, each carrying five human fibroblasts were incubated in 0.3 µl incubation volumes in PMC's. Because enzyme activities are cell-cycle dependent, each measurement had to be done in fivefold to get a reliable average. By measuring at five different pH values, an indication of the pH optima of ADA (fig.1), NP-Hx (fig. 2) and NP-Ino (fig.3) with in total 125 cells only for each enzyme could be obtained.

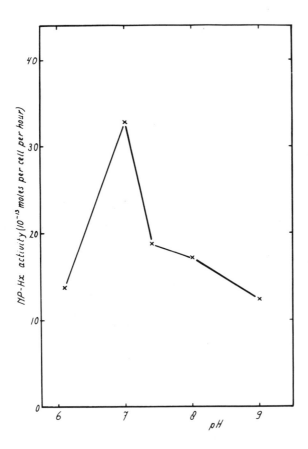

Figure 2.
pH dependence of NP-Hx from normal human fibroblasts. Each point represents an average of five measurements.

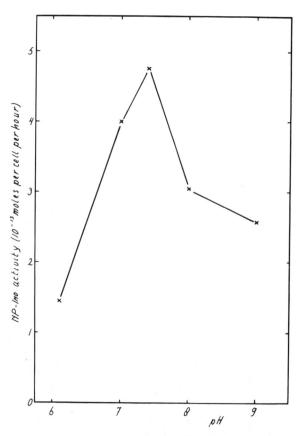

Figure 3.
pH dependence of NP-Ino from normal human fibroblasts. Each point
respresents an average of five measurements.

The pH optima found in this way were in accordance with literature values, e.g. Agarwal et al.(7) and Meyskens and Williams (8)
found a broad pH optimum, respectively, between pH 6 and 8 and betwe
pH 5.5 and 8 for ADA from human erythrocytes. Fig.1 shows a flat pea
in the same area for ADA from human fibroblasts.

By assaying at five different substrate concentrations K_M value.
of adenosine for ADA, of inosine for NP-Ino and of hypoxanthine and
ribose-1-phosphate for NP-Hx could be determined with again only 125
cells for each K_M. The K_M of adenosine for ADA was about 33 uM. Previous studies from other groups using conventional techniques gave
values of 30 uM (9), 25 uM (7) and 40 uM (8) in human erythrocytes.
A K_M of inosine for NP-Ino from fibroblasts of 66 uM could be calculated. This is in good agreement with the K_M values reported for NP-
Ino measured with macromethods: 58 uM (10), 67 uM (11) and 61 uM (12)

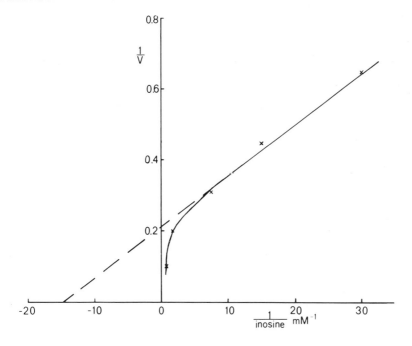

Figure 4.
Lineweaver-Burk plot of NP-Ino activity vs. inosine concentration.

In addition, the ultramicrochemical method could also detect
the activation of NP-Ino at high inosine concentrations (fig.4)
reported by other groups (11,12,13,14).

The K_M values of hypoxanthine and ribose-1-phosphate for NP-Hx
were found to be 16 µM and 0.3 mM, respectively.

The composition of the standard reaction mixtures of NP-Hx, NP-
Ino and ADA is based on the results mentioned above. Assays for NP-
Hx, NP-Ino and ADA performed in this way are completely quantitative
as can be seen in fig.5 and table I. Fig.5 shows a linear dependence
between the activity of NP-Ino and the number of cells incubated and
table I gives ADA activity per cell per hour determined with different
numbers of cells per incubation volumes and with different incubation
times.

As can be seen from the comparison between the results reported
here and enzyme characteristics known from literature, the method
presented in this study offers a rapid and accurate way to get an
indication of enzyme kinetics, consuming only an extremely little
amount of cell material.

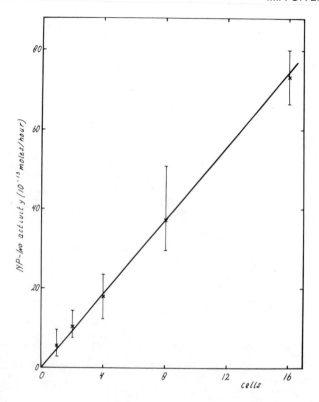

Figure 5.
Relation between NP-Ino activity and the number of cells per assay
volume. Each point represents an average of ten measurements. The
vertical bars indicate the range of the values measured.

These types of ultramicrochemical enzyme characterisations
are of course not restricted to the enzymes mentioned above but
can be applied to any enzyme assay making use of radioactive (1)
or fluorogenic substrate (15). This technique can be very useful
when there is only a small number of cells available e.g. when
several enzymes have to be tested in human biopsies or in biochemical research where prolonged cultivation of a cell clone to
obtain more material would take too much time.

Table I.

ADA activity of human control fibroblasts, expressed in 10^{-13}moles/ cell/hr., determined with 5 and 10 cell incubations using different incubation times.

number of cells per incubation volume	incubation time	
	110 min	230 min
5 cells	7.94	7.83
10 cells	8.67	not tested

ACKNOWLEDGEMENTS

This study was supported by a grant from FUNGO, Foundation for Medical Scientific Research in the Netherlands. The authors gratefully acknowledge the expert assistance of Miss E.Vogt (cell culture) and Mr.F.Oerlemans (enzyme assays).

REFERENCES

1. de Bruyn C.H.M.M., Oei T.L. and Hösli P. (1976).
   Biochem.Biophys.Res.Commun. 68, 483-488.

2. Giblett E.R., Anderson J.E., Cohen F., Pollara B. and Meuwissen H.J. (1972).
   Lancet II, 1067-1069.

3. Smyth J.F. and Harrap K.R. (1975).
   Br.J.Cancer 31, 544-549.

4. Zimmer J., Khalifa A.S. and Lightbody J.J. (1975).
   Cancer Res. 35, 68-70.

5. Giblett E.R., Amman A.J., Wara D.W., Sandman R. and Diamond L.K. (1975).
   Lancet I, 1010-1013.

6. Hösli P. (1972).
   Tissue cultivation on plastic film (Tecnomara,Zürich).

7. Agarwal R.P., Sagar S.M. and Parks R.E.Jr. (1975).
   Biochem.Pharmacol. 24, 693-701.

8. Meyskens F.L. and Williams H.E. (1971).
   Biochem.Biophys. Acta 240, 170-179.

9. Osborne W.R.A. and Spencer N. (1973).
   Biochem.J. 133, 117-123.

10. Kim B.K., Cha S. and Parks R.E.Jr. (1968).
    J.Biol.Chem. 243, 1771-1776.

11. Agarwal R.P. and Parks R.E.Jr. (1971).
    J.Biol.Chem. 246, 3763-3768.

12. Turner B.T., Fisher R.A. and Harris H. (1971).
    Eur.J.Biochem. 24, 288-295.

13. Kim B.K., Cha S. and Parks R.E.Jr. (1968).
    J.Biol.Chem. 243, 1763-1770.

14. Agarwal, K.C., Agarwal R.P., Stoeckler J.D. and Parks R.E.Jr.
    (1975).
    Biochemistry 14, 79-84.

15. Hösli P. (1976).
    In: Current Trends in Sphyngolipidosis Research and Allied Dis-
    orders. pp. 1-13. Plenum Press, New York.

LOCALIZATION OF XANTHINE OXIDASE ACTIVITY IN HEPATIC TISSUE.

A NEW HISTOCHEMICAL METHOD

C.Auscher, N.Amory, C.Pasquier and F.Delbarre

Institut de Rhumatologie. Centre de Recherches sur les
Maladies osteoarticulaires U.5 INSERM;ERA 337 CNRS.
27 rue du Fg St Jacques 75014 Paris, France

A histochemical method of localization of xantine oxidase
activiy in glutaraldehyde-fixed rat jejunum had been reported
by Sackler (1). The reaction is carried out by the reduction
of nitro-blue tetrazolium (NBT) into insoluble formazan, in
phosphate buffer with hypoxanthine as substrate. Though Picklett
et Al (2) have introduced a modification in fixing the sections,
that histochemical technique is ineffective to demonstrate xanthine
oxidase activity in rat liver. On the opposite side, biochemical
methods can show this enzyme activity. The inability to demonstrate
the hepatic enzyme is believed to depend on its lower concentration
in liver as to compare to that of the jejunal mucosa (1).

In the present report, new technical conditions will be described
They permit to locate xanthine oxidase (XO) activity in hepatic
tissues by the reduction of nitro-blue tetrazolium into formazan
and to enhance the staining on jejunal mucosa sections.

The method is based on the polyvinyl alcohol (PVA) techniques
developped by Altman (3) and by Chayen et Al.(4) in connection with
tetrazolium method for oxidative enzymes. The main following
conditions were used :
1 - rapid and low freezing temperature of the tissue,
2 - unfixed sections,
3 - polyvinyl alcohol as stabiliser,
4 - soluble menadione as intermediate hydrogen carrier.

605

MATERIALS AND METHODS

Female Sprague Dawley rats weighting 150 g were used in all experiments. The rats were fed ad libidum on a complete laboratory diet and had free access to water. The animals were killed by stunning.

The liver and the jejunum were removed at once and placed at + 4°C. Jejunum was carefully washed in cold saline to evacuate its content. The tissues were then blotted dry and small pieces of liver (5x5x5 mm) or of jejunum (0.5 cm) were immediately chilled by precipitate immersion in n-isopentane (2-methyl butane OSI), cooled by liquid nitrogen. Temperatures were carefully controled. The pieces of liver were supercooled for 5 seconds at - 60°C and that of jejunum for 10 seconds at - 70°C. Blocks were kept at - 20°C for about 18 hours. They were sectioned at 10 µ for liver and at 4 µ for jejunum in a IEC cryostat in which the ambient temperature was maintained at - 20°C. The sections were picked off the knife on a cover-slide and, to avoid drying, they were maintained in a humidity chamber at + 4°C.

The incubation medium was prepared as follows: polyviol M 05/140 (Wacker-Chemie GMBH) was dissolved at 22 % (w/v) in phosphate buffer 200mM, pH 7.4 and hypoxanthine 0.25 mM in a boiling water-bath. The PVA solution was kept during the night at + 4°C. The next morning, just before sectioning the tissues, menadione sodium bisulfite (Merck) 1 mM and p-NBT (Calbiochem) 1 mg/ml were added to the medium which was kept in darkness. At the time of the reaction media were equilibrated at 37°C and saturated with nitrogen. For control reaction allopurinol (10^{-5}M) was included in the medium.

The reaction was carried out in darkness at 37°C in oxygen-free nitrogen, water vapour saturated chamber. A small perpex ring was placed over the section. About 0.5 ml of the incubation medium was dropped on the section, inside the ring.
When incubation was over, after 30 minutes for liver and 15 minutes for jejunum sections the slides were immediately rinsed in distilled water in staining jar, in order to eliminate the medium and to stop the reaction. The slides were then cleared in successive rinses of alcohol and toluene and mounted on glass-slides with permount.

The control reaction. In all assays, serial sections were incubated in the same way in medium including allopurinol.

RESULTS

There was no disruption of tissue either in liver or in jejunum mucosa sections. The sections though unfixed were quite intact.

The liver sections showed an important purple deposit of forma-zan in the cytoplasm of hepatic cells. There was a good diffusion of the staining. No reaction appeared on the control sections. The nuclei are faintly pink-colored in the reaction as in control (fig.1).

The jejunal sections showed an intense diffuse staining of the apical portion of the epithelium of villi that did not appear in control sections. The chorionis were stained in both the control and the reaction sections (fig.2).

There was no modification when hypoxanthine was replaced by xanthine (10^{-5} M) as a substrate or when pH of phosphate buffer was increased to pH 7.8 ; but, at pH 8.2 in glycyl-glycyl buffer, there was disruption of the tissues though the staining was about the same. Other serial sections were done in order to control the effects of different factors (fixation, gas phase, concentration of substrate) on the reaction. It seems that the complete medium was needed to perform the reaction. Menadione was strickly essential.

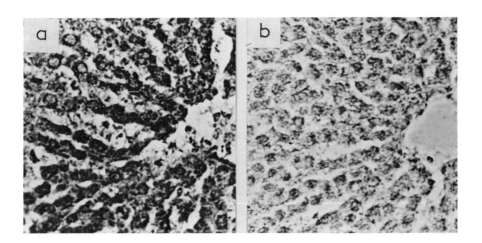

Figure 1.   10 μ unfixed rat liver sections. 30 min. anaerobic
             incubation in the medium (see text)
             a) reaction   b) control reaction

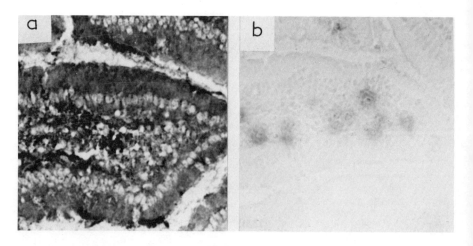

Figure 2.   4 μ rat jejunum unfixed sections.15 min. anaerobic
            incubation in the medium (see text)
            a) reaction b) control reaction

     Serial sections were incubated in the same medium in which
phenazine methosulfate (PMS) was substituted to menadione as to
compare their relative values. PMS also appeared to enhance the
reaction but the diffusion of the staining was not as good as with
menadione. Furthermore, PMS is a very unstable coumpound especially
in light.It is also a substrate for aldehyde oxidase (5) while menadione is an inhibitor of aldehyde oxidase (6).

                              CONCLUSION

     Rapid supercooled tissues and PVA techniques on unfixed sec-
tions developped by Altman (3) for the quantitative dehydrogenase
histochemistry by the reduction of tetrazolium salts are suitable
to locate xanthine oxidase activity in rat liver and to enhance it
in jejunal mucosa sections. However, the intermediate hydrogen
carrier, menadione, appears to be a critical factor for the histochemistry performance of xanthine oxidase activity. Preliminary
studies have shown that this method should be effective for the
localization of the enzyme in human liver sections.

This histochemical method preserves the quantity of enzyme in a native state inside the tissue and does allow to locate and maintain xanthine oxidase activity in situ.

The conventional method (1) was inefficient to show this enzyme activity in liver sections. The histochemical method we have developped allows as well as the biochemical methods to localize xanthine oxidase activity in hepatic tissue (7). It should allowed also to investigate XO activity in other tissues, especially in the kidney and to develop the method in a quantitative way.

ACKNOWLEDGEMENTS

We are indebted to Dr H.Farkas for her helpful advice. (laboratoire de neuro-pathologie, Hopital St Vincent de Paul Paris, France.)

REFERENCES

(1). Sackler M.L. Xanthine oxidase from liver and duodenum of the rat. Histochemical localization and electrophoretic heterogeneity. J. Histochem. Cytochem. 1966, 14, 326-333.

(2). Picklett J.P, Pendergrass R.E, Bradford W.D and Elchepp J.G. Localization of xanthine oxidase in rat duodenum. Fixation of section instead of blocks. Stain Techn. 1970, 45, 35-36.

(3). Altman F.P. Quantitative dehydrogenase histochemistry with special reference to the pentose shunt pathway. Prog. Histochem. Cytochem. 1972, 4, 225-273.

(4). Chayen J. Loveridge N. Ubhi G.S. The use of menadione as an intermediate hydrogen-carrier for measuring cytoplasmic dehydrogenating enzyme activities. Histochemie 1973 , 35, 75-80.

(5). Rajagopolan K.V. and Handler P. Oxidation of phenazine methosulfate by hepatic aldehyde oxidase. Biochem. Biophys. Research. Comm. 1962. 8, N°1, 43-47.

(6). Johns D.J. Human liver aldehyde oxidase : differential inhibition of oxidation of charged and uncharged substrates. The J. of Clin. Invest. 1967, 46, N°9, 1492-1505.

(7). Auscher C. and AMORY N. The histochemical localization of xanthine oxidase in the rat liver. Biomedicine Express. 1976, 25, 37-38

ANALYSIS OF PURINES AND THEIR NUCLEOSIDES BY THE REVERSE PHASE PARTITION MODE OF HIGH PRESSURE LIQUID CHROMATOGRAPHY

P. R. Brown, A. M. Krstulović and R. A. Hartwick

Department of Chemistry
University of Rhode Island
Kingston, Rhode Island 02881

INTRODUCTION

Although the analysis of nucleotides in blood is now being carried out routinely by high pressure liquid chromatography (HPLC) (1-4), and work is in progress to determine alterations casued by various disease states of free nucleotide concentrations in physiological fluids and cell extracts (5,6), less attention has been paid to free nucleoside levels in cells. Recent investigations, however, have focused attention on methylated nucleosides in cells as biological markers in cancer (7,8). Moreover, nucleoside concentrations are thought to be important in cardiac disease and birth defects. The intracellular level of adenosine, one of the physiological regulators of coronary blood flow, has been found to increase in cardiac hypertrophy and after brief periods of myocardial ischemia and hypoxia (9,10). In abnormalities in purine metabolic pathways caused by disease, the concentrations of purine bases and/or their nucleosides often are involved. For example in patients with adenosine deaminase deficiency, it has been observed that excess adenosine in cultured mammalian cells is toxic (11) and it has been postulated that accumulated adenosine is associated with severe immunological defects in children (12).

In metabolic studies, a change in the concentration of a specific nucleoside may be accompanied by a concomitant change in its base. Thus it is important to be able to monitor the nucleosides simultaneously with their bases in small samples of tissue or physiological fluids. For example, in a study of the Lesch Nyhan Syndrome, it was shown that when erythrocytes from patients with this disease are incubated with guanosine, the guanosine is

quantitatively converted to guanine. The guanine accumulated because it cannot be converted to GMP (12) due to the absence of
hypoxanthine guanine phosphoribosyl transferase. In addition,
nucleosides and their bases in cell extracts may be present in
very small concentrations while the nucleotides are present in far
larger concentrations. Thus either the nucleotides must be removed
prior to the HPLC analysis or they must have different retention
times so that they do not interfere with the analyses of the
nucleosides and bases.

While ion exchange chromatography has been the method of
choice for the separation of nucleic acid components (3) this
technique has certain limitations. Since nucleosides and bases do
not possess ionic phosphate groups, the partition mode of HPLC can
be utilized for the analysis of these compounds. The development
of microparticle chemically-bonded reverse phase partition packings
for high pressure liquid chromatography (HPLC) has opened new possibilities for these analyses. Various applications of the partition mode of LLC have been applied to the nucleosides and bases.
However, prior to the development of chemically-bonded packings,
the partition mode could be difficult to use because of column
instability caused by the liquid phase bleeding off the support.
Packings in which the liquid stationary phase is chemically-bonded
to an inert support have an advantage because the loading of the
liquid phase on the inert support is stable and permits reproducible separations over long periods of time.

Furthermore, since it has been predicted by theory (16,17)
that a decrease in particle size increases column efficiency, the
use of microparticle reverse phase packings should have advantages
over similar packings with larger particle size.

Therefore, microparticle chemically-bonded reverse phase packings were investigated for the development of a rapid, sensitive,
reliable method for the quantitative HPLC analysis of the naturally
occurring nucleosides and their bases which could be used in investigations of normal and abnormal purine metabolism. Conditions
were optimized so that levels of purines could be determined simultaneously with their nucleosides in the presence of nucleotides in
cell extracts and plasma.

EXPERIMENTAL

A Waters ALC 202 liquid chromatograph with a micro UV detector
with a fixed wavelength at 254 nm was used. Gradients were generated by a solvent programmer accessory. Peak areas were electron-
ically integrated using a Hewlett-Packard Model 3380-A integrator.
A Model SF-770 Spectroflow Monitor variable wavelength detector

was used to obtain the absorbance ratios used in peak identifications.

The columns were μBondapack C_{18} from Waters Associates, Inc. or Partisil 10 ODS from Whatman, Inc. The dimensions of the stainless-steel columns were 4 mm ID x 30 cm for the micro C_{18}, and 4 mm ID x 25 cm for the Partisil 10 ODS. The packing material in both columns (prepacked), consisted of a totally porous silica backbone to which octadecyl groups are bonded through a silane bond.

Aqueous buffers were prepared using reagent grade potassium dihydrogen phosphate from Mallinckrodt Chemical Works and distilled water, which was filtered through a membrane filter (Whatman Inc.). Unless otherwise indicated, the buffers were prepared in the formal concentrations of 0.01 F, after which the pH was adjusted to 5.8 using dilute potassium hydroxide.

Standards of the nucleosides adenosine, guanosine, inosine, xanthosine, and the bases adenine, guanine, hypoxanthine, xanthine were obtained from the Sigma Chemical Co. Standard solutions were prepared at a concentration of 1 mM in a solution of 0.005 F KH_2PO_4, and pH adjusted to 4.8. Before analysis, all cell extracts were filtered using a membrane filter, pore size GS, from the Millipore Co.

The extraction procedure described by Khym (18) was used to prepare the cell extracts prior to chromatography. The proteineous material was precipitated by the addition of 2 volumes of cold TCA (6% by weight). The solution was centrifuged and then filtered. To one ml of the supernatant fluid two ml of a 0.5 F solution of tri-octyl amine in freon was added. After vortexing and centrifuging the solution, the top layer was withdrawn and stored at -4°C.

Recovery experiments were carried out by adding known amounts of adenosine to a plasma matrix. These solutions were extracted according to the procedure outlined above, and recovery was quantitative.

Initial peak identification was made on the basis of retention times. Standard solutions were run daily before and after samples to determine reproducibility of retention times. The standard addition method was also used in which known amounts of pure compounds were added. A quantitative increase in peak area was taken as further identification of a peak.

Some cell extract peaks were identified by the enzymatic peak shift technique. The enzyme used in this study was xanthine oxidase which catalyzes the conversion of hypoxanthine and xanthine to uric acid. The xanthine oxidase, E.C. No. 1.2.3.2 (grade IV) was obtained from the Sigma Chemical Co. In the standard assay, the sample was incubated with the enzyme at 25° and a pH of 9.2 for 5 minutes. The reaction of the enzyme was stopped by the addition of TCA, which was subsequently removed by extraction with water-saturated ether.

Area ratios of peaks at various wavelengths were also employed to aid in identification. In such instances, triplicate runs were made at 260, 250 and 230 nm, and the ratios of the unknown peaks were correlated to the area ratios of pure compounds eluted with similar retention times.

RESULTS

Optimal resolution of the purine bases and their nucleosides under study was achieved using gradient elution, with 0.01 F KH_2PO_4, pH 5.8 as a low strength eluent and a solution of 100% CH_3OH as a high concentration eluent. The slope of the linear gradient was from 0% methanol to 55% over a period of 30 minutes. The optimal flow rate was found to be 2.0 ml/min. Column temperature was ambient.

The separation of the four bases (in order of elution) xanthine, hypoxanthine, guanine and adenine and their nucleosides is shown in Fig. 1A, these nucleosides and bases in the presence of the pyrimidines thymine, uracil and cytosine and their nucleosides in Fig. 1B, and all these compounds in the presence of the mono-, di-, and triphosphate nucleotides of both the purines and pyrimidines in Fig. 1C. As can be seen from the chromatogram, the nucleotides were not retained on the column and all the pyrimidine compounds except thymidine which eluted at early retention times (3.3-5.0 min)   did not interfere with the purines. Thymidine eluted at a retention of 16.7 minutes which was clearly separated from all other peaks. However, thymine and xanthine had close retention times (7.9 and 8.3 min respectively).

The retention times, peak areas and shapes were reproducible. They remained constant for sample sizes up to 200 nanomoles and the lower limits of detection was approximately 20 picomoles (Fig. 2). The pH of the system must be carefully controlled using reverse phase partition separations if reproducible retention times are to be obtained as the retention times of xanthosine, adenine and adenosine are affected by slight changes in pH (19).

This HPLC method was used in several biomedical studies. An application of this technique was a pharmacological study in which

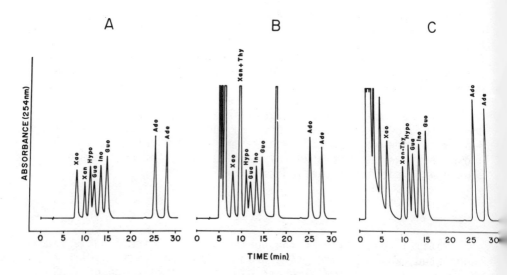

FIGURE 1A. Purines and their bases separated by HPLC using the conditions described in the text; 1B. in the presence of pyrimidines and their nucleosides; 1C. in the presence of pyrimidines, their nucleosides and the purine and pyrimidine mono-, di-, and triphosphate nucleotides.

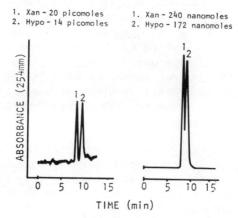

FIGURE 2. Limits of detection of xanthine (1) and hypoxanthine (2). A. Lower limits ∿20 picomoles; B. upper limits ∿200 nanomoles.

nucleosides and/or bases were monitored in human plasma. It was
found in this sample, that xanthine was present in significant
quantities. In order to identify unambiguously the peak tentatively identified as xanthine, the standard addition method, the
enzymic peak-shift, and peak area ratios at two wavelengths were
used. Co-injection of the sample first with hypoxanthine, then
xanthine, indicated that the peak was xanthine (Fig. 3). Further
identification was made on the basis of ratios of peak areas at
different wavelengths. The area ratios of standard solutions of
hypoxanthine and xanthine were obtained at 260 nm/250 nm and
250/230 nm. As shown in Table 1 the ratio of the peak areas at
260/230 nm and 250/230 nm of the peak in the cell extract was similar to that of xanthine, and dissimilar to that of hypoxanthine;
thus, further evidence was obtained that the identity of the extract peak is that of xanthine. In addition, the enzymic peak-
shift technique was used, utilizing xanthine oxidase to convert
either xanthine or hypoxanthine to uric acid (Fig. 3B).

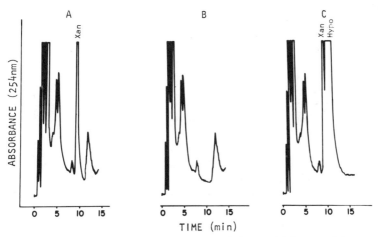

FIGURE 3. Enzymatic peak shift with xanthine oxidase.
A. Human plasma sample; B. Human plasma sample after treatment with
xanthine oxidase; C. Human plasma coinjected with hypoxanthine.

TABLE I

| | Standards | | |
	Hypoxanthine	Xanthine	Peak of Interest
260/230 nm	0.67±0.05	2.65±0.04	2.71±0.01
250/230 nm	1.14±0.03	1.98±0.02	1.96±0.05

DISCUSSION

The reverse phase partition mode for the separation of nucleosides and bases offers several advantages over ion-exchange techniques used previously in HPLC. The naturally occurring purines in cell extracts and plasma can be analyzed simultaneously with their nucleosides in the presence of pyrimidines, their nucleosides and nucleotides in less than 30 minutes.

The analyses are achieved at ambient temperatures with improved sensitivity, selectivity and efficiency. Furthermore, the eluents are dilute salts and methanol, which can be readily removed after fraction collections. The chromatograms are highly reproducible and column stability is good. Columns have been used continuously in our laboratory for six months with no loss in efficiency. Peaks can be identified unambiguously by a combination of methods; standard addition method, enzymic peak-shift technique and use of peak area ratios at two different wavelengths.

Proper care is especially necessary to obtain long life from microparticle columns. Particulate matter must be removed from samples and solvents by membrane filtration. Mechanical shock and extremes of pH and temperature must be avoided. The recommended pH range for silica based packings is from 2.0 to 7.5. The column should be flushed periodically with a moderately polar eluent such as ethanol or acetonitrile to prevent accumulation on the column of trace organics from the solvents or high molecular weight organics from the samples.

It should be noted that retention characteristics may vary in columns of this type from batch to batch or from different manufacturers. Although the conditions for this analysis may be used as a guideline, conditions must be optimized for each column to obtain the best separations.

ACKNOWLEDGMENTS

We thank Waters Associates Inc. for the use of the Waters ALC 202 liquid chromatograph and columns; Whatman Inc. for columns; the Schoeffel Instrument Co. for use of the Spectroflow 770 Detector; Dr. Hilaire Meuwissen for the samples of plasma; and Roberta Caldwell for her help with the manuscript.

This work is supported by Grant No. CA-GM-17603-01 from the United States Public Health Service.

REFERENCES

1. N. G. Anderson, J. G. Green, M. L. Barber and L. C. Ladd, Anal. Biochem., 6, 153 (1963).

2. C. G. Horvath, B. A. Preiss and S. R. Lipsky, Anal. Chem., 39, 1422 (1967).

3. P. R. Brown, J. Chrom., 52, 257 (1970).

4. P. R. Brown, "High Pressure Liquid Chromatography, Biochemical and Biomedical Applications," Academic Press, N. J., 1973.

5. C. D. Scott, J. Attril and N. G. Anderson, Proc. Soc. Expt. Biol. Med., 125, 181 (1967).

6. C. A. Burtis and K. S. Warren, Clin. Chem., 14, 290 (1968).

7. J. E. Mrochek, D. D. Chilcote and S. R. Dinsmore, "Liquid Chromatographic Analysis of Urinary Nucleosides in Normal and Malignant States," ACS Meeting, August, 1973.

8. C. W. Gehrke, R. W. Zumwalt, K. C. Kuo and T. P. Walker, Symposium on Chromatographic Analysis of Biologically Important Compounds, ACS Meeting, Div. of Anal. Chem., August, 1975.

9. R. M. Berne, R. Rubio, Supplement III to Circ. Res., 34 & 35, 109 (1974).

10. R. Rubio, R. M. Berne, Circ. Res., 25, 407 (1969).

11. K. Ishii and H. Green, J. Cell. Sci., 13, 1 (1973).

12. H. Green in "Combined Immunodeficiency Disease and Adenosine Deaminase Deficiency (H. J. Meuwissen, et al., eds.), Academic Press, Inc., N. Y., 1975.

13. P. R. Brown and R. E. Parks, Anal. Chem., 45, 948 (1973).

14. J. J. Van Deemter, F. J. Zuiderweg and A. Klinkenberg, Chem. Eng. Sci., 5, 271 (1956).

15. J. C. Giddings, Dynamics of Chromatography, Marcel Dekker, N. Y., 1965.

16. L. R. Snyder, J. Chromatogr. Sci., 7, 352 (1969).

17. R. A. Hartwick and P. R. Brown, J. Chromatogr., 112, 651 (1975).

18. J. X. Khym, Clin. Chem., 21, 1245 (1975).

19. R. A. Hartwick and P. R. Brown, J. Chromatogr. (in press).

ENZYMATIC DETERMINATION OF URATE, CHOLESTEROL AND TRIGLYCERIDES IN THE SYNOVIAL FLUID WITH THE CENTRIFUGAL FAST ANALYZER

J. van Wersch, M. Peuckert, H. Greiling

Department of Clinical Chemistry

Technical University Aachen, FRG

The synovial fluid may be understood as a dialysate of the blood serum, because the synovial fluid has the same composition with respect to the electrolytes and other molecular organic substances as the blood serum. The main difference between the synovial fluid and the blood serum exists in the composition of proteins and the hyaluronate. Hyaluronate is the active secretion product of the lining cells from the synovial membrane. Hyaluronate is synthesized by specific enzymes of the lining cells, from UDP-N-acetylglucosamine and UDP-glucuronic acid as shown in figure 1.

All carbohydrate components of the hyaluronate origin from the glucose, which is present in normal synovial fluid, and has the same concentration as in blood serum. In the synovial fluid of patients with various joint diseases for instance rheumatoid arthritis, osteoarthritis and also gout, there are similar concentrations of potassium, sodium, phosphate, urate, urea, creatinine and bilirubin but lower concentrations of proteins, cholesterol and the triglycerides. Several attempts have been made for using the synovia analysis in the differentiation of the joint diseases. Great differences exist between the urate concentrations in the synovial fluids of gouty patients and of patients with other diseases. For synovial fluid analyses we have only a small amount of fluid. For this purpose we had to develope a micromethod and we found that the fast analyzer is convenient for this purpose.

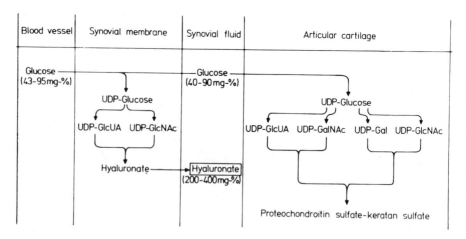

Figure 1:   The compartiments of the synovial system and the
            biosynthesis of glycosaminoglycans from glucose

In figure 2 is shown the scheme for the enzymatic determination
of cholesterol (1), triglycerides (2) and urate (3) in the synovial
fluid with the **GEMSAEC FAST ANALYZER.**

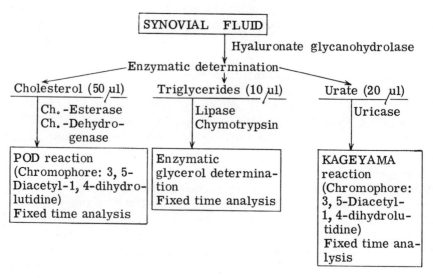

Fig. 2: Scheme for the analytical procedure and the determination
        of cholesterol, triglycerides and urate in synovial fluid.

The determination of the various components is not possible in the authentic synovial fluid, because its viscosity is too high. Therefore we must depolymerize the hyaluronate by hyaluronate glycanohydrolase. The precision for all methods is comparable with other methods in use (table 1).

		Urate [μmol/1] Europe control (Hyl.)	Triglycerides [mmol/1] Precilip (Boehringer)	Cholesterol [mmol/1] Europe control (Hyl.)
precision within rotors	$\bar{x} \pm 2s$	506 $\pm$ 20	1.32 $\pm$ 0.07	3.32 $\pm$ 0.17
	CV (%)	2.1	2.7	2.6
	n	15	15	15
precision between rotors	$\bar{x} \pm 2s$	506 $\pm$ 26.7	1.28 $\pm$ 0.08	3.37 $\pm$ 0.23
	CV (%)	2.7	3.2	3.4
	n	15	15	15
precision from day to day	$\bar{x} \pm 2s$	506 $\pm$ 35.6	1.3 $\pm$ 0.11	3.27 $\pm$ 0.25
	CV (%)	3.5	4.2	4.4
	n	15	15	15

Table 1: Precision of the enzymatic determination, of urate, triglycerides and cholesterol

We found a linearity for urate to 714 μmol/1, for cholesterol to 7,77 mmol/1 and for the triglyceride determination to 3,96 mmol/1. In table 2 you see the synovial values for urate, triglycerides and cholesterol concentration in patients with several joint diseases. The urate concentrations in rheumatoid arthritis and osteoarthritis are similar to them in the blood serum. In table 2 the maximum values are in the first line, the average value in the middle and third line shows the minimum value. The values for cholesterol and triglycerides are far lower then the values in the serum of the same patients. The cholesterol concentrations of all chronic joint diseases are similar. The highest triglyceride values are found in arthritis urica.

	Urate [μmol/l]	Cholesterol [mmol/l]	Triglycerides [mmol/l]
Rheumatoid arthritis (n = 54)	464 286 131	3, 47 2, 48 1, 19	1, 19 0, 74 0, 23
Osteoarthritis (n = 38)	387 286 155	4, 07 2, 19 1, 04	0, 75 0, 24 0, 14
Arthritis urica	785 696 613	2, 59 1, 89 1, 06	2, 61 1, 56 1, 15
Normal values in serum	3, 4 - 7, 0 (200 - 420)	150 - 300 (2 - 8)	74 - 172 (0, 84 - 1, 94)

Table 2: The uric acid-, cholesterol-, and triglycerides- concentration in the synovial fluid of patients with different joint diseases

There is a high correlation between the synovial fluid urate and triglycerides concentration in gouty patients with a correlation coefficient of 0, 96. The lowest triglyceride values are found in osteoarthritis. The cause for the lower cholesterol and triglyceride concentrations in comparison to blood serum values might be the reduced permeability of the synovial membrane for alpha- and ß-lipoproteins.

Nettelbladt, Sundblad and Jonsson (4, 5) showed that after pressing serum through a hyaluronate gel by repeated centrifugations the protein profile was similar to that of the normal synovial fluid. Immunoelectrophoretic analysis of the fluid showed the absence of ß$_2$-macroglobuline, ß-lipoproteine and alpha-lipoproteins. These results support the theory, that hyaluronate of the perisynovial connective tissue regulates the passage into the joint cavity.

Perhaps is also the size of the hyaluronate molecule an important factor for this effect. In the synovial fluid of osteoarthritis with the lowest values for triglycerides concentration a high hyaluronate concentration and a high viscosity is found. This is due to a high molecular weight of the hyaluronate.

The high coefficient of correlation between urate and triglyceride in the synovial fluid of gouty patients may be explained by a interrelationship between the urate and lipid metabolism in the synovial tissue (6).

References:

1. Röschlau, P. et al., Z. klin. Chem. klin. Biochem., 12, 403 (1974)

2. Bucolo, G., H. David, Clin. Chem., 19, 476 (1973)

3. Kageyama, N., Klin. Chim. Acta, 31, 421 (1971)

4. Nettelbladt, E., Sundblad, L., Jonsson, E., Acta Rheumatol. Scand., 9 (28), 249 (1963)

5. Sundblad, L. in: The Amino Sugars, Academic Press, New York, Vol. II A, 249 (1965) Ed.: E. A. Balazs and R. W. Jeanloz

6. H. Greiling, Therapiewoche, 25, 4358 (1975)

AUTHOR INDEX